COLOR ORDERED

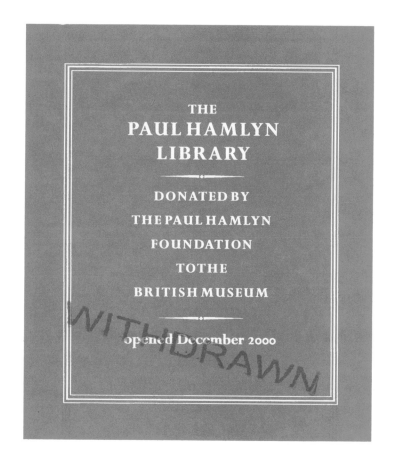

COLOR ORDERED

A Survey of Color Order Systems
from Antiquity to the Present

Rolf G. Kuehni and Andreas Schwarz

OXFORD
UNIVERSITY PRESS

2008

OXFORD
UNIVERSITY PRESS

Oxford University Press, Inc., publishes works that further
Oxford University's objective of excellence
in research, scholarship, and education.

Oxford   New York
Auckland   Cape Town   Dar es Salaam   Hong Kong   Karachi
Kuala Lumpur   Madrid   Melbourne   Mexico City   Nairobi
New Delhi   Shanghai   Taipei   Toronto

With offices in
Argentina   Austria   Brazil   Chile   Czech Republic   France   Greece
Guatemala   Hungary   Italy   Japan   Poland   Portugal   Singapore
South Korea   Switzerland   Thailand   Turkey   Ukraine   Vietnam

Copyright © 2008 by Oxford University Press, Inc.

Published by Oxford University Press, Inc.
198 Madison Avenue, New York, New York 10016

www.oup.com

Oxford is a registered trademark of Oxford University Press

Library of Congress Cataloging-in-Publication Data
Kuehni, Rolf G.
Color ordered: a survey of color order systems from antiquity to the present / Rolf G. Kuehni
and Andreas Schwarz.
p. ; cm.   Includes bibliographical references and index.
ISBN 978-0-19-518968-1
1. Color vision — History. 2. Colors — Classification — History. 3. Colorimetry — History.
I. Schwarz, Andreas.
II. Title.
[DNLM: 1. Color Perception. 2. Color. 3. Colorimetry — history. WW 150 K95c 2007]
QP483.K82 2007
152.14'5 — dc22   2007011343

9 8 7 6 5 4 3 2 1

Printed in China
on acid-free paper

Dedicated to the memory of Dorothy Nickerson,
Deane B. Judd, and Manfred Richter;
and also dedicated to Friedrich Schmuck

PREFACE

Color plays a very important part in the world we experience visually. Not only does color help us to distinguish between objects having different light-reflecting properties, but naturally or artificially caused color also gives us aesthetic pleasure. Color helps us to categorize objects and experiences.

Assigning names is the result of a categorization process. Among the first named color categories were red and "grue" (a combination of green and blue), white, and black. From this stage, we progressed to one or two dozen named categories recognized by most humans. More complex and detailed verbal categorization required the development of color reference collections with material samples representing (in some fashion) color perceptions. The color order systems implicit in such collections represent not only growing insight into the nature of color experiences but also idiosyncratic expressions and cultural achievements. Color order is an attempt to systematize color experiences.

This book offers an extensive if incomplete survey of color order systems, illustrated mostly with original figures. It places color order systems into categories that are based in part on history, that is, the gradual growth in understanding of the logical arrangement of color experiences, and in part on science and technology. It is arranged according to the following plan:

Chapter 1 gives a brief introduction into specific aspects of vision and color vision to help the reader establish a foundation for the remaining chapters.

Chapter 2 discusses linear color order systems beginning with Aristotle and reaching into the twentieth century.

Chapter 3 presents the realization of a second dimension in color order, resulting in color diagrams and color circles.

Chapter 4 demonstrates the systematic development of the third dimension in color order, beginning in the eighteenth century with the work of Tobias Mayer, Johann Heinrich Lambert, and others.

Chapter 5 describes psychological color order systems, with a basis of more or less detailed psychological scaling.

Chapter 6 is filled with scientifically and technically important systems derived from psychophysical or neurobiological descriptions of color stimuli.

Chapter 7 covers systems of the twentieth century that attempt to connect empirical perceptual data with psychophysical scaling data.

Chapter 8 is a brief introduction to physical order systems that place color stimuli as a result of mathematical manipulation of purely physical measurements.

Chapter 9 presents technical color order systems where the samples of the system are arranged according to systematic mixture of usually three primary colorants (dyes or pigments, typically in form of printing inks) or three lights.

Chapter 10 contains groups of systems that represent mixtures of ideas not easily fit into the other chapters. There are systems with a basis in early psychophysical results and a group of systems based on disk mixture data. There is also a group of systems based on the form of the cube but not directly related to the halftone printing and display systems of chapter 9.

Chapter 11 contains miscellaneous systems: a group containing systems of an incomplete nature, and another group based on more or less unconventional ideas.

Chapter 12 is a concluding synthesis.

An ***appendix*** shows 29 larger reproductions of rare and unusual original color figures of color order systems, several not previously reproduced. The figures that appear in the appendix are identified in the main text with figure numbers highlighted in red.

We describe systems, particularly older ones, conservatively. In recent years a few authors have interpreted certain systems as three dimensional, even though there is no direct evidence to support such views.

Entries in the book contain brief descriptions of the color systems and their inventors, with original or schematic illustrations. Chapters 2–11 begin by discussing the ideas behind the systems and the relationships among them. The book also contains an alphabetical and chronological listing of all entries, a select bibliography of general literature, and a glossary. Explanatory comments within quotations have been placed in square brackets.

We are both much interested in color order, so writing this book has been a labor of love. We hope that it will be useful to scientists and technicians involved with color, designers and artists, historians, students, and any reader with an interest in the historical path of human thought on color order.

R.G.K. expresses his appreciation to friends and colleagues interested in the history of color order, to fellow members of the Inter-Society Color Council, researchers who have shared their knowledge and opinions, color philosophers, and several helpful librarians around the world. Specifically, I thank Werner Spillmann in Basel, Switzerland, for generously sharing his great collection of historical color order systems. I thank Roy Berns, Robert Feller, Erich Fischer, Larry Hardin, David Hinks, Tarow Indow, Sarah Lowengard, Joy Turner Luke, Claudio Oleari, Dale Purves, Alan Robertson, Ralph Stanziola, and Françoise Viénot for supplying information on various subjects. I particularly thank Charles di Perna of the Charlotte-Mecklenburg Public Library system for much help with interlibrary loans, Christine de Valet of the Yale Art Library for important help with images, and Gary Field, from whose knowledge the print-related portions of chapter 9 profited. Last but not least, I thank Margret for generous understanding of this minor obsession.

A.S. thanks all his friends, colleagues, organizations, and institutions that supported him in various ways in this project. My thanks go to the Wilhelm Ostwald Forschungs- und Gedenkstätte in Grossbothen for generous help with materials and extended support regarding inquiries; the Fachhochschule Köln, present host to the Collection Friedrich Schmuck; the Color Reference Library in London; the Bayerische Staatsbibliothek München; and the Staatsbibliothek in Berlin. Thanks are due to Hartmut Adam, Eckhard Bendin, Robert L. Dillon, Günter Döring, Heinwig Lang, Doris Oltrogge, Klaus Palm, Albrecht Pohlmann, Bud Plochere, and Gerhard Zeugner for images and text materials, together with pleasant correspondence. I give heartfelt thanks to my colleague Silvia Danilieva for translations from Russian. Our designer, Michael Marschhauser, deserves thanks for design work, the schematic drawings, and his patience in dealing with many special requests. I also thank the late Manfred Richter, who not only supplied valuable information but also brought to my attention and made available several rare color order systems. Many thanks go to my wife, Iris, for always showing understanding for my passion and for providing support during difficult phases. My largest gratitude, however, is owed to Friedrich Schmuck, always reachable and ready with valuable information and providing me with unlimited access to his important collection. Without its many treasures, this book in its present form would not have been possible.

We both express our appreciation to the editorial personnel at Oxford University Press, in particular, Catharine Carlin, for helpful and unstinting support of the project and for letting us design the book according to our wishes.

SPECIAL ACKNOWLEDGMENTS

We are particularly grateful to Datacolor International in Lawrenceville, New Jersey, and its president, Terry Downes, for a generous contribution toward the cost of the design process for the book.

We also would like to acknowledge a generous financial contribution toward the design process from CAPAROL in Ober-Ramstadt, Germany and its managing partner, Dr. Ralf Murjahn, as well as from RAL e.V. in St.Augustin, Germany and its general manager, Dr. Wolf D. Karl.

CONTENTS

COLOR ORDERED

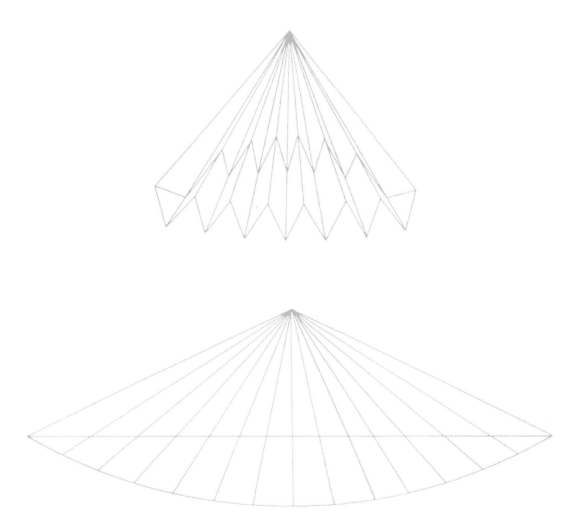

THE UNIVERSE OF HUMAN COLOR EXPERIENCES

A brief introduction into specific aspects of vision and color vision to help the reader establish a foundation for the remaining chapters.

Vision and color vision

Vision is an important capability of all animals, helping them to interact in a meaningful way with the environment. In humans, in particular, vision is mediated by complex mechanisms in the eye and the brain. Bees, butterflies, and hummingbirds, despite having a very small brain volume, are also capable of intricate, vision-based behavior. A special capability of many types of animals is some kind of color vision. Certain areas in the field of vision can be distinguished from others by what humans experience as colors. What the corresponding experiences are for different kinds of animals not believed to be conscious is unknown.

Color vision is an aspect of general vision. Eyes have been invented by evolution several times and in different forms. Mammals have camera-type eyes consisting of a hollow spherical body with an opening of variable size covered by a lens whose curvature can be changed by muscular action. The lens focuses light energy onto a highly sensitive spot on the retina that lines the interior wall of the eye. There, light interacts with different kinds of light-sensitive cells. The number and types of cells in the retina vary by species as well as within a species. Most mammals have three types of cells, one responsible for colorless night vision and two for color vision in daylight.

Many primates, including humans, have four cell types: rods that mediate night vision and three types of cones mediating daylight color vision. As a result, our color vision is called "trichromatic vision." Comparatively recently, it was discovered that a number of human females have the genetic capability for four or in some cases even five cone types. The extent to which there are individuals with four or five expressed cone types and the consequences of this for their color vision are not yet known. Approximately 10% of all humans (mostly males) have more or less impaired trichromatic color vision. This can range from modifications in cone sensitivity to absence of one, two, or all three cone types. Even so-called color-normal individuals have considerable variation in color vision that depends on their own specific expression of the color-vision system.

Trichromatic or even pentachromatic vision is not the pinnacle of chromatic discrimination. Mantis shrimp have at least 11 types of photosensitive cells varying in their spectral response (it is not known how exactly the resulting information is used). Bees have a trichromatic system but their spectral sensitivity range is partially shifted into the ultraviolet region, with a range of approximately 320–600 nanometers (abbreviated nm, a billionth of a meter), compared to that of

humans, with a range of approximately 400–700 nm. The red-eared turtle (*Pseudemys scripta elegans*) has six different cone types, some with differently colored oil droplets in front to modify their spectral sensitivity.

Light is a form of electromagnetic energy, and the energy content can be expressed either in terms of wavelength or in terms of electron volts. Electromagnetic energy that can be sensed by animals' eyes falls in the range of approximately 300–800 nm. Energy in this range is provided in abundance by the sun. The three standard cone types in humans vary in their sensitivity along the spectrum. The relative sensitivity curves of the three average cone types (named S, M, and L for short-, medium-, and long-wavelength bands) are shown in figure 1.1. It is apparent that the ranges of the three kinds of cones are fairly broad and overlapping. As a result, across the visible spectrum, light of single wavelengths activates at least two, and for many wavelengths all three, cone types, but to different degrees.

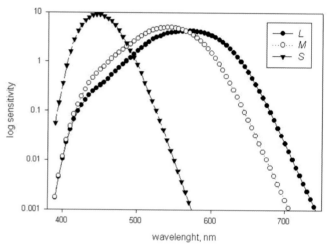

Figure 1.1.
Average spectral sensitivity curves of the three human cone types. Stockman-Sharpe 10° observer cone fundamentals based on the Stiles and Burch 10° color matching functions (data from www.cvrl.org).

Light enters our eyes through the lens, rarely directly from a light source, most often scattered or reflected from objects. Objects have the property of reflecting or scattering light in qualitatively and quantitatively different ways. The amount of light as a function of wavelength reflected from a given object can be determined using physical measurement (using a reflectance spectrophotometer). The quality and quantity of light entering the eye coming from a particular object depend on the spectral distribution and quantity of the light striking the object and the portion of it that is reflected.

Daylight varies in quality and quantity to a considerable extent depending on the time of day and the weather. It can be modified further, for example, by filtering through a tree canopy in woods or by reflection from mountainous terrain in a steep valley. Artificial light sources add many more and widely differing spectral power distributions to the set of possible sources of illumination. And it is readily apparent that the light entering the eye reflected from, for example, a tomato can vary widely in terms of quality and quantity depending on the light it is seen in. Yet in virtually all cases, we recognize the fruit without any problem. Complex mechanisms in the brain determine from the information generated in the eyes the most likely nature of the objects and activities in our field of view.

At the front end, so to speak, of the human visual system are the cone types. They transform electromagnetic energy of light into chemical energy. The resulting signals are already extensively modulated in the retina of the eye by cells connected to the cone cells. From there, they travel along the visual pathway to the main visual processing center at the back of the brain. After the brain manipulates the signals (in ways that are not yet fully understood), they disappear into a "black box," out of which our subconscious and then our conscious color experiences emerge.

How old is color vision?

In recent years, the genetics of opsins, the light-sensing proteins in the eyes, have been clarified to a considerable extent. We can now estimate when genetic modification responsible for a given property of the vision system occurred. The history of the development of human color vision is rather complex. The ability to sense light is estimated to have first occurred some two billion years ago. Several basic light-sensitive molecules are known, but most vertebrates only employ one, called retinal. The spectral sensitivity of this substance is modified in different species by five different kinds of protein compounds, forming the opsins, two resulting in ultraviolet or short-wave sensitivity (SWS1 and SWS2), two in mid-wave sensitivity (RH1 and RH2), and one that can be expressed to have either mid-wave or long-wave sensitivity (LWS/MWS). Some birds (e.g., chickens) and some fish (e.g., goldfish) have all five kinds of opsins. Dinosaurs likely had all five opsins.

Mammals began to develop about 200 million years ago, and during the Oligocene, they were spreading widely over the earth, with primates beginning to develop some 60 million years ago. Early mammals, and with them early human ancestors, lost the ultraviolet-sensitive SWS1 as well as RH2 opsin sometime in their history. In humans, RH1 in the form of the rod cell type supports night vision and is not believed to be involved in color vision in regular daylight. Our so-called "ancient system" consists of the SWS2 opsin, expressed as the S cone type (figure 1.1), and the LWS opsin, expressed as L. This system is estimated to have developed perhaps 100 million years ago. In addition to distinction between degrees of light and dark, it allowed qualitative distinction of some sort between short- and long-wave light.

The separate expression of an MWS cone type, M, developed only in the Oligocene, some 35 million years ago. The family of Hominidae is only about four million years old, indicating that our typical trichromatic system developed in primates, but long before the appearance of hominids.

What has been the driving force behind the evolutionary development of color-vision systems? Scientists since Darwin have wrestled with this question. It is now quite evident that, in a given species, the spectral sensitivity of the color-vision system has adapted itself in support of the viability of that species. In bees, butterflies, and certain birds, the specific forms of spectral sensitivity may have developed as a result of the loosely symbiotic relationship they have with flowering plants. The latter trade nectar for pollen dispersal. For the purpose of effective distinction in the natural environment, flowers developed particular reflectance properties (in the ultraviolet and/or visible range), and the guest animals developed appropriate visual systems to detect them. Thus, from the point of view of the animals, color vision developed to make detection of food sources easier.

In recent years, researchers in Africa have studied the relationship between apes' and monkeys' color vision and their food's coloration: fruits and plant leaves as viewed in varied natural greenish-brownish surroundings (e.g., Regan et al. 2001). During ripening, the fruits change from green to yellow, to reddish orange, or to red. Very young edible leaves are reddish and more easily digested by the animals than are older, green ones.

The research showed that trichromatic color vision nearly optimally distinguishes between unripe and ripe fruit or younger and older leaves, and between food and the general surroundings, and that it supports "judgment" about the ripeness and edibility of both fruit and leaves. So it is likely that our color-vision abilities are, to an important degree, the result of our primate ancestors' need for efficient feeding from plants.

In addition, there is little doubt that trichromatic color vision also developed as a useful tool for detecting hidden predators and potential sexual partners. The ancient system effectively distinguished between lights of short and long wavelength. The later system added effective distinction between lights in the mid-wavelength central region of the spectrum.

Vision: reconstruction or construction?

Vision provides much more than efficient distinction between food and surroundings. It is a primary tool for animals to pursue their goals (instinctive or rational) efficiently in the environment in which they exist. An unresolved issue is whether the brain reconstructs veridical images of objects or constructs sufficiently informative, but different, images from the energy patterns it receives through the eyes. The issue is complex. Hummingbirds have a very small brain, yet they operate very effectively in their environment, identifying food sources rapidly, defending them fiercely, flying safely with great speed in complex environments, and every year making two extended trips of hundreds of miles. It is unlikely that their brain provides them with a veridical reconstruction of the world. The human system is much more complex but unlikely to be radically different. Differences likely are of a quantitative, not qualitative, nature.

An important function of the human visual system is to find "nuggets of gold" in mounds of dross. The eyes take in visual streams that have an estimated information content of 10 million bits/second, which is believed to exceed the information content that can be processed in consciousness by a factor of about 250,000:1. The visual system contains mechanisms that automatically extract from the stream the information presumed most important to the individual at that moment. Knowing what information to extract has been learned adaptively over millions of years. It is information that other neural systems use to generate actions to deal with it. Do I reach for what may be a ripe fruit, or could it be just a yellowed leaf? Do the yellowish patches indicate a tiger or just dried leaves? All but the final interpretation of this processing is done subconsciously. It is now well established that in humans and higher primates, visual information passes along two separate paths to centers in the brain that have much different purposes. The resulting two ways of seeing have been named "vision for action" and "vision for perception" (Goodale and Milner 2004). Most kinds of animals have more or less elaborate versions of the vision-for-action system. This system produces appropriate bodily movements from visual stimuli, for example, prey capturing or evasive actions. In humans, the results of this system's operation are not available to consciousness.

The vision-for-perception system, fully expressed only in higher primates, results in our ability to contemplate and compare objects for recognition and planning actions, involving both form and color. The substance of our visual memories, including objects and their colors, is available in consciousness.

The two systems smoothly interact under normal conditions. When Jane perceives a basket with apples in front of her and reaches out to grasp one, experience has taught her that the apples are actually there in front of her. The first system, without Jane's conscious support, smoothly grasps an apple for closer inspection. The second system perceives the apple as ripe and edible, with the first system guiding (again without conscious thought) the movement of the apple to the mouth.

The idea of construction also receives support from the following several points:

Visual illusions: It is well known that in many situations, the human visual system produces results that don't agree with physical facts. These results can be viewed as aberrations, or as normal results of the interpretation of certain photon flow data by the less than all-powerful visual system. Doubts regarding the veridical reconstruction theory have a long philosophical tradition. Nicholas Malebranche wrote in 1674: "It can be predicted at the outset that few people will not be taken aback by the following general proposition, viz., that we have no sensation of external objects that does not involve one or more false judgments" (p. 76). Despite such insights, the developing view in the Enlightenment and Industrial periods was one of a relatively simple relationship between physical measurement data and resulting experiences. This came to mean that if one knew the reflectance properties of a material, one also knew its color. And that idea resulted in two competing theories: the objectivist, realist theory (colors are out in the world and the brain apprehends them) and the subjectivist theory (our brain constructs color experiences, rather than reconstructs colors). The two theories continue to be debated.

Metamerism: One of the striking facts about color vision is the phenomenon of metamerism. When a given reflecting material is viewed under given conditions of surround and illumination, the effect of the photon flow from the object on the three cone types can be expressed with three numbers (representing the degrees of activation of the three cone types). Thus, the complex, many-dimensional photon-flow data are abbreviated and reduced to three pieces of information. It is a natural law that if this kind of dimension reduction involves continuous filter functions, there are many different input functions that result in the same three numbers in the corresponding three-dimensional space. That means that materials with many different spectral reflectance functions can look exactly the same. Spectral distributions with this property form a set of *metamers*.

Metamers are of great practical importance in the modern world. On the one hand, they have made possible efficient technologies of color photography based on three dyes, color television based on three lights, and color printing based on three or four pigments. In all of these technologies, all color stimuli possible can be reproduced using three primary lights or colorants. On the other hand, metamers have made life difficult for industrial colorists working with textiles, paints, and other materials: A sample matched with colorants different from those used in the reference material will likely be metameric to the reference sample. Under the standard light (e.g., standardized daylight), it will match the appearance of the reference sample. Under another light, say, that from an incandescent light bulb or a fluorescent light, it will often appear as a mismatch.

Depending on the colorants in use, it may be possible to find a recipe resulting in an acceptable match under two or three different light sources. However, the larger the number of light sources, the smaller the likelihood that a good match for all lights is technically achievable. Whether or not two lights arriving at the cones match is a well-known phenomenon with a well-fitting mathematical model, because it is controlled at the level of the relative sensitivities of the cones in the eye.

Color constancy: The phenomenon of color constancy (or lack thereof) is connected with metamerism. It refers to the fact that a material with a given reflectance function (the spectral function of the reflectance properties) in a given surround and light either maintains its color appearance when viewed in lights from different sources or changes (possibly drastically) in appearance. Thus, in a metameric pair, one sample may be color constant and the other not constant, or both may be not constant but in different ways. It is a fact that natural objects, when viewed in natural conditions, generally maintain their color appearance reasonably well regardless of the spectral power distribution of daylight (excluding colorful sunrises and sunsets).

For our early primate ancestors, it seems to have been important to be able to unfailingly recognize edible fruit and leaves in widely different natural illumination conditions, despite large differences in the photon flow arriving from them at the eyes. The visual system elaborated a (not yet fully understood) process of maintaining relatively constant appearance for natural products despite possibly large changes in photon flow to the eyes. But this built-in constancy does not generally extend to artificial materials and light sources. Our species has been exposed to most of these for only some 200 years or less.

Surround: A further process with a potentially strong effect on an object's appearance is related to the object's surround. Objects, if they can be recognized, are always seen in a surround. Surrounds can be of extreme simplicity (uniform neutral gray) or very complex (the floor of a deciduous woods in the fall). Any kind of surround has an effect on the appearance of objects, based on their lightness. Figure 1.2 illustrates the effect of a varying neutral surround on perceived lightness of five identical neutral patches. Figure 1.3 illustrates the fact that the apparent location of the light source illuminating the object plays a major role in what is perceived. The measured luminous reflectance values of the areas highlighted in the gray sketch at the bottom right are identical despite the large difference in appearance in the color figure. Figure 1.4 demonstrates the effect of a varying chromatic surround on the appearance of identical chromatic patches. (For the skeptical, a paper mask with a cutout will provide proof that they are identical.)

In complex chromatic surrounds, objects may change their appearance within a wide range depending on the local surround, according to processes that are not yet understood. An object with low reflectance across the visible spectrum can be made to appear from white to black depending on

illumination and surround. Here, there is again a lack of agreement between reflectance properties and related appearance. Mathematical models to predict these effects remain comparatively crude, with reasonable results only for limited conditions.

Figure 1.2
Effect of surround on perceived lightness of five identical squares.

Contrast and assimilation: A connected process is that of contrast and assimilation. Under certain conditions, the perceived difference between an object and its surround is enhanced; in other cases, it is reduced. The argument can be made that there is no color vision without contrast. In conditions where any possibility of contrast is eliminated, color appearance rapidly fades. An important perceptual task is to distinguish an object from a surround of similar reflectance properties. The relative change in photon flow required for perception of a difference is small. Pairs of objects with similar reflectance properties are most readily distinguished if the reflectance of the surround falls between the reflectances of the two objects. If the surround reflectance is much different, the apparent difference between the two objects is reduced or absent.

Figure 1.3.

Lightness perception as a function of spatial arrangement and apparent lighting. The small sketch at the bottom right indicates the fields in the figure that have identical luminous reflectance. From Purves and Lotto (2003), reprinted with permission.

An example of another effect, assimilation, is shown in figure 1.5. The stripes have uniformly the same reflectance even though some appear yellowish red and others bluish red. Also in this case, there is no close relationship between reflectance and perceived color.

Figure 1.4

Effect of varying surround on the appearance of five identical color chips.

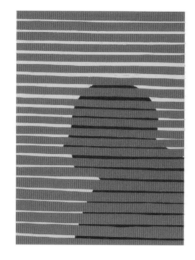

Figure 1.5

Color assimilation: stripes of uniform red appear much yellower against the yellow than against the blue background.

Perceptual variability of observers: Observers with so-called normal color vision also contribute to variation in perceived color. This can be demonstrated by determining the personal "unique hue" stimuli of an observer in standard observation conditions. Unique hues (as defined by Ewald Hering; see entry in chapter 5) are those four hues—yellow, red, blue, and green—that observers perceive as not having admixtures from other hues. Unique blue, for example, is blue that observers see as having neither a reddish nor a greenish component; its appearance is just blue. Unique hue perceptions can occur either in the spectrum or, more naturally, from an array of color chips changing smoothly in hue without changing noticeably in lightness or intensity.

When determining stimuli that result in unique hue perceptions under identical observation conditions, a surprisingly large variation among observers is obtained (Kuehni 2004). It is evident from the result that one color-normal observer's color experience when viewing a given painted chip (with a specific reflectance function) can differ significantly from that of another color-normal observer.

Fifty or more years ago, such effects were conventionally seen as exceptions to the imputed rule of a close relationship between reflection and appearance. Color technology, by relying on strictly controlled "standard" conditions of surround, illumination, and viewing, continues on this path,

with a good measure of success. A more fruitful way of viewing the situation is that specific perceived colors of objects under any set of conditions are the symbols our brains elaborate for the constituting complex energy signatures impinging on the retina. A given perceived color is the result of construction by the brain based on interpretation of the information available to it. That the perceived colors of an object can change depending on conditions is the result of ambiguity in the available information and the limited capabilities of the visual apparatus. The visual system is only as complex as it needed to be for our primate ancestors to be able to lead successful lives. To assume that the colors we experience are the real colors of objects imposes a narrow, normative, anthropocentric viewpoint.

The failure of a simple reductive theory to explain the colors of objects that individuals experience under any given set of circumstances is complicating development of broadly meaningful color order systems and mathematical models of color vision.

How many colors are there?

Aristotle's view on this question was that the number is infinite. But based on the preceding discussion, the question needs to be posed differently. For purposes of simplification, we will limit the question to human trichromats: How many chromatic stimuli can the average trichromat identify without error on an absolute basis, that is, without reference to any other stimulus except a gray surround? The answer is, surprisingly few. When considering stimuli giving rise to different hue perceptions, the answer is six to eight. When considering in addition variations in lightness and chromatic intensity, the number doubles for each situation. Add three achromatic percepts (white, gray, black) for a total of perhaps 30.

The issue becomes one of distinction rather than identification. The minimal degree of distinction is the "just noticeable difference" (JND). Perceiving a JND between two colored fields requires a difference in stimuli from the two fields of a magnitude that depends on several factors. In theory, all possible color stimuli can be mapped in such a fashion. Up to now, only small regions have been mapped in this manner, usually in differing conditions, and the results have been extrapolated.

Given a standard condition of viewing and surround, the number of stimuli from objects that are distinguishable is estimated to be about one million. An additional large number is distinguishable in the form of mixtures of spectral lights only. The visual system accepts an infinitely large number of photon streams of different qualitative and quantitative nature through three filters (represented by the cone sensitivities). It constructs from the result perhaps two million perceptually different color experiences. The problem that color order systems have attempted to solve is how to place these experiences into a systematic order.

To have controlled experimental conditions for determining psychological spacing of color products, color scientists usually employ extreme minimalist visual fields and very short exposure time of observers to such fields. Some psychologists and philosophers have questioned whether the minimalist visual fields produce results akin to those that might be obtained in natural environments that presumably formed our visual system as it exists. Others question the existence of a natural psychological color space in which all color experiences are systematically ordered, and believe these to be cultural artifacts. Such questions strike at the basis of efforts to create color order systems. They also raise the question of nature versus nurture in color order. It seems reasonable to assume that nature and culture are jointly involved. To enter into the details of these controversies is outside the scope of this text.[1]

A conclusion that can be drawn so far is that color perception is a complex subject. Most of us see the world in color, which helps us to identify and distinguish objects in visual fields of various levels of complexity. Many objects appear to have about the same color when seen in different circumstances. But some artificial objects can change apparent color quite dramatically. The possibility of about a million different object color experiences allows for fine-grained distinctions. But, as discussed above, there is no singular relationship between reflectance functions of objects and the resulting color experience.

It is not possible to specify colors for fine-grained work with words because categories around names are large (discussed below). And it is not universally satisfactory to specify color experiences by physical measurement. A sample chip has the same reflectance function regardless of the conditions it is viewed in. But the viewing experience depends on the observer and the viewing conditions. Color specification is possible only in the sense of color stimulus specification, and as a result, chip collections representing color order are of limited value.

Color categories and names

Wikipedia defines a category as "a group, often named or numbered, to which items are assigned based on similarity or defined criteria." It is the human attempt to bring a kind of order into the chaos of the world. The innumerable flow of sensory experiences and feelings we can have requires categorical classification to efficiently deal with them. This is particularly important for communicating with other people (Harnad 2003). Stevan Harnad thinks that language is largely a categorization activity. Just about every noun, verb, adjective, and adverb represents categorization

of some kind. The specific words used for a given category are assigned either by some brain cognitive processes or by cultural agreement.

On the other hand, color experiences are private and cannot be shared directly with anyone. Our age-old means of communicating with fellow humans about color experiences is with gestures and words. Primitive color names (referring to categories of stimuli), as Berlin and Kay (1969) and their colleagues have shown, appear to some degree to be universal; that is, some (e.g., Hering's six *Urfarben*, discussed further below) are likely generated in essence by the visual system. Berlin and Kay think that there are 11 of these universal simple color categories: white, black, gray, red, orange, yellow, green, blue, purple, pink, and brown. Aside from the first three achromatic (hueless) categories, there are six categories representing hues, and only two categories where lightness and intensity of coloration are important additional aspects: brown and pink.

The six primitive hue categories correspond to our ability to absolutely identify colors according to hue. It seems reasonable to assume that their number is influenced to a degree by this ability. When moving beyond the simple category level, color categorization becomes arbitrary and largely culturally determined. Exemplars recognized by smaller or larger cultural groups of humans often inspire categories, such as olive, lime, peach, violet, beige, and navy. At a higher level, large collections of color names with representative stimulus chips, such as the Maerz and Paul (1939) color dictionary (see below), are attempts to codify regional, temporal, and cultural color naming (e.g., the culture of American interior designers or of ornithologists), with many meanings often understood only by small subgroups.

It is interesting to note that, of the six hue-related basic category names, four refer to those for which unique hues can be defined. The other two, orange and purple, are taken to fall halfway between two unique hues, both involving red. The relationship between a given color stimulus and the resulting perceptual category, such as red, is, as mentioned above, vaguely defined. The category "red" includes some 60,000 just noticeably different object color stimuli, but which of these are included in the category varies significantly among color-normal observers. The stimulus representing the color experience that is paradigmatic for a given category varies widely among individuals.

In prehistory, color names began to be attached to widely known exemplars, such as red for blood, carnelian stones, or some types of ochre. At some point, humans realized that the color of certain stones or earth types has more or less resemblance to that of blood, and "red" became an idea. The original meaning of red and its cognates in other languages is lost because the terms are very old. The same applies to the other three unique hues, as well as to white, black, and brown. But the original meanings of the exemplars for the basic terms orange, purple, and rose (pink) are still known: the color of citrus fruit, crocus, or saffron; the color of dyes made with bodily fluids of certain types of sea snail (purple); and the prototypical color of roses. In the English language, the term "pink," taken from the name for the *Dianthus* flower, began to be used as a categorical color term only in the eighteenth century, even though the idea of pink as a color existed in England before then. Homeric Greek had a term for pink and at least four terms for red, and one of these, *erythron*, by some cultural process became the basis for the category name in many languages.

The meaning of color words in ancient texts is often uncertain and seems to have varied considerably. That could be because basic category formation was still in progress, and the exemplars were not generally known to later translators. Color terms are relatively sparse in early written documents. The Homeric epics include fewer than 25 different color words. Different translators have often used different words for colors.

A curious discussion about Greek and Latin color words was recorded in the second century A.D. by the Roman writer Aulus Gellius in his collection *Noctes atticae* (Gellius 1493/1927). The conversation is between Gellius, the orator Cornelius Fronto, and the sophist philosopher Favorinus. Among Latin names for red, Fronto mentions *flavus* (commonly taken to mean yellow) and *luteus* (saffron-colored, orange). About *fulvus* (taken to mean tawny brown, the color of lions), he comments: "[It] seems to be a mixture of red and green, in which sometimes green predominates, sometimes red"(Book II, p. 215). He continues: "Virgil, when he wished to indicate the green color of a horse, could perfectly well have called the horse *caeruleus* rather than *glaucus*" (p. 217) (*caeruleus* is usually taken to mean sky blue and *glaucus* [from the classic Greek word *glaukon*], blue-gray).

This raises the strange case of the meaning of the Greek term *glaukon*, used in the Homeric epics to describe the color of Athene's eyes (*glaukopis*), variously translated as bluish gray, gray, light blue, grayish green, and flashy. In the later Middle Ages, the meaning of the derived Latin term *glaucus* changed dramatically from blue-gray to yellow. In his listing of colors, the English monk and philosopher Roger Bacon (ca. 1214–1292; see Parkhurst 1990) used *glaucitas* to refer to yellowness and *ceruleus* (wax color) as a yellow color falling between *glaucus* and *citrinus* (yellow and orange, respectively). *Glaucus* was used with the same meaning in the fourteenth century by Theodoric of Freiberg (see entry in chapter 2). In the mid-sixteenth century, Scaliger (1557) consigned it again to the blue category, as did François d'Aguilon (see entry in chapter 2) at the beginning of the seventeenth century.

An early attempt at relating specific color stimuli to specific names (in multiple languages) was made by Richard Waller in the seventeenth century (see entry in chapter 3). A more modern example is Maerz and Paul's *Dictionary of Color* (1939), with 7,056 printed and mostly named color samples. Many of the names were derived from use in fashion, and today a large percentage is without generally recognizable meaning.

An attempt at a "universal" color language was made in the mid-1950s by the American Inter-Society Color Council in cooperation with the (then) U.S. National Bureau of Standards (Kelly and Judd 1976). The object color space as represented by the Munsell system (see Albert Henry Munsell, chapter 5) was divided into 29 major regions (categories) in an 18-hue circle, with 267 subregions around central colors. The subregions are identified by modifiers, such as vivid, strong, deep, dark, and so on, attached to color names. Most of the 267 central colors are illustrated in a chart in the form of glossy paint chips. They are attached to pages printed in a continuous gray scale from white to dark gray at the corresponding level of lightness. Thus, the samples are seen in an immediate surround of comparable lightness, making lightness differences between different samples more easily visible. This is a pragmatic approach to eliminate the impact of a single surround on the appearance of the chips, but it also highlights the tenuousness of trying to specify color experiences with color chips and the names connected to them.

The absence or vagueness of materials paradigmatic for color names of the classical period makes assignment to one of the 29 major categories a subjective matter. Different directions of cultural development and languages as well as translation issues create additional problems for a full understanding of the intentions of early color system developers.

Color and music

A presumed connection between music and color dates back at least to Pythagoras (ca. 580–500 B.C.). Among his intellectual achievements is the well-known geometric theorem bearing his name and the discovery of musical consonances represented by strings of certain ratios: 2:1, 3:2, and 4:3, corresponding to the octave, the fifth, and the fourth. The important ratios are displayed in the symbolic triangle of the *tetractys* (the holy fourfoldness, the triangle generated from dots representing the numbers 1, 2, 3, and 4). The religious Pythagorean society considered him a "divine man," and the *tetractys* was taken to be a universal symbol of harmony. Plato was deeply influenced by Pythagorean thought, and what is known today about Pythagoreanism derives mainly from Plato's interpretations. Aristotle reported that, for the Pythagoreans, all things are numbers or relate as numbers do. Implied is the idea that all senses have fundamental similarities. Mentioning specifically the Pythagorean musical ratios as examples (see Aristotle, chapter 2), Aristotle described harmonious color combinations as those resulting from mixture of basic colors in simple

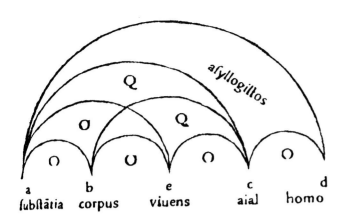

Figure 1.6.

Arc diagram illustrating relationships between logical categories, from a printed edition of Boethius's translation of Aristotle's Libri logicorum *(1503).*

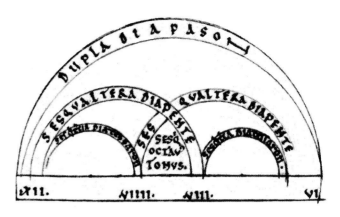

Figure 1.7.

Illustration of the musical diapason from a twelfth-century manuscript of Boethius's De musica *(see Boethius 1492/1989).*

ratios. He believed in a fundamental relationship between color, sound, and taste. One of the musical tone scales of ancient Greece is known as the chromatic scale. Aristotle's authority assured that such views remained received knowledge until after the time of Isaac Newton.

After Pythagoras, several other Greek philosophers wrote on music theory, Aristotle, Euclid, Nichomacus, and Gaudentius among them. Some of these writings were transmitted to the Middle Ages by Anicius Manlius Severinus Boethius (ca. 480–524) of *De consolatione philosophiae* fame. Boethius also translated and commented on Aristotle's influential treatises on logic, the *Libri logicorum* (Aristotle 1503). With his text on music, *De institutione musica* (see Boethius 1492/1989), Boethius helped to assure a place for music in the *quadrivium*, the upper division of the seven liberal arts taught in medieval universities. Boethius believed some of the logical and musical relations described in these works to be difficult for readers to understand, so he developed the idea of using graphical illustrations to aid comprehension. Boethius is quoted as having said in a commentary to his translation of a mathematics manuscript to have sometimes used paradigms and diagrams to make things clearer (Boethius 1492/1989). These are perhaps the first examples outside of geometry and geography of using graphics in aid of perspicuity.

In both works, Boethius used different variants of graphical figures. In the *Libri logicorum*, the figures show relationships between logical entities; in *Musica*, they show the division of the diapason, the compass of musical notes, in the form of arc loops. Figure 1.6 is from the 1503 edition of Aristotle's *Libri logicorum*, and figure 1.7 is from a twelfth-century manuscript of *Musica*. Boethius commented on the relationship between color and music:

Just as when a rainbow is observed, the colors are so close to one another that no definite line separates one color from the other – rather it changes from red to yellow, for example, in such a way that continuous mutation into the following color occurs with no clearly defined median falling between them – so also this may often occur in pitches. (Boethius 1492/1989, p. 103)

The basic idea of Boethius's graphical representation was used in the fourteenth century by the French mathematician Nicole Oresme to compare ratios of regular polygons, and by English mathematician Richard Swinehead to illustrate changes in force-resistance as a function of changes in speed. Robert Fludd (see entry in chapter 2) also used the idea in his alchemical text *Utriusque cosmi* of 1621 to show God's descent by three harmonic octaves to imbue man with spirit.

It is not surprising that diagrams of the same nature should result from considerations of color order. As described in his entry in chapter 2, the first known such diagram is by d'Aguilon, a Belgian Jesuit, found in his 1613 book *Opticorum libri sex*. A somewhat modified version was offered in 1646 by the polymath and Jesuit Athanasius Kircher, who attached to the color diapason a wide-ranging series of correspondences. Another Jesuit, the Austrian Zacharias Traber, modified Kircher's diagram slightly in his 1690 book *Nervus opticus*. In 1702, German abbot Johannes Zahn converted the semicircle into a triangle (but maintained Kircher's correspondences).

Philosopher René Descartes (1596–1650) was also a theoretician of music. In his book *Musicae compendium* of 1650, he used numbers derived from the Boethian system to express the relationship of major and minor tones in a circle. It is believed that this figure influenced the form of Newton's color circle (see entry in chapter 6). Newton, being the foremost rationalist of his time, is known to also have been an alchemist and admirer of Johannes Kepler's treatment of a harmonic world, *Harmonices mundi* (1619). Newton acknowledged the influence of musical order on the diagram in his seminal work *Opticks* (1704): "[D]escribe a Circle and distinguish its circumference into seven parts, proportional to the seven musical Tones or Intervals of the eight Sounds . . . contained in an Eight" (p. 114).

The implied relationship between music and color continued to occupy the minds of pan-harmonists until the present, a story that exceeds the bounds of the current text.

Color spaces: dimensions, attributes, and scales

Color space

The term "color space" has been used in different senses. For the purposes of this text, we define it as a conjectural geometric space in which all possible human color experiences are arranged in some kind of rational order, forming a color solid. A three-dimensional solid is implicit in the reduction of stimulus dimensionality from a number such as 31 (one point every 10 nm from 400 to 700 nm) to 3 when the light stimulus is converted to electrochemical energy by three cone types. The conversion of such signals to color percepts proceeds according to an unknown and likely very complex process, as has been touched on above. But it is obvious that the resulting experiences can also be ordered into a three-dimensional solid. Color experiences can be sorted into a color solid only by mental process. Material color solids, on the other hand, are filled with color stimuli: mixed lights in the case of a display on a color monitor or painted chips in the case of atlases.

The shape of the solid depends on the meaning of the distances between individual color percepts. It can be imposed at the outset. Color solids can also be expressed in many different ways as sets of mathematical equations or geometric models. Both imply dimensions. Chapter 2 presents early one-dimensional color order systems. With the invention of the color circle (or color "wheel") in the early eighteenth century, color systems became two dimensional and, not long thereafter, three dimensional. Around the color circle, color experiences are ordered according to the attribute hue. Hues are ordered according to the sequence in which they appear in the spectrum. The order is supplemented by mixtures of lights at different ratios from both ends of the spectrum to form red and purple colors that do not exist in the spectrum. This completes the arrangement of the spectral colors into a perceptually continuous hue circle. The color circle can be generated in the same order by mixing pairs of

three or more selected primary colorants (dyes or pigments). The hue circle is implicit in the mathematical dimension reduction referred to above. Thus, there is a natural, closed hue order that encompasses all known hues.

Two important color experiences from objects are white and black. When white and black pigments are mixed in different proportions in a medium, the result is colorations appearing as grays of different lightness. Somewhat comparably, lights can be of different brightness. The question arises how the brightness/lightness dimension is to be integrated with the hue circle. Mixtures of lights and of colorants provided generally similar answers, if different in detail. If in a yellow paint, for example, increasing amounts of the yellow pigment (in a medium) are replaced by an appropriate blue pigment, the resulting colorations will appear as an increasingly dull, grayish yellow until at a given ratio the color experience is that of a more or less neutral gray (depending on the pigments used; see figure 1.8).

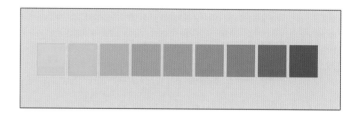

Figure 1.8.

Desaturation of yellow and blue in four steps in the direction of neutral gray.

As the relative concentration of the second pigment increases, the color experience begins to be of bluish gray with increasing blue intensity, until 100% of the blue pigment is reached, with the resulting experience of a saturated blue color. (Achieving a smooth scale of this kind is technically complex, however, because of the specific optical properties of the pigments involved.) A similar situation occurs when spectral lights (or mixtures) placed opposite on a specific type of hue circle (so-called complementary colors) are combined. But at the neutral point, the resulting light when reflected from a white surface appears whitish.

We will refer to chromatic object colors of less than highest possible saturation, such as produced in the manner just described, as *tonal* colors. These are colors that perceptually consist of mixtures of chromatic as well as gray (white and black) experiences. They do not include achromatic grays. If the opposing pigments are mixes to form a mid-level neutral gray (in reality impossible without adding white), it is evident that the appearance of gray from a particular mixture of white and black pigments and that of the mixture of the two complementary chromatic pigments (and white) is the same. The two grays would very likely be members of a metameric set (see "Vision: reconstruction or construction?" above). The painter Phillip Otto Runge first explicitly explained this phenomenon (see entry in chapter 4),

The plane on which the hue circle resides can be filled with tonal chips in the manner just described. Such planes can be repeated for different lightness levels of the central gray. It is logical to stack the resulting planes so that the central grays form a continuous gray scale from white to black, placed at a right angle to the plane of the color circle. The result is a solid formed by the stacked planes. A corresponding geometric solid is obtained with mixtures of the lights of the hue circle. At one vertical extreme, there will be total darkness; at the other, unbearably brilliant white light.

Many details complicate the matter of color order. One of the questions concerns how many colorants or chromatic lights are required to construct a space containing all possible color experiences. Mixtures of three spectral lights of varying intensity, from the beginning, middle, and end of the spectrum, allow generation of stimuli producing perhaps 80% of all experiences that can be generated directly with lights. As mentioned above, because of the broad reflectance functions of colorants and the usually broad spectral power of light in which the colorations are viewed, such reflected light is almost always absorbed to a smaller or larger degree by all three cone types. This results in veiling or desaturation, the generation of tonal colors. While all hues of a hue circle can be generated by binary mixtures of three pigments, all will be desaturated compared to those created from spectral lights, particularly strongly halfway between two pigments.

Desaturation is generally unsatisfactory for painters, who since antiquity have used palettes with multiple pigments to minimize this problem. Rather than live with the dull result of mixing, for example, a red and a blue pigment, many preferred to use a pure violet pigment of the same hue. Desaturation also has been a problem in four-color printing with yellow, red, blue, and black primary pigments, as demonstrated in the eighteenth century by Jacob Christoph Le Blon (1667–1741). Today, up to eight chromatic pigments are used in highest quality halftone color printing to obtain the widest possible range (gamut) of color experiences (see chapter 9 for more discussion of this subject). Only a limited range of the theoretical space containing all possible object colors can be filled with pigments. The color samples of the modern Munsell system (see Munsell Renotations, chapter 7) fill only approximately 60% of the optimal object color space.

The idea of three colors being sufficient to match all other colors has a long history, beginning in written documents around 325 A.D. with Chalcidius (see entry in chapter 2). In the early thirteenth century, Bacon posited three fundamental chromatic color categories (*glaucitas* believed to mean yellowness, *rubedo*, and *viriditas* believed to mean bluegreenness) from which all other colors are derived. A later writer making a similar claim was the Italian sixteenth-cen-

tury theologian Filippo Mocenigo, who designated *flavus*, *hyacinthinus*, and *ruber* as the three fundamental chromatic colors (Mocenigo 1581). *Hyacinthinus* is usually taken as having the meaning of blue or reddish blue. Mocenigo indicated that a mixture of the first two of his primaries forms green (*viridis*) and the second two, purple. A few years later, the same basic idea was put into writing and a diagram by d'Aguilon and, a few years later, by Kircher (see entries in chapter 2). Robert Boyle, in his book *Experiments and considerations touching colour* (1664), stated:

> [T]here are but few Simple and Primary Colours (if I may so call them) from whose various compositions all the rest do as it were Result . . . I have not yet found, that to exhibit this strange Variety they [painters] need imploy any more than White, and Black, and Red, and Blew, and Yellow; these five, Variously Compounded, . . . being sufficient to exhibit a Variety and Number of Colours, such as those that are altogether Strangers to the Painters Pallets, can hardly imagine . . . by these simple compositions again Compounded among themselves, the Skilfull Painter can produce what kind of Colour he pleases, and a great many more than we have yet Names for. (pp. 219–220)

In 1677, the English physician Francis Glisson introduced a color specification system consisting of a yellow, a red, a blue, and a gray scale with which any color can be described, implying that they can all be matched with these colors. In 1708, an anonymous author stated in a French book on miniature painting: "Properly there are only three primitive colours . . . yellow, red, and blue. White and black are not properly colors, white being nothing than the representation of light and black the privation of that light." From these primitives, additional colors are formed by mixture, such as orange from yellow and fire red: "[T]hey make a suite or circle" (Anonymous 1708, p. 152, translated by R.G.K.). Additional colors are mixed from primitives and secondaries, such as purple from carmine red and violet. "All these colors are lively, but other mixtures, for example orange with violet, fire red with blue, violet with green, or green with orange or fire red produces dirty and disagreeable colors" (p. 154).

A similar statement is found in the book *Coloritto* (ca. 1723) by Le Blon, the inventor of four-color printing: "Painting can represent all *visible* Objects, with three Colours, *Yellow*, *Red*, and *Blue*; for all other Colours can be compos'd of these *Three*, which I shall call *Primitive*" (Le Blon ca. 1723, p. 6; italics original). All hues, indeed, can be obtained from three primary colorants or lights, but at a cost in color intensity, particularly in the case of pigments.

On color television and color monitors, as well, the ranges of obtainable colors, the so-called color gamuts, are less than complete. Here the primary colors are located near the beginning, middle, and end of the spectrum and are usually named red, green, and blue (see Color Display Solids, chapter 9). The corresponding lights have spectra significantly broader than monochromatic spectral lights, and thereby they are desaturated. The degree of desaturation differs somewhat according to the technology employed. The theory of why three primary lights are sufficient is due to James Clerk Maxwell, Hermann von Helmholtz, and Günter Grassmann (see entries in chapter 6). As a result (in part), human color vision came to be termed "trichromatic."

Opponent color systems

A different viewpoint also has deep roots. The Italian Renaissance painter, architect, and art theorist Leon Battista Alberti (see entry in chapter 2) proposed in 1435 four primary chromatic colors: red, blue, green, and a not well-defined color he named *bigio et cenericio*. It is generally interpreted to mean a light, dull yellow of some kind. Some 65 years later, Leonardo da Vinci (see entry in chapter 2), having extensive painting experience, described the four primary chromatic colors unhesitatingly as yellow, green, blue, and red, in a sequence that considers the spectrum and suggests the (then not yet existing) hue circle.

The idea of four psychological primary colors was proposed in the second half of the nineteenth century by Hering (see entry in chapter 5). Based on experiment and observation, he concluded that there are four perceptual chromatic *Urfarben* (fundamental colors). These presumably are the most pure and intense versions that we can imagine of yellow, red, blue, and green and are, he thought, fixed in our minds. The hues of these *Urfarben* are unique; that is, in each no components of any of the other three are present. All intermediate hues are obtained by mental mixture of two fundamental colors.

The color solid is filled out by mentally replacing increments of fundamental chromatic colors with the other two *Urfarben*, white, black, or both. Hering named colors containing white and/or black "veiled," a term we adopt in this text. The resulting color solid has the form of a double cone (see Ewald Hering, chapter 5). Each color perception represented in it is the sum, adding to 100, of fundamental chromatic color(s), black, and white. Hering's "natural color system" is based on the idea of opponency. There are three pairs of opposing colors: white–black, yellow–blue, and green–red. Chromatic color perceptions that are simultaneously yellowish and bluish, or greenish and reddish are not possible. The Natural Color System (see entry in chapter 5) implements Hering's ideas.

The opponent color theory has considerable explanatory power for our color experiences. As mentioned above, there are also indications that our physiological color-vision system has an opponent organization of some kind. Efforts to determine fundamental colors, or at least unique hues, have resulted in the interesting facts briefly mentioned above. Individual color-normal observers show considerable variation in choosing color chips that they see as having unique hues. Individual selections of color chips for the four unique hues encompass about two-thirds of the complete hue circle. There is as yet no viable psychophysical model based on cone functions that offers an explanation for unique hues. It is evident that the idea, mentioned above, of simply subtracting one cone signal from another to create a model of perceptual opponency is simplistic.

General rules of ordering

The question of how the human mind orders objects of which it is aware has not yet been convincingly answered. What does it take for us, for example, to recognize a sign as a letter A? A general rule is that we place objects into categories, as discussed above. Within a category, there can be different levels of order. The first of these is known as *nominal* order. In the case of members of a basketball team (as a category), we can order players according to their name or the number on their jersey. It makes no difference if some names or numbers are changed in the next game—the identity of the team remains the same. The second ordering principle is ordinal order. This is a rank order. Basketball players can be ranked according to height or to their average score per game. Colored materials can be rank-ordered, for example, according to attributes lightness, redness, or blackness. In ordinal order, nothing has been expressed about the absolute magnitude of, for example, redness or the difference in redness between two samples of adjacent rank.

When concerned with the magnitude of differences between members of a class, we speak of *interval* order. There can be interval scales of hue, or lightness, or other attributes. Intervals don't say anything about absolute magnitude of the attribute, but the size of the differences is now specified. The resulting numbers can be multiplied by any constant, or any constant can be (uniformly) added to the numbers of interval order, and the relative results remain the same.

Finally, we come to *ratio* order, the order that expresses absolute magnitude. A certain noise is heard as being twice as loud as another, or a distance is measured as being half as long as another. Such scales must minimally have an absolute zero point. In ratio scales, ratios, fractions, and multiples have quantitative meaning. Ratio scales of colors are controversial, however. The meaning of "twice as red" is not obvious to many people.

As will become apparent in the course of this book, all historical color order systems are, to a greater or lesser extent, only ordinal in nature (in terms of perceptual order). Systems that specify color stimuli can be exactly uniform in terms of stimulus increments but will not be perceived as being uniform by observers. Some systems are approximations of perceptual uniformity, however. Because of perceptual variability in observers and issues relating to surrounds, an interval system can be highly accurate only for a small percentage of observers and one set of viewing conditions.

Dimensions and attributes

As this book shows, many kinds of color spaces and many forms of color solids have been proposed. On an ordinal basis, and with the space dimensions undefined, most any geometric space form can be used for the color solid. One may consider all possible color experiences, each represented by a point, to form a "color cloud." This cloud can be shaped into many different geometric forms without losing its ordinal order. Geometric distances between individual color points vary as a function of the geometrical form in which the cloud is presented.

Spaces, and solids placed into the spaces, are either mathematically consistent or not. A mathematically consistent solid fills a portion of a space whose dimensions are directly related to what the solid represents. A cube in an axis system has an internal structure of distances that are related to the distances on the axes. When the color solid represents a continuum of perceptual experiences, a mathematically consistent relationship is much more difficult to achieve, as demonstrated in chapter 7. The perceptual experiences must be in the form of some kind of general attributes. Many different kinds of color attributes are possible. For the color experiences to fit into a three-dimensional space in a mathematically meaningful manner, there must be three independent attributes.

In the late nineteenth and the twentieth centuries, two kinds of perceptual color attributes became prominent:

Hering attributes	Helmholtz attributes	
	Lights	*Objects*
Unique hue content	Hue	Hue
White content	Saturation	Chroma
Black content	Brightness	Lightness, value

Hue, value, and chroma are the attributes of the Munsell system of object colors (see Albert Henry Munsell, chapter 5). With the related attributes hue, saturation, and brightness for lights, they form the most commonly used attributes in color technology and science. In the Cartesian version of such a space, a hue circle is defined by pairs of *x*- and *y*-dimension values of the diagram (these are either negative or positive). The radial angle in the chromatic plane can represent the hue of a chip. The radial distance from the origin can represent the intensity of coloration (chroma) and the vertical dimension can express lightness.

But as discussed in more detail further below, a system in which all internal distances are directly related to corresponding perceptual distances cannot be represented in Euclidean space. On the other hand, if the perceptual attributes are Hering's, mathematical consistency between the axes and the solid is lacking, because the vertical dimension cannot represent lightness in an absolute sense. In addition, equal-judged incremental distances in unique hue, whiteness, and blackness content are not identical with, but distinctly different from, perceptually equal differences (figure 1.9).

As demonstrated above, the stimuli's surround has a significant effect on the perceived color resulting from given stimuli (see, e.g., figures 1.2–1.4). If we look at a group of color chips, we probably look them all in the same set of general surround and lighting conditions. If the general surround is light, the darker samples will look significantly darker than if they are viewed against a dark background, and vice versa. Appearance of the chips changes in different ways with a single chromatic color or a multicolored surround (recall figures 1.2 and 1.4). Color experiences obtained from given chips are standard (for the average observer) only if the surround and lighting conditions are specified.

The question arises of how to define sample chips that are considered standards. Defining them by substrate and colorant recipe provides data of only limited value, due to inconsistencies in colorants, substrates, and processing. Beginning in the early twentieth century, sample chips began to be defined by physical measurement. The applicable

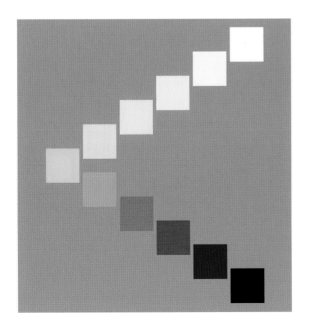

Figure 1.9
Perceptually equal steps from full yellow to white and black. The steps from yellow to white are perceptually smaller than those from yellow to black.

measurement is reflectance factor, the ratio of light incident on the sample to that reflected from the sample. Such measurements have variability of their own. Once all samples are measured, there's still the question of whether the physical measurements relate in some manner to the geometric model of average perceptual differences (ideally in a manner resulting in a linear relationship). This has been a goal of color science and technology for the last 100 years, and the results remain unsatisfactory, as will be shown.

Perceptual scaling

In the case of the Munsell system, the three Helmholtz attributes were scaled individually, with unit differences that differed in magnitude in all three attributes, as discussed below. Thus, the system is only approximately perceptually uniform along attributes, but not throughout (i.e., it is not perfectly uniform, or isotropic).

Earlier systems typically were not even uniform within an attribute. As already mentioned, spacing of a scale from paradigmatic (unique) yellow to red and from red to blue, each with, say, four intermediate equal-percentage steps, does not result in steps of equal perceptual magnitude in both scales. In the color chart of Matthias Klotz, for instance (see entry in chapter 4), hue steps around the hue circle are not uniform, and saturation (tonal) steps toward gray may be reasonably uniform within a hue but differ in perceptual magnitude among hues. Achieving a high level of general uniformity is a difficult technical problem, and a measure of success was not achieved until the middle of the twentieth century.

The question of the relationship between physical measurements of stimuli and the resulting perceptual responses is the domain of psychophysics, invented in the nineteenth century. Psychophysics is an empirical science that as yet lacks a deep basis. But it has proven useful to a degree in many areas. There are many different psychophysical scaling methods, and different methods usually produce different results.[2] In vision, important scales relate to brightness and lightness. Among the possible methods for determining a scale are adding up JNDs and perceptually first dividing in half the step between white and black and then successively halving the remaining differences. The two methods produce somewhat different results. They are also a function of the surround lightness. The results are even more complicated for complex (e.g., natural) surround and viewing situations.

Only in the late 1930s was it discovered that, if the Munsell hue and chroma scales are adjusted to units of equal magnitude, it is not possible to display on a plane the resulting chromatic diagram that appears approximately uniform. The total hue angle of such a system is not 360°, as expected, but approximately twice that size. The American physicist

Deane Brewster Judd named the phenomenon "hue superimportance" and derived a fan-folded geometric model to represent it (figure 1.10). The phenomenon is present regardless of the method of scaling color differences. This is why a perceptually isotropic color solid cannot be represented in Euclidean space.

From the 1940s to the 1970s, a committee of the Optical Society of America (OSA) worked on developing a Euclidean color space of maximum achievable uniformity in which, in a crystalline structure, every color in the interior of the solid was to have 12 perceptually equally different neighbors. The result is the OSA Uniform Color Scales (OSA-UCS; see OSA Uniform Color Scales, chapter 7). However, in the preparatory scaling experiments, the committee also found the hue superimportance effect. As a result, the committee changed its goal to the development of color samples showing the closest a Euclidean space can come to represent an isotropic color space, as well as a corresponding formula. OSA-UCS, therefore, is not isotropic but more uniform than the Munsell space. In formulas for calculation of the average perceptual size of small differences from differences in reflectance values of samples, hue superimportance was dealt with by separate weighting of implied (calculated) hue and chroma differences (see CIEDE2000, chapter 7). Such formulas predict the judgments of an average observer with an accuracy of about 65%. The industries involved have found this useful if not satisfactory. It is not yet known what is required to improve the accuracy, if it is possible.

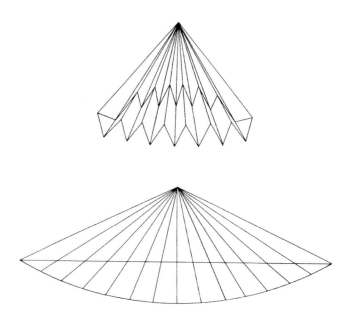

Figure 1.10.
Judd's (1969) folded fan illustration of hue superimportance.

Scaling by colorant weight

A different approach to scaling has been practiced since the late nineteenth century, initially by dyestuff and pigment manufacturers who wanted to demonstrate the range of colors that their products could achieve. Rather than perceptual scaling, they used scaling by colorant weight or volume. Figure 1.11 illustrates a triangle with three commercial dyes at given concentrations in the corners and regularly changing concentrations elsewhere. In this case, the total concentration of dye on the weight of fabric was maintained with only the ratios changing in regular intervals. As is evident from figure 1.11, perceptual differences between the grades vary widely in magnitude. Such triangles were often prepared at different total dye concentrations (making the charts on average lighter or darker), and the color matcher was able to find approximate dye recipes by interpolation of concentrations in the charts. These triangles encompass only a comparatively small portion of possible object color experiences.

Figure 1.11.

Dye mixing triangle with pure dyes in the corners and regularly changing weight ratios along the sides and in the interior. The total dye concentration of dye on weight of fabric is the same for all patches.

The growth in halftone printing with three chromatic pigments, and the identification and industrial standardization of the optimal process inks cyan, magenta, yellow, and black (CMYK) created an interest in systematically demonstrating the colors that can be achieved as a function of dot size, printing sequence, paper quality, and other technical parameters. Such systems are usually presented in cube form with yellow, magenta, and cyan in three corners and black in the fourth (white is obtained from the paper itself). Between primary colorants as well as in the interior of the

cube, "concentrations" vary in a regular manner by changing dot size and degree of over-print. The resulting space is perceptually not uniform and is limited to color experiences achievable with the primary inks, the printing process, and the quality and whiteness of the paper. Examples of such systems are those by Alfred Hickethier and M. and P. Rogondino (see entries in chapter 9).

Similar in nature are spaces implicit in computer monitors. Here the primaries are "red," "green," and "blue" lights generated with phosphors (RGB system). The lights are not spectral, so the achievable gamut (range) of colors is limited but larger than that achievable with print process pigments. A unit component of the image (pixel) is formed by three dots, one each for the three primary lights. The relative light output of each dot can be varied on a scale from 0 to 255. The corresponding scales form the axes of the RGB space.

Many software programs allow the choice of colors within the gamut as a function of output intensity. Because of the linear nature of the representation in RGB (or, derived from it, HSB for hue, saturation, and brightness), hue in a given window is usually not perceptually uniform. Color values typically can be selected not only in the form of RGB but also in HSB, an imitation of process colors (CMYK), and CIELAB-like L, a, b values (see CIELAB, chapter 7). The display changes as a function of the scale system used.

Ordering without reference to the human visual system: physical spaces

As briefly touched on above, the "filters" of cone sensitivities reduce to three dimensions the data arriving at the eye in the spectral power signatures of light. High complexity or dimensionality of information is in this manner reduced to lower complexity (three dimensions). (In the case of colors, high dimensionality might mean, say, 31 dimensions: one data point for every tenth wavelength from 400 to 700 nm.) When using standard cone sensitivity data as filters, the dimensionality reduction of the data represents that from an observer considered average. Dimensionality reduction in this case is psychophysical. Metamers (see "Vision: reconstruction or construction?" above) in this system are seen as approximately equal in color by real observers. Other, purely physical/mathematical dimensionality reduction techniques are also possible (e.g., principal component analysis) and have been applied to color stimulus data, such as the reflectance functions of Munsell chips.

The treatment of data is here at the physical level only, not involving information about an observer in the form of cone sensitivities. The resulting functions describe different three-dimensional spaces (depending on the mathematical technique used) in which every Munsell chip is also located at a point (see chapter 7). The functions depend on the collection of reflectance data used in their calculation. As a result, the locations occupied by given chips vary according to method and data set used. Members of metamer sets related to a standard observer occupy different positions in spectral spaces (they are no longer metamers), and metamers related to a given spectral space occupy different positions in the cone space. For this reason, even though spectral spaces generally place Munsell chips into the same ordinal order (the same sequence) as in the cone space, they should not be called color spaces, given the human, psychological meaning of the term "color."

In this section we have presented some of the intuitive and rational reasons for color order developed over the last four centuries. Color spaces always imply divisions, and the space division offering the most information is the one where geometrically equal steps in all directions represent perceptual steps of equal size. We have briefly mentioned geometric as well as several perceptual facts, but not all, that make it impossible to achieve an isotropic space. In this quest, most has been achieved for the limited case of small color differences evaluated in highly simplified conditions of viewing and surround, of interest to industries manufacturing colorants and colored goods.

Mixing lights and mixing paints

A subject causing much confusion for more than 2,000 years is the seemingly contradictory perceptual results obtained from mixing light stimuli and mixing colorants. In the former case, mixing three primary lights in appropriate ratio results in "white" light; in the latter case, mixing three primary colorants in appropriate amounts results in black. The issue was resolved conclusively only in the second half of the nineteenth century. It continues to be a source of confusion today in the wider population. One of the difficulties is that, due to subject's complexity, clear explanation in a few sentences is difficult.

Fundamentally, normal color stimuli are lights. Sometimes these lights come directly from a light source, for example, the sun, a traffic light, a laser, or a prism. But in most situations we encounter, the light is reflected from an object and thus is usually modified, for example, moonlight or light reflected from a banana, a flower, or a color chip. Surfaces of objects generally have spectrally distinct absorption properties; that is, while light at some wavelengths may be more or less reflected, light at other wavelengths may be more or less absorbed. The absorbed light is missing in the stream reaching our eyes. In this respect, as mentioned above, objects are defined and physically measured in terms of their reflectance properties. The results are expressed as spectral reflectance factor, the ratio of reflected to incoming light. While most real light sources have relatively complex spectral signatures, for purposes of discussion we will use a light with an equal energy (*EE*) spectrum: a light that has

a value of 100 at any wavelength of the visible range. The appearance of this light is colorless and is considered "white."

Individual spectral lights and *EE* light are at this point "filtered" with International Commission on Illumination (CIE) color matching functions, which contain the same information as the cone sensitivity functions shown in figure 1.1, but in a different format (different axes of the implicit space). The result describes the light, transmitted through the lens of the eye and absorbed in the retina, with three numbers: the tristimulus values named *X*, *Y*, and *Z* (see CIE X, Y, Z Color Stimulus Space, chapter 6). The tristimulus values imply a three-dimensional space, but the chromatic aspects of the results can be expressed in the CIE chromaticity diagram. This diagram does not represent brightness information (it would be plotted perpendicularly to the chromaticity diagram).

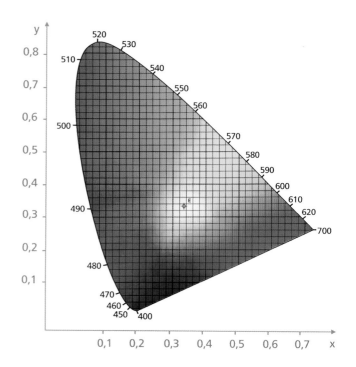

Figure 1.12.

CIE chromaticity diagram with the trace of spectral colors (horseshoe-shaped line) and the purple line connecting its ends. Colors are approximate and illustrative only.

Figure 1.12 is a rough approximation of the appearance of different lights in this diagram. Colorless (*EE*) light is in the center of the diagram. Lights resulting in color perception of approximately identical hue, but varying saturation are on radial lines from the center to the periphery. Spectrally different lights that are metameric (i.e., those that have different spectral functions but the same tristimulus values) plot in identical locations in the diagram. Individual spectral lights form the horseshoe-shaped outline, closed off at the bottom with nonspectral lights that can be generated by mixing in appropriate ratios the spectral lights at 400 and 700 nm. This line segment is known as the "purple line" because lights are located on it that cause people to have red to purple perceptions. This operation discloses the linear nature of the diagram. We can find the implicit chromatic results of mixing any two lights by connecting their locations in the diagram with a straight line. The chromaticity of the mixed light is located on that line and depends on the ratio of the two component lights.

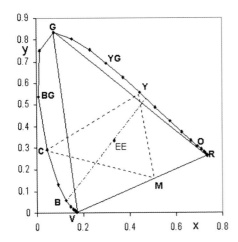

Figure 1.13.

CIE chromaticity diagram with certain spectral wavelengths associated with color names (R, red; O, orange; Y, yellow; YG, yellow-green; G, green; BG, blue-green; C, cyan; B, blue; V, violet; M magenta). EE is the location of the equal energy (colorless) light.

The chromaticity diagram demonstrates clearly why early researchers, such as Helmholtz and Maxwell, used violet, green, and red lights as primary lights in their color mixture experiments. In the version of the diagram in figure 1.13, certain wavelengths have color names attached.[3] The figure shows that there is no combination of three spectral lights with which all other spectral lights can be matched exactly, but violet, green, and red lights come closest. Mixtures of the violet and green lights lie on the line connecting them; the same is true of mixtures of red and green. The graph also demonstrates the empirical fact that appropriate mixtures of green and red lights (as defined in the diagram) have hues from yellowish green through yellow to orange, depending on the mixing ratio.

On and within the triangle *VGR* are the chromaticities of any light that can be mixed with these three lights (we can call them primary lights). In this diagram, hues can be identified by the angle of the line from point *EE* to the point representing a given light to a chosen reference line. The distance along radial lines from *EE* through the chromaticity of a given color to the spectral (and purple) line is a (nonlinearly related) measure of perceived chromatic intensity or saturation. Spectral lights are seen as maximally saturated, which means that the points defining all possible lights must fall on or within the outline of the diagram.

Three other spectral lights are identified as yellow (*Y*), cyan (*C*), and magenta (*M*). Dashed lines on the diagram form the triangle containing the chromaticities of all possible mixtures of these lights. From these lights we can also create lights that have all the hues of the hue circle. However, *saturation* levels are generally much more limited than in the case of primary lights *VGR*, thus this general rule: A large number of appropriately selected spectral triples can generate hues, but some can result in more highly saturated appearances than others.

Any such triple of lights can be called primaries when they have been used as a basis for mixtures. The appearance of colorless light *EE* can be matched (among others) with appropriate pairs of spectral and purple line colors (connected with a line passing through *EE*). The dash-dot line from blue *B* through *EE* ends up near yellow *Y*, indicating that light *B* and a light with slightly higher wavelength than *Y* is required to match the appearance of *EE* (i.e., they will desaturate each other completely). Light *EE* can also be matched with triples, for example, the two sets already mentioned, or by violet *V*, yellow-green *YG*, and magenta *M*, or an endless number of additional combinations. We can use four, five, ten, or more wavelengths for the mixture, and we know that if we mix all wavelengths appropriately, the result is also colorless light *EE*. The diagram is a fully quantitative tool that shows the outcome of light mixtures in terms of chromaticity for one particular observer, the CIE standard observer. Individual color-normal observers differ in their results.

The first semiquantitative form of this kind of diagram was produced by Newton at the end of the seventeenth century (see figure 6.1 in chapter 6). Two important issues not considered in this discussion are brightness of lights and slit width of the apparatus producing spectral lights. Both are complications the full consideration of which is beyond this

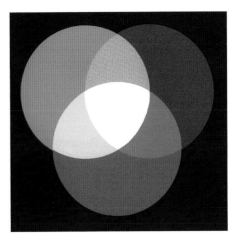

Figure 1.14.

Illustrative depiction of the results of addition of red, green, and violet light beams.

level of discussion. Regarding brightness, the following applies: As mentioned above, appropriate mixture of *R* and *G* spectral lights form near-spectral *Y* light, a light that is, as is known from experience, brighter than the two component lights. When *Y* light is appropriately mixed with *B* light, the result is the colorless light *EE* that is even brighter than *Y* light. Mixing primary lights *G* and *V* in appropriate ratio results in a slightly desaturated *C* light and the appropriate mixture of *V* and *R* light results in *M* light. *C*, *M*, and *Y* are usually described as the result of binary mixture of *R*, *G*, and *V*, with the appropriate triple mixture resulting in *EE*. Because the lights are added together, this kind of mixture is known as "additive color mixture." It is often illustrated with an image such as that of figure 1.14, where red-, green-, and violet-appearing light beams are projected, partially overlapping, onto a white screen.

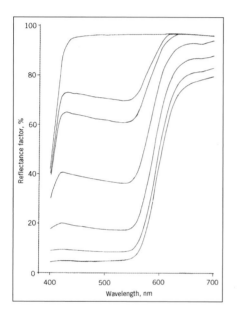

Figure 1.15.
Spectral reflectance functions of a red pigment (bottom curve) diluted with increasing amounts of a white pigment (top curve). From Berger-Schunn (1994).

But what about mixing dyes or pigments? As mentioned above, some artists consider yellow, red, and blue to be the primaries of pigment mixture. While it is true that any hue of the hue circle can be obtained with these pigments, the chroma and, particularly, the lightness of binary mixtures are often much reduced. At the same time, yellow, red, and blue pigments can be matched in hue from binary mixtures of pigments with other hues. (For an example of a "true" red from a yellowish and a bluish red, see Anonymous, chapter 3; however, the farther apart the parent pigments, the darker and less saturated the result will be.)

For this reason, painters who want high chroma use multiple pigments, each at the highest chroma available. While spectral lights can be of a single wavelength, colorants always

have broad and irregular spectral reflectance properties. Nonfluorescent colorants do not produce light, but absorb light at certain wavelengths; that is, they subtract from the light that the light source throws at them. Figure 1.15 illustrates typical reflectance functions of a red pigment mixed in different ratios with a white pigment, in a lacquer medium. The bottom curve represents the red pigment only; the curves above represent increasing additions of the white pigment (shown in pure form in the top curve). The lower the amount of white pigment, the more light is absorbed and

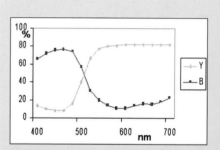

Figure 1.16.
Spectral functions of the lights and pigments used in the example. Reprinted with permission.

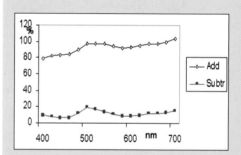

Figure 1.17 *Spectral functions of the additive mixture (top) and the subtractive mixture (bottom) of the spectral functions of figure 1.16.*

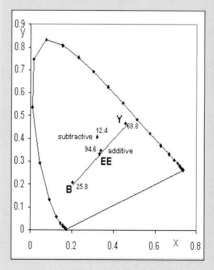

Figure 1.18.
Results of the additive and subtractive mixture of the spectral functions of figure 1.16. EE is the location of the equal energy illuminant. The numbers next to the locations of the stimuli and mixture results represent their luminance and luminous reflectance values, respectively.

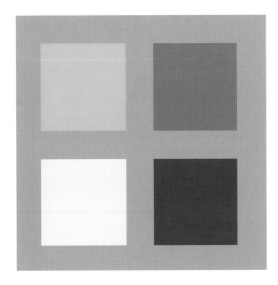

Figure 1.19.

Perceptual results of the additive and subtractive mixture. Top left, yellow light or paint; top right, blue light or paint; bottom left, result of additive mixture as lights; bottom right, result as subtractive mixture of paints.

the darker and (up to a point) the more saturated the color appears. The rules to learn are that (1) colorants are broadband absorbers of light, and (2) the smaller the ratio of white pigment to colored pigment, the more light is absorbed.

To demonstrate the difference in results between mixing lights and mixing colorants, we use a simplified case of binary mixture where the spectral functions of light as well as colorants are identical. In figure 1.16, the blue curve represents the reflectance function of a blue pigment in a given ratio with white pigment, and the yellow curve comparably for a yellow pigment. If the observer views the corresponding colorations in colorless *EE* light, the two functions can also be considered to represent the light reflected at each wavelength from the painted samples.

For the next step, imagine a piece of equipment where the total light reflected from each sample can be collected and the two lights can be mixed together (superimposed). The location of this additive mixture of two lights in the chromaticity diagram is then compared to that of the light reflected from a painted sample where both pigments have been applied in a mixture in the same concentrations as individually. To calculate the reflectance function resulting from this pigment mixture, it is necessary to apply the Kubelka-Munk law.[4] Figure 1.17 illustrates the two resulting spectral functions. The top curve shows the result of the addition of the two lights reflected from the individual painted samples, and the bottom curve is the reflectance function of the pigment mixture and thereby represents the amount of *EE* light reflected from the combined sample.

The corresponding plot in the CIE chromaticity diagram is shown in figure 1.18. Because lights and painted samples before mixture have the same functions, they plot at the same location in the diagram. Mixing the lights is additive, and the result must plot on the straight line connecting the two individual lights. Mixing the two pigments and viewing the result in *EE* light produces a chromaticity that is nonlinearly related to those of the individual pigments. It is the result of so-called "subtractive color mixture." The light mixture is nearly colorless, with a slight greenish tinge compared to *EE* light. The result of the pigment mixture is a more saturated green.

The much larger difference, however, is in the perceived brightness of the light mixture versus the perceived lightness of the painted pigments. This is expressed in terms of luminance for the lights and luminous reflectance for the painted sample (both identical to the CIE tristimulus value *Y*). The *Y* values are given in the figures next to the lights and painted samples in the chromaticity diagram. The result for the light mixture is the sum of the individual *Y* values. For the pigment mixture, the result is less than half that of the darker pigment, the result of broad light absorption across the spectrum by one or the other of the two pigments. The appearance is near white in one case (light reflected from a white screen) and a dark olive in the other, approximated in figure 1.19.

The saturation of the mixture color depends on the saturation of the two component colors and the nature of the overlap of the two reflectance functions. Mixing high-saturation yellow, red, and blue colorants in appropriate ratios results in black coloration, as Johann Heinrich Lambert (see entry in chapter 4) found to his surprise, because most light is absorbed by one or another of the three pigments. Compared to colorless light, black paint can be produced with endless combinations of pigments, from two to many. (In practice, economics usually prevents mixtures of pigments from being used for blacks.) This is illustrated for two different sets of pigment triples in the centers of Moses Harris's prismatic and compound color charts (see entry in chapter 3).

As indicated in the beginning of this section, the difference between the results of light mixture and colorant mixture was not understood until the second half of the nineteenth century and before that caused endless confusion. To come to a full understanding required an ability to make spectral measurements of lights and objects and to have a system, such as the CIE colorimetric system, in which to express the results quantitatively. Particularly difficult to grasp is the nonlinear nature of colorant mixture. Early developers of color systems struggled endlessly to achieve colorations that expressed their intent in the system. Only with computer-supported colorant formulation was it possible to obtain desired results easily.

Coloring the solid

How to practically scale the attributes of their lines, circles, planes, or three-dimensional systems has been a great problem for color system developers since the days of Theophilus (see entry in chapter 2). The tools available to system designers are dyes or pigments (more recently also lights on display units). Changes in concentrations of colorants do not directly relate to changes in perceptual attributes. It was therefore necessary to have in mind some perceptual ordering principle and to mix colorants by trial and error until the desired color experience was obtained.

Coloring systematic samples of an object color solid requires matching of imaginary or real reference samples, and knowledge of the optical behavior of the chosen colorants (pigments or dyes) on the selected substrate. Another requirement is the development of a clear concept of coloration in the interior of the color solid. All real dyes and pigments are imperfect colorants. Their absorption of light across the spectrum is irregular and when they are applied in different quantities, the signatures of the reflectance curves change (see, e.g., figure 1.15). When viewed under a given set of conditions, the color of the resulting chips usually varies in all three attributes, but not in a simple relationship. One would expect lightness and chroma changes, but in most cases hue also changes. And chroma usually approaches and reaches a maximum or often even reverses direction, making perceptually systematic coloration difficult.

Another important matter is the gamut that can be covered with a set of colorants. The gamut indicates the range of color stimuli that can be generated with a series of colorants or lights. Because of their imperfect nature, no set of three colorants can fill the entire range of possible stimuli. The largest gamut of material colorants is obtained from dyes used in transparency color film. Pigments on paper have a reduced gamut, while that of dyes on paper falls in between. But dyes are less fast to light, so have rarely been used for color systems. (An exception is the Aemilius Müller atlases; see entry in chapter 5.)

Because of the nature of the cone sensitivity curves, it is not possible to fill the maximal gamut with three spectral lights, either. But choice of lights from the beginning, middle, and end of the spectrum results in a gamut of more than 80% of the maximum. Color television and computer monitors have a reduced gamut because of the relatively broad emission curves of three phosphor compounds that generate the primary lights of the systems. Gamuts of coloration systems are particularly important when transferring color stimuli from one system to another, such as from a computer monitor to the color printer. Figure 1.20 illustrates the respective gamuts. They are not in agreement, and complex calculations are required to produce perceptually acceptable, smooth results when transferring stimuli from one gamut to the other.

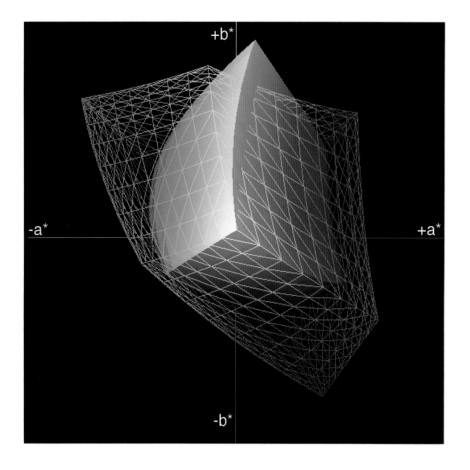

Figure 1.20.
Three-dimensional representations of the gamuts of two different coloration systems in the CIELAB a*, b* diagram. The solid outline represents the gamut of conventional four-color paper printing (CMYK); the wire frame, the gamut of a typical cathode ray tube computer display unit (RBG) (figure reprinted with permission).

The role of practical coloration in the development of color order systems

In the history of color system development, a path from one-dimensional via two- to three-dimensional systems is discernible, as this book demonstrates. The following examples can be taken as paradigmatic for early forms of the systems: the one-dimensional, strictly lightness-oriented scale of Girolamo Cardano (1550), Newton's chromatic color plane of 1704, and Runge's 1810 color sphere (see entries in chapters 2, 6, and 4, respectively). The necessity for three dimensions was clearly understood by the later eighteenth century, but the discussion about the shape of the solid and its interior order has continued to the present. As discussed above, Helmholtz and Hering identified two kinds of three attributes of color perception, resulting in a systematically structured space. But only at the beginning of the twentieth century, with the work of Munsell and Wilhelm Ostwald (see entry in chapter 10), were such ideas converted to extensively colored models and atlases. Only early in the twentieth century was clear understanding gained that most color chips are simultaneously members of at least three separate scales (and as many as six, in OSA-UCS) within a color solid.

That understanding was won not only from thinking about the problem, but also to a very significant extent from practical coloration efforts. The two methods were mutually reinforcing. To understand the role of coloration in this process, it is necessary to distinguish between (1) coloration, referring to exactly separated areas painted in a given color (e.g., color chips or samples that have a clearly specified place in the system and can be used for comparison), and (2) illustration, referring to continuous painted color transitions covering extended areas of the system without having locally homogeneous or clearly defined color fields in areas suitable for comparison.

A few examples given below show important steps in this struggle for developing a clear understanding of the inner structure of the solid, on the one hand, and a way to color samples to conform to the structure, on the other.

Tobias Mayer and Johann Heinrich Lambert

Problems appeared immediately in the eighteenth century with the first attempts at a color solid. Tobias Mayer's concept (see entry in chapter 4) was of a double tetrahedral space with the primary colors yellow, red, and blue at the corners of the central triangle. The outline of the triangle was to be filled with mixture grades so that 12 perceptually equal steps resulted (figure 1.21).

Mayer believed such grades to be just perceptually distinguishable but did not realize that, if the goal was to have just noticeably different grades, the number of grades between

Figure 1.21.
Mayer's (1758/1775) sketch of the basis triangle of his double tetrahedral color solid.

the three primary colorants could not be the same. He also planned 12 steps from that plane toward white and 12 steps toward black, in the erroneous belief that the central color of the central plane, mixed from his primaries, would be a medium gray. Further, he did not realize that 12 equal steps from pure yellow to white or black would be of different perceptual magnitude than comparable steps from pure red and pure blue. Mayer died before making extensive attempts to color his system, and it was Lambert who experienced the practical problems in detail.

Lambert (see entry in chapter 4) noted that, while Mayer identified his individual colors with a systematic numbering system based on amounts of primary colors (see Tobias Mayer, chapter 4), he never explicitly defined the meaning of the numbers attached to the color symbols in terms of colorants. Lambert knew that the numbers could not refer to their weights because of varying coloration strength. So he hired the painter Benjamin Calau to help color Mayer's system.

After some tests, the two settled on gamboge (yellow), carmine, and Prussian blue as the colorants representing neutral yellow, red, and blue most closely. Working in a watercolor system on white paper, Lambert determined the strength of these colorants by finding the relative weights in mixtures of two that resulted in a color perceived as halfway between those of the pure colorants. Ratios varied by combination, and a compromise ratio was calculated that was believed to apply also to mixtures of all three colorants. When mixing the three primaries in full concentration, Lambert and Calau found the result not to be a middle gray but a black, and as a result, Lambert decided there was no need for the lower tetrahedron (figure 1.22).

Figure 1.22.
Lambert's color pyramid of 1772.

Lambert selected seven grades between each of his three primaries in the basis triangle. He noted: "It is immediately obvious that the eight steps from red to blue are less different than those from red or from blue to yellow" (Lambert 1772, pp. 81–82, trans. R.G.K.). The triangular system could not have uniformly equidistant steps based on his primary colorants. To create lighter colors, Lambert and Calau did not add white pigment, but prepared more diluted mixtures.

Lambert described how he selected the colors for the seven triangles constituting his tetrahedron as follows:

In the seventh [top] triangle we have the only uncolored square representing whiteness or light. . . . In the sixth triangle we have light dissolved into its three basic components red, blue, and yellow out of the mixture of which, prismatically as well as in paints, all others are formed. . . . In the fifth triangle . . . we have in addition to the three basic colors the medium mixtures between always two of them. (Lambert 1772, pp. 82–83)

The fourth triangle consists of the three basic colorants with always two grades between two of them:

In the center there is a mixture consisting of equal portions of the three basic colors. When applied in dilution, a somewhat reddish-brown gray color results, the first indication of shadow in paint colors. . . . In the third triangle three main types of such shaded colors are found. On the outline of the triangle there are 12 prismatic colors consisting only of mixtures of two each of the basic colors. . . . These outer colors encompass the three in the middle representing the same number of main classes of darker paint colors. The one adjoining blue [is a slightly brownish gray]. The other two are brown colors, the first with a yellowish cast, the other with a reddish one. . . . When applied less diluted, the same colors are found in the lowest triangle. In the second triangle there are [always five grades between] the primary colors. . . . Here copper and olive colors are beginning to appear. By the way, between the second and the third triangle, as well as between the second and the first, there should be one more triangle if space would have allowed placing them or if it would have been necessary to also show them. . . . There remains the lowest triangle in which the colors . . . are mostly quite dark except where they border on yellow. Yellow cannot be dark without turning to brown. . . . [Four of the colors] appear completely black. (Lambert 1772, pp. 83–84)

It is apparent that Lambert was aware of many of the problems with his tetrahedron. In order to show it on a single page, he had to draw it from a point located above it. So as not to overlap triangles 6 and 7 in the view of the tetrahedron (figure 1.22), he left out intermediate triangles. The dilution levels between the triangles were selected based on trial and error. Triple mixtures make their appearance only on triangle 4. The perceptual differences between grades vary not only along the three sides of the triangles but also along the lines between the three primary colors and white, as well as within the triangles. The fact that two levels of relative lightness were left out results in a lack of systematic relationship in the vertical direction. There is no systematic development of shade colors, and there is no central gray axis. The experimental strength relationships of the colorants proved not to be valid in all combinations and dilutions.

Ignaz Schiffermüller

In 1771, Ignaz Schiffermüller achieved a color circle and three blue tint/shade scales (Schiffermüller 1771; see entry in chapter 3). The circle has engraved partition lines for different colors but (in two inspected copies) is painted in an attempt to be continuous. On the other hand, the three blue scales have spatially defined color samples with attempted perceptually uniform spacing. Initially, Schiffermüller intended to include grays in his blue "plane" but did not penetrate the interior of his implicit solid, remaining instead on its surface. After completing the blue scales, he discontinued his efforts.

Phillip Otto Runge

In his small book *Farben-Kugel* of 1810, Runge (see entry in chapter 4) illustrated his color sphere with two surface views and horizontal and vertical sections. The surface of the sphere is divided by lines of longitude and latitude, the resulting enclosed fields having one color (figure 1.23). Because the grades are colored by hand, they are less than uniform. The equatorial cross section illustrates the problems of mixing imperfect primary colorants: Violet and green are much duller than the primaries. Runge did not realize that mixture of the three primary colorants would result in near-black, and the central gray is noticeably lighter than most adjacent more chromatic grades, resulting in discontinuity. (Ninety years later, Munsell demonstrated the solution for a sphere where all colors on the same latitudinal plane have the same lightness.) The least successful illustration is the polar cross section, where full green is very desaturated, resulting in an excessively grayish half. Runge prematurely died in the year his book was published, and it is not known if he had plans for more systematic coloration of the sphere.

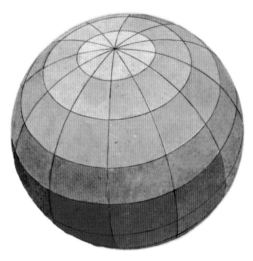

Figure 1.23.
One of the two surface views of Runge's color sphere (1810).

Michel-Eugène Chevreul

A chemist and textile-dyeing expert, Michel-Eugène Chevreul (see entry in chapter 4) had surprising difficulties in creating a three-dimensional color solid, given his background and experience. The difficulties are mainly due to his choice of chromatic plane (reversed from that of Harris). The colors that in Runge's sphere cover the surface are shown by Chevreul on circle segments on a plane, with white in the center, the full color at grade 10, and black beyond grade 20 outside of the circle (see figure 4.26). Each circular slice represents a 21-step tint-shade ladder of constant hue with attempted perceptually uniform steps (within a ladder).

Now the problem arose of how to represent the desaturation steps toward gray. Chevreul decided to raise a hemisphere above the plane with a central, vertical gray scale axis, with black on top. He added increasing amounts of black in 10 steps to the colorations of the plane. The central vertical axis is a gray axis starting with white at the bottom and ending in black at the top (figure 1.24). Chevreul discovered that having all full colors in the tenth position results in a different step magnitude when adding black to yellow than when adding black to blue.

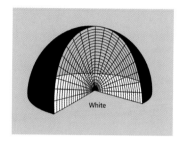

Figure 1.24
Sketch of Chevreul's hemispheric color order system.

In his publications, Chevreul did not illustrate a coloration of his conceptual color solid. Instead, he published 10 hue circles in which the first contains 72 full colors and the other nine contain the same 72 colors with increasing amounts of black added in 10% (perceptual) steps. In addition to a gray scale, he also illustrated 12 tint/shade scales of major hues in 22 grades from white to black. In the plane, subscales are often reasonably uniform, but also here the perceptual magnitude of steps varies from subscale to subscale. Had he colored the proposed blackish colors of the hemisphere, he would have discovered that they do not form perceptually uniform subscales. Chevreul had considerable problems in filling the interior of his hemisphere in a way that was systematic and perceptually uniform, and he never published a complete proposal.

Robert Ridgway

Robert Ridgway (see entry in chapter 10) was an ornithologist whose interest in an ordered color arrangement was for specifying birds' colors. His color atlas illustrates and names 1,115 colors. Ridgway's approach borrowed (implicitly) from Chevreul: He also developed a 72-grade hue circle, with only half of them expressed as samples. Each of the hues is shown in nine-grade tint/shade scales from white to black, with the full color at grade 5.

For the interior of the implicit solid, Ridgway prepared several series of decreasing saturation with four (sometimes five) intermediate grades from the saturated color toward neutral gray. All grades were determined using disk mixture (see glossary), employing for the disks the most highly saturated dyes and pigments he could obtain. Experienced colorists matched the appearance of the disk mixtures with watercolor paints on chips using the same colorants used for the disks. However, there appears to be some discrepancy between Ridgway's description of his scales and the actual scales. It is evident that Ridgway developed linear scales in several directions, without seeking to connect them in an internally consistent manner.

The purpose of this brief discourse is to show that the development of systematic three-dimensional color order systems was beset by problems of how to color the interior. Pioneering efforts by Lambert and Runge fell short in this respect. As said of Ridgway, essentially all of the early developers thought only in terms of linear scales between given endpoints without concern for general uniformity. It was only Munsell and Ostwald (see entries in chapters 5 and 10, respectively) who, in the early twentieth century, offered the first extensive systematic, smoothly varying three-dimensional color systems (albeit conceptually different) in which most color samples are simultaneously a member of at least three color scales.

Colorant selection

All system developers have faced the issue of colorant selection. One aspect is fastness properties. Colorants should have excellent fastness to light and atmospheric contaminants, and the medium in which they are applied should be resistant to soiling so that color chips remain unchanged for extended periods. Another aspect is the appearance of chips when viewed in different "white" lights. The appearance of the chips should change minimally for light sources other than the primary light for which the system has been developed. That means that the resulting colorations must be color constant (i.e., not change in appearance as a function of light source). It is impossible to develop colorations that are color constant under all light sources, but differences should be minimized. Surprising irregularities within a scale can be seen today in early systems when viewing them under certain fluorescent daylight simulators.

Why color atlases?

Creating a color atlas is an undertaking of considerable complexity, and over the centuries, different kinds of justification have been offered for such an effort. The color circles in *Traité de la peinture en mignature* (Anonymous 1708) were published in a book on miniature painting and served the purpose of instructing painters in the results of mixing primary colorants (see Anonymous, chapter 3).

Lambert's primary purpose was to show what a systematic color arrangement looks like. But he also had practical uses in mind. He saw it as a standard collection of colors for tailors and fabric stores to use. The owner could make sure he had fabrics of all colors at hand, and customers could point to specific colors they desired. Lambert also saw it as a useful tool for dyers. If they could relate the coloring strength of their dyes to those of the pigments he used, he thought, dyers could precalculate formulas for colors they had to match and interpolate to obtain intermediate colors. Johann Christof Frisch, Runge, and Klotz were painters, and the primary purpose of their color systems was didactic, including having systematic color collections to develop and demonstrate presumed laws of color harmony. Rules of harmony also were high on the list of reasons for Chevreul, Munsell, Ostwald, Alexander Pope, and others.

Schiffermüller, Harris, and Ridgway were entomologists or ornithologists, and their efforts, in part or whole, were to have systematic color collections for identifying colors of insects and birds. Specialty color charts, limited in scope, were also developed to help identify and specify colors of various kinds of materials.

System, space, atlas: some definitions

Following are definitions of terms used in the succeeding chapters for the various constructs of systematic arrangement of color experiences:

Color (order) system is a general term used for any effort to systematically arrange color experiences. It is applied to early one-dimensional scales sorted according to some concept, such as lightness; to two-dimensional scales, such as Waller's color chart; and to three-dimensional efforts of many kinds, such as Lambert's tetrahedron, Runge's sphere, and the RGB system of color monitors.

Color space is a three-dimensional axis frame, such as in an opponent color space, defining (in a mathematically consistent system) the dimensions and their divisions of the related color solid.

Color solid is the volume occupied in a color space by symbols of all possible color experiences. For a color system such as Munsell's, color chips fill the solid in the corresponding perceptual color space to the extent achievable with available pigments. A MacAdam color solid (see David Lewis MacAdam, chapter 6) represents all possible object color stimuli as viewed by an average observer under a given light source in CIE x, y, Y color space (see CIE X, Y, Z Color Stimulus Space, chapter 6).

Color atlas is a systematic collection of color chips or color prints encompassing a large systematic range of possible experiences. Examples are atlases of the Munsell system, OSA-UCS, Natural Color System, DIN 6164, Colorcurve, and many others.

Color collection is a more or less systematic but incomplete selection of color chips. Examples are the Pantone and the RAL incomplete collections or the typical offerings in a paint store.

Notes

1. For an extended discussion of the subjective or objective nature of color, see Byrne and Hilbert (2003). For a defense of subjectivism, see Hardin (2004). For a sharply critical view of the enterprise of color science, see Saunders and van Brakel (2003).

2. In recent years, deeper doubts about the possibility of a reasoned basis for the relationship between stimulus and response have developed as a result of analysis of experimental data. Laming (1997) has concluded that there is no simple relationship and that judgments of magnitude and, implicitly, of difference are primarily the result of the judged magnitude of the previously viewed stimulus or stimulus difference. As a result, reliability of judgment deteriorates if the stimulus difference from one judgment to the next varies strongly such as in randomized presentation of sample pairs.

3. The color names selected are used only categorically. The relationship between spectral wavelengths and perceived color is quite complex and depends on light intensity and the individual observer.

4. The relationship between chromatic pigment concentration and reflectance is not linear, as figure 1.15 implies. The empirical law that linearizes the relationship so that the results of pigment mixtures can be calculated with relative ease was found in the 1930s by the two German physicists, P. Kubelka and F. Munk.

References

Anonymous. 1708. *Traité de la peinture en mignature,* The Hague: van Dole.

Aristotle. 1503. *Libri logicorum, ad archetypos rocogniti, cum novis at litteram commentaries, Boethius Severinus interpretatus est,* Paris: Hopilius & Stephani.

Berger-Schunn, A. 1994. Practical color measurement, New York: Wiley.

Berlin, B., and P. Kay. 1969. *Basic color terms,* Berkley: University of California Press.

Boethius, A. M. S. 1492. *Arithmetica, geometria et musica Boetii,* Venice: Gregorius. Reprint: *Fundamentals of music,* Bower, C. M., trans., New Haven, CT: Yale University Press, 1989.

Boyle, R. 1664. *Experiments and considerations touching colours,* London: Herringman.

Byrne, A., and D. R. Hilbert. 2003. Color realism and color science, *Behavioral and Brain Sciences* 6:3–64.

Cardano, H. 1550. *Hieronymi Cardani medici mediolanensis de subtilitate libri XXI,* published simultaneously in Nürnberg, Germany, and Lyon and Paris, France.

Descartes, R. 1650. *Renati Des-Cartes musicae compendium,* Utrecht: Zyll & Ackendyck.

Fludd, R. 1621. *Utriusque cosmi,* Oppenheim, Germany: de Bry.

Gellius, A. 1493. *Auli Gellii noctes atticae,* Venice: Lazaroni. Reprint: *The attic nights of Aulus Gellius,* 3 vols., Rolfe, J. C., trans., London: Heinemann, 1927.

Goodale, M., and D. Milner. 2004. *Sight unseen: an exploration of conscious and unconscious vision,* Oxford: Oxford University Press.

Hardin, C. L. 2004. A green thought in a green shade, *Harvard Review of Philosophy* 12:29–39.

Harnad, S. 2003. *To cognize is to categorize: cognition is categorization,* paper presented at UQàM Summer Institute in Cognitive Categorisation, June 30, 2003, Montreal, Canada. Available at http://www.ecs.soton.ac.uk/~harnad/Temp/catconf.html.

Judd, D. B. 1969. Ideal color space, *Palette* 29:25–31, 30:21–28, 3:23–29.

Kelly K. L., and D. B. Judd. 1976. *Color: universal language and dictionary of names,* NBS Special Publication 440, Washington, DC: U.S. Government Printing Office.

Kepler, J. 1619. *Harmonices mundi libri V,* Linz, Austria: Plancus.

Kuehni, R. 2004. Variability in unique hue selection: a surprising phenomenon, *Color Research and Application* 29:158–162.

Lambert, J. H. 1772. *Beschreibung einer mit dem Calauischen Wachse ausgemalten Farbenpyramide,* Berlin: Haude und Spener.

Laming, D. 1997. *The measurement of sensation,* Oxford: Oxford University Press.

Le Blon, J. C. ca. 1723. *Coloritto; or the harmony of colouring in painting: reduced to mechanical practice,* London.

Malebranche, N. 1674. *De la recherche de la verité,* Paris: Pralard.

Maerz, A., and M. R. Paul. 1939. *The dictionary of color,* New York: McGraw-Hill.

Mayer, T. 1758. De affinitate colorum commentatio, in *Opera inedita Tobiae Mayeri,* Lichtenberg, G. C., ed., Göttingen, 1775.

Mocenigo, F. (Philippus Mocenicus). 1581. *Universales institutions ad hominum perfectionem,* Venice: Manutius.

Newton, I. 1704. *Opticks,* London: Smith and Walford.

Parkhurst, C. 1990. Roger Bacon on color: sources, theories & influence, in *The verbal and the visual,* Selig, K.-L., Heckscher, W. S., eds., New York: Italica Press.

Purves, D., and R. B. Lotto. 2003. *Why we see what we do,* Sunderland, MA: Sinauer.

Regan, B. C., C. Juillot, B. Simmen, F. Viénot, P. Charles-Dominique, and J. D. Mollon. 2001. Fruits, foliage and the evolution of primate colour vision, *Philosophical Transactions of the Royal Society London* B 356:229–283.

Runge, P. O. 1810. *Die Farben-Kugel oder Construction des Verhältnisses aller Mischungen der Farben zueinander,* Hamburg: Perthes.

Saunders, B. A. C., and J. van Brakel. 2003. The trajectory of color, *Perspectives on Science* 10.3:302–355.

Scaliger, J. C. 1557. *Iulii Caeasaris Scaligeri exotericarum exercitaionum liber quintus decimus de subtilitate ad Hieronum Cardanum,* Paris: Vascosani.

Schiffermüller, I. 1771. *Versuch eines Farbensystems,* Wien: Augustin Bernardi.

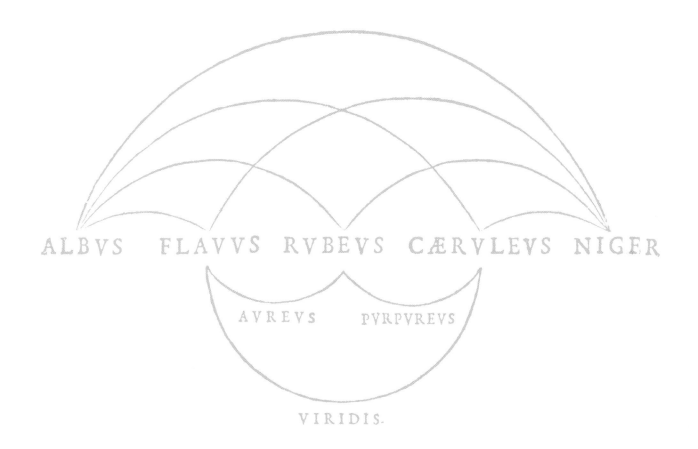

ALBVS   FLAVVS RVBEVS CÆRVLEVS NIGER

AVREVS   PVRPVREVS

VIRIDIS.

CHAPTER 2

LINEAR SYSTEMS

A description of linear (one-dimensional) color order systems beginning with Aristotle in classical Greece and reaching into the twentieth century.

The Cro-Magnon (the early direct ancestors of Europeans) brought cognitive awareness of forms and colors (and a primitive language) on their path out of Africa. This is witnessed impressively in the Chauvet cave discovered in 1994 in southern France, with 32,000-year-old artwork. The most widely used colorants were charcoal and ochre.

Beginning approximately 7,000 - 8,000 years ago, use of multiple colorants to achieve given colors is known from wall paintings in Catal Höyük, Turkey, and some 5,000 years ago from Egyptian statues and paintings of Dynasty IV. Writing is believed to have begun soon after, as demonstrated by Egyptian hieroglyphs and Mesopotamian clay tablets.

From classical Greece to the Renaissance

While we can imagine that people in many places gave thought to the mysteries of the world, our knowledge of early thinking on such matters is limited to transmitted writings, primarily from the Greeks. Pythagoras, mystic and philosopher living in the sixth century B.C., believed the universe to be based on numbers. None of his writings is extant, but some of his ideas are known from the writings of disciples. Pythagoras believed in universal harmony and established principles of musical harmony.

According to Pythagoras's disciple Philolaus, he equated colors with the number 5, the first four numbers presumably equating to the four classical elements of the universe. According to Plutarch, the Pythagoreans named the basic species of color white, black, red, and yellow. The Pythagorean color categories became influential in early Greek philosophy and art and were related to the four elements of nature.

Some 200 years later, Aristotle extended the number of basic color categories from four to seven by adding (what we take to mean) blue, green, and purple. His categories were arranged in a linear, lightness-dependent order from white to black that was influential for the next 2,000 years. Aristotle's ordering system (see entry in this chapter) may have been meant to be a lightness scale, given his belief that all colors are created from light and darkness. From today's viewpoint, one might think that *alourgon* (purple) is a darker color than *prasinon*, the green generally translated as leek green, but we have no direct knowledge of the typical appearance of purple dyeings in his time. The workings of the cognitive process of category generation are largely unknown, and it is not known whether Aristotle simply reported the additional categories that had established themselves in the Greek population (out of a much larger number

[some 140] of color names in the Greek language of the time) or if he added them out of his own insight.

Aristotle chose the number 7 for his simple colors, perhaps in part because seven has connoted a magical number since ancient times. As some commentators, for example, Nicolaus Leonicus Thomaeus (1523), have pointed out, depending on the treatment of yellow and gray in Aristotle's statement, there are either six or eight simple colors. Translators and commentators also continue to discuss the identity, not just the number, of Aristotle's simple colors. Table 2.1 gives an idea of the breadth of interpretation.

The English thirteenth-century encyclopedist Bartholomaeus Anglicus (see Robert Grosseteste, this chapter) used the term *glaucus* for yellow (as did Theodoric of Freiberg and, later, Roger Bacon). He also placed red in the center and replaced green with purple. To achieve symmetry around red, he added orange (*"puniceus, id est citrinus"*) and dropped blue. Bartholomaeus's work was widely copied, and these changes became broadly accepted, as shown by his influence on Lodovico Dolce and Isaac Vossius.

Thomaeus's Latin color words, on the other hand, reasonably agree with Aristotle's Greek ones. Dolce (1565) and Vossius (1662) do not specifically refer to Aristotle, but it is evident that their lists interpret Bartholomaeus's version.

Bartholomaeus clearly sought to improve on Aristotle. Red's placement between white and black appears seemingly for the first time in Bartholomaeus's encyclopedia. He interpolated this idea from various writings of Aristotle. According to Bartholomaeus:

> [I]*nter album et rubeum erit glaucus a parte albi puniceus a parte rubei. Inter nigrum autem et rubeum purpureus a parte rubei, et viridis a parte nigri. Bec nomina alio modo vocatur i Greco ut kianus/purpure/karapos/glaucus. Sed de nominibus grecis non est vis, sed nomina latinorum attendentur.*

> [B]etweene white and red, the yeolow is towarde the white, and the citrine towarde the red, betweene blacke and redde, purple is toward the red, and greene towarde the blacke. These names bee otherwise called in Greeke, for Purple is called *Kyanos* in Greeke, and yeolowe is called *Karapos*, but of names of Greeke is no charge, but we take heede to Latine names. (Batman 1582, pp. 388–389)

Aristotle used the term *charopon* in *De anima* to describe the eye color of certain animals, with its meaning unclear. Other authors liken *charopon* to the color of camel hair or lion skin.

Bartholomaeus went on to describe his logic behind the re-arrangement:

> Citrine and Purple compasseth the redde colour, for either of them hath more of redde then of white or of blacke, but Citrine is farther from blacke then is Purple, as Aristotle meneth, in secundo de Somno & Vigilia, where he speketh of curruption of these colours, and turning into blacke. And he sayeth that Citrine passeth by Purple into blacke. And therefore Purple must be between redde and blacke, and citrine between white and blacke. (p. 389)

Phoinikoun is now often translated as crimson. Antonio Telesio gave *puniceus* its own chapter that describes purples in his small book on color names (Telesio 1528). He explained the name as deriving from the purplish brown fruit of the Greek palm tree named after Phoenix. *Halourgon* is often translated as violet (by others as purple) and was used by Aristotle to describe the color of short-wavelength light of the rainbow. Some 700 years later, Chalcidius (see entry in this chapter), the fourth-century translator of Plato, reduced the simple chromatic colors to three, perhaps from knowledge of painting practices. The colors were (a form of) yellow, red, and blue.

The five chromatic colors of Aristotle denote chromatic categories. Classical Greeks were well aware of the existence of chromatic colors having a given hue but differing in lightness or intensity, but there is no known attempt to classify them into a system. Systematic consideration of such "veiled" colors (to use Ewald Hering's term; see the glossary) began only in the Middle Ages. Knowledge of veiled colors began to be reflected in color theory only in the eleventh century, in the three scales of Avicenna, the Persian

Aristotle	Modern Translation	Bartholomaeus Anglicus	Thomaeus	Dolce	Vossius
1. xanthon	yellow	glaucus	flavus	pallido over violaceo	pale green
2. phoinikoun	crimson	puniceus, id est citrinus	puniceus	croceo o giallo	yellow
3. halourgon	violet	rubeus	purpureus	rosso	red
4. prasinon	leek green	purpureus	viride	purpureo	purple
5. kyanoun	blue	viride	caeruleus	verdo	green

Table 2.1

Aristotle's terms for simple chromatic color and translations/modifications

Century	System	Simple colors
4th BC	Aristotle	2 AC 5 C
5th AD	Chalcidius	2 AC 3 C
11th	Avicenna	proto tint/shade scales
12th	Theophilus	tint/shade scales
13th	Grosseteste	2 AC 14 C plus tint/shade scales
15th	Alberti, Leonardo	4 C, compared to elements
	Ficino	5AC 7 C ordered by lightness
16th	Cardano	2 AC 7 C ordered by estimated lightness degree
17th	Forsius	1 AC and 4 C tint/shade scales ending in common W&B
	d'Aguilon, Kircher	graphical representation of colorant mixture
	Glisson	color specification with 1 AC (gray) and 3 C tint scales

Table 2.2

Historical development of linear color order (AC refers to achromatic, C to chromatic)

commentator of Aristotle. Veiled colors were also acknowledged in the extended Aristotelian linear color scale of the urine circles, a fact also mentioned by Bartholomaeus.

Theophilus, German monk and artisan addressing fellow artisans, described tint/shade scales that were systematic to a smaller or larger degree, but without considering a more encompassing color order.

In the early thirteenth century, English philosopher Robert Grosseteste believed in the Aristotelian concept of colors generated from white and black. It is not clear what perceptions were included in a scale of 14 chromatic colors he described, but each of these were to have an "infinite" number of tint/shade grades toward white and black. His views on color became widely known by the liberal quotations in Bartholomaeus's encyclopedia, without identifying him as the author.

At the beginning of the fourteenth century, Theodoric of Freiberg used empirical knowledge of the rainbow to reorder the Aristotelian linear color scale. In the following centuries, several artists and art theoreticians began to frame the question of color order.

The views of Leon Battista Alberti (ca. 1435) and Leonardo da Vinci (ca. 1550) were influenced by their experiences as painters. Alberti described four basic chromatic colors (red, blue, green, and ash gray or dull light yellow) and compared them to the four classical elements. Leonardo's views changed over time but always were a mixture of classical and pragmatic knowledge.

By mid-sixteenth century, art theoreticians Dolce (mentioned in table 2.1) and Giovanni Paolo Lomazzo appear to have borrowed, with minor changes, Bartholomaeus's description of the Aristotelian scale. Lomazzo provided an even stronger interpretation of the role of red: He claimed that it is possible to generate all chromatic colors from white, red, and black.

Philosophers Girolamo Cardano and Marsilio Ficino were much interested in brightness and lightness as they integrated Roman philosopher Plotinus's neoplatonic thinking into Christian doctrine. Similar views have been offered earlier by Grosseteste. Cardano's scale with estimated numerical values has five chromatic grades. Ficino's brightness scale has five achromatic and seven chromatic grades.

Linear scales from the seventeenth to the twentieth centuries

At the beginning of the seventeenth century, several new ideas appeared. Finnish astronomer Sigfrid Aronus Forsius (1611) conceptually built on Theophilus's tint/shade

scales and formalized them. His graphical interpretation of classical color order appears to be derived from Grosseteste, or perhaps rather from Grosseteste's interpreter Bartholomaeus. But then he posited four tint/shade scales of primary hues and a gray scale, connecting all to common white and black points. Each scale has five intermediate grades, with the full color and medium gray in the middle, in continuation of the classical number of colors in a scale. Forsius concentrated on tint/shade scales without considering hue scaling. His color order system does not seem to have been known outside Sweden until the twentieth century.

But with Aristotle's works firmly placed in the educational canon of the time, his ideas on color order persisted. In English alchemist Robert Fludd's 1629 version of the Aristotelian scale, red is in the center of a circular scale and is considered to generate, together with white, both yellow and orange. On the other hand, colors darker than red are generated from white and black.

Meanwhile, painters' knowledge that they could mix all hues from just yellow, red, and blue continued to assert itself. Those primary colors, along with additional hues, were considered to be generated from black and white. In 1613, the Flemish physicist François d'Aguilon was the first to express this idea by using the Boethian method of visualizing logical relationships. Mixing the three chromatic primaries in combinations of two resulted in three additional categories (green, purple, and orange), and the five Aristotelian chromatic categories became six by the addition of orange.

This idea was embellished later in the century by Germans Athanasius Kircher and Johannes Zahn, as well as Austrian Zacharias Traber. D'Aguilon named only the middle colors of his chromatic mixtures, while Kircher and Traber also named the middle colors of the tint/shade mixtures, if with minor variation, and moved the mixed chromatic colors to intersections of the upper semicircles, as shown in the respective entries. In the second edition of his book, Zahn changed the semicircles to equilateral triangles. A Chinese version of such diagrams was published in mid-nineteenth century by Zheng Funguang.

In the same time period in which Isaac Newton performed his classic prism experiments, the aging English physician Francis Glisson developed a system of object color specifications. In the process, he quantitatively defined the pigment weights needed to produce what he believed to be a visually equidistant gray and red-to-white scale. Interested users were invited to construct yellow-to-white and blue-to-white scales according to the same methodology. Just as in Forsius's system, the full scales end in common white and black.

In the later eighteenth century, Portuguese Diogo de Carvalho Sampayo, perhaps not familiar with developments in color order elsewhere in Europe, made three-grade mixtures of six basic colors in all possible combinations. He did not proceed beyond these simple linear scales even though a double-pyramidal system is implicit in them.

Linear scales of an entirely different nature were offered in the early nineteenth century by the Italian physicist Leopoldo Nobili. Using methods of electrochemistry, he deposited metals atoms on metal plates in a manner resulting in the appearance of interference colors. He believed his method to create colors in natural order and called the resulting scale "metallochromic."

The historical development of linear color order scales is summarized in table 2.2. (see page 29)

As mentioned above, the power of Aristotelian thought influenced thinking on color order as late as the eighteenth century, as illustrated by d'Aguilon and his followers. At the same time, painters and craftsmen knew about tint/shade scales; the documents of Avicenna and Theophilus are among the first to reflect it. Strict ordering by lightness assumed importance briefly as a result of the integration of neoplatonic thought into Christianity. Beginning with Theodoric of Freiberg, several authors anticipated Newton in describing simple colors in spectral order. Forsius formalized tint/shade scales by making them end in common white and black. Glisson, finally, used tint/shade achromatic and chromatic scales to define any object color.

Linear scales are the least complex and easiest-to-comprehend ways to view color order, which is why they have persisted. This can be seen in the tint/shade scales of Michel-Eugène Chevreul, Robert Ridgway, and Jiri Paclt, for example (see entries in chapters 4, 10, 11).

ARISTOTLE   circa 330 B.C.

Aristotle, *Peri Aistheseos kai aistheton, Meteorologia*
Aristotelian School, *Peri khromaton*

Aristotle (384-322 B.C.), born in Stagira in Macedonia, a student of Plato, biologist, philosopher, teacher of Alexander the Great, and founder of the Lyceum school, was the universal scholar of Greek antiquity. In connection with his interest in color, he visited painters' studios, experimented with colored glass, and attentively observed natural phenomena such as rainbows, sunrises, and sunsets.

In the work on psychology, *Peri Aistheseos kai aistheton* (Sense and Sensibilia, part III), Aristotle explained chromatic colors to be generated from mixtures of white and black in Pythagorean harmonic ratios:

> It is conceivable that the white and the black should be juxtaposed in quantities so minute that either separately would be invisible, though the joint product would be vis-

ible; and that they should thus have the other colors for resultants. Their product could, at all events, appear neither white nor black; and, as it must have some color, and can have neither of these, this color must be of a mixed character—in fact, a species of color different from either. Such then is a possible way of conceiving the existence of a plurality of colors besides the white and black; and we may suppose that [of this plurality] many are the result of a [numerical] ratio; for they may be juxtaposed in the ratio of 3 to 2, or of 3 to 4, or in ratios expressible by other numbers; while some may be juxtaposed according to no numerically expressible ratio, but according to some incommensurable relation of excess or defect; and, accordingly, we may regard all these colors as analogous to concords, and suppose that those involving numerical ratios, like the concords in music may be those generally regarded as most agreeable, as for example, purple, crimson, and some few such colors their fewness being due to the same causes which render the concords few. Or it may be that, while all colors whatever [except black and white] are based on numbers, some are regular in this respect, others irregular; and that the latter, whenever they are not pure, owe this character to a corresponding impurity in their numerical ratios. This then is one way to explain the genesis of intermediate colors. . . .

> Colors will thus, too be many in number on account of the fact that the ingredients may be combined with one another in a multitude of ratios; some will be based on determinate numerical ratios, while others again will have as their basis a relation of quantitative excess or defect not expressible in integers. (translated in Barnes 1984, vol. 1, pp. 698–700, reprinted with permission)

Later in the same work (part IV), he stated:

> Savors and colors, it will be observed, contain respectively about the same number of species. For there are seven species of each, if, as is reasonable, we regard dun [gray, *phaion*] as a variety of black [*melanon*] (for the alternative is that yellow [*xanthon*] should be classed with white [*leukon*], as rich with sweet); while [the irreducible colors] crimson [*phoinikoun*], violet [*alourgon*], leek-green [*prasinon*] and deep blue [*kyanoun*], come between white and black, and from these all others are derived by mixture. (p. 702; see figure 2.1)

In Greece, before Aristotle, only four basic colors – white, black, yellow, and red, aligned with the four classical elements – were recognized. Aristotle refers to the seven names as representing species, of which five are hue related. Except for the missing orange, they are the hue-based simple color categories still used today. The sequence recognizes the lightness of yellow (nearness to white) and the relative darkness of object color blue. Whether it should be considered ordered according to lightness in all respects is a matter of opinion.

Meteorologia is a text dealing with the heavens, the earth, and weather. In book 3, there is a description of halos and the rainbow:

> When a cloud is close to the sun, when we look directly at it, it appears to have no colour but to be white, but when we look at its reflection in water it seems to be partially rainbow-coloured. The reason is clearly that, just as our vision when reflected through an angle and so weakened makes a dark colour appear still darker, so also it makes white appear less white and approach nearer to black. When the sight is fairly strong the colour changes to [*phoinikoun*, scarlet], when it is less strong to green [*prasinon*], and when it is weaker still to [*alourgon*, violet]. (Lee 1978, p. 261)

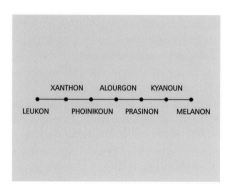

Figure 2.1.
Aristotle's order of seven simple colors.

In a preceding passage, Aristotle wrote: "So if our assumptions about the appearance of colors are correct, the rainbow must be three-colored, and its only colors must be these three. The yellow color [*xanthon*] that appears in the rainbow is due to the contrast of the two others." There is an implicit scale of colors as follows: white, scarlet, (yellow), green, violet, black. Of the three chromatic colors, he wrote: "These colours are almost the only ones that painters cannot manufacture; for they produce some colours by a mixture of others, but [red], green and [blue] cannot be produced in this way" (p. 243), thus affirming their fundamental nature. This scale, obtained from direct observation of nature, is the predecessor of spectral scales of color.

Peri khromaton (On Colors) is believed to represent Aristotle's final thinking on the subject, even though some writers attribute it to Theophrastus, Aristotle's successor at the Lyceum. In sections 1–3 of that text, color generation is treated, and there are only two primary colors, white and yellow:

> Simple colors are those which belong to the elements, i.e., to fire, air, water and earth. Air and water in themselves are by nature white, fire (and the sun) yellow, and earth is naturally white. The variety of hues which earth assumes is due to dyeing, as is shown by the fact that ashes turn white when the moisture that tinged them is burned out. ... Black is the proper color of elements in the process of transmu-

tation. The remaining colors, it may easily be seen arise from blending by mixture of these. . . . This then is the list of simple colors. From these the rest are derived in all their variety of chromatic effects by blending of them and by their presence in varying strength. The different shades of crimson and violet depend on differences in the strength of their constituents, whilst blending is exemplified by mixture of white with black, which gives gray. . . . For it is after this fashion that we ought to proceed in treating of the blending of colors as our basis and making mixtures with it. . . . So we must start from a color previously established, and observe what happens when it is blended. Thus we find that wine color results from blending airy rays with pure lustrous black, as may be seen in grapes on the bunch, which grow wine-colored as they ripen; for, as they blacken, their crimson turns to violet. . . . But we must not proceed in this inquiry by blending pigments as painters do, but rather by comparing the rays reflected from the aforesaid known colors, this being the best way of investigating the true nature of color-blends. . . . We must not omit to consider the several conditions which give rise to the manifold tints and infinite variety of colors. It will be found that variations of tint occur either because colors are possessed by varying and irregular strengths of light and shade (for both light and shade may be present in very different strengths, and so whether pure or already mixed with colors they alter the tints of the colors); or because the colors blended vary in fullness or in powers; or because they are blended in different proportions. Thus violet and crimson and white and all colors vary much both in strength and in intermixture and purity. . . . We never see a color in absolute purity: it is always blended, if not with another color, then with rays of light or with shadows, and so it assumes a tint other than its own. (translated in Barnes 1984, vol. 1, p. 1219, reprinted with permission)

Aristotle's handed-down texts have been contemplated, interpreted, and commented by many people in the past 2,300 years. These texts have monopolized Western thinking on color order into the sixteenth century and continue to have an impact today.

CHALCIDIUS   circa 325
Chalcidii Timaei Platonis traductio et eiusdem argutissima explanatio, 1520

Little is known about the person of **Chalcidius** or his life dates. But we do know this: In the fourth century, he translated (with commentary) Plato's cosmological work *Timaeus* from Greek sources into Latin. Until the twelfth century, this translation was the only source of access to Plato's work for Western Europeans. The first printed version of the extant fragments of the translation appeared in Paris in 1520. *Timaeus* contains Plato's most extensive statement on color. Chalcidius's translation of Plato's statements on color is not extant, but he commented on the subject of color in his section 333 as follows (in part):

> Those [things] most distant from each other are contrary, although they are of the same kind; pure whiteness and, likewise, pure blackness are of the same kind, and called colors. But these two are most distant from each other. The color with the smallest distance from white [*candidus*] is what is called pale yellow [*pallidus*],[1] a little further away is red [*rubeus*], even further is the color blue [*cyaneus*], and at greatest distance is true blackness [*nigredo*], therefore white and black are not different but contrary, with an extended interval between them. (Wrobel 1976, pp. 357–358, translated by R.G.K.)

It is not evident how Chalcidius arrived at this scale of five simple colors (figure 2.2). It appears to be a model for the sixteenth- and seventeenth-century scales of d'Aguilon, Kircher, Zahn, and Traber (see entries in this chapter) and presages the painter's primaries described in 1664 by the chemist Robert Boyle.[2]

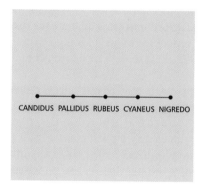

Figure 2.2.
Chalcidius's order of five simple colors.

URINE COLOR SCALES   *Medieval*

These were an early version of a (more or less) linear scale in circular form. In the opinion of John Gage (1993), medieval physicians developed color wheels of this kind to help diagnose digestive diseases. Several such diagrams are found in illuminated manuscripts dating at least to the twelfth century and copied in early printed medical texts (figures 2.3 and 2.4). A tenth-century anecdote describes Notker, a Swiss monk and physician, predicting a miracle from the presumed urine sample of the Bavarian king: that the king would soon give birth to a baby boy. The sample was in reality from one of the king's female servants.

In these scales, urine flasks containing samples of various colors are arranged in a circle, from white to black. It is not known who developed the first such scale. In five versions of the scale from manuscripts and books by different authors, the

sequence of usually 20 colors is found to be identical: white, *glaucus* (yellow), *lacteus* (milky), *charopon* (camel hair color), two versions of *pallidus* (light grayish brown), *subcitrinus* (dark orange), *citrinus* (orange), *subrufus* (dark ruddy), *rufus* (ruddy), *subrubeus* (dark red), *rubeus* (red), *subrubicundus* and *rubicundus* (versions of deep red), an unidentified blackish red color, *kyanos* (blue), *viridis* (green), *lividus* (yellowish gray, lead color), and two versions of black.

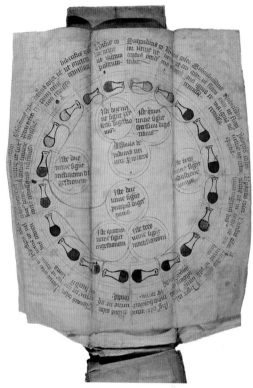

Figure 2.3.
Urine circle from a fourteenth-century physician's belt book. Rosenbach Museum and Library, Philadelphia; Rosenbach manuscript MS 1004/29 fol. 9. Reprinted with permission.

Figure 2.4.
Urine circle from Pinder's Epiphanie medicorum (1506).

In the Rosenbach manuscript (figure 2.3) the sequence begins at the lower left; in Pinder's version (figure 2.4), it begins at the top. The scale is thus an expanded version of Aristotle's scale with the usual ambiguities as to the exact meaning of the colors. Lighter and darker versions of hues are placed next to each other. Depending on the color, the patient's digestion is either judged poor, good, or excessive. Urine color charts of new design remain in use in our time.

AVICENNA   circa 1015

Avicenna, *Kitab al-Shifa, Liber de anima seu sextus de naturalibus*, circa 1015

The Persian physician, philosopher, encyclopedist, mathematician, and astronomer **Abd Ali al Hosain ibn Abdallah ibn Sina** (980-1037) was known in the Latin world by the name **Avicenna**. He wrote more than 90 books on a wide range of topics. But he is best known for his medical work *Al-Qanun fi al-Tibb* (The Canon), an encyclopedic work on medicine that remained highly important for six centuries. Avicenna also translated and commented on the works of Aristotle and other Greek philosophers. In the philosophic encyclopedic work *Kitab al-Shifa*, he included a treatise on the soul, based in part on Aristotle's work on the same subject. It was translated in the twelfth century in Spain into Latin under the name *Liber de anima* and survived in several manuscript copies. *Liber de anima* consists of five sections with several chapters each. The third section discusses vision, and the fourth chapter, color. Avicenna does not dispute Aristotle's seven-color scale directly but argues that there is more than one transition from white to black:

> Moreover, if whiteness does not exist without light and blackness not in ways already discussed then whiteness and blackness cannot only be joined in one manner. A manifestation of this is the fact that white gradually passes to black by three paths. The first is via light yellow-green (*subpallidus*) and its progression is pure: it will indeed be of pure progression, at first it progresses to light yellow-green, from there to yellow-green (*pallidus*), and

continuing in this manner until black is obtained, because thus proceeding to its limit it does not veer from gradually stretching toward blackness, until it becomes pure black. There is also another path proceeding [from whiteness] toward light red (*subrubeus*), and from there to red (*rubeus*), thereafter to black. The third path is the one going to blue-greenness (*viriditas*), from there to indigo (*indicus*), thereafter to blackness. And in these ways not all color diversity can exist, neither can they be the source of the diversity of [Aristotelian] median colors. [A] sort of black smoke mixed with fire will be red if blackness is abundant, or it will be orange (*citrinus*) if blackness is subdued and there is a superabundance of resplendent whiteness; if orange is mixed with blackness and splendor is not present greenness will occur; and altogether with blackness it will be more hidden and with whiteness more palpable. When in the first case blackness is abundant the color will be dark (*fuscus*, dark color of undefined hue), if blackness is abundant in the second the color will be that of leek leaves; on the other hand if it [greenness] becomes more and more it will be intense greenness, for which there is no name. But if whiteness is added to it will be grayish green; on the other hand if blackness with a small amount of redness is mixed into grayish green the result will be indigo; but if redness is mixed to indigo carmine will result. (Van Riet 1972, pp. 205–206, trans. by R.G.K.; see figure 2.5)

Avicenna's subject matter is light rays, rather than object colors. His conclusions are drawn from empirical observations, such as flames that turn red in the presence of black smoke. Avicenna clearly argued that the path from white to black includes additional paths, such as tints and shades, not only directly through the five Aristotelian median colors. As mentioned by Gage (1993), Avicenna's thirteenth-century commentator Al Tusi added yellow and blue tint/shade scales. Similar scales were described in the thirteenth century also by Theophilus (see entry in this chapter). Avicenna's scales are referred to in the thirteenth-century encyclopedic work *Speculum majus* by Vincent de Beauvais.

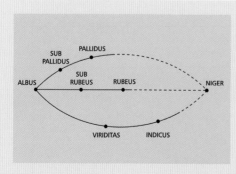

Figure 2.5.
Conceptual sketch of Avicenna's linear scales from white (albus) *to black* (niger).

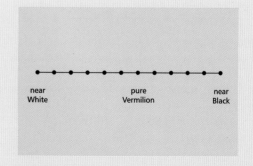

Figure 2.6.
Conceptual sketch of a 12-grade tint/shade scale as described by Theophilus.

THEOPHILUS   circa 1120

Theophilus, *De diversis artibus*, circa 1120

Theophilus, also known as Theophilus Presbyter, is believed to be the same person as Roger von Helmarshausen (1080?-1125?), a German Benedictine monk and priest, goldsmith, and creator of an important portable altar. In approximately 1120, Theophilus wrote a treatise, *De diversis artibus* (The various arts), in which he described technical details for painting, glass making, and metalworking. In book 1 of the manuscript, he described how to mix colorants for painting flesh and hair tones with their shadows and highlights. He also described how to mix colorants for painting draperies and for imitating the rainbow.

For flesh color, Theophilus started with lead white that turned yellowish, to which he added unheated lead white and vermilion (mercuric sulfide, red pigment). For the color of light faces, he added more white; for that of unwell persons he replaced vermilion with green earth. Theophilus described first and second shadow colors, progressively darker. He described first and second rose colors and a dark red for cheeks, first and second highlight colors on the cheeks, as well as a dark gray for eyes. Thus, he described scales of colors centered on average (Caucasian) flesh color and moving in lighter, darker, redder, and greener directions. For example, he described how to get the second shadow color for flesh:

> Afterwards take the [first] shadow color for flesh which has been referred to above, and mix with it more green earth and burnt ochre so that it is a darker shade of the former color. Then fill the middle space between the eyebrows and eyes, under the middle of the eyes, near the nose, between the mouth and chin, on the down or beards of young men, on the half-palms toward the thumb, on the feet above the smaller areas of relief, and on the faces of children and women from the chin right up to the temples. (Theophilus ca. 1120/1986, p. 7, trans. by R.G.K.).

Similarly, Theophilus described various tint/shade scales for painting drapery, for example for a greenish yellow hue:

> Mix pure viridian (green) with yellow ochre so that the yellow ochre predominates, and fill the drapery. Add to this color a little sap green and a little burnt ochre and make the drawing. Mix white with the ground-color and paint the first light areas. Add more white, and paint the lighter areas on top. Mix with the above shadow-color more sap green and burnt ochre and a little viridian and make the shadow on the outside. . . . Mix dark blue with white in the above way. Similarly mix black with white. In the same way mix yellow ochre with white and for its shadow add a little burnt ochre. (pp. 10–11)

He described the colors of the rainbow (in part) as follows:

> The band which looks like a rainbow is composed of various colors: namely vermilion and viridian, also vermilion and dark blue, viridian and yellow ochre, and also vermilion and folium [a vegetable red lake]. . . . Then mix from vermilion and white whatever tones you please so that the first contains a little vermilion, the second more, the third still more, the fourth yet more, until you reach pure vermilion. Then mix with this a little burnt ochre, then burnt ochre mixed with black and finally black. . . . You can never have more than twelve of these strokes in each color range. And if you want these many so arrange your combinations that you place a plain color in the seventh row. (pp. 14–15)

In the rainbow section, Theophilus clearly described 12-grade tint/shade scales for various pure or mixed color pigments that represent the most intense color available for a given hue (figure 2.6). Presumably, he meant the grades to be mixed so that the steps appear approximately even. He did not offer a formal arrangement in which to place the scales. However, it is evident that his proposal results in a two-dimensional arrangement of tint/shade scales from white via full colors to black, such as they are described later by Forsius (see entry in this chapter).

ROBERT GROSSETESTE   circa 1230

Grosseteste, *De colore*, circa 1230
Bartholomaeus Anglicus, *De proprietatibus rerum*, circa 1245

Robert Grosseteste (ca. 1170-1253), an English philosopher, natural scientist, and theologian, translator into Latin and commentator of Aristotle, was active in the Franciscan school in Oxford and served as Bishop of Lincoln. Grosseteste was interested in astrology and mysticism and he translated the complete works of pseudo-Dionysius, the Aeropagite (active in the fifth century A.D.),[3] a text of considerable influence on mystical thought in the later Middle Ages and the Renaissance period. With pseudo-Dionysius, he believed light to be the basis of all matter and that natural phenomena can be described mathematically.

Grosseteste authored a brief text on color, *De colore* (circa 1230), clearly influenced by Aristotle. Most of this text was used nearly verbatim by the contemporary English encyclopedist Bartholomaeus Anglicus. Bartholomaeus's encyclopedia *De rerum naturalis* (Of natural things) attempted to provide an overview of all that was known and written about the universe and man in it.

Completed about 1245, the encyclopedia became hugely successful: Dozens of manuscript copies are extant. John of Trevisa's translation into English at the end of the fourteenth century was first printed in the fifteenth century by William Caxton. It saw at least 50 editions in several languages, the last one in the seventeenth century. In (the final) chapter 19, Bartholomaeus discussed the senses, and in section 8, "Of the opinion of them which would have light, to be of the substance of colour," he reprinted almost verbatim approximately two-thirds of Grosseteste's brief text on color. It includes:

> Some men deeme or suppose, that light is of the substaunce of colour, and they saye, that colour is in cleane and cleere matter. . . . [B]lackness is privation of clearnesse, and for to speake in this wise, hee followeth, that there be seaven colours that stretch from white, toward blacke. And this is known, and three thinges maketh whitenesse, brightnesse of light, and plentye thereof, and pureness of cleare matter. And while meane [intermediate] colour may abate, then in this wise is generation of three colours, if one abideth alone, the other two abate: and so of white cometh 7 colours and stretch from the white toward the blacke: also from black to white stretcheth 7. And by this consideratio colours be 16, two principall, blacke and white, and 14 meane, for 7 stretch from white toward blacke, and 7 from blacke toward white, and in the stretching, the first 7 abate in whitenes, and the other 7, abate in blacknes, and meeteth in the middle. In every meane colour, be as it were endles meane degrees of deep colour and of lyght, as they be farre from white or blacke or nigh thereto. (quoted in Batman 1582, p. 389)

The meaning of Grosseteste's translated words is somewhat controversial, because he did not name the intermediate 14 colors or provide a sketch of his arrangement (figure 2.7 shows an interpretation). The simplest explanation is that it is a description of a 16-color, brightness-ordered linear scale: The lighter colors "stretch" from white and become progressively darker, while the dark colors stretch from black and become progressively lighter, the two scales meeting in the middle. Each of these 14 colors is chromatic, because all of them have associated "endless intermediate

degrees" of lightness and darkness, depending on the content of white or black.

It is very likely that Forsius's first circle (see entry in this chapter) is another interpretation of Grosseteste's idea, but without consideration of the multitude of variation of intermediate chromatic colors

THEODORIC OF FREIBERG   circa 1310

Theodoricus Teutonicus de Vriberg, *Tractatus de coloribus*, circa 1310

Theodoric of Freiberg (ca. 1250-after 1310), philosopher, mystic, scientist, and member of the Dominican order, received higher education in Paris and was an influential church official in Germany. In scientific circles, he is best known for his work on the rainbow, *De iride* (ca. 1305). He was the first to place the cause of rainbows in individual raindrops and investigated the path of light in water-filled glass spheres as models for raindrops. From the results, he was able to explain the formation of the primary and secondary rainbows.

In his short tract on color, *Tractatus de coloribus*, Theodoric discussed, among other matters, white and black as the Aristotelian extreme colors as well as the number and sequence of the simple middle colors. His list is clearly influenced by the earlier work on the rainbow (figure 2.8):

> [T]he second kind [middle colors] differ thus that in red and in yellow [*glaucus*] whiteness prevails, while blackness predominates in green as well as in blue . . . in the first combinations whiteness predominates, but red is nearer to the extreme, which is white, and yellow farther. On the other hand, in the combinations where blackness dominates, blue is closer to black and green is less so. The middle colors stretch in this order to form the rainbow. (Rehn 1985, p. 281, trans. by R.G.K.)

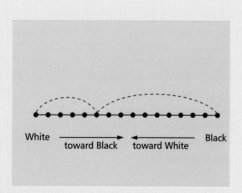

Figure 2.7.
Conceptual sketch of Grosseteste's 16-color scale. The "endless intermediate degrees" between middle colors and black and white are indicated for only one of the middle colors.

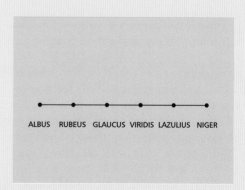

Figure 2.8.
Conceptual sketch of Theodoric of Freiberg's system of six simple colors, ordered according to the spectrum.

Theodoric was not prepared to give up Aristotle's idea that the simple middle colors are formed between white and black. White is placed next to red and black next to blue. He arranged the middle colors in the sequence he had seen in rainbow experiments, and reduced them to four, leaving out purple because it does not exist in the spectrum. (Aristotle had described the colors of the rainbow as being three, *phoinikoun*, *prasinon*, and *alourgon* [red, green, and violet].) Typical for his time period, Theodoric used the term *glaucus* for yellow.

LEON BATTISTA ALBERTI   circa 1435
L. B. Alberti, *Della pictura libri tre*, circa 1435

The Italian architect **Leon Battista Alberti** (1404-1472) was also a painter, poet, and musician, a paradigm of the Renaissance man. Perhaps his most famous achievement as an architect is the Tempio Malatestiano in Rimini. When he wrote his book on painting, he was well aware that at the beginning of the high Renaissance, he was the first author to write a treatise on art theory. Of Alberti as a painter, Giorgio Vasari said: "In painting Leon Batista did not do great or very beautiful work . . . nor is this to be wondered at, seeing that he devoted himself more to his studies than to draughtsmanship" (Vasari 1979, vol. 1, p. 497).

Several Latin and Italian fifteenth-century manuscripts of Alberti's text on painting are extant, although none in his own writing. The first printed Latin edition appeared in 1540. In the first of three books, he discussed the elements of painting, lines, planes, areas, and colors. The second covers object outlines, planes, and color use when painting historical subjects. The third deals with painting's importance and the painter's role.

Alberti viewed colors from a phenomenological viewpoint. He was aware that the apparent color of objects changes, depending on the light in which they are viewed. There is no metaphysical aspect to Alberti's view of colors. His book is a primer on painting and an aid for progressive painters of his time with an interest in sciences.

On colors he wrote as follows:

> It seems obvious to me that colors take their variations from light, because all colors put in the shade appear different from what they are in the light. . . . The philosophers say that nothing can be seen which is not illuminated and colored. . . . Let us omit the debate of philosophers where the original source of colors is to be investigated. . . . However, I do not despise those philosophers who thus dispute about colors, and establish the kinds of colors at seven. . . . It is enough for the painter to know what the colors are and how to use them in painting. I do not wish to be contradicted by the experts, who, while they follow the philosophers, assert that there are only two colors in nature, white and black, and there are others created from mixtures of these two. . . . From a mixture of colors almost infinite others are created. I speak here as a painter.
>
> Through the mixing of colors infinite other colors are born, but there are only four true colors—as there are four elements—from which more and more other kinds of colors may be thus created. Red is the color of fire, blue of the air, green of water, and of the earth gray and ash [*bigia et cenericia*]. . . . Therefore, there are four genera of colors, and these make their species according to the addition of dark and light, black or white. They are thus almost innumerable. Therefore the mixing of white does not change the genus of colors but forms the species. Black contains a similar force in its mixing to make almost infinite species of color. . . . For this reason painters ought to be persuaded that white and black are not true colors but are alterations of other colors. . . . I should like to add that one will never find black and white unless they are [mixed] with one of these four colors. (quoted from Spencer 1966, pp. 49–50)

It is evident that Alberti, on empirical grounds, rejected Aristotle's view of chromatic colors generated from white and black. His four "true" colors are analogs of the four classical elements. Some debate has resulted from his choice of earth's color analog, *bigia et cenericia*.[4] His color order scheme can be loosely interpreted as two-dimensional (figure 2.9).

LEONARDO DA VINCI   circa 1500
L. da Vinci, *Trattato della pittura*, circa 1500

During his life as a painter, architect, sculptor, engineer, and scientist, **Leonardo da Vinci** (1452-1519) kept notebooks with observations and insights connected with his activities. After his death, the notes were sorted into themes; one collection became the *Trattato della pittura* (Treatise on painting). In his wide-ranging career, Leonardo was acquainted with such luminaries as Verrochio (in whose workshop

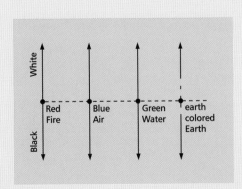

Figure 2.9.
Alberti's comparison of four basic colors with the classical elements and conceptual representation of his color genera and species.

37

he was an apprentice), Botticelli, Ghirlandaio, Lorenzo di Credi, and Machiavelli. He is credited with inventing the painting style of *sfumato*, subtle tonal transitions of colors of comparatively low intensity.

To a base of traditional thinking, Leonardo added broad insights about the generation of shadows and their colors, the effect of reflected light on the color of objects, and effect of the atmosphere on coloration of natural scenes. Leonardo's views on color were influenced by Aristotle and Alberti, but acquired more layers and complexity. What he wrote on color order, however, is slim and not always consistent. The notes indicate Leonardo's plans for a more extensive statement on color that was never written. He held ambivalent views about white and black. At times, he did not recognize them as colors. At other times, he included them because they were needed for *rilievo* (the illusion of volume in painting) and *chiaroscuro* (the achromatic drawing of figure versus ground). He also did not define a set number of simple colors. His most extended statement on color order is as follows:

The simple colors are six, of which the first is white, although some philosophers do not accept white or black in the number of colors, because one is the origin of all colors and the other the absence of them. However, because the painter cannot do without them, we place them in the number of the others, and we say that, in this order white is the first among the simple colors, and yellow the second, green the third of them, blue is the fourth, red is fifth, and black is the sixth. And white is given by light without which no color may be seen, yellow by earth, green by water, blue by air, and red by fire, because there is no substance or dimension on which the rays are able to percuss

and accordingly to illuminate it. (Leonardo da Vinci 1956, p. 176; see figure 2.10)

Somewhat later, though, Leonardo explained: "[B]lue and green are not in themselves simple colors, because blue is composed of light and darkness, like that of air. . . . Green is composed of a simple [yellow] and compound color [blue], that is to say, composed from blue and yellow" (p. 177). Here, color names no longer seem to represent purely perceptual categories, but in part also the results of colorant mixture; the number of simple colors shrinks to the pre-Socratic four. The unusual statement about blue is based on Leonardo's empirical knowledge that a glaze of "thin and transparent white" over the "finest black . . . will exhibit no other color than the most beautiful azure." Leonardo had found some truth in Aristotle's view that chromatic colors are generated from white and black. It is interesting to note that Leonardo's order of chromatic colors relates to the four elements in the sequence of earth, water, air, and fire (sun) and at the same time is in spectral order. It appears that Leonardo did not see value in the classical lightness order.

LODOVICO DOLCE   1565

L. Dolce, *Dialogo nel quale si ragiona della qualità, diversità, e proprietà dei colori*, 1565

Lodovico Dolce (1508-1568) was a Venetian art theorist, playwright, translator, and commentator. He translated classical writers such as Seneca, published a commented version of Dante's *Divina commedia* (he was the first to call it *divina*), and wrote several plays. He is the author of a small book on color published in 1565, and a book on painting (*Dialo-*

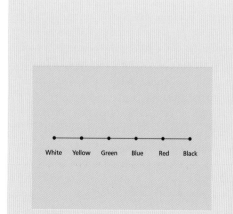

Figure 2.10.
Leonardo's system of six simple colors.

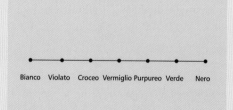

Figure 2.11.
Conceptual sketch of Dolce's (1565) scale of simple colors.

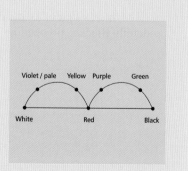

Figure 2.12.
Conceptual sketch of Lomazzo's (1584) second scale.

go della pittura, 1557). Dolce believed that all of creation, including colors, was generated out of light and darkness.

In his work on color, Dolce described the seven-grade Aristotelian color scale, with a translation of color names identical to those reported in the encyclopedia of Bartholomaeus Anglicus (see Robert Grosseteste, this chapter), except for the second color, *glaucus* in the encyclopedia, and the specific use of *vermiglio* for red.

Bianco	white
Violato, pallido	violet, pale
Croceo o giallo	saffron or yellow
Vermiglio	vermilion, red
Purpureo	purple
Verde	green
Nero	black

The selection of *violato*, *pallido* as the color nearest to white is unusual.

GIOVANNI PAOLO LOMAZZO   1584

G. P. Lomazzo, *Trattato dell'arte de la pittura, scultura ed architectura*, 1584

Giovanni Paolo Lomazzo (1538-1600) was a painter in Milan until he became blind when he was 33 years old. As a result, he began to concentrate on art theory and wrote two treatises, *Trattato dell' arte* (1584, described later as the "Bible of Mannerism," a style of painting of the late Renaissance) and *Idea del tempio della pittura* (1590). In the *Trattato*, color is the heading of the third of seven chapters. Here Lomazzo interpreted the generation of colors from various effects of cold and heat:

> Now there be 7 sortes of simple colours, from which all the rest arise. Of these 2 are extreames, as white and Blacke; and 5 middle, as light yeallow, redde, purple, Blewe and Greene. Now concerning the generation of colours, Colde produceth white, whereonto much light is required. Heate engendereth Blacke, proceeding from a smale quantity of light and much heate. Redde is made by the mixture of white and Blacke. Violet or pale of much white and a little redde; saffron colour or yeallow of much redde and little white. Purple of much redde and a little Blacke, and greene of a little blacke and much redde. And this may suffice for the foundation and originall of colours. (from the first English translation, Lomazzo 1584/1598, p. 98)

Lomazzo described two different scales, the first reasonably in line with Aristotle's, the second without blue and interpreting all members in terms of cold and heat, a subject on which Aristotle also commented. In this scale, "pale" and "saffron" are generated from mixture of white and red, and "purple" and "green" from black and red (figure 2.12).

GIROLAMO CARDANO   1550, 1558, 1563

Cardanus, *De subtilitate*, 1550
De rerum varietate, 1558
De gemmis et coloribus, 1563

Girolamo Cardano (Hieronimus Cardanus, 1501-1576) was a highly acclaimed Italian physician and mathematician who first conceptualized negative numbers. As was typical during this age, he was also deeply interested in astrology (in later life, he was imprisoned by the Inquisition for casting Jesus' horoscope). Cardano discussed the subject of color in three of his works: (1) *De Subtilitate* (On subtleness), first published in 1550, a book on many aspects of natural science, including optics, in which Cardano described a biconvex lens as a component of a camera obscura and how it sharpens images; (2) *De rerum varietate* (On the variety of things), a book first published in 1558; and (3) the essay *De gemmis et coloribus* (On gems and colors) in a collection of shorter essays first published in 1563.

Cardano's view on color was conventional, largely according to Aristotle as expressed in the latter's *De sensu*. His list of "principal" colors in *De subtilitate* consists of nine, but with both white and black having double terms, as Aristotle had alluded to: *albus/flavus, croceus, puniceus, purpureus, viridis, caeruleus, niger/fuscus*. Thus, Cardano included yellow with white (as gray with black), and he included the color saffron (orange). Cardano also offered a list of nine tastes and seven planets related to colors. (Interestingly, he related Mars – known since antiquity as the red planet – with blue, not red).

In *De rerum varietate*, Cardano described red as an equal mixture of white and black. In *De gemmis et coloribus*, Cardano went further and assigned values to the brightness of colors, thus producing the first (pseudo-) quantitative scale of brightness. White and black are at the ends of the scale: "[W]e assume that white contains a hundred parts of light, scarlet fifty, black nothing." It is not known how Cardano arrived at these estimates of reflectance. He also listed six additional intermediate colors, resulting in a nine-grade brightness scale (figure 2.13).

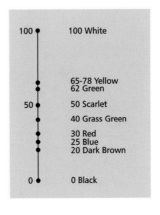

Figure 2.13.

Conceptual sketch of Cardano's (1563) linear scale of nine colors sorted according to lightness.

MARSILIO FICINO   circa 1480

M. Ficino, *Opera*, Vol. 1, 1561

Marsilio Ficino (1433-1499) is regarded as the preeminent humanist scholar and philosopher of the Italian Renaissance. He prepared an influential new translation of Plato's works and helped integrate platonic thinking into Christian religion. Attached to the court of Cosimo de Medici, Ficino was head of the Platonic Academy in Florence and a canon of Florence Cathedral. He also had considerable interest in astrology and mysticism. In the neoplatonic tradition of the third-century philosopher Plotinus, Ficino believed light to be an image of spirit and darkness to be an image of matter. He visualized a hierarchy of brightness levels from God as pure spirit to black as pure matter. He offered a 12-grade color scale as demonstration of this hierarchy:

> In light there are many ideas of colors as there are colors in objects. At the lowest degree where it is communicable, there is the idea of black [*niger*], at the second the idea of brown [*fuscus*], at the third dark yellow [*flavus*], at the fourth dark blue and green [*caeruleus et viridis*] at the fifth sky blue and bluish gray [*caelestis et glaucus*], at the sixth full red [*rubeus plenior*], at the seventh light red [*rubeus clarior*], at the eighth saffron yellow [*croceus*], at the ninth white [*albus*], at the tenth the transparent or the shining [*limpidus sive nitidus*], at the eleventh the brilliant [*splendidus*], and finally there is the idea of splendor [*splendor*]. (Barasch 1978, pp. 177–178)

This scale (figure 2.14) is an exemplification of Plotinus's neoplatonic emanation theory. The theory has the supreme principle (One) overflowing to create intelligence [*nous*], in turn creating psyche which, by degree, becomes less and less perfect and more and more numerous. The emanation from psyche is matter. Notable aspects of the scale are the three levels above white, usually already equated with light, and the absence of purple and violet.

Figure 2.14.

Conceptual sketch of Ficino's (1561) brightness-based scale of lights and colors.

FRANÇOIS D'AGUILON   1613

F. Aguilonius, *Opticorum libri sex*, 1613

François d'Aguilon (Franciscus Aguilonius, 1566-1617) was a Flemish mathematician/philosopher and rector of the Jesuit College in Antwerp, where he taught mathematics and natural sciences. His extensive work on optics (in six parts) is notable for containing the principles of stereographic projection. It is also notable for its title page and six illustrations designed by his countryman, the painter Peter Paul Rubens (1577–1640) (see figure 2.15). Among many other illustrations, it contains what appears to be the first printed color mixture diagram (figure 2.16). Its form derives from the Boethian illustration of relationships in logic and music (see chapter 1) and is explained as follows:

> Among the primary colors the two having extreme positions are whiteness and blackness, they are maximally separated: of them whiteness is superior, it is similar to light; black is truly inferior, being near to darkness. . . . Of intermediate colors we number not more than three, yellow, red and blue. They complete with whiteness and blackness the five simple colors. Further, all others are generated from combination of the three median colors. To be sure, golden [*aureus*] is created from yellow and red, purple [*purpureus*] from red and blue, and lastly green [*viridis*] from yellow and cyan. Some unpleasant colors are generated from mixture of the three truly simple colors, ghastly yellow, lurid blackish blue, and even the color of cadavers. Friendly association results from the mixture of the extreme colors with all the intermediates. Here is found a large number of diverse colors belonging to a particular species: under yellow we find yellowish gray [*luteus*], lemon color [*citrinus*], *ruffus*, weasel color [*mustelinus*], rust color [*ferugineus*], yellowish black [*pullus*], roan color [*roanus*], *tanalus*, royal color [*regius*], lion color [*leonatus*]. Under red there are rose color [*roseus*], reddishness [*rubidus*], ruddy [*rubicundus*], auburn [*rutilus*], blood red [*sangineus, giluus*], chestnut color [*spadix*], glowing color [*igneus*], flame color [*flammeus*]. Under blue there is bluish gray [*caesius, glaucus*], lead color [*plumbeus*], Venetian blue [*venetus*].[5]

And, just as the simple colors, the mixed ones have many that fall under them. For under golden fall orange [*arantius*], saffron [*croceus*], yellowish brown [*fulvus*]. To green belong verdigris [*aerugineus*], leaf color [*herbaceus*], leek color [*prasinus*], sea color [*cymatilis, marinus*]. Purple is accompanied by purplish pink [*rosaceus*], *balasius*, amethyst color [*amethystinus*], crimson [*puniceus*], violet [*violaceus*]: that which comes from conches must be judged first among them, it produces a most lively color. . . . The collection of different colors results not only from admixture in various ratios of whiteness and blackness but also from mixtures among them in various unequal proportions. Leaf color or *vi-*

ror has more yellow while verdigris and crimson have more blue, amethyst color and saffron have more red. (d'Aguilon 1613, pp. 39–40, trans. by R.G.K.)

D'Aguilon, possibly influenced by Roger Bacon, listed five simple colors. From the three chromatic colors, he mixed three more to result in six hue-related categories. D'Aguilon distinguished between three kinds of color mixture: (1) mixture of colorants (*realis compositio colorum*), (2) intentional mixture: a blue object seen in the yellow light of a candle appears greenish, and (3) notional mixture: the mixture in the eye of individual colors "sprinkled" onto the retina.

D'Aguilon was in close contact with Rubens, who himself wrote a now-lost treatise on light and color. At the same time that d'Aguilon worked on his book on optics, Rubens created the mythological painting, *Juno and Argus* (1611, Wallraf-Richards Museum, Cologne). It is usually understood to be an allegory of vision and looks like the illustration of a color diagram with the simple chromatic colors yellow (gold of the carriage), red (dress of Juno), and blue (dress of Iris) in one line, together forming the cadaverous color of Argus's head (Parkhurst 1961). Juno and Iris are busy inserting the 200 eyes of Argus into a peacock's tail (a peacock also is featured prominently in Rubens's title page for d'Aguilon's book).

Figure 2.15.
Title page of Opticorum libri sex *(d'Aguilon 1613).*

D'Aguilon's color diagram proved influential in the seventeenth century, resulting in several copies and modifications.

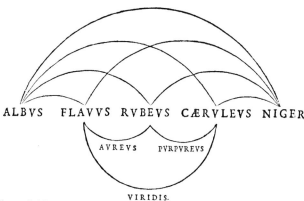

Figure 2.16.
D'Aguilon's basic color scale and mixture diagram. The upper part of the arc diagram illustrates the tint and shade scales of the three simple chromatic colors; the lower part, the mixtures between them (d'Aguilon 1613).

ATHANASIUS KIRCHER   1646

A. Kircher, *Ars magna lucis et umbrae*, 1646

Athanasius Kircher (1602?-1680), a polymath of German birth and education and a member of the Jesuit order, was ordained as a priest in 1628. He taught philosophy and mathematics in Würzburg and Avignon. In 1633, the pope called him to Rome to be professor of mathematics, physics, and oriental languages. Among his wide-ranging interests were magnetism, music, and ancient Egyptian culture. Kircher claimed to be fluent in 12 languages, including hieroglyphic Egyptian. Among his 40-odd works is a book on light and shade and their offspring, color (figure 2.17). The book contains discussions not only of the planetary system, light and color, and magic lanterns, but also of the color of angels. In chapter 2 of the part III, he presented, in a conservative manner, "the manifold variety of colors":

Of the true colors two are extreme, three are intermediate, three are mixed from these. . . . White and black are of the first kind, they are contrary . . . white is most similar to light, it is well known that black is in the vicinity of darkness. . . . There are three intermediate colors, yellow, red and blue. From the mixture of the two extreme colors, each with the three intermediate, all remaining colors emanate. Of the intermediate three real colors yellow and red form the color of gold, red and blue purple, and finally from yellow and blue green is composed, it is of all the mixture ratios the most perfect, everybody gives it the highest praise; it is seen as causing in the eye as well as in the audible diapason the most agreeable and satisfying harmony. . . . The two extreme colors easily undergo mixture with the intermediates, but they do not change them, rather they stretch them and make them yield. White elevates the others, black depresses them by obfuscation. . .

Lights and colors also recall other kinds of entities, each grade is shown under the attached figure, each has a certain kind of analogy to lights and colors, each is disposed toward the other in comparable manner. (Kircher 1646, pp. 48–49, trans. by R.G.K.)

Figure 2.17.

Title page of Kircher's Ars magna lucis et umbrae *(1646).*

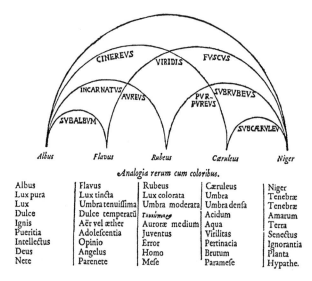

Figure 2.18.

Kircher's (1646) color diagram. The middle colors of the mixtures of the three chromatic primaries are identified near the intersections of their tint/shade scales. Below the diagram are lists of color analogs.

Kircher employed the Boethian arc form for his diagram (figure 2.18), already seen in the work of d'Aguilon (see entry in this chapter). However, he dispensed with the lower arcs and placed, with some minor changes, the secondary chromatic colors at the appropriate intersections of the upper arcs.[6] Attached to the figure is a fascinating table containing analogies between lights and color and other "things." It is reminiscent of similar figures in Boethius's commentary on Aristotle's *Organon*. The last row lists the strings of the Greek lyre.

ROBERT FLUDD   1629

Fludd, *Medicina catholica*, 1629–1631

Robert Fludd (1574 -1637) was an English hermetic philosopher, Rosicrucian, and physician who attempted to reconcile the mystical tradition of Paracelsus with the then-beginning scientific views of the human body. In his major two-volume work, *Utriusque cosmi* (1617–1621), Fludd presented several versions of a Boethian arc figure showing universal harmony. In another work, *Medicina catholica*, he illustrated a color circle (figure 2.19).

Figure 2.19.

Fludd's (1629) circular form of the modified Aristotelian scale of simple colors.

Fludd's circle is an Aristotelian seven-grade scale from black to white, but he used a different selection of five chromatic colors: yellow, orange, red, green, and blue. Following Bartholomaeus Anglicus, he placed red in the center. The composition of each of the colors is given on the spokes:

White	No blackness
Yellow	Equal in whiteness and redness
Orange	More redness, less whiteness
Red	Intermediate between white and black
Green	Equality of light and blackness
Blue	More blackness, less light
Black	Zero light

Notably, the first three colors are expressed in terms of whiteness, blackness, and redness, while the last three are expressed in terms of blackness and light. It is not clear whether Fludd viewed red and green as having identical lightness.

Having established his lightness-based color scale, Fludd arranged it in the form of a ring with black at bottom center and white and blue adjoining it. The circular form does not have an explicit purpose and may express hermetic symbolism: the *ouroboros*, the snake that bites its tail, emblematic for the battle between lightness and darkness, or good and evil.

ZACHARIAS TRABER   1675

Z. Traber, Nervus opticus, 1675

Like d'Aguilon and Kircher, Austrian **Zacharias Traber** (1611-1679) was a member of the Jesuit order. In 1675, Traber wrote a book on vision, *Nervus opticus* (figure 2.20). In the book, he also discussed color and used Kircher's form (see figure 2.18) of the color order diagram (figure 2.21), but with changes in terminology. The mixture between blue and black that Kircher called *subcaeruleus* (below blue, dark blue), was called *violaceus* (violet) by Traber. The mixture between red and black, *subrubeus* (dark red) to Kircher, was *cupreus* (copper colored) to Traber. Kircher's mixture of white and blue is *cinereus* (ashen), while Traber's is *lucido caeruleus* (light blue). Traber used the term *cinereus* for the mixture of white and black, a mixture not named by Kircher.

Traber refers to results obtained by painters: "[T]he summit of the semi-circles indicates the creation of one color from two. When two are mixed unevenly, wherein more or less is used of one or the other, all possible and conceivable colors are generated, best created by painters in the execution of their works" (Traber 1675, p. 17, trans. by R.G.K.). But his scheme does not consider desaturated or grayish mixtures from three colorants, used by painters at the time.

Figure 2.20.

Title page of Nervus Opticus *(Traber 1675).*

JOHANNES ZAHN   1685

J. Zahn, Oculus artificialis teledioptricus sive telescopium, 1685

Johannes Zahn (1641-1707) was provost of a women's convent near Würzburg, Germany. In his spare time, he occupied himself with mathematics, physics, and mysticism. His book on optical equipment includes the first description of binoculars and what may be the first description of an apparatus projecting a moving image. The book also includes a conventional chapter on the nature of color.

Zahn demonstrates the relationships between colors with a modification of Kircher's diagram (published 39 years earlier and printed in several editions; see entry in this chapter). The lines connecting mixed colors are triangles rather than arcs, such as those of d'Aguilon and Kircher (figure 2.22). The simple colors are located on the base corners and the mixed color is at the apex of the mixture triangle. Kircher's mixture of white and blue, called *cinereus*, is replaced by a color Zahn called *aqueus* (water-colored).

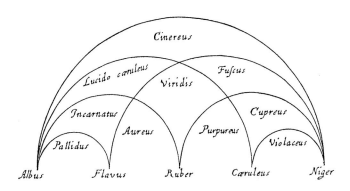

Figure 2.21.

Traber's (1675) version of the Kircher-style color diagram.

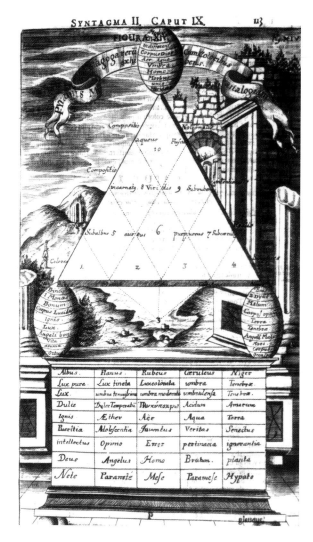

Figure 2.22.
Depiction of Zahn's (1685) fundamental color triangle.

In other respects, Zahn's diagram is identical to Kircher's, including the table of color analogs below the triangle. Zahn distinguished between the two extreme colors white and black and the intermediate simple colors yellow, red, and blue. White is taken as analog of light and black as that of darkness. Red is "exactly intermediate" to white and black;

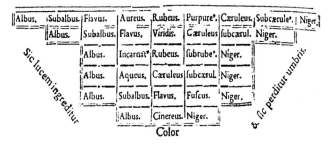

Figure 2.23.
Tabular representation of various lightness scales (Zahn 1685). The angled text reads: Here light increases (left) and here it disappears into shade (right).

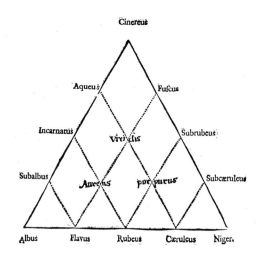

Figure 2.24.
Zahn's fundamental triangle (Zahn 1685/1702).

yellow and blue are "not intermediate." Zahn understood real colors (compared to apparent and phantastical) to be an integral, nonperishable part of materials; thus, "Such kinds of colors are whiteness in chalk and lead white, redness in red lead, cinnabar and similar materials. Yellowness is found in saffron and corn, blackness in coal, the color blue in indigo, etc." (Zahn 1685, p. 111, trans. by R.G.K.).

Zahn described his diagram as a mystagogical (i.e., indicating knowledge of the mysterium, the ways of God) concept where various things and ideas are related to colors. The triangle is taken as the proper form for the diagram because it is the constituent of the simplest of the platonic solids, the tetrahedron. Zahn quoted Aristotle from *On the Heavens* as further support for this shape: "For, as the Pythagoreans say, the universe and all that is in it is determined by the number 3" (p. 114) as further support for this shape. The ellipsoids that support the white and black base corners and the one on top of the triangle (figure 2.22) bear the following analogs:

White	**Apex**	**Black**
Beginning	Middle	End
Unity	Triality	Duality
Goodness	Indifference	Badness
Transparent body	Translucent body	Opaque body
Fire	Air, water	Earth
Light	Shade	Darkness
Good angels	Humans	Bad angels
Life	Illness	Death
Soul	Union	Body

In a separate – tabular– diagram (figure 2.23) Zahn described various lightness-dependent paths across the diagram. The first scale from the bottom is a simple gray scale with three grades. The second scale (from bottom) is a sort of constant-hue lightness scale from white via yellow and brown to black. The third and fourth scales are comparable to the second one but for blue, and red, respectively. The fifth is a lightness scale based on mixture of two chromatic fundamentals: from white to off-white, yellow, green, blue, dark blue to black. The final scale combines lightness with mixtures of the adjacent chromatic primaries: white, off-white, yellow, orange, red, purple, blue, dark blue, black.[7]

Zahn was one of the last explicit representatives of the classical tradition. In the second edition of his work (two years before the publication of Newton's *Opticks* in 1704; see entry in chapter 6), the mystagogical apparatus has been removed from the figure which was now a plain triangle with the logical color gray (*cinereus*) placed on top (figure 2.24). In this form, the figure is an early version of a two-dimensional color order diagram.

SIGFRIDUS ARONUS FORSIUS   1611
S. A. Forsius, *Physica*, 1611

Sigfridus Aronus Forsius (1560-1624) was a Finnish mathematician, astronomer, and clergyman. He was named Royal Astronomer at the Swedish court and had exclusive rights to issue almanacs and cast horoscopes.

In 1611, while in Stockholm, Forsius wrote a manuscript on physics, which was never printed as a book. In chapter 7, titled "On Vision," he presented two color order diagrams that are an important bridge between classical and more modern ideas on color order. Both diagrams are circular but represent linear scales.

The first diagram has a range of seven "light" colors descending on the left-hand side from white (white gold, gold, burnt gold, red, purple, brown and violet brown) and an equal number of "dark" colors ascending from black on the right (blackish green, green, blue green, blue, sky blue, gray, dapple). Forsius described the scale as follows: "Among the colors there are two prime colors, white and black, from which all others have their origin. . . . Gold between white and red, . . . brown between red and black. . . . Then on the other side between white and blue is gray. . . . And on the lower part green between blue and black" (translation from Feller and Stenius 1970, p. 50).

The diagram is emblematic of Forsius's understanding of classical color order: "In the middle between these colors [white and black], red since ancient times has been placed on one side and blue on the other" (p. 50; figure 2.25). In the writings of antiquity, there are no evident antecedents for

this claim, however. It likely derives from Forsius's reading of Grosseteste's linear scale, as interpreted by Bartholomaeus (see Robert Grosseteste, this chapter). Forsius may have had access to Grosseteste's text in Stockholm and quite certainly had access to Bartholomaeus's encyclopedia.

If Forsius depended on Grosseteste, he overlooked or was not able to interpret the latter's remarks concerning the meeting of the two scales in the middle. In any case, Forsius did not find this arrangement satisfactory and continued with a second proposal that resulted in his own contribution to color order:

> But if you want right to consider the origin and relations of the colors, you should start from the five principle middle colors which are red, blue, green, gold, and gray of white and black. And their gradings, they rise either closer to white by their paleness or to black by their darkness; albeit they are (as above has been made known) related to one another as previously shown. Because red rises to white through pale red (pink) and skin color; to black through purple, brown, violet brown and black brown. Similarly gold relates toward white through pale gold, wooden and wheat color; to black through burnt gold and blackish brown. Equally blue rises to white through sky blue and pale blue, like Dutch cloth; and to black through dark blue like indigo color that has some brownish to it. So rises also green toward white through verdigris and pale green; to black through blackish green. Gray approaches white by the color of light gray, dapple gray and lime: to black by mouse gray, black gray and pale black. And this is the correct relationship of colors that in their number agree with that of the planets as do the lower colors with the five membranes of the eye, and with the five senses. All this can be seen from the accompanying figure. (p. 50)

The resulting diagram also has circular form (figure 2.26). In this diagram, Forsius used four chromatic and two achromatic categories. The result is four tint/shade scales of simple chromatic colors (of the kind described by Theophilus; see entry in this chapter) as well as a gray scale, placed between a common white and black.

On the left side of the diagram is a red tint/shade scale that descends via purple and violet-brown to black, perhaps the result of mixture of particular pigments. It is followed by the gold scale. In the center is the gray scale. Here, lime is to be understood as chalk white. On the far right side is the blue scale, and on the near right side the green scale. The arrangement maintains the hue order on the left and right of his first diagram, thus keeping some continuity with the classical order as he understood it.

Some commentators (e.g., Feller and Stenius 1970, Gage 1993) have interpreted the second of Forsius's circles as a color sphere, with white and black at the poles and four primary hues, gold (yellow) and blue, and red and green, in the proper sequence of the spectrum and opposing each other on the equator. The left semicircle would be paired with the first right curved line to form a circle and vice versa. But it is an implausible interpretation for two reasons: First, how to properly draw a transparent sphere was well known in the seventeenth century from several earlier books on perspective and geometry. Second, the text does not indicate that Forsius had a three-dimensional arrangement in mind. Forsius quite certainly knew that there are many intermediates between his primary hues and with them many more curved lines. How he would have arranged these is not clear.

Forsius's manuscript was rediscovered only in the twentieth century, and his arrangement of linear tint/shade scales between common white and black did not have noticeable influence on developments elsewhere. A somewhat similar arrangement was published in 1677 by Glisson (see entry in this chapter).

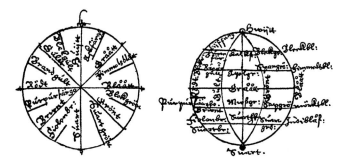

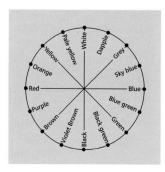

Figure 2.25.

Forsius's (1611) interpretation of classicalcolor order and schematic representation with translated color names.

Figure 2.26.

Four tint/shade and a gray scale connecting in common white and black locations in a circular arrangement (Forsius 1611), and schematic representation with translated color names.

FRANCIS GLISSON   1677

F. Glisson, *Tractatus de ventriculo et intestinis*, 1677

Francis Glisson (1597?-1677) was an English physician and Regius professor of medicine at Cambridge. He wrote several books on medicine, among them *Tractatus de ventriculo et intestinis* (Treatise of ventricle and intestines), published in London in the last year of his life. Its chapter 9

is titled "De coloribus pilorum" (On the colors of hair). The main purpose of the chapter seems to have been to present Glisson's ideas about a quantitative color specification system that could be used to specify not only the color of hair but also of any other object.

Perhaps influenced by d'Aguilon, Kircher, or Boyle's book of 1664, Glisson believed any color could be specified with five simple colors, white, black, yellow, red, and blue, represented by four visually equally spaced scales: a gray scale and tint scales (from white to full color) of yellow, red, and blue. While originally envisaging the chromatic scales to be complete tint/shade scales (running from white via the full color to black), he argued that chromatic scales from white to the full color were sufficient because he could specify the black content of darker colors with the help of the gray scale. He sketched the placement of the full scales in an arrangement (figure 2.27) not unlike Forsius's (figure 2.26), with common white and black points (it is unlikely, though, that Glisson knew Forsius's *Physica*).

Glisson used a novel approach to obtain scales with what he took to be perceptually equidistant steps. He determined the "equivalent strength" of his white and black pigments (lead white, carbon black) by making a mixture that perceptually fell halfway between the two extremes, that is, middle gray. He found that he needed a weight ratio of 50:1 to achieve this. Because he wanted to have a scale of 24 steps and the middle gray was to be step 12, he multiplied the weights of his two pigments by 12 to arrive at 600 grains of white and 12 grains of black for the middle grade. For the next darker grade, he used 1 grain more of black and 50 grains less of white, that is, 13 and 550. In this fashion the scale was completed in both directions (figure 2.28).

As an example of a chromatic scale, Glisson specified the redness scale (figure 2.29). His example of color specification using the system consists of grade values for the golden yellow color of certain blossoms; their color is given to be the equivalent of the sum of grade 11 of the yellowness scale, grade 3 of the redness scale, and grade 2 of the gray scale. If necessary, Glisson said, colors could be specified to half-steps between his grades.

Glisson described the gray scale (in his terminology, blackness scale) as being straight and the three chromatic scales as rounded and sideways from or oblique to (*tres . . . scalae obliquae sunt*) the gray scale, but ending in common white and black (see figure 2.27). The midpoints, indicated in the figure by marks, are the locations of the pure chromatic pigments he used: orpiment, vermilion, and azurite or bice.

A reconstruction with the specified pigments and weights of Glisson's gray and red scales resulted in larger steps near both ends of the scales (Kuehni and Stanziola 2002). In the middle region, they are surprisingly even, but they gradually decline in size toward black and full red, respectively. Only extensive visual scaling would have disclosed this fact.

Glisson believed he could specify any object color with the scales. The horizontal scale arrangement does not indicate extensive thought on how all specifiable colors could be systematically arranged.

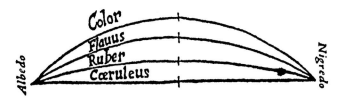

Figure 2.27.

Arrangement of the linear gray and three chromatic scales with common endpoints in white and black (Glisson 1677). The vertical dashes denote the location of middle gray and the three chromatic full colors.

Scala Nigredinis.

Gradus ejus.	Grana ceruffæ.	Grana a-tramenti fuliginei.	Utriufque proportio minima.
13us.	Simplex Nigredo.		
22us.	100.	gr. XXII.	C. 4 6/11 F. I.
21us.	150.	gr. XXI.	C. 7 1/2 F. I.
20us.	200.	gr. XX.	C. 10. F. I.
19us.	250.	gr. XIX.	C. 13 1/3 F. I.
18us.	300.	gr. XVIII.	C. 16 2/3 F. I.
17us.	350.	gr. XVII.	C. 20 6/19 F. I.
16us.	400.	gr. XVI.	C. 25. F. I.
15us.	450.	gr. XV.	C. 30. F. I.
14us.	500.	gr. XIV.	C. 35 5/7 F. I.
13us.	550.	g. XIII.	C. 42 4/11 F. I.
12us.	600.	gr. XII.	C. 5. F. 1/10
11us.	650.	gr. XI.	C. 5 10/11 F. 1/10
10us.	700.	gr. X	C. 7. F. 1/10
9us.	750.	gr. IX.	C. 8 1/3 F. 1/10
8us.	800.	gr. VIII.	C. 10. F. 1/10
7us.	850.	gr. VII.	C. 12 1/10 F. 1/10
6us.	900.	gr. VI.	C. 15. F. 1/10
5us.	950.	gr. V.	C. 19. F. 1/10
4us.	1000.	gr. IV.	C. 25. F. 1/10
3us.	1050.	gr. III.	C. 35. F. 1/10
2us.	1100.	gr. II.	C. 55. F. 1/10
1us.	1150.	gr. I.	C. 115. F. 1/10
Simplex Albedo, bafis fcalæ.			

Figure 2.28.

Specification of the pigment composition for the blackness (gray) scale. The leftmost column contains the designation of the grade, the next the amount of lead white in grains, the third the amount of carbon black, and the last the pigment ratio in reduced format (Glisson 1677).

Scala Rubedinis.

Gradus ejus.	Grana ceruffæ.	Grana Cinna-baris.	Utriufque proportio minima.
11us.	Satura Rubedo.		
10us.	gr. 40.	gr. X	C. 4. Ci. gr. I.
9us.	gr. 60.	gr. IX.	C. 6 1/7 Ci. gr. I.
8us.	gr. 80.	gr. VIII.	C. 10. Ci. gr. I.
7us.	gr. 100.	gr. VII.	C. 14 2/7 Ci. gr. I.
6us.	gr. 120.	gr. VI.	C. 20. Ci. gr. I.
5us.	gr. 140.	gr. V.	C. 28. Ci. gr. I.
4us.	gr. 160.	gr. IV.	C. 4. Ci. 1/10 gr.
3us.	gr. 180.	gr. III.	C. 6. Ci. 1/10 gr.
2us.	gr. 200.	gr. II.	C. 10. Ci. 1/10 gr.
1us.	gr. 220.	gr. I.	C. 22. Ci. 1/10 gr.
Simplex albedo, bafis fcalæ.			

Figure 2.29.

Specification of the pigment composition for the redness scale (Glisson 1677).

DIOGO DE CARVALHO E SAMPAYO   1788

D. de Carvalho e Sampayo, *Dissertaçâo sobre as cores primitives*, 1788

Diogo de Carvalho e Sampayo (1750–1807) was a Portuguese nobleman and diplomat, and a knight of the Maltese order. During his stay in Malta, he worked on color problems and wrote two books in rapid succession. The first, *Tratado das cores* (Treatise on colors), published in 1787 in Malta, demonstrated the results of double and triple mixtures of pigments. While he quoted Newton liberally, he denied simple color status to yellow and blue (in the manner of Leonardo da Vinci, who had earlier denied it to blue). But he soon changed his mind, and in 1788 he published his discourse on simple colors that contains a color order system based on mixtures of six simple colors. Despite the contemporaneous development from two- to three-dimensional systems elsewhere, Sampayo's system is based on one-dimensional color scales.

Sampayo's, six simple colors are white, black, yellow, red, blue, and green. Every color is combined in turn with the other five colors, and each combination has three intermediate grades, resulting in four steps. In this manner, Sampayo generated 15 linear scales, illustrated on five pages with a total of 51 samples (figure 2.30). Figure 2.31 shows a table identifying six generic colors and their "species." Sampayo's system is the first (if late) fully colored system of one-dimensional simple hue and tint/shade scales as well as a gray scale (figure 2.32). Following classical examples Sampayo called the simple colors generic and the mixed colors species.

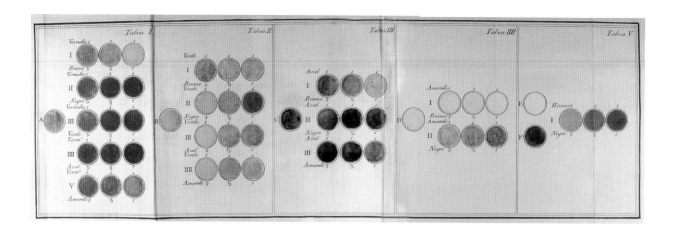

Figure 2.30.
Sampayo's (1788) five plates with 15 linear scales.

SCHEMA DAS CORES GENERICAS COM AS SUAS RESPECTIVAS ESPECIES.

Cores genericas.	Cores efpecificas.	Cores genericas.	Cores efpecificas.
Vermelho.	Verm-claro. Verm-efcuro. Verm-Verde. Verm-azul. Verm-amar.	Amarello.	Amar-claro. Amar-efcuro. Amar-Verm. Amar-Verde Amar-azul.
Verde.	Varde-claro. Verde-efcuro. Verde-Verm. Verde-azul. Verde-amar.	Branco.	Branc-efcur. Branc-Verm. Branc-Verd. Branc-azul. Branc-amar.
Azul.	Azul-claro. Azul-efcuro. Azul-Verm. Azul-Verde. Azul-amar.	Negro.	Negro-claro. Negro-Verm Negro-Verde Negro-azul Negro-amar.

Figure 2.31.
Table of the six generic colors and their species (Sampayo 1788).

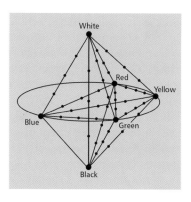

Figure 2.32.
Schematic representation of the position of the 15 scales in the implicit double cone space.

ZHENG FUNGUANG   1847

Zheng Funguang, *Jingjig Lingchi*, 1847

In the book *Jingjig Lingchi* (Introduction to lens and mirror), Chinese natural scientist **Zheng** (1780–1864) published a diagram that agrees in all essential details with d'Aguilon's linear color order diagram (figure 2.33). The five simple colors in the large circles are, from top to bottom, white, yellow, red, blue, and black. They coincide with the five classical Chinese elements metal, earth, fire, wood, and water. Next to the circle segments are the names of the binary mixtures:

White and yellow make pale yellow

White and red make pale red

White and blue make moon white

White and black make pale black

Yellow and black make brown

Red and black make purple

Blue and black make dark blue

Yellow and red make orange

Red and blue make purple

Yellow and blue make green

Historically, it is likely the final example of an Aristotelian color scale with component mixture.

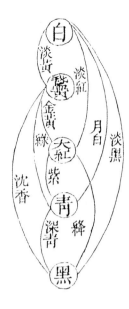

Figure 2.33.
Zheng's (1847) color order diagram, derived from that of d'Aguilon.

LEOPOLDO NOBILI   1830

L. Nobili, *Sui colori in generale ed in particolare supra una nuova scala chromatica dedotta dalla metallochromia ad uso delle scienze e delle arti*, 1830

Italian physicist **Leopoldo Nobili** (1784–1835) invented the thermopile for measuring weak sources of infrared radiation. He also invented a type of galvanometer for measuring electric current. He studied electrochemical effects and produced a large series of displays of interference colors resulting from electrochemical metal deposits (figure 2.34). In 1830, he published an article about a "novel chromatic scale obtained by means of metallochromia."

Such deposits are obtained by various methods, for example by attaching a pointed platinum electrode to the center of a steel or silver plate that has been placed into a solution of a metal salt, such as lead acetate. After connecting the negative pole of a battery to the electrode and the positive to the metal plate metal, deposits in various patterns and thicknesses appear on the plate. The pattern design and deposit thickness (and thereby the resulting apparent colors) depends on the experimental conditions.

Nobili noted that the colors generated when viewing such plates tended to appear in four rings. He compared these to the rings Newton generated by bringing a curved glass plate into contact with a flat plate. In Nobili's rings, the apparent colors often represent mixtures, rather than being purely spectral. In addition, colors in a ring tend to be close together so that their separation involves a degree of subjectivity. Nobili identified 44 colors in these rings and considered them to form a "natural chromatic scale" (figure 2.35). The complete scale is the sum of four linear scales with a small degree of repetition.

The vividness of many of the apparent colors led Nobili to call the scale "voluptuous." He described a scale formed from larger uniform color fields (not extant). He also described efforts to reproduce the scale in oil and watercolors, but judged the result to be poor.

The scale of figure 2.35 is separated into four sections that he compared to the first four Newton diffraction rings (Newton 1704, book 2, p. 37). However, the apparent colors do not closely agree due to the different methods of producing them. Starting from the bottom of the scale, Nobili's first group ranges from blond through copper-color to bluish red. He described the appearance of these colors as "metallic and foggy."

The second and largest group passes from violet through blues, yellows, oranges, and reds to deep carmine (a spectral scale without green). This is "the most beautiful of the four groups."

The third and fourth groups are somewhat similar, with the colors described as saturated. Both groups run from purplish carmine through various greens to pinkish carmine. In the third group, we encounter unusual color descriptions, such as "carmine with a hint of turquoise," or "violetish-green." Such colors are normally considered impossible and difficult to imagine. Another way in which Nobili's scale does not follow common perceptual rules of color order: Some of the colors are spectral in nature, while others are tonally reduced.

The ability to produce the displays repetitively by quasi-natural procedures led Nobili to claim natural status for his scale. It is likely that the resulting color experiences can be predicted from detailed knowledge of the surface structure of the metal deposits. Today, such color effects are sometimes seen in fashion jewelry.

Figure 2.34.

Example of a "metallochromic" plate produced by Nobili showing Newton-ring-like interference colors. Photography courtesy of University of Padua.

Nobili's color list (figure 2.35), with translations, is as follows:[8]

44	lacca rosea	pinkish carmine
43	verde-giallo rossio	reddish yellow-green
42	verde-giallo	yellow green
41	verde	green
40	violaceo-verdognolo	violetish-green
39	lacca-violacea	violetish carmine
38	lacca-rosea	pinkish carmine
37	rancio-roseo	pinkish orange
36	rancio verde	greenish orange
35	verde-rancio	orange-green
34	verde-giallo	yellow-green
33	verde-giallognolo	yellowish green
32	verde	green
31	porpora-verdognola	greenish purple
30	lacca turchiniccia	carmine with a hint of turquoise
29	lacca purpurea	purplish carmine
28	lacca accesa	vivid carmine
27	lacca	carmine
26	lacca-rancia	orange-carmine
25	rosso rancio	orange-red
24	rancio-rosso	red-orange
23	rancio rossicio	reddish orange
22	rancio	orange
21	giallo-rancio	orange-yellow
20	giallo acceso	vivid yellow
19	giallo	yellow
18	giallo chiarissimo	very light yellow
17	celeste-gialognolo	sky blue with a hint of yellow
16	celeste	sky blue
15	bleu chiaro	light blue
14	bleu	blue
13	bleu carico	intense blue
12	indaco	indigo
11	violetto	violet
10	rosso-violaceo	violetish red
9	ocria-violacea	violetish ochre
8	ocria	ochre
7	rosso di rame	copper red
6	fulvo acceso	vivid fulvous
5	fulvo	fulvous
4	biondo acceso	vivid blond
3	biondo d'oro	gold blond
2	biondo	blond
1	biondo argentine	silver blond

Figure 2.35.
Nobili's (1830) "natural chromatic scale."

Notes

1. The meaning of the term *pallidus* is uncertain and varies for different classical Latin authors. It is the source of the current word "pale," and *Cassell's Latin Dictionary* (fifth ed., 1968) translates it as "pale, wan, sallow."

Classical usage seems to indicate that it was at times considered to have meaning equivalent to the Greek term *chloros* (yellow-green). It was a term used in the Latin translation of Avicenna's work (see entry in this chapter), where he discussed three tint/shade scales from white to the full color and on to dark colors. The meaning here may be yellow or gray. Roger Bacon, in the thirteenth century, described it as between wax color and orange in one location, and between orange and red in another.

In the fifteenth century, the compiler of manuscripts on colorant technology Jehan le Begue declared "*pallidus est color non proprie albus, sed declinans aliquantulum ad obscuratem*" (pallidus is a color not white in a proper sense but declining about half way toward darkness) (Merrifield 1967).

The sixteenth-century Italian Coronato Occolti described *pallidus* in his treatise on color (1568) as "*colore brutto e vile, genera di color bianco debile, accompagnato da un poco di negro e rosso*" (a brutish and vile color, of the genus of off-white, accompanied by a little black and red). A few years later (1595), his countryman Antonio Calli described it as "*Sono tra esso bianco e 'l nero il pallido, detto squalido, color de gli amanti . . . de gli impauriti et de morti*" (Between white and black is pallid, called squalid, the color of lovers, the poor and the dead). In the same time period, the Italian art historian Raffaello Borghini (1584) called *pallido* a yellowish greenish gray.

2. Another early description of a system with three chromatic colors is that by Filippo Mocenigo (1581), copied by Vitus Antonius Scarmilionius (1601), with the unusual sequence white, yellow, blue (*hyazinthinus*), red, and black. A system with two chromatic primary colors, red and blue, intermediate to black and white, was described by Louis Savot (1609).

3. Pseudo-Dionysius the Aeropagite refers to an anonymous fifth-century philosopher and author of *Corpus Aeropagiticum*, a philosophical work falsely ascribed to the real Dionysius the Aeropagite, a first-century Bishop of Athens.

4. Gavel (1979) interpreted it as yellow-brown, thus implying a Hering-like order of chromatic colors.

5. Early lists of color names sorted into categories from which d'Aguilon may have profited are found in Bacon (thirteenth century), Telesio (1528), Cardano (1550), and Scaliger (1557).

6. Johann Scheffer (1621–1679), professor of law and history at Uppsala University and mapmaker, copied Kircher's diagram unchanged in his treatise on painting, *Graphica, id est arte pingendi* (Scheffer 1669).

7. This scale was reproduced nearly unchanged by Johann Kaspar Funk (1680–1729) in *Liber de coloribus coeli* (1716).

8. Translation of color names by Claudio Oleari.

References

Alberti, L. B. 1540. *De pictura*, Basel: Westheimer.

Alberti, L. B. 1956. *Leon Battista Alberti on painting*, Spenser, J., ed., New Haven, CT: Yale University Press.

Barasch, M. 1978. *Light and color in the Italian Renaissance theory of art*, New York: New York University Press.

Barnes, J. (ed). 1984. *The complete works of Aristotle*, 2 vols., Princeton, NJ: Princeton University Press.

Bartholomaeus Anglicus. 1245. *De proprietatibus rerum*, English translation, fourteenth century: *On the properties of things*, John of Trevisa, trans., Oxford: Clarendon Press, 1975.

Batman, S. 1582. *Batman upon Bartholomew*, London: East.

Borghini, R. 1584. *Il riposo di Rafaello Borghini*, Florence: Marescotti.

Boyle, R. 1664. *Experiments and considerations touching colours*, London: Herringman.

Calli, A. 1595. *Discorso de' colori d'Antonio Calli*, Padua: Pasquati.

Cardano, G. 1550. *Hieronymi Cardani medici mediolanensis de subtilitate libri XXI*, published simultaneously in Nürnberg, Germany, and Lyon and Paris, France.

Cardano, G. 1563. De gemmis et coloribus, in *Hieronimi Cardani opera omnia*, vol. 2, Lyon, France.

Cassell's Latin dictionary (Simpson, D. P., fifth ed.). 1968. New York: Macmillan.

d'Aguilon, F. (Aguilonius). 1613. *Opticorum libri sex*, Antwerp: Plantin.

da Vinci, L. 1956. Treatise on Painting "Codex Urbinas latinus 1270, translated and annotated by A. P. McMahon, 2 vols., Princeton, New Jersey: Princeton University Press.

Dolce, L. 1565. *Dialogo nel quale si ragiona della qualità, diversità, e proprietà dei colori*, Venice: Sessa.

Feller, R. L., and Å. S Stenius. 1970. On the color space of Sigfrid Forsius, 1611, *Color Engineering*, June, 48–51.

Ficino, M. 1561. *Opera*, vol. 1, Basel: Henricpetrus.

Fludd, R. 1626. *Medicina catholica*, vol. 1, Frankfurt: Rötelli.

Forsius, S. A. 1611. *Physica*, manuscript, Stockholm: Royal Library.

Gage, J. 1993. *Color and culture: practice and meaning from antiquity to abstraction*, Boston: Little, Brown.

Gavel, J. 1979. *Colour, a study of its position in the art theory of the Quattro- & Cinquecento*, Stockholm: Almquist and Wiksell.

Glisson, F. 1677. *Tractatus de Ventriculo et Intestinis*, London: Brome.

Kircher, A. 1646. *Athanasii Kircheri Fuldensis Ars magna lucis et umbrae in decem libros digesta,* Rome: Scheus and Grignani.

Kuehni, R. G., and R. Stanziola. 2002. Francis Glisson's color specification system of 1677, *Color Research and Application* 27:15–19.

Lee, H. D. P. (trans.). 1978. *Aristotle VII meteorologica*, Cambridge, MA: Harvard University Press.

Lomazzo, G. P. 1584. *Trattato dell'arte de la pittura, scultura ed architectura*, Milan: Pontio. First English translation, *A tracte containing the artes of curious paintinge, carvinge and buildinge*, Oxford: Barnes, 1598.

Merrifield, M. P. 1967. *Medieval and Renaissance treatises on the arts of painting*, Mineola, NY: Dover.

Mocenigo, F. (Philippus Mocenicus). 1581. *Universales institutions ad hominum perfectionem*, Venice: Manutius.

Newton, I. 1704. *Opticks*, London: Smith and Walford.

Nobili, L. 1830. Sui colori in generale ed in particolare supra una nuova scala chromatica dedotta dalla metallochromia ad uso delle scienze e delle arti, *Antologia* 117:1–39.

Occolti, C. 1568. *Trattato de colori di M. Coronato Occolti da Canedolo*, Parma: Viotto.

Parkhurst, C. 1961. Aguilonius' optics and Rubens' color, *Nederlands Kunst-Historisch Jaarboek*, vol. 12, Zwolle, The Netherlands: Waanders.

Parkhurst, C. 1990. Roger Bacon on color: sources, theories & influence, in *The verbal and the visual*, Selig, K.-L., Spiers, E., eds., New York: Italica Press.

Pinder, U. 1506. *Epiphanie medicorum*, Nürnberg: Sodalitas Celtica.

Rehn, R. (ed.). 1985. Magistri Theodorici ordinis fratrum praedicatorum tractatus de coloribus, in Opera omnia, Schriften zur Naturwissenschaft, Briefe, in *Corpus Philosophorum Teutonicorum Medii Aevi*, vol. 4, Pagnoni-Sturlese, M. R., Rehn, R., Sturlese, L., Wallace, W. A., eds., Hamburg: Meiner.

Sampayo, D. de C. 1788. *Dissertaçâo sobre as cores primitives*, Lisbon: Regia Officina Typographica.

Savot, L. 1609. *Nova, seu verius nova-antiqua de causis colorum sententia*, Paris: Plantin.

Scaliger, J. C. 1557. *Iulii Caesaris Scaligeri exotericarum exercitationum liber quintus decimus de subtilitate ad Hyeronymum Cardanum,* Lutetia (Paris): Vascosanus.

Scarmilionius, V. A. 1601. *De coloribus*, Marburg, Germany: Egenolphus.

Scheffer, J. 1669. *Graphica, id est arte pingendi,* Nürnberg, Germany.

Spencer, J. R. 1966. *Leon Battista Alberti on painting,* New Haven, CT: Yale University Press.

Telesio, A. 1528. *Antonii Thylesii Cosentini libellus de coloribus.* Venice: Vitalis.

Theophilus. ca. 1120. *De diversis artibus*, manuscript. *The various Arts, De diversis artibus*, Dodwell, C. R., ed. and trans., Oxford: Clarendon Press, 1986.

Thomaeus, N. L. 1523. *Aristotelis Stagiritae parva naturalia, omnia in latinum conversa, & antiquorum more explicata a N. Leonico Thomaeo*, Venice: Vitalis.

Traber, Z. 1675. *Nervus opticus,* Vienna: Cosmerovius; 2nd ed., 1690.

Van Riet, S. (ed.). 1972. *Avicenna latinus, Liber de anima seu sextus de naturalibus*, Louvain, Belgium: Peeters.

Vasari, G. 1979. *Lives of the most eminent painters, sculptors, and architects*, 3 vols., De Vere, G. duC. trans., New York: Abrams.

Vossius, I. 1662. *De lucis natura e proprietate*, Amsterdam: Elzevier.

Wrobel, I. 1976. *Platonis Timaeus interprete Chalcidio cum eiusdem commentario*, Leipzig: Teubner.

Zahn, J. 1685. *Oculus artificialis teledioptricus sive telescopium.* Würzburg, Germany: Heyl; 2nd ed., Nürnberg: Lochner, 1702.

Zheng Funguang. 1847. *Jingjig Lingchi.*

ORANGE
by union of Yellow & Red

Olive & Brown
mixes here

mixtures
by the two prev.

OLI.&BRO.
Neuter

YELLOW
A Primitive Colour

OLIVE
by Green & Orange

mixture of the
two preceding compounds

OLIVE
Neuter

Neuter

Central
UNION.
Blackness

Neuter

OLI.& SLA.
Neuter

BRO.& SLA.
Neuter

BROWN
Neuter

RED
A Primitive Colour

BROWN
by Orange & Purple

mixture of the
two preceding Compounds

Neuter

mixture by the two
previous

SLATE
Neuter

GREEN
by union of Yellow & Blue

Olive and Slate

mixture by the two
previous

Slate and Brown

PURPLE
by union of Blue & Red

mixture of the two
preceding Compounds

SLATE
by Green and Purple

BLUE
Primitive Colour

CHAPTER 3

COLOR DIAGRAMS AND COLOR CIRCLES

A description of the realization of a second geometric dimension in color order, resulting in color diagrams and color circles.

The Aristotelian linear color scale paradigm held sway into the eighteenth century, as described in chapter 2. There, the circular form of the systems of Sigfrid Aronus Forsius and Robert Fludd was arbitrary. François d'Aguilon's figure indicated the complexity of a more complete color order for which more than one dimension was required, without offering a system in which geometry was meaningfully employed to demonstrate color relationships. And it was Isaac Newton who paved the way.

Two-dimensional color diagrams as the result of empirical knowledge

Some members of the early (scientific) English Royal Society realized that color phenomena were not clearly understood, as expressed in Robert Boyle's *Experiments and considerations touching colour* (1664). Various authors attempted to help resolve the conundrum.

At the beginning of the eighteenth century, Newton published his experimental findings on the composition of sunlight in English as well as Latin, thus making them available to a wide audience. His book *Opticks* (1704) contains a (semiquantitative) circular diagram in which he graphically represented the results of mixture of spectral lights (see entry in chapter 6). Spectral stimuli are placed on the circle, which represents a (thereby incomplete) hue circle. The pan-harmonist Newton used musical ratios to place seven primary colors along the circumference of his circle, with "white" light placed at its center. In this manner, the circle represented not only hue (limited to spectral lights) but also saturation along lines from the circumference to the center. It did so in a manner that used geometry meaningfully to represent perceptual facts.

In the early and mid-Renaissance, pigment mixing was generally frowned upon as a corruption of pure colors. But in the Baroque period, pigment mixing became a standard procedure. Many painters arranged pigments in specific order on the palette as an aid to systematic mixture. Thus, they learned empirically the results of mixing specific pigments.

On the one hand, the painters learned they could achieve all hues with only three properly selected pigments. On the other hand, they found that the larger the perceptual distance between colors from the two generating pigments, the more subdued the mixed middle hues tended to be. The idea of three basic pigments being sufficient was correct if only hue, but not saturation, was considered. But in painting practice, the idea is unsatisfactory, because painters cannot ignore the effects of saturation.

Color mixture from three primary pigments raised the question of how to systematically organize the results. In 1686, English botanist Richard Waller published the systematic mixture of pigment pairs in two-dimensional tabular form. His expressed purpose was to provide a degree of standardization by attaching color names in multiple languages to specific pigments and their binary 1:1 mixtures. The rectangular two-dimensional chart placed blue and black pigments on the abscissa and yellow and red colorants on the ordinate.

The identity of the first person who published a complete series of hues placed systematically in circular form is unknown. But four years after publication of Newton's *Opticks* (1704), such a (hand-illuminated) circle was published in a Dutch edition of a French book on miniature painting, *Traité de la peinture en mignature* (Treatise on miniature painting; see entry Anonymous in this chapter). It soon proved to be paradigmatic for the idea of the hue circle. The circle's primary purpose is to demonstrate the results of mixture in various ratios of pairs of three primary pigments, yellow, red, and blue.

Although French Jesuit Louis-Bertrand Castel came to oppose Newton, in 1740 he proposed a color spiral on basis of an octave design, in line with Newton, except for the choice of the seven primary hues.

From hue circles to chromatic color circles

Some 30 years later, English engraver Moses Harris was the first to design a pigment mixture hue circle with tint/shade grades for each hue. In this manner, his "prismatic" circle represented three attributes in two dimensions (without apparent awareness of this fact). The prismatic circle, as well as a second, "compound" circle, shows that he had not advanced to the idea of a three-dimensional color order system.

The same applies to Austrian botanist Ignaz Schiffermüller, who in 1772 published a 12-grade hue circle, illustrated in approximately continuous form. Unlike the anonymous author's and Harris's circles, which have the three primaries in geometrically equidistant positions, Schiffermüller's 12 color (hue) classes consist of four grades between red and blue, one between blue and green, one between green and yellow, and two between yellow and red. Such a distribution is more in line with perceptual distances.

Schiffermüller also envisaged tint/shade scales for his 12 hues, but executed only three. It is not known if he was aware of Tobias Mayer's 1756 lecture on the double tetrahedron space (see entry in chapter 4). In the year of publication of Schiffermüller's book, German physicist Johann Heinrich Lambert published his own on the *Farbenpyramide* (color tetrahedron; see entry in chapter 4), thus presenting for the first time an illustrated three-dimensional color solid based on yellow, red, and blue primary pigments.

Harris is the first system designer who specifically referred to pairs of hues diagonally opposed on his prismatic chart as contrasting to a maximum degree. The usefulness of contrast in painting is a subject with a long history.[1] The hues that most strongly contrast with Harris's primary hues are purple with yellow, green with red, and orange with blue. At about the same time, the neoclassical painter Anton Raphael Mengs (1762) promoted the same choices, but it is unlikely that the two were acquainted. The Harris chart allows the determination of an additional six pairs of highly contrasting colors.

In 1788 German painter and painting instructor Johann Christoph Frisch described another version of a tint/shade color circle. He realized that not all 32 hues of his saturated hue circle could be distinguished as blackish colors, so he reduced their number to two near black. The number of grades increased steadily from two with reduction in black content to the maximum for pure pigments. Frisch envisaged tint shades of increasing lightness for the next five rings but never illustrated them. To achieve perceptual equidistance between his grades, he placed more grades between primaries yellow and blue and red and blue than between red and yellow.

Three-dimensional data projected onto two dimensions

Since the work of Mayer and Lambert in the second half of the eighteenth century (see entries in chapter 4), it had become evident that a complete geometric representation of color stimuli requires three dimensions. But solids can be represented only imperfectly in two dimensions (on a sheet of paper). For this reason, there were attempts to find a solution limited to two dimensions.

A curious version of a two-dimensional colorant mixture system was offered in 1809 by English engraver James Sowerby, perhaps influenced to a degree by previous-generation entomologist and engraver Harris. Sowerby chose a triangular rather than a circular format, which let him illustrate binary colorant mixture by overlay. The triangle contains only pure primaries and their binary mixtures. To illustrate ternary mixtures, he attached three rhombi with varying levels of yellow and red but only one level of blue in each. It is apparent that this system is far from complete.

Incompleteness also applies to the work of portrait painter Charles Hayter. In 1826, he developed a system based on Harris's but with clearer ideas regarding tonal colors in the direction of white as well as black. He also improved on Harris's system by considering the third color attribute:

saturation. However, he used different types of diagrams to demonstrate the relations of colors; like his predecessors, he was unable to develop a closed system.

German poet and natural scientist Johann Wolfgang von Goethe extensively studied the extant color literature from antiquity to his time. He designed a color circle from perceived psychological principles, attempting to avoid errors he saw in his predecessors' work. Based on his theory of intensification, Goethe placed red at the 12 o'clock position. The circle consists of only three primary and three secondary hues because he believed that six hue categories were sufficient. From his studies of color literature and his friendship with the painter Phillip Otto Runge (see entry in chapter 4), he was aware of the three-dimensional nature of a complete color solid. But he believed that Runge had solved this problem to the degree required. Among some philosophers and artists, Goethe's color circle has remained influential into the present.

Attempts to color various forms of two-dimensional diagrams began to raise serious issues. In the seventeenth and eighteenth centuries, there was only very limited colorant standardization. Depending on the quality of raw materials, the manufacturing process used, and the clarification steps taken, a specific painter's pigments could vary significantly in the resulting hue, saturation, and lightness (and thereby in price). Certain dyes and pigments used in painting, such as gamboge, indigo, and ochre, are natural products, and others are manufactured, all with the potential for considerable variation.

Waller identified his colorants by name, but this indicates little about their quality. None of the other system developers discussed in this chapter identified the colorants used or provided any colorant quality information. Hand coloration of plates in books is somewhat haphazard. One can imagine the book publisher employing inexpensive labor to color the plates after an example provided by the author. Considerable variation can be expected depending on the care of the illustrator.

The four extant copies of hand-colored plates of Harris's work may have been painted somewhat differently, perhaps at different times and by different people. The same applies to hand-colored figures in different copies of C.B.'s book (see entry Anonymous in this chapter). We can at best take the coloring of these systems as approximate illustrations of the author's intent. But the posthumous second edition of Hayter's work takes advantage of color printing and thereby achieved a higher level of standardization.

RICHARD WALLER 1686

R. Waller, A catalogue of simple and mixt colours with a specimen of each colour prefixt to its proper name, 1686

Richard Waller (1647?–1715) studied botany at Christ College in Cambridge, England. He was elected a Fellow of the Royal Society in 1681 and served as its secretary and *Philosophical Transactions* editor from 1694 to 1713. In the early 1680s, Waller read *Nomenclatura et species colorum* (Nomenclature and species of colors), a booklet by the Swedish painter and archeologist Elias Brenner, published in Stockholm in 1680. In the booklet, Brenner presented samples and short descriptions of 31 colorants in six groups, as an aid to miniature painters. Waller improved on Brenner by placing not only colorants but also one-to-one mixtures into a systematic order.

Waller defended limitation to one-to-one mixtures as follows:

> Not that I pretend to give the *Shades* of all the *mixt Colours*, which were indeed infinite as the Compositions and Proportions of them may be unlimited; but I have mixt each of the *Simple Yellows* and *Reds* with each of the *Simple Blews*, and these Mixtures give most of the *mean* Colours, *viz.* Greens, Purples &c. (Waller 1686, p. 24, italics original)

In Waller's table (figure 3.1), the column header colorants are Spanish white, azurite, ultramarine, smalt, litmus, indigo, and ink black. The row label yellow colorants are (lead white), Naples yellow, gamboge, ochre, orpiment, and umbra; the red colorants are minium, burnt ochre, vermilion, carmine, red lake, dragon's blood, red ochre, (carbon black). Regarding mixtures, Waller explained:

Poppinjay-green is made of *Blew Bice* and *Cambodia*, an equal *weight* of each. . . . I have added the *Latin, Greek, French* and *English* Names that I knew, which the more skilful Reader may supply where wanting. I propose to my self that this *Table* will be of some use and advantage in the describing of the Colours of Natural Bodies, . . . which may be done by this *Table*, and represented more nearly to the Reader provided with one of the same *Tables*, with less ambiguity, I think, than is usual: *A Standard of Colours* being yet a thing wanting in *Philosophy*. (p. 25, italics original)

Appended to the table is a short description of the colorants representing Waller's simple colors and their sources.

Waller was fully aware of Newton's findings, published some 20 years earlier in the *Philosophical Transactions*, and the opposition they raised in the Royal Society. He attempted to standardize a limited number of color perceptions resulting from colorants and colorant mixtures, as well as their names in four languages.

ANONYMOUS 1708

Traité de la peinture en mignature, 1708

Although the author of the first illustrated modern hue circle is not known, its publishing history is. An influential self-help book on miniature painting was first published in 1673 in Rouen, France, under the title *École de la mignature dans laquelle on peut aisément apprendre a peintre sans maître* (School of miniature painting wherein one can easily learn how to paint without a teacher). The author, listed only as C.B., was later identified as Claude Boutet.[2] The work was

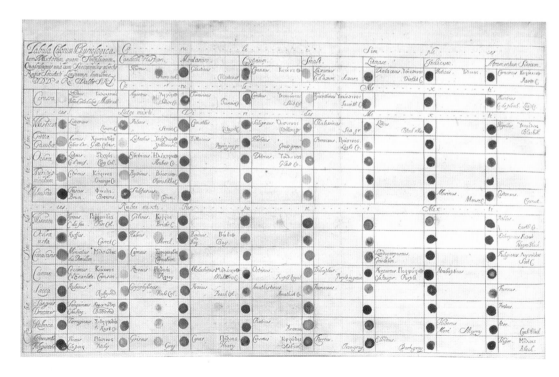

Figure 3.1.
Waller's (1686) colorant and colorant mixture table.

issued in several editions and languages (often with additions) by various publishers; by the year 1800, according to Parkhurst and Feller (1982), there were at least 33 editions.

The 1708 edition by the Dutch publisher van Dole is titled *Traité de la Peinture en Mignature* (Treatise on miniature painting) and is considerably larger than the original work. Aside from additional chapters, it includes a new, separate section on pastel painting. According to the publisher's advertisement for the book: "One finds there [in the section on pastel painting] something rather curious concerning primitive colors and the generation of their composites." After saying that successful work requires pastel crayons of many different colors, the section's (unknown) author launches into a description of primitive and composite colors:

> Properly there are only three primitive colours . . . yellow, red, and blue. White and black are not properly colors, white being nothing than the representation of light and black the privation of that light. But there are two kinds of red primitives, one tending toward yellow, such as fire red or vermilion, the other in direction of blue, like carmine red or lac. (p. 152, translated by R.G.K.)

From these four primitives, additional colors are formed by mixture, such as orange from yellow and fire red: "[T]hey make a suite or circle of seven colors" (p. 153). Additional colors are mixed from primitives and secondaries, such as purple from carmine red and violet. "All these colors are lively, but other mixtures, for example orange with violet, fire red with blue, violet with green, or green with orange or fire red produces dirty and disagreeable colors" (p. 154). The author illustrated the hue circle in two figures, the first with seven samples and the second with 12.

> Here are two circles by which one will be able to see how the primitive colors, yellow, fire red, crimson red and blue generate the other colors, and which one might call *Encyclopedia of Colors*. The first figure includes the four primitive colors and three composed from them, and the second includes those same colors with five others which are produced as much from the primitives as of their composites. (p. 154; see figure 3.2)

Coloring of the fields was by hand, with noticeable differences between copies.[3]

The two circles show that their author was familiar with the idea of pure hue primaries, for example, a red of neither yellowish nor bluish cast. Because he did not have a pigment for such a red, he used a mixture of "fire red" (perhaps vermilion) and "carmine red" to obtain what for him constituted pure red hue. He located the primaries at equal angular intervals in the 12-grade circle. The circles represent only the attribute of hue, and they are in essence one-dimensional, even though represented in two geometrical dimensions. These colorant mixture circles for painters are the first known illustrated systematic hue circles. The only other editions of the *Traité* with copies of the hue circles were Dutch translations of the van Dole edition (Verly 1744/1759).

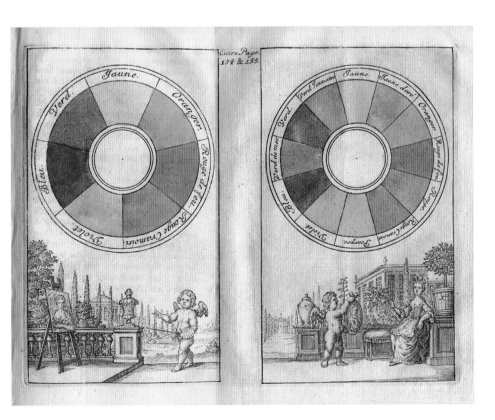

Figure 3.2.
Pages 154 and 155 of Traité de la peinture en mignature *(Anonymous 1708). Page 154 shows the circle of four primitive colors and three intermediate mixtures; page 155, the same seven colors plus five additional binary mixtures of adjacent colors. Here, the two reds of page 154 can be taken as mixtures of the primary red with yellow and blue, respectively.*

LOUIS-BERTRAND CASTEL   1740

L.-B. Castel, *Optique des couleurs*, 1740

Louis-Bertrand Castel (1688–1757) was a French math-ematician and member of the Jesuit order. He is best known for his proposal of a *clavecin oculaire* (piano for the eyes). Castel saw a strong analogy between music and colors, both produced by vibrations. He attributed to every tone and half-tone of the chromatic scale a color taken from the spectrum, resulting in a scale of 12 colors. In contradiction to Newton, these he claimed to be the 12 major hues identifiable in the spectrum (see figure 3.3).

While visiting Castel, German composer Georg Friedrich Telemann observed a prototype of Castel's *clavecin* in op-eration and described it in a booklet in 1739. Pressing keys resulted in color chips being moved and becoming visible to the viewer. Voltaire called the result "music for the eyes" and reported that Castel had painted "minuets and beautiful sarabandes."

Castel was an early supporter of Newton, but later strongly opposed him, as Goethe (1810) gleefully reported. Com-pared to Newton's seven, Castel claimed to have distin-guished several thousand different colors in the spectrum (modern research results limit this number to about 200). In 1740, Castel published *Optique des couleurs* (Color op-tics; see figure 3.4), in which he described his music-inspired 12-color scale, in the form of a circle segment, not a circle. Such segments repeat themselves in consecutively lighter and darker colors toward white and toward black, forming a spiral: "It must not form a completely round circle but it should be called circular, like a spiral spring" (p. 132, trans. by R.G.K.). Of the 12 spiral segments, Castell said:

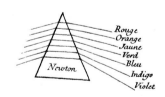

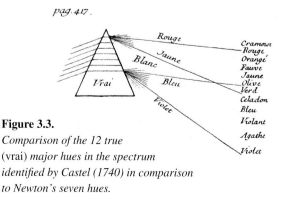

Figure 3.3.

Comparison of the 12 true (vrai) major hues in the spectrum identified by Castel (1740) in comparison to Newton's seven hues.

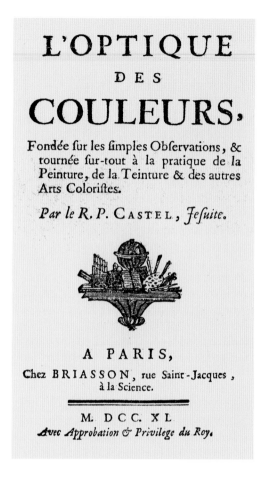

Figure 3.4.

Title page of Optique des couleurs *(Castel 1740).*

> I also determined the rise and decline of the colors to have twelve to thirteen grades; but I confess that their determi-nation is halfway accidental and useful in the arts rather than determined by nature. . . . The colors . . . distinguished alone by their intensity and lightness or darkness cannot be more than 144 or 145, possibly 146. The reason is not difficult to find. Because 12 times 12 is 144. (p. 217)

Castel did not offer a graphical image of his spiral color or-der system, and his description is difficult or perhaps impos-sible to comprehend. Elsewhere, Castel described all colors to be mixable from three primaries, identified as yellow: *stil de grain* (reddish yellow), red: fire red (yellowish red), and blue: sky blue (light blue).

Castel, having knowledge of the craft of dyeing, worked on a woven band to illustrate all hues of the spectrum. His book contains a sample table of binary mixtures of the three ba-sic colors, showing gradual transition from one to another. The ratios, for unexplained reasons, vary irregularly. Castel planned to fill one or more rooms with dyed color bands to produce a *cabinet universel de coloris* (universal color cabi-net), but there is no report that he did so.

IGNAZ SCHIFFERMÜLLER 1772

I. Schiffermüller, *Versuch eines Farbensystems*, 1772

Ignaz Schiffermüller (1727–1806/9), Austrian entomologist and member of the Jesuit order, is best known for his book on butterflies (*Ankündigung eines systematischen Werkes von den Schmetterlingen der Wienergegend*, 1775). In connection with his interests in entomology and in rules of harmony, and in keeping with the Jesuit tradition, he also worked on a color system.

In 1772, Schiffermüller published *Versuch eines Farbensystems* (An attempt toward a color system). He desired a systematic way of naming insect colors. He mentioned that he was inspired by a wish to derive rules for harmonic combinations for several artist friends.

Unlike Castel (see entry in this chapter), Schiffermüller had little interest in a link between color and music, but used Castel's 12-color hue spiral segment (but not his color selections) as a basis for a closed hue circle. Each of the identified hues is thought to represent a genus in the class of colors, numbered with a Roman numeral (figure 3.5). The colors of the circle are designated as *blühend* (florid); that is, they are

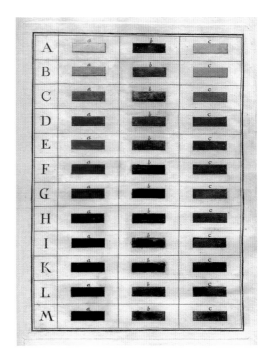

Figure 3.6.
Three tint/shade scales of blue colors (Schiffermüller 1772).

full colors. The circle moves from the 12 o'clock position in 30° segments beginning with blue as genus I, followed by sea green, green, olive green, yellow, orange yellow, fire red, red, crimson red, violet red, violet blue, and fire blue.

As was the case for Castel's color spiral, Schiffermüller's scale is an attempt at perceptual uniformity with two steps between blue and green and between green and yellow, three steps between yellow and red, and five steps between red and blue. In the matter of desaturated colors, Schiffermüller considered only tint/shade ranges from white via the full color to black without attempting to fill the interior of the implied double cone. For his book, he produced only three such scales, all involving bluish colors and containing 12 grades each, not including white and black, with the full color in the center (figure 3.6).

Schiffermüller's work is one of several incomplete attempts of the time to develop a rational color circle and place the tint/shade colors derived from it into systematic order.

MOSES HARRIS circa 1772

M. Harris, *The natural system of colours*, circa 1772

Moses Harris (1731?–1785?) was an English entomologist who engraved his own copperplate illustrations. Among his best-known works are a book named *Aurelian: a natural history of English moths and butterflies, together with the plants on which they feed* (1766), *Exposition of English Insects*

Figure 3.5.
Illustration of Schiffermüller's (1772) 12 classes of colors in the form of a hue circle.

(1776), and a brief work on color order (eight printed pages and three charts) that was published sometime after 1769.

The book's full title is *The Natural System of Colours, wherein is displayed the regular and beautiful Order and Arrangement from the Three Premitives, Red, Blue, and Yellow, the manner in which each Colour is formed, and its Composition, the Dependence they have on each other, and by their harmonious Connections are produced the Teints or Colours, of every Object in the Creation and those Teints, tho' so numerous as 660, are all comprised in Thirty Three Terms, only.*

The book is dedicated to the painter Joshua Reynolds, at the time president of the Royal Academy, indicating that Harris had good contacts in the artistic community. Only four copies (three incomplete) of the book are known to exist.

Harris's color system consists of two analogously built color charts (figures 3.7 and 3.8), identified as "prismatic" and "compound." Harris supported the three-primary-color theory, with all colors composed of them "except . . . white, which is the term for total privation or absence of colour" and black "a compound of Red, Blue, and Yellow in equal force and of the strongest powers" (Harris 1772, p. 4). Black is located in the center of both diagrams, demonstrated by superimposed triangles. In the so-called prismatic chart, the colors of the triangles are yellow, red, and blue. In the compound chart, they are orange, green, and purple ("mediate" colors). He identified the colorants used as the primitives as vermilion, kings yellow, and ultramarine, and placed them at 120° angles, with red at the top of the triangle.

Between primitives and mediates are two more grades ("divisions"), resulting in an 18-hue circle. Harris justified the circle form as implicit in the hue continuity of spectral colors. Each of the hues (none generated from more than two others)

> is divided into twenty parts or degrees of power, from the deepest or strongest, to the weakest; or from the outermost circle to the innermost. They are called teints, of which the whole circle contains, 360, so that each of the colours in the innermost or smallest circle contains 20 degrees of power, but each of the outermost but one. (p. 5)

Harris illustrated only 10 "teints." The teints are scaled in two ways: (1) There is saturation scaling using more or less diluted watercolors, and (2) they are also scaled in terms of blackness by addition of increasing numbers of black lines beginning with zero in the outermost circle, continuing with three in the second circle and increasing after that by one line in each circle. The latter is a standard engraver's technique for shading with black. Harris's teints are tonal color scales that begin with a tint color and proceed toward more saturated and at the same time more and more blackened colors. But as a result of the technique used, full colors are not shown anywhere.

The compound chart has prismatic mediates as its primaries. The mediates of these primaries are brown, "olave," and slate, with always two additional intermediate grades. These 18 hues (three duplicates from the "prismatic" chart) also are described as having 20 teints (with only 10 illustrated). Harris calculated a total of 33 hues (18 + 15) and 660 teints. He determined that more teints or divisions would not "render it more useful but rather tend to create confusions between the teints, which is now but sufficiently conspicuous" (p. 6). Harris considered the charts useful for painters to find the most contrasting colors directly opposite any color on the charts. He mentioned that all opposing colors, when mixed at equal "power," form black.

The third chart (figure 3.9), extant in only one of the four existing copies, illustrates the result of superimposing (more or less mixing) watercolor layers of various colors. Examples 1–3 are pairwise mixtures of the primitives; examples 4–6 are mixtures of primitives and opposing mediates. In example 9, all three primitives are mixed. Because of the over-painting process, the results do not support his theory as much as Harris seemed to wish. He explicitly indicated: "It must be observed here that the author treats on colour in the abstract." Predictions of the results of mixtures are difficult because "colours . . . being made of various substances, as animal, vegitable, and mineral maketh the colouring part extremely difficult if not impossible to be done, with any degree of perfection" (p. 7).

Harris briefly mentioned some perceptual color effects: bluing yellowish textiles for greater apparent whiteness, adapting to green eyeglasses that, when removed, make the world appear reddish for a short time, and viewing colored shadows. He also commented that in European languages, color names of primitives and mediates refer only to themselves and have no other meaning (he claimed that the fruit was named after the color orange, not vice versa).

Harris's color circles are peculiar in that they represent a compromise between two ideas about additive mixture, and the reality of subtractive mixture. The first chart (figure 3.7) purports to show 20 powers of prismatic colors, but all but one are blackened with lines. The colors of the compound chart are mixtures of mediates (except for the basic mediates themselves) and, as such, represent ternary mixtures, which he counted as new hues. In an idealized color circle, they would fall on lines connecting mediates, in most cases not representing additional hues. In a color solid, all colors illustrated by Harris (perhaps with the exception of the outermost circle) fall into its interior.

The plates' coloring in the second, posthumous edition of Harris's book (1811) is much different from that of the first edition, suggesting that the second edition's publisher may not have had access to a colored example of the first (Spillmann 2004).

Harris also included a color circle in his *Exposition of English Insects* (1776; figure 3.10). Based on the concept of the prismatic chart, its purpose is to give some idea of the meaning of color terms Harris used to describe the illustrated insects. Circle I consists of full colors with names based on red, orange, yellow, green, blue, and purple, and combinations thereof. In circle II, they are lightened to illustrate, for example, "cream colour," light blue, or "rose colour." Circle III has triple mixtures of the three primaries to result in colors such as nut brown, greenish olive, or purple slate. In circle IV, this selection is lightened to result in colors such as light olive-brown, or light greenish slate.

Harris's color charts represent only a small step forward in the effort to understand color order. Charts of a somewhat similar implied organization were described later by Frisch (see entry in this chapter) and produced by Michel-Eugène Chevreul (see entry in chapter 4).

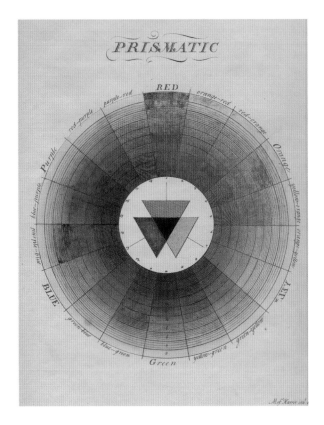

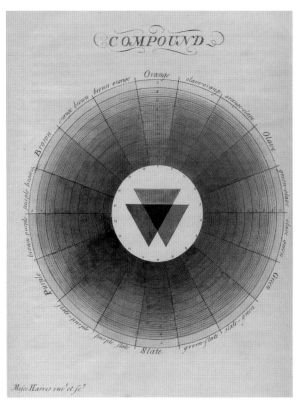

Figure 3.7.
"Prismatic" color chart (Harris ca. 1772; Spillmann copy).

Figure 3.8.
"Compound" color chart (Harris ca. 1772; Spillmann copy).

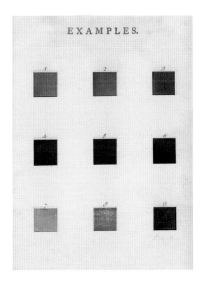

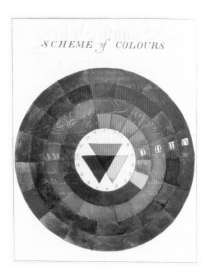

Figure 3.9.
"Examples" color chart (Harris ca. 1772; Spillmann copy).

Figure 3.10.
Color chart from Exposition of English Insects *(Harris 1776).*

JOHANN WOLFGANG VON GOETHE   1808–1810
J. W. von Goethe, *Farbenlehre*, 1808–1810

German poet, novelist, and natural philosopher **Johann Wolfgang von Goethe** (1749–1832) believed that there were many poets of his rank but that, in color knowledge, he was unique in his time. In 1808–1810, Goethe published an extensive *Farbenlehre* (color theory) that includes the unique and detailed *Materialien zur Geschichte der Farbenlehre* (Materials toward a history of color theory, 1810) in which he demonstrated that exceptional knowledge.

Wanting to personally experience the famous Newtonian experiments, Goethe reported that he placed a borrowed prism directly in front of his eyes. Based on the observed phenomena and other reasons, Goethe came to disagree violently with Newton. He devoted about a quarter of the *Farbenlehre* to an extended (less than successful) critique of Newton and his analysis of the composition of sunlight.

In *Farbenlehre*, Goethe included several color plates, the first shown in figure 3.11. In addition to his color circle (section 1), the plate comprises illustrations of several of his phenomenological findings.[4] The circle is described in the "Explanations of the charts belonging to the color theory" as follows:

> It is the simple but entirely sufficient scheme to explain the general facts of color. Yellow, blue, and red form a triad, as do the intermediate, mixed, or derived colors. The advantage of this scheme is that any diameter drawn in the circle immediately indicates the physiologically demanded color [the color appearing as a result of successive contrast]. If a devotee wants to expand on it by continuous and careful coloration it will be possible to make even clearer what I have here illustrated only in a schematic manner. (Goethe 1979, vol. 2, p. 233, trans. by R.G.K.)

In concept, Goethe's color circle is based on complementary colors. At the same time, he used it to imply psychological-philosophical aspects of color. The left semicircle is considered the plus side, and the right the minus side. He believed that colors are shadows between white and black. Both sides gradually intensify, culminating in "red in the zenith" and intensification is stronger on the plus side than on the minus side.

Concerning color mixture Goethe said:

> All colors mixed together maintain their general character . . . and because they are no longer individually visible there is no totality, no harmony and therefore gray results that, as any visible color, is always somewhat darker than white and somewhat lighter than black. . . . That all colors mixed together form white is an absurdity, repeated since a century with other absurdities, and is contrary to appearances. (p. 218)

Figure 3.11.
Color plate I from Goethe's Farbenlehre *(1808–1810) with the color circle in the upper left corner, section 1.*

It is evident that Goethe did not distinguish between additive and subtractive mixture. He had little interest in color order beyond the hue circle and considered Runge's effort (see entry in chapter 4) to be as extensive as needed.

JOHANN CHRISTOPH FRISCH   1788
J. C. Frisch, *Über eine harmonische Farben-Tonleiter und die Wirkungen und Verhältnisse der Farben im Colorite*, 1788

Johann Christoph Frisch (1738–1815) was a German painter of portraits and monumental canvases with historical themes, and a painting instructor. In 1786, he rose to the position of rector at the *Berliner Kunstakademie* (Berlin Academy of Art). Frisch described a color order system in his article "On a harmonic color scale and the actions and relations of colors in coloring." The colors he included represent the taste of the time; that is, they agree with the muted light and grayish colors of neoclassicist painting. Frisch considered Lambert's pyramid (see entry in chapter 4) insufficient because of the limited number of colors. As a painter and painting instructor, Frisch wanted to be able to see all selected colors simultaneously, thus limiting himself to a two-dimensional system. He produced only one illustrated copy, which was located at the time in a lecture hall of the academy.

The described color diagram (figure 3.12) consists of 20 concentric rings surrounding the central black color (described as having a reddish cast). The first nine rings represent tint/shade scales, proceeding from black via the full color to white in the tenth ring. From there, a second series of grayed colors was to proceed through the following nine rings, ending in ring 20 again in black. Only two colors are located in the innermost ring: dark red and dark blue. In the second ring, where the two colors of the first ring are slightly lightened, Frisch added a brownish yellow.

The third ring contains eight hues: yellow, orange, red, purple, violet, blue, sea green, and leaf green. Frisch chose these hues based on his argument that more hues are distinguishable between red and blue, and between yellow and blue, than between yellow and red. By adding intermediate hues, the number of hues always doubles in the fourth and fifth ring, that is, to 16 and 32, respectively. Here his description ends, with his comment that this many hues were barely distinguishable. It is not known how many colors the system was to display or the quality of coloration of the now-lost color chart.

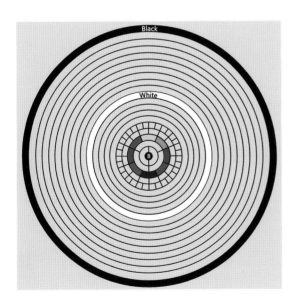

Figure 3.12.
Conceptual sketch of Frisch's color diagram.

JAMES SOWERBY   1809

J. Sowerby, *A new elucidation of colours, original prismatic and material; showing their concordance in the three primitives, yellow, red and blue: and the means of producing, measuring and mixing them: with some observations on the accuracy of Sir Isaac Newton, 1809*

James Sowerby (1757–1822) was the preeminent illustrator of English works of natural science of his time. After studying art at the Royal Academy in London, he contributed hand-colored copperplates to the important botanical work *Flora Londinensis*, published between 1777 and 1789. Sowerby and his children illustrated and wrote more than 100 works of natural science, among them *British Mineralogy* (1804–1817) and *Exotic Mineralogy* (1811–1820).

As an illustrator, Sowerby had a natural interest in color and color order. Simultaneously with his first work on mineralogy, Sowerby wrote a small book on color containing a color order system. Its purpose was to demonstrate color mixture as well as to supply color standards for scientific communication:

> The use of a true original for colours, and a regularity of arrangement, is almost infinite; for to the artist in any line it will be a solid satisfaction to know when he treads on a sure foundation, laid by unerring Nature. The mineralogist, the botanist and the zoologist may in future agree in their descriptions and ideas, so as to identify them to all parts of the world, and remotest ages. (Sowerby 1809, p. 5)

Sowerby's color theory does not distinguish between light and colorant mixture. He was familiar with Newton's work and with Thomas Young's publications of 1802 (see entries in chapter 6) but, as an artist, chose to use Young's initial set of primitives, yellow, red, and blue. Sowerby's color order system (figures 3.13 and 3.14) bears some resemblance to that of Harris (see entry in this chapter), as well as Edmé-Gilles Guyot's (see entry in chapter 11). It contains binary and ternary mixtures of the primitives, executed in watercolors.

The form is triangular, and the pure primitives are located midway between the corners. They are illustrated as "full tints," "middle tints," and "light tints." The central small triangle shows the white of the paper, also representing the white of light. Binary mixtures at three tint levels result where the bands are superimposed.

The secondary color orange is located at the bottom of the triangle (color 15 at full tint). Here the triangle expands to show ternary mixtures in three rhombi. These are colors Sowerby found missing in optical works. The four small rhombi forming the large rhombus on the bottom mostly differ in blue content, with the uppermost small rhombus containing no blue. In the left rhombus, fields vary in yellow and red content but have blue uniformly at the light tint. Fields in the right rhombus have blue uniformly at the middle tint, and those in the bottom one at the full tint level.

Figure 3.13.
Sowerby's (1809) color diagram with the three primary colors in three saturation levels on the sides of the main triangle and secondary mixtures in the corners. The three rhombi added on the bottom are ternary mixtures with three different levels of blue.

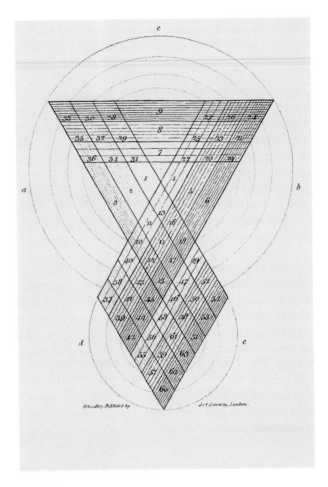

Figure 3.14.
Schematic drawing of Sowerby's (1809) diagram with identification of the colors.

CHARLES HAYTER   1826

C. Hayter, *A new practical treatise on the three primitive colours*, 1826

Sowerby defined black as supersaturated versions of each of the three primitives (as Matthias Klotz also did; see entry in chapter 4), not as a mixture of them. In the uncolored version (figure 3.14), Sowerby used dots to represent yellow, wavy lines for red, and straight lines for blue.

The diagram contains 63 colors plus white. All colors are identified by a number and a letter code. The colorants used as primitives are not explicitly identified, but his book names Lambert's primaries gamboge, carmine, and Prussian blue (see entry in chapter 4) as the most perfect colorants for the three primitives. Created more than 30 years after Lambert's color pyramid, Sowerby's color order remains in the two-dimensional world.

In 1826, English portrait painter **Charles Hayter** (1761–1835) published a color order system with considerable resemblance and reference to that of Harris (see entry in this chapter). Thirteen years earlier, Hayter published *An introduction to perspective, practical geometry, drawing and painting* (1813), an introductory instructional manual primarily for use by women. It appeared in several editions, the sixth and last published posthumously in 1845.

Unlike the previous editions, the posthumous edition included an appendix containing the complete text of Hayter's *Practical treatise*. In the 1826 edition, the color charts were hand-colored; those of the 1845 appendix were color-printed under the supervision of the famous English architect and ornamental designer Owen Jones.

Like Harris, Hayter did not distinguish between colors of light and of objects, that is, between additive and subtractive color mixture. To explain discrepancies, he formulated six axioms:

Axiom 1:

Yellow, red, and blue are the primary, primitive colors.

Axiom 2:

All other colors are obtained from mixtures of the primary, primitive colors yellow, red, and blue.

Axiom 3:

When the primaries are mixed in appropriate ratio, black is obtained.

Axiom 4:

"[E]very practical degree of light" can be obtained by dilution of any color (pigment) by addition of "white paint."

Axiom 5:

All color appearances of lights ("all transient or prismatic effects") can be imitated with colorants, but only to the degree white paint can imitate light.

Axiom 6

supports the special nature of the three primitives by stating, "There are no other materials, in which colour is found, that are possessed of any of the foregoing perfection." (Hayter 1826, p. 13)

Figure 3.16.

Strength and darkness variations of the three primary, secondary, and tertiary colors. On the bottom is a scheme of a color mixture disk where, from a combination of 90° each of yellow and red and 180° of blue, white is supposed to result when the disk is rapidly rotated (Hayter 1826).

Hayter developed a three-part color diagram (figure 3.15), built in nearly analog form to Harris's diagrams. He called each part a "color compass." In the center of the first part, the three primary colors yellow, red, and blue are shown in overlapping circles of gamboge, red lake, and Prussian blue. In the first circle (from the center), pairs of the primitives are mixed, in all cases in five grades. Toward the periphery, the resulting 18-grade hue circle is tinted in two steps.

In the second compass, similar to Harris's diagrams, the primitives are replaced by the secondaries orange, purple, and green. Because of the reduced perceptual distinction, the number of intermediate grades is reduced to three, resulting in a circle of 12 colors. These are also whitened in two grades. In the third compass, the key positions are held

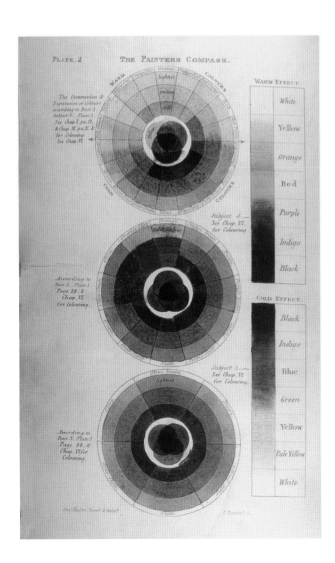

Figure 3.15.

Hayter's (1826) color compass in three parts, with separate scales of colors between black and white with warm and cold effects.

by tertiaries olive, brown, and slate, with the circle further reduced to six colors, again whitened in two grades. Hayter pointed out that he limited the whitening steps to two to save publication cost, and recommended that his readers attempt as many whitening tonal steps as possible in their own work. The three compasses contain only full and whitened tonal colors; Hayter demonstrated darkened colors on a separate chart (figure 3.16). Here the three primitives, the three secondaries, and the three tertiaries of the color compasses are modified "in different degrees of strength" pale, medium, and full, and then darkened in two grades with black, resulting in

81 colors. Together with the colors of the three compasses, Hayter's system contains a total of 189 (162 different) colors. Hayter expanded Harris's work by clearly defining the grades of a particular hue made by adding white and black. By mixing (in figure 3.16) dilution and adding black, Hayter also took a step in the direction of defining the dimension of chromatic intensity. In the sixth edition of *An introduction to perspective, practical geometry, drawing and painting* (1845; see figure 3.17), Hayter attempted to show in a single diagram (*Compendium of colours*) the relationship of all major colors of his three compasses.

THE COMPENDIUM OF COLOURS.

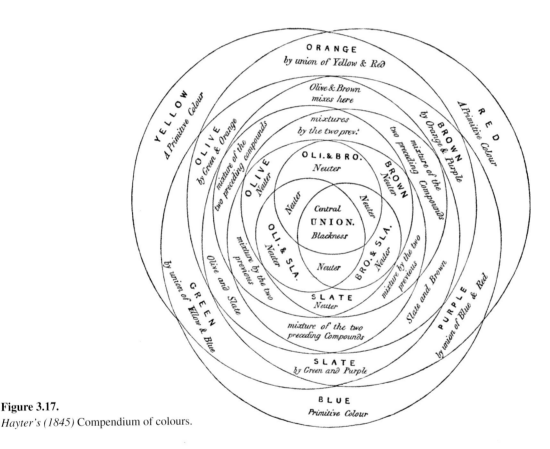

Figure 3.17.
Hayter's (1845) Compendium of colours.

Notes

1. The naturalness of contrasts already was commented on by Aristotle (*On the Universe*): "It may perhaps be that nature has a liking for contraries and evolves harmony out of them and not out of similarities. . . . The arts, too, apparently imitate nature in this respect" (Barnes 1984, vol. 1, p. 633). Contrast and its relation to harmony continued to be a much-discussed subject in music and painting through the Renaissance and beyond. Leonardo da Vinci commented that colors that go together harmoniously are green with red or purple, or violet, and yellow with blue. What hues should be considered maximally contrasting is a subjective matter. The issue became commingled in the nineteenth century with that of complementary or compensating colors.

2. The identification is documented in Quérard's *La France Littéraire* (1827–1864). Boutet was a minor seventeenth-century French miniature painter.

3. Each of four copies inspected by Parkhurst (Parkhurst and Feller 1982) are colored somewhat differently and perhaps with different pigments.

4. In figure 3.11, sections 2, 8, and 11 are illustrations of what Goethe believed a person with blue blindness (*alkyanoblepsy*) experiences compared to a color-normal observer. Sections 5 and 6 are experimental setups for seeing colored shadows. Section 7 is an image of the colors of an alcohol spirit lamp. Sections 3 and 4 represent the reflection of lamp light from a wall, and section 3 shows intensification into red at the outer edge. This effect is used to discuss the (seen as related) phenomenon of *aureoles* shown in section 4. Section 9 is an illustration of contrast colors appearing when viewing a white paper mask over colored water. Section 10 shows the change in apparent color of dazzling images after the eye is redirected onto a white or black surface.

References

Anonymous (C. B.). 1673. *École de la mignature dans laquelle on peut aisément apprendre à peintre sans maître,* Rouen, France: Le Brun.

Anonymous (C. B.). 1708. *Traité de la peinture en mignature,* The Hague: van Dole.

Barnes, J. (ed.). 1984. *The complete works of Aristotle,* 2 vols., Princeton, NJ: Princeton University Press.

Brenner, E. 1680. *Nomenclatura et species colorum,* Stockholm.

Castel, L.-B. 1740. *L'optique des couleurs,* Paris: Briasson.

Frisch, J. C. 1788. Über eine harmonische Farben-Tonleiter und die Wirkungen und Verhältnisse der Farben im Colorite, *Monatsschrift der Akademie der Künste und mechanischen Wissenschaften zu Berlin* II, 8. Stück, 58–77.

Goethe, J. W. von. 1808–1810. *Farbenlehre,* Tübingen: Cotta. English translation, *Theory of colours,* Eastlake, C. L., trans., London: Murray, 1840.

Goethe, J. W. von. 1810. *Materialien zur Geschichte der Farbenlehre,* Tübingen: Cotta.

Goethe, J. W. von. 1979. *Farbenlehre,* Ott, G., Proskauer, H. O., eds., 3 vols., Stuttgart: Verlag Freies Geistesleben.

Harris, M. ca. 1772. *The natural system of colours,* London: Laidler.

Harris, M. 1776. *Exposition of English insects, with curious remarks.* London.

Hayter, C. 1826. *A new practical treatise on the three primitive colours,* London: Booth.

Hayter, C. 1845. *An introduction to perspective, practical geometry, drawing and painting (including: A new and perfect explanation of the mixture of colours),* 6th ed., London: Bagster.

Mengs, A. R. 1762. *Gedanken über die Schönheit und den Geschmack in der Malerey,* Zürich.

Newton, I. 1704. *Opticks,* London: Smith and Walford.

Parkhurst, C., and R. L. Feller. 1982. Who invented the color wheel? *Color Research and Application* 7:217–230.

Quérard, J.-M. 1827–1864. *La France Littéraire ou Dictionnaire Bibliographique,* Paris: Didot.

Schiffermüller, I. 1772. *Versuch eines Farbensystems,* Vienna.

Sowerby, J. 1809. *A new elucidation of colours, original prismatic and material; showing their concordance in the three primitives, yellow, red and blue: and the means of producing, measuring and mixing them: with some observations on the accuracy of Sir Isaac Newton,* London: Taylor.

Spillmann, W. 2004. Moses Harris's *The natural system of colours* and its later representations, *Color Research and Application* 29:333–341.

Verly, P. J. 1744. *Verhandeling van de Schilderkonst in Miniatuur,* Utrecht: Lobedanius; 2nd ed., Amsterdam: de Groot, 1759.

Waller, R. W. 1686. A catalogue of simple and mixt colours with a specimen of each colour prefixt to its proper name, *Philosophical Transactions of the Royal Society* 26:24–32.

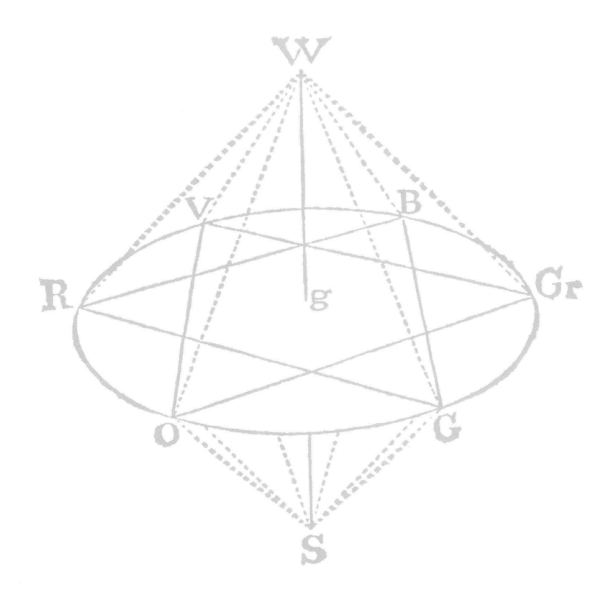

A description of the systematic development of the third geometric dimension in color order, beginning in the second half of the eighteenth century, as demonstrated in the work of Tobias Mayer, Johann Heinrich Lambert, and others.

The step from simple, linear, more-or-less lightness-based color order systems to hue circles and related tint/shade scales took some 2,000 years. At that point, progress toward three-dimensional systems came relatively rapidly, but not without difficulties.

In the eighteenth century, it became evident that the geometric representation of all colors known in art and the natural sciences could not fit on a plane. Moses Harris had already gone to a second diagram, and James Sowerby added rhombi to his triangle to represent ternary mixtures (see entries in chapter 3). The third dimension was needed to guarantee the geometric placement of all possible color perceptions into a logical system. Common to all color order pioneers was a gap between the theoretical structure of the system and the actual placement of color samples. Bringing the conceptual order into agreement with coloration of the system was a key problem. The third dimension also requires considering not just two, but three attributes. At the time, the idea of color attributes was not well understood or developed. In some cases, this resulted in unusual or partial solutions.

Also common was a solid conviction that yellow, red, and blue are the fundamental colors in light mixtures, not only pigment mixtures. The resolution of the difference between additive and subtractive color mixture would have to wait another hundred years. As late as 1846, Christian Doppler (see entry in chapter 10) used yellow, red, and blue as the primaries for the first three-dimensional system for colored lights, a prototype of Edwin Schrödinger's object color solid (see entry in chapter 6). The authors featured in this chapter developed their color solids through practical experience with subtractive mixture. Although they correlated with sensory attributes such as hue, lightness, and saturation, the practical outcome was only partially successful.

The development of three-dimensional systems began in 1758 with the work of German astronomer Tobias Mayer, the first entry in this chapter. It continued with the work of Swiss mathematician Johann Heinrich Lambert. But stepping into a third dimension does not mean that the two fully understood the totality of color experiences. The meaning of the spatial dimensions, the geometrical distances in the resulting space, and the technological problems of producing appropriate color samples were difficult issues that interfered with correctly placing specific samples in the system. That indicates a lack of understanding of the theoretical order or placement system.

Such problems were so great that often only partial coloration of a system was attempted. For simple collections, one could produce large numbers of different samples, with no need to place them appropriately in a three-dimensional system. Examples of such collections are C. F. Prange's color lexicon (1782) with 4,600 samples, and the Vienna *Farbenkabinet* (1784), based on Prange's work, with 5,400 samples (figure 4.1). Compared to these numbers, Lambert's 108 samples, Matthias Klotz's 104, and even David Ramsay Hay's 228 are lamentably few.

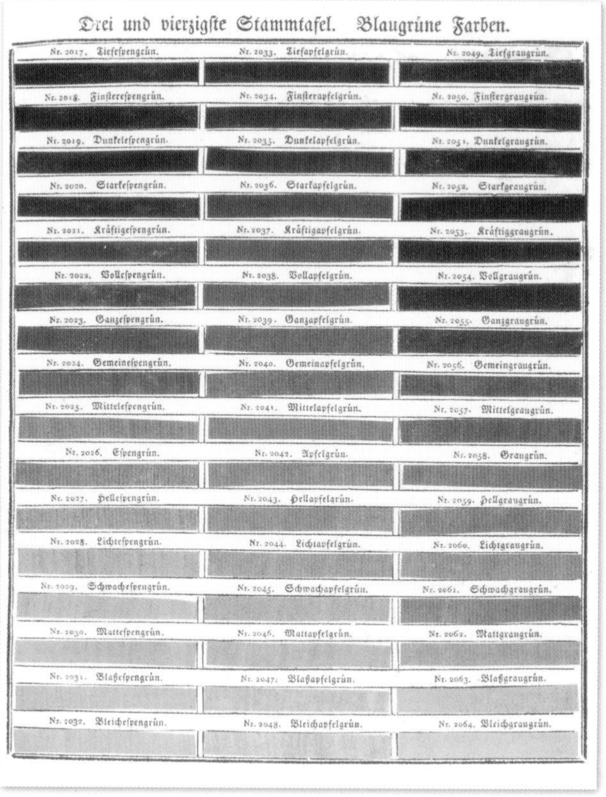

Figure 4.1.
Chart 43 with blue-green colors from Wiener Farbenkabinet *(Anonymous 1794).*

More impressive in number is Michel-Eugène Chevreul's collection of 950 samples, but it is only a partial sampling of perceptual color space and does not agree in all respects with his conceptual system. On the other hand, issues that arose during attempts to fill a system with samples at times resulted in modifying the conceptual system itself. This was the case with Lambert's interpretation of Mayer's system.

Mayer and Lambert

Mayer pursued the idea of developing all possible hues from mixtures of yellow, red, and blue. He attempted to have perceptually equidistant hue differences, which he believed to result from regular weight ratios of the colorants in the mixtures. The basis triangle of hues was to be expanded upward toward white and downward toward black, the resulting geometric solid being a double tetrahedron. This pioneering effort resulted in the first spatial color system, if only on a conceptual, qualitative basis. It was Lambert who undertook to prepare appropriate color samples to populate the system. He selected colorants different from those suggested by Mayer and reduced the number of mixture grades. He also discovered that mixing his three primary colorants in an appropriate ratio resulted in black, not the gray Mayer had anticipated.

As a result, Lambert discarded the lower of Mayer's two tetrahedra, resulting in a single tetrahedron with white on top. Even though Lambert used a "scientific" method to determine the relative coloristic strength of his primary colorants, the color black does not fall on the gravimetric center of the triangle. In addition, only the samples of the basis triangle are colored according to systematic weight ratios. All dilutions with white were obtained on visual basis alone, thus having only the character of an illustration.

German colorant manufacturer August Ludewig Pfannenschmid knew Mayer's and Lambert's efforts but limited his system to the basis triangle; that is, he dropped the third dimension. However, his triangle is expanded to 64 colors from Lambert's 45. Pfannenschmid's main interest and concern were to provide colorant standards for his colors, offering 64 pigment mixtures that required only dispersion in water. This effort represented, for its time, the highest level of reproducibility of coloration.

Spherical and cylindrical systems

German Romantic painter Phillip Otto Runge was using his knowledge of desaturation as theoretical basis for developing a sphere out of Mayer's double tetrahedron. In Runge's view, the sphere is an ideal solid, where all chromaticity dissolves in the center (red + yellow + blue = black + white). He wished to model the sphere with transparent media, believing that, from any viewing angle, desaturation would make it look gray. But the illustrations of the sphere do not make it clear that Runge (who was sometimes called the developer of the first modern color order system) had a clear concept of its internal structure. Perhaps for that reason, but more likely due to Runge's premature death, he produced no sphere with samples.

A somewhat clearer vision of the color attributes used to structure color order was offered by French silk merchant and inventor Gaspard Grégoire. He conceived a system with three independent attributes: hue (*teinte*), chroma-like saturation (*ton*), and lightness (*nuance*). The hue scale consists (in theory) of 24 grades. The gray scale consists of five grades between white and black, and the tonal scale, seven saturation grades between the full hue and the gray judged

to be of comparable lightness. He did not discuss a geometrical model even though the cylindrical form is implicit. Grégoire's final theoretical concept required 962 color samples, considerably fewer than the 1,351 samples of the atlas he originally produced. He also produced a small atlas of 103 samples, the only one of which copies have survived.

The situation is similar in the case of German painter Klotz. He named his attributes hue, desaturation, and light/dark modification and considered them to be the cornerstones of his color canon. Like Grégoire's attributes, Klotz's clearly point toward the modern attributes hue, chroma, and lightness. Klotz prepared a colored circular chart with the three primary colorants and their pairwise mixtures on the periphery. The center is occupied by a gray of light/dark level 4, presumably the result mixture of any opposing pair in the circle. There are three desaturation grades between gray and full colors. He stopped there, without offering a geometrical model (schematic drawing) of the implicit cylindrical solid. It is unknown whether he would have been able to complete his system because he equated light and object colors and tried to correlate his attributes with pigment properties and application techniques. It is not clear how he would have resolved the issues surrounding very light and very dark colors, so his system concept must be considered incomplete. However, both Grégoire's and Klotz's approaches prefigure the system of Albert Henry Munsell (see entry in chapter 5) of some 90 years later.

French chemist Chevreul's system is a paradigm for discrepancy between theoretical concept and population with color samples. He changed his mind more than once about the internal structure of the conceptual semisphere (in a backward step not clearly based on three independent color attributes). In addition, the color samples in different publications reflect his conceptual design only to a limited extent. Of the 72 constant-hue scales of the conceptual basis plane, 12 have been published, for a total of 240 samples. But they vary considerably in different publications and editions, ranging from hand-painting to printing. In addition, there are 10 charts with 720 samples of full colors gradually declining to black. Many of these samples fall outside Chevreul's conceptual system.

Chevreul's example demonstrates that full understanding of a given color solid's interior structure in terms of color samples was still lacking. Understanding of internal relationships consistent with three-dimensionality—each color is a member of three independent scales—was not yet fully developed. Also lacking was comprehension of the relationship between colorant properties and color perceptions resulting from mixtures. It may have been natural to assume this relationship to be linear, but it is an assumption that is generally invalid.

Implicitly three-dimensional systems

Hay, an English interior decorator, mixed colorants for practical purposes and avoided spatial ordering. He prepared his samples in consequent fashion according to primary, secondary, and tertiary colors, expanding the sample selection in the direction of white and black. This implies a spatial order that might be represented in a double tetrahedron. However, the internal order would be different from Mayer's.

Tint/shade scales according to Chevreul's example were also prepared by German publisher Otto Radde. Of his 42 scales, 30 are based on full colors, and 10 derive from grayed colors. In addition, there is a brown and a gray scale. Radde organized the interior of his implicit color solid but left large gaps. His scales are independent and one dimensional. Although developing lightness relationships between the scales may be possible, developing saturation relationships is not. In the late nineteenth century, French physiologist Charles Henry developed a scientific aesthetic of both color and form. His work was much discussed by postimpressionist painters, and Georges Seurat used it in several of his later paintings. Henry published a continuous color circle based on the spectrum, related to Chevreul's basis plane. It can be interpreted as an infinite number of tint/shade scales with white in the center, the full colors in the middle ring and black at the periphery.

One-dimensional tint/shade scales continued in the early twentieth century in the work of Robert Ridgway and Jiri Paclt (see entries in chapters 10 and 11, respectively). While Ridgway somewhat systematically populated a double-cone color solid, he was interested in neither developing relationships between the scales nor clarifying or demonstrating his work's three-dimensional nature.

It was Munsell and Wilhelm Ostwald who early in the twentieth century finally demonstrated clear understanding of their respective concepts and populated them with samples as fully as they technically could (see entries in chapter 5 and 10, respectively). Both also took advantage of all representations implicit in their systems: circles of constant chroma/saturation, constant lightness, or constant-hue planes, and so on. The sample population was no longer dominated by colorant properties; rather, colorants were used to demonstrate attribute sequences. Only then was the third dimension not just passively used, but fully understood in terms of the postulated color attributes.

TOBIAS MAYER   1758
T. Mayer, De affinitate colorum commentatio, 1758 (1775)

German **Tobias Mayer** (1723–1762), after only limited formal education, joined the firm of mapmaker Hohmann in Nürnberg, where he learned to use astronomical and surveying equipment. Based on these capabilities and his acquaintance with some leading astronomers, he became a professor in Göttingen. There he directed the newly built astronomical observatory until his untimely death at age 39. Mayer's lunar navigational tables that demonstrated a method of longitude estimation earned him, posthumously, half of a £10,000 British Admiralty prize.

When Mayer began work on color order, it is likely that he knew of Isaac Newton's writings and Jakob Christof Le Blon's three-color printing technique (see chapter 9), which was only 30 years old. Mayer's is the first concept of an object color solid based on mixture of three chromatic primaries yellow, red, and blue, as well as black and white.

Mayer's 1758 Latin essay *De affinitate colorum commentatio* (Commentary on the relationship of colors) was published posthumously in a collection of unpublished work (*Opera inedita Tobiae Mayeri* 1775) by his Göttingen colleague physics professor Georg Christoph Lichtenberg (for translations into German and English see, respectively, Lang 1980; Fiorentini and Lee 2000).

Also in 1758, Mayer gave a public lecture on his color order system, a report of which was published in the *Göttingische Anzeigen für gelehrte Sachen* (Göttingen reports on learned matters). The report was read by Lambert, who attempted a physical implementation of the system (see entry in this chapter).

Like his contemporaries, Mayer did not understand the difference between the addition of colored lights and of colorants, and his system is inconsistent in this regard. He commented:

> There are three simple or basic colors and no more than that, all others can be generated from their mixture; they themselves cannot be generated in any way from others, in whatever ratio they might be mixed: red, yellow, and blue. We see them in rainbows, but even more distinctly in rays of the sun captured by a glass prism, though there they are accompanied and surrounded by secondary colors. (Lang 1980, p. 9, translated by R.G.K.)

Based on arguments of discriminability as well as general aesthetics, Mayer used 12 steps between pairs of the three chromatic primaries as well as 12 steps each from the basis plane to white and to black: "One has to employ such ratio between colors to be mixed as can be expressed with numbers which are not very large. . . . Neither in architecture nor music are proportions greater than twelve generally accepted, unaided senses only barely perceive differences between such mixture ratios" (p. 12).

Mayer understood the principle of threshold differences and believed that 12 steps would produce them: "In this way, there will be between two simple colors eleven intermediate mixed colors distinguishable by the eye, while the colors that fall in between these must be considered indistinguishable" (p. 12). Mayer placed the three primary chromatic colors at the corners of an equilateral triangle and designated them with the index 12. By filling the interior of the triangle in systematic ratios, he arrived at a total of 91 colors for the basis plane (figure 4.2).

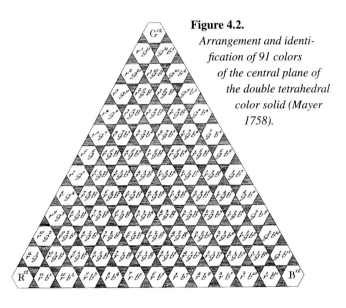

Figure 4.2.

Arrangement and identi-fication of 91 colors of the central plane of the double tetrahedral color solid (Mayer 1758).

On the next lighter and darker levels are 10 grades between the corners, resulting in 78 colors each, and so on. The complete double tetrahedron contains 819 defined colors (figure 4.3). For the intermediate colors, Mayer used a system of designation where two and three primary colors, respectively, are indexed with what Mayer called *partientes* (coefficients) to indicate their positions.

These coefficients must be taken as representing not pigment weights or volumes but amounts of yellow, red, and blue that result in perceptually equidistant steps. Mayer did not identify any pigments as representing the primary colorants. However, the lecture report identifies the Mayer coordinates of several pigments, including, with the index 12 (i.e., primary colorant), orpiment (king's yellow), vermilion, and azurite (*Bergblau*). These are the pigments previously used by Francis Glisson (see entry in chapter 2).

Among Mayer's contributions is that he understood gray to be composed either of white and black or of the three primary chromatic pigments. As a result, he considered black to be composed of the three chromatic primaries, not to be a primary color. By assigning values of primary components to each mixed color, he developed an arithmetic of color mixture in that any three-component color can (in theory) be mixed from other colors, the sum of which have the same total indices as the reference color. This is valid in a linear system, but inaccurate when using pigments due to the nonlinearity of the relationship between weights and resulting perceptual color of mixtures, as Lambert later discovered.

Mayer appears not to have proceeded past rudimentary experiments toward populating his system with color samples. Lichtenberg, in a commentary to the published essay, introduced a modified version of Newton's center of gravity principle to Mayer's system.[1] He also included a color figure of a simplified basis triangle painted with five intermediate grades using gamboge, cinnabar, and Prussian blue as primary colorants; he commented on the difficulties of producing many copies of this figure (figure 4.4). Lichtenberg also mentioned Mayer's idea for color reproduction, developed using the color order system.

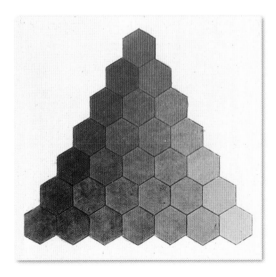

Figure 4.4.
Lichtenberg's 1775 abbreviated color sketch of Mayer's 1758 basis plane.

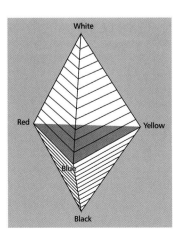

Figure 4.3.
Schematic representation of Mayer's (1758) color solid.

Mayer incorporated finely ground pigments into a solid wax medium according to ratios agreeing with the system's color specifications, and formed elongated bars. An artist cut thin layers sliced from the bars into appropriate shapes. When assembled, they resulted in "paintings" of which multiple copies can be produced in exactly identical colors. Mayer obtained financial support to commercialize the idea, but died without seeing it through.

As Pfannenschmid later wrote (see entry in this chapter), the fact that Mayer wrote his essay in Latin kept it from people who probably would have the most interest in its message: painters, dyers, and other craftsmen. His pioneering work represents an important first step in three-dimensional color order.

JOHANN HEINRICH LAMBERT   1772

J. H. Lambert, Beschreibung einer mit dem Calauischen Wachse ausgemalten Farbpyramide wo die Mischung jeder Farbe aus Weiss und drey Grundfarben angeordnet, dargelegt und derselben Berechnung und vielfacher Gebrauch gewiesen wird, 1772

Johann Heinrich Lambert (1728–1777), born in a part of Alsace then belonging to Switzerland, left home at age 17 to work as a private teacher in various European cities as well as in Tunisia. In Berlin in 1764, several compatriots teaching at the Prussian Academy arranged for him to become a member of the Berlin Academy of Sciences. He wrote books on mathematics, architectonics, photometry, and pyrometry. Lambert's lengthy book title translates to "Description of a color pyramid painted with Calau's wax, in which the mixture of every color from white and the three basic colors is arranged, explained, and their calculation and various uses indicated." After his death, a four-volume collection of letters between him and a wide range of academics, including Immanuel Kant, was published.

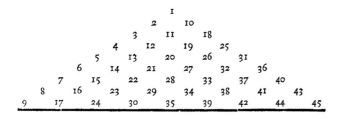

Figure 4.5.
Identification scheme of primary colors (1, 9, and 45) and color mixes used for the basis plane of Lambert's (1772) tetrahedral color solid.

Lambert read and was influenced by the report about Mayer's 1758 Göttingen lecture on color order (see entry in this chapter). The subject of light and color interested him, and in 1760 he published his important work on light measurement, *Photometria*, in which he cited Mayer's essay.

Lambert decided to build a colored model of Mayer's system. With Prussian court painter Benjamin Calau, Lambert set out to find the best available primary colorants: a neutral yellow neither reddish nor greenish, and comparably a red and a blue neutral pigment. For yellow, they chose the natural dye gamboge; for red, carmine lake, and for blue, the pigment Prussian blue. Wanting primary colorants to produce intensities resembling spectral colors as closely as possible, Lambert commented: "I leave it undecided if in the future colorants will be found which approach the spectral colors even closer" (Lambert 1772, p. 58, trans. by R.G.K.).

Lambert realized the varying coloristic power of the three selected colorants. He set out to gauge their relative strength by finding the weight ratio resulting in a color halfway between each pair of the basic colorants (in a manner comparable to Glisson; see entry in chapter 2). He determined them to be (on average) 3 parts Prussian blue to 2 parts carmine to 12 parts gamboge.

Lambert and Calau applied the colorants in a medium of Calau's invention, a waxy water-dispersible plant substance that, with gum, added gloss and durability to the painted papers. When they applied a full-strength mix of the three normalized primary colorants on white paper, they obtained a black rather than the medium gray predicted by Mayer. Lambert, realizing that Mayer's basis plane would not look as Mayer had imagined it, eliminated the lower part of Mayer's double tetrahedron. All three colorants have a degree of transparency, so instead of using white pigment for lighter colors, Lambert diluted the colorant mixtures of the basis plane appropriately with medium, thus making use of the paper's whiteness. Because the pigments selected as primaries behaved far from ideally, the practical work toward coloration resulted in system design changes.

Perhaps the 91 colors of Mayer's basis diagram were too many for Lambert's patience. His basis diagram contains only 45 colors, that is, seven grades between the primaries (figure 4.5). As a result, none of the color samples falls on the center of gravity of the basis triangle. Valid determination of the coloristic strength ratios of the three primary colorants might be expected to result in a near-achromatic color when the colorants are mixed in ratios corresponding to their strengths.

Lambert found that black does not fall near the center but closer to primary blue, that is, colors 11, 12, 19, and 20 in figure 4.5. He found the most neutral gray to be represented by color 51 on the second level, consisting of three parts blue and one part each of yellow and red, but did not further comment on this empirical fact. He colored the planes toward white (i.e., appropriate dilution of the colorant dispersions used for the basis plane, as well as new combinations) on visual basis only, and made several compromises.

Figure 4.6.
Image of Lambert's Farbenpyramide *(color pyramid, 1772).*

peared to be correct. While the basis plane represents systematic colorant mixture, the higher planes are essentially only illustrations.

In Lambert's image of the pyramid, there are 108 colored samples on six planes. Including the two planes left out, the model contains 164 chromatic samples. Lambert wrote that he proposed that Calau place "twelve of the more beautiful painter's colors on the foundation of the pyramid so that they can be compared with the colors in the pyramid" (p. 117). They are, from left to right, Naples yellow, king's yellow, orpiment, azurite, smalt, indigo, lamp black, sap green, chrysocolla, verdigris, vermilion, and Florentine lake.

Lambert envisaged several user groups for the pyramid:

1. Merchants could use it to determine if they have goods in all appropriate colors.

2. Consumers could use it to determine what colors of material they desire: "Caroline wants to have a dress like Selinda's. She memorizes the color number from the pyramid and will be sure to have the same color. In case the color need to be darker or go more in direction of another color, this will not pose a problem" (p. 109).

3. Dyers, he believed, would find it most valuable. After determining the coloristic strength of three primary dyes relative to Lambert's primaries, they could (in theory) use it to easily calculate formulations for any of the samples in the system.

4. Artists could use it to be able to reproduce in the studio the colors they sketched from nature.

Lambert also briefly discussed Jacques Fabian Gautier d'Agoty's (see chapter 9) three-color printing process as being easy to understand when considering all the colors produced from three primary colorants in his pyramid. Although some of these ideas are conceptually interesting, nonlinearities between colorant concentrations and perceived color make them problematical.

Despite its many shortcomings, Lambert's pyramid represents a valiant first attempt to solve the problems of systematic coloring of a conceptual color solid. But the practical impact of Lambert's pyramid was limited. It took another 30 years for the next, even more conceptual and less complete attempt by Runge (see entry in this chapter).

On the sixth level above the basis plane, there is only white. On the fifth level are the three diluted primary colorants alone. On the fourth level, two (diluted) primaries have been mixed to result in the intermediate pure colors. Ternary mixtures show up only on the third level, with a single grayish color. The second level has just three ternary mixtures. Here (as well as between the basis and the first planes), Lambert left out a plane to avoid hiding any of the color samples of the chosen planes in his image of the pyramid (figure 4.6). He defended this choice by saying, "[I]t is not good to use a separate page for each triangle because the system of colors presents itself better as a *whole*, when it can be viewed on *one* sheet at *one* glance" (p. 74, emphasis original).

A total of 36 different hues are illustrated on the six chromatic planes, many appearing only on two planes, a consequence of the system's design. Given the Mayer/Lambert design, no systematic uniformity is possible in the planes' interior. The scales along the sides of the triangles are approximately uniform within a scale, but differ in unit size between scales. Without a measuring tool to aid in the dilutions, Calau experimented with the planes until they ap-

AUGUST LUDEWIG PFANNENSCHMID 1781

A. L. Pfannenschmid and E. R. Schulz, *Versuch einer Anleitung zum Mischen aller Farben [aus] blau, gelb und roth, nach beiliegendem Triangel*, 1781

August Ludewig Pfannenschmid, whose life dates are not known, was active in or near Hannover, Germany. He was what the English called a colorman, a producer of standardized colorants for artists and technicians. The work cited here was written and published by his friend Ernst Rudolph Schulz, a pastor in a town near Hannover, because Pfannenschmid "was too occupied with his business." The book appeared in two German and two French editions. Its title translates as "Essay of instructions for mixing all colors from blue, yellow, and red, according to the included triangle."

Pfannenschmid, familiar with Mayer's and Lambert's works on color order (see entries in this chapter), did not fundamentally advance their efforts, but concerned himself with its practical implementation. The purpose of the book was to help artists, apprentices, and house painters in the task of matching a large number of colors with three primary color-ants. On the number of required primary colorants, Pfannenschmid conclud-ed: "[O]ne has to admit that only blue, yellow and red are the true unmixed primary colors, and as a consequence it should be

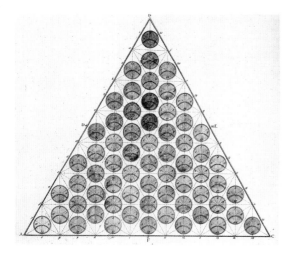

possible to mix from these three all other existing colors" (Pfannenschmid and Schulz 1781, p. 28, trans. by R.G.K.).

White was not regarded as primary because, Pfannenschmid argued, it does not change the hue of chromatic colors. He agreed with Lambert that mixing the three primaries made black. Using his knowledge of hue and purity of colorants, he selected gamboge, carmine, and ultramarine as primaries (deploring the high cost of the red and blue pigments). He determined their relative strength according to Lambert's method. In a Mayer/Lambert type of triangle, he placed five perceptually equidistant binary mixtures between the primaries.

But unlike Lambert, whom he faults for this shortcoming, Pfannenschmid filled the interior space of the triangle with 46 ternary mixtures (21 in Lambert's base plane), arguing "this is in better agreement with nature." The interior colors were mixed in ternary combinations of simple ratios. The triangle printed in the book (figure 4.7) is not colored, but the book offered 12 watercolors from which interested readers could mix the 64 colors of the triangle. (For those not inclined to do the mixing, he even offered 64 watercolors.) Pfannenschmid's business guaranteed standard quality of these watercolors, an important condition for a claim of universality. Pfannenschmid was one of the first colorant suppliers offering an extensive range of products of standardized quality.

Pfannenschmid's effort in color order did not leave a discernible trace beyond his native Germany. What he contributed was greater detail in the interior of Lambert's color solid (without resolving any of the basic problems of that approach) and products for standardized colorants to be used for color order systems.

DAVID RAMSAY HAY 1845

D. R. Hay, *A nomenclature of colours applicable to the arts and natural sciences, to manufacturers, and other purposes of general utility*, 1845

David Ramsay Hay (1770–1840), a renowned Scottish interior and exterior decorator and painter, was named Decorator to the Queen. He was personally and professionally interested in colors and color harmony. In 1828, he published *The laws of harmonious colouring adapted to interior decorations*, a text that appeared in five editions. Seventeen years later, he added *A nomenclature of colours*, a color sample collection Hay aimed to replace with a system.

Hay was influenced by the writings of Johann Wolfgang von Goethe, the English colorant manufacturer George Field, and his countryman Patrick Syme, who edited *Werner's Nomenclature of colours* (Werner 1814). With his contemporary the Scottish physicist and inventor David Brewster, Hay believed the theory of yellow, red, and blue as the three primary colors to have been generally proven: "It may, therefore, now be confidently assumed that there are in the

Figure 4.7.
Pfannenschmid's version of the Mayer triangle as published and hand-colored with Pfannenschmid inks (Pfannenschmid and Schulz 1781).

scientific theory, as in that of the artist, only three primary homogeneous colours, of which all others are compounds" (Hay 1845, p. 11). According to Hay:

> White and black are representatives of light and darkness; and yellow, red, and blue, the primary elements of colour, out of which, by commixture and union amongst themselves, every conceivable variety of colour and hue arises. White and black are not colors themselves but are, as the representatives of light and darkness, simply the modifiers of colours, in reducing them, and the hues arising from them, by their attenuating and neutralizing effects, to tints and shades respectively. (pp. 6–7)

But Hay did not accept the idea of white light being produced from the colors yellow, red, and blue. He describes red "pre-eminent amongst colours, and may be justly termed the life-blood of every chromatic composition. The other two primary colours may be termed links between colour and the principles of absolute light and darkness, represented by white and black, while absolute colour is represented by red" (p. 7).

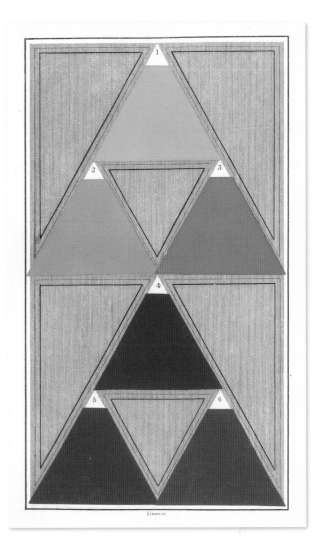

Figure 4.9.
Plate 13 from Nomenclature of colours *(Hay 1845).*

Secondary and tertiary compound colors can be mixed from the simple colors. All simple mixtures can be lightened (tints) and darkened (shades): "[T]hey can be multiplied in tint and shade, as well as in hue, almost to infinity. . . . By tint is meant every gradation of colour in lightness, from its most perfect or intense state up to white. . . . By shade is meant every gradation of a colour in depth, from its perfect state down to black" (p. 24).

In *Nomenclature of colours*, Hay displayed 240 (228 different) color samples (see figures 4.8 and 4.9). He named chrome yellow, carmine, and lapis lazuli as examples of primary pigments that, when mixed in their "strength ratio 1:2:3," produce a neutral gray. He did not ascribe a geometric form to his system, but an idealized form, a triangle, was later suggested to him by James David Forbes (see entry in chapter 11).

The three secondary mixtures are located perceptually halfway between primaries. Hay placed three additional steps

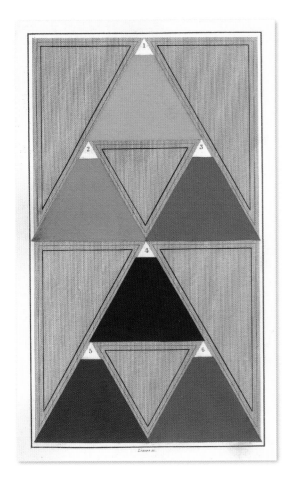

Figure 4.8.
Plate 8 from Hay's (1845) Nomenclature of colours.

Figure 4.10.

Conceptual sketch of the placement of colors mixed from the three primaries and located in the interior of the primary color triangle in Hay's system.

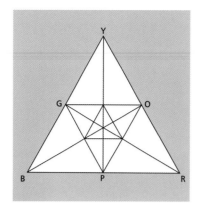

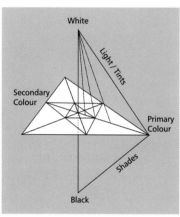

Figure 4.11.

Conceptual sketch with the placement of additional light tint colors in Hay's system.

between each primary and secondary hue sample. Between a primary and the opposed secondary color, Hay placed a total of 14 tertiary steps in irregular intervals favoring grayish colors (see figure 4.10), resulting in a total of 66 samples in the basis plane.

Shade samples at three levels were produced of only 12 colors(every second one). But tint samples were produced at three levels for every second color of the basis plane, including the primary and middle secondary colors. Hay saw a practical need for "light tints," additionally whitened light colors, useful for decorative purposes. Such additional light tints later became a standard feature of systematic paint-sample collections (see, e.g., Colorcurve, chapter 9). In Hay's system, there are 12 tints each at three levels (see figure 4.11). He explained that there are many more tints than shades, a judgment that may have been influenced by a light surround color for his samples.

Hay arranged color samples in *Nomenclature* according to aesthetic criteria, not neighboring relationships. Each chart contains two triples that are always contrasting (on opposite sides in the conceptual diagram). The numerical ratios of white, black, and the three primary chromatic colorants identify each color.

Unlike Mayer, Lambert, and other authors, Hay did not proceed from a preconceived geometric space with resulting problems in how to structure the interior organization. In-

stead, he used simple mixing as well as colorant strength ratios to color his samples in a systematic (if not perceptually uniform) manner. His contribution is that of an artisan with much practical experience in selecting pleasing color combinations for which he attempted to provide a systematic approach.

PHILLIP OTTO RUNGE   1810

P. O. Runge, *Farben-Kugel, oder Construction des Verhältnisses aller Mischungen der Farben zu einander, und ihrer vollständigen Affinität, mit angehängtem Versuch einer Ableitung der Harmonie in den Zusammenstellungen der Farben* 1810

German painter **Phillip Otto Runge** (1777–1810), born by the Baltic Sea, studied art in Copenhagen. With his friend Caspar David Friedrich, he became a chief representative of German romanticist painting with its belief that emotion and intuition rival reason and logic. One of Runge's spiritual sources was the mystic Jakob Böhme. Among his contemporaries are the painters Delacroix, Blake, Turner, and Goya.

Runge's interest in color and its symbolic values was an important aspect of his art. He was sympathetic to the views of Goethe (see entry in chapter 3), and they kept up a correspondence in the last years of Runge's short life. Goethe's *Farbenlehre* and Runge's *Farben-Kugel* were both published in 1810, the year of Runge's death.

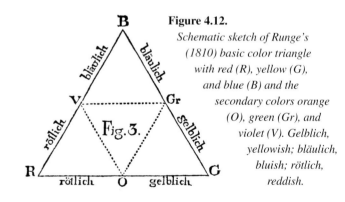

Figure 4.12.

Schematic sketch of Runge's (1810) basic color triangle with red (R), yellow (G), and blue (B) and the secondary colors orange (O), green (Gr), and violet (V). Gelblich, yellowish; bläulich, bluish; rötlich, reddish.

Figure 4.13.

Color star formed from two overlapping triangles connected by a hexagon and expanded to a circle (Runge 1810).

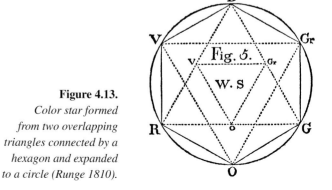

Runge's interest in color order derived from his wish for colors to have well-founded concepts similar to those for perspective, to "discover the relationships of given colors, as pure colors as well as according to the law from which their combinations appear to proceed, and to predict with certainty the changed appearances, generated from their mixtures and have the ability to produce them unfailingly with our materials" (Runge 1810, p. 2, trans. by R.G.K.). Having in hand a systematic arrangement of colors, Runge also attempted to derive the harmony of color combinations, as announced in the subtitle of his book.

Runge developed the sphere model (*Farbenkugel*) from the equilateral (Mayer-like) triangle of the basic colors yellow, red, and blue, with, as in Ignaz Schiffermüller's circle (see entry in chapter 3), blue on top (Matile 1979). The secondary colors are placed in a second equilateral triangle midway on the sides of the triangle (figure 4.12), and the whole is expanded into a hue circle (figure 4.13). Placing white on top and black on the bottom of a central axis perpendicular to the plane of the hue circle resulted in a double cone (figure 4.14).

Using the same argument that produces a circle from the double triangle, Runge described the ideal form of a color solid to be a sphere (figure 4.15). He did not identify primary colorants, but took them to be idealized materials. As a result, he placed middle gray as the central color of the sphere. As Lambert did before him, he postulated middle gray as derived from equal amounts of white and black, on the one hand, and, on the other, from colors opposing each other diametrically on the hue circle or, indeed, elsewhere in the sphere:

> Since all three colors, blue, yellow and red stand at the same distance from white and black, therefore, the center of the color disk in which those three have lost their individuality through equal activity must be in the same relationship and in the same distance to white and black as those three. Both of these points (the center point between white and black and that of the triangle Blue, Yellow, Red) coincide mathematically and it follows that both must be one and the same . . . and that from such identical difference a complete indifference results into which all individual qualities have dissolved. . . . This point, since it is in equal distance to all five elements, is therefore to be seen as the general center of them all. . . . All mixtures resulting from inclination of a point on the complete color circle toward white or black (a tendency common to all these points) will slowly loose themselves toward white and toward black. . . . [T]he differences of all points of inclination toward white or black from the central point being radii the points form nothing but circle segments ending in the poles white and black. . . . Thereby, the complete relationship of all five elements, by difference and inclination, results in a perfect sphere. Its surface contains all five elements and those of their mixtures generated in

friendly inclination of their qualities, and toward its center all colors of the surface dissolve in equal steps into a balanced gray. . . . Every color is placed in its proper relationship to all pure elements as well as all mixtures and in this manner the sphere is to be seen as a general table by which he who requires various tables in his business, can always find the relationship connecting the totality of all colors. It now must be evident to the attentive reader that it is not possible to find a plane figure that is a complete table of all mixtures; the relationship can only be presented as a solid. (pp. 12–14)

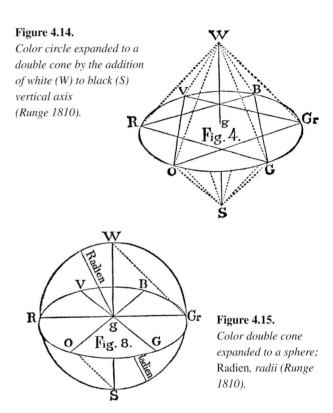

Figure 4.14.
Color circle expanded to a double cone by the addition of white (W) to black (S) vertical axis (Runge 1810).

Figure 4.15.
Color double cone expanded to a sphere; Radien, *radii (Runge 1810).*

Unlike Lambert (see entry in this chapter), who developed a more or less systematic mixture space from his three chromatic primary colorants and white paper, Runge had in mind a perceptually uniform space. His colorants were idealized to have equal strength (*Kraft*) and were expected, in equal mixture ratios, to be perceptually intermediate. Equal mixtures of the three primary colors (or their appropriately selected mixtures) were taken to form neutral gray.

Efforts toward coloration of the sphere are restricted to illustrative vertical and horizontal central slices and views of the surface (figure 4.16). Had Runge lived longer, he might have attempted detailed coloration and thereby confronted the problems with which Lambert wrestled. The hand-colored copperplate indicates limited uniformity, perhaps due to the need for mass production (coloration is known to vary in different copies of the original edition). The equatorial chart contains 12 hues, each with four desaturation steps toward the central neutral gray. The intermediate colors indicate the problems of maintaining saturation when mixing them from three primary colorants (e.g., green and violet). Nine grades of the central gray scale come between white and black, but the steps do not appear to be perceptually uniform.

Although his color order system was the most complete produced up to that time, it is evident that Runge's proposal still suffers from several unresolved issues. Yet Goethe thought that Runge's sphere "has successfully concluded this kind of effort."

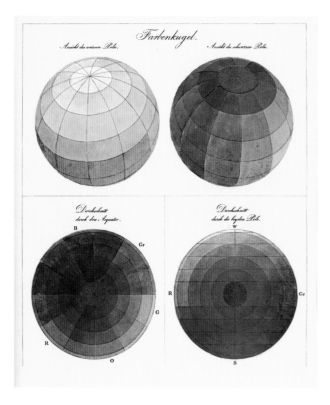

Figure 4.16.
Illustrated Farben-Kugel *(Runge 1810). On the top are views toward the white and black poles. On the bottom, to the left is the equatorial, and to the right, the polar cross section.*

MATTHIAS KLOTZ   1816

M. Klotz, *Gründliche Farbenlehre*, 1816

German **Matthias Klotz** (1748–1821) was born in Strassbourg and became court and theater painter at the Bavarian royal court. He developed his *Farbenlehre* in serious critical engagement with that of Goethe (see entry in chapter 3). Klotz proposed to Goethe the joint development of a *Farbenlehre*, an offer the latter ignored. Klotz then became more and more critical of Goethe's efforts, a position that did not endear him to the public.

Klotz spent 20 years developing *Gründliche Farbenlehre* (Thorough theory of colors). In attempts to keep the public interested while he continued to refine his work based on new insights and practical experience gained as a painter, he published three other works: *Aussicht auf eine Farbenlehre* (Prospect for a theory of color, 1797), *Meldung einer Farbenlehre* (Report of a theory of color, 1806), and *Erklärende Ankündigung einer Farbenlehre* (Explanatory announcement of a theory of color, 1810). Finally, in 1816, at his own expense, he published *Gründliche Farbenlehre*, but only 300 copies, the first 48 of which he hand-illuminated.

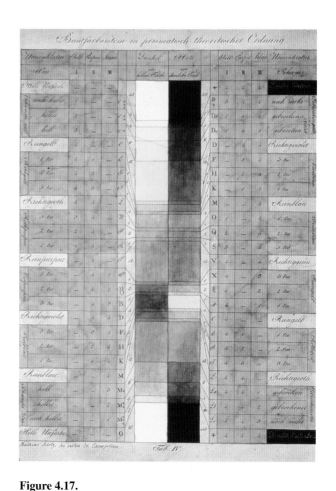

Figure 4.17.
System of chromatic colors in their theoretical prismatic order (Klotz 1816).

Klotz's work centers on a color order based on Klotz's view of a natural grammar of color that, due to its clarity, demonstrates the true relationship between colors. As a result of learning the system, artists will gain the ability to select harmonic color combinations and avoid those that are disharmonic. A biographical note about Klotz states that after completion of the *Farbenlehre*, his paintings were much improved.

The text is divided into "chromatics" and "prismatics." The former treats object colors as a "theory of colors in art"; the latter treats lights as "natural theory of colors." Klotz conducted prismatic experiments similar to those by Goethe and concluded that three primary colors apply to both lights and objects. Of these, pure yellow and pure blue are defined as, respectively, neither greenish nor reddish, and neither violetish nor greenish. Interestingly, the red primary is a "pure purple," defined as neither violetish nor reddish.

In partial agreement with Goethe, Klotz sees colors interact with white and black "by expansion from the inner dark into the outer light . . . [or from] expansion of the outer dark to the inner light" (Klotz 1816, p. 23, trans. by R.G.K.). This is illustrated in systematic manner with named colors in figure 4.17. The prismatic order transitions from light, via light yellow to pure yellow, on to red, pure purple, violet, pure blue, and blue, fading into white. The order of prismatic colors bounded by black is black, violet, pure blue, green, pure yellow, and red transitioning to black. These scales are in qualitative agreement with those given by Goethe.

Klotz viewed colors as pure perceptions:

> Here it is called color only in the sense that its stimulus influences our perception. This color theory is completely indifferent to physical, optical, or any and all other stimuli of our color sense. Similarly, the basic composition of all colorants does not have the least importance for the theory, they may be purely natural products or the result of chemistry. (p. 14)

Klotz published what may be the first illustrated gray scale with seven grades between white and black (figure 4.18, left) and established the lightness of the (unidentified) primary colorants against it. Interestingly, it is a logarithmic scale. The primary colorants are also applied at seven different intensity levels, with grade 8 called "oversaturated."

Klotz also generated a color chart, named *Buntfarbenkanon* (chromatic color canon; figure 4.19), with the three primary colors at the 2, 6, and 10 o'clock positions and seven inter-

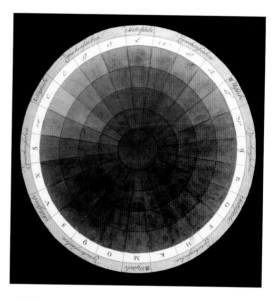

Figure 4.19.
Buntfarbenkanon *chromatic chart at lightness level 4 (Klotz 1816).*
Figures 4.17. to 4.19. the Bayerische Staatsbibliothek München

Figure 4.18.
Klotz's 1816 nine-grade (logarithmic) gray scale (left) and "light-dark images" (right) for the purpose of obtaining color effects when viewed through a prism (as also described by Goethe in his Farbenlehre).

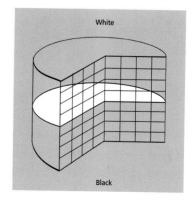

Figure 4.20.
Conceptual sketch of Klotz's implied color solid.

mediate grades between the primaries. Klotz chose grades such that, in his view, compensating colors are located diametrically opposite in the chart. At the same time, Klotz attempted perceptually uniform spacing but, due to the incompatibility of the two principles, succeeded only to a modest degree. Between the full colors and the central "dead gray," there are three hue scales of "broken color nuances" (tonal colors) with tonal declines toward middle gray of lightness grade 4. The chart has a total of 97 colors. Klotz identified the lightness grade of his samples. In the chart they vary as a function of hue.

Klotz described three color attributes for the samples, *Buntmodifikation* (hue variation), *Brechungsmodifikation* (degree of desaturation), and *Hell/Dunkelmodifikation* (lightness variation). Together with Grégoire (see entry in this chapter), he was the first to do so. The color solid implicit in these attributes is cylindrical (figure 4.20). It should be noted that the samples of the *Buntfarbenkanon* do not fall on a horizontal plane in this solid.

Klotz was seemingly not aware of the geometric implications of his three attributes. But his color chart is the most thoroughly developed of its time.

GASPARD GRÉGOIRE ca. 1810–1820

G. Grégoire, *Table des couleurs* (1,351 samples) [ca. 1810]
Théorie des couleurs, contenant explication de la table des couleurs [ca. 1820]
Table des couleurs sur trois feuilles, précédée d'une planche indicative des couleurs [ca. 1820]

Gaspard Grégoire (1751–1846) was born in the south of France into a family of silk merchants and became active in their silk business. At age 26, he conceived the idea of making naturalistic pictures using the medium of silk velvet. He produced several dozen works, of approximately 30 × 30 cm size, among them portrait images of Napoleon, Louis XVIII, and Pope Pius VII, reproductions of famous paintings by Raphael, and other subjects. To achieve the great detail necessary for such work, Grégoire invented a special weaving technique (Jacquard weaving did not yet exist at the time) (Algoud 1908).

Grégoire became interested in color order as a result of working with silk. Around 1810, he published a large-format color atlas (*Table des couleurs*) with 1,351 samples. The French minister for manufacturing and commerce arranged to send copies to royal manufacturing operations. In approximately 1820, Grégoire published an atlas describing his

Figure 4.21.
Depiction of Grégoire's ca. 1820 color order system concept (1820b).

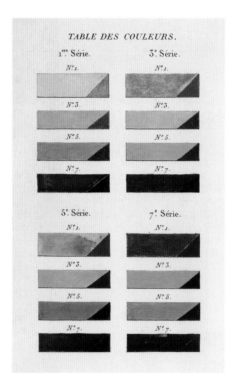

Figure 4.22.
Tonal grades of four hue series. The lighter fields represent the grades at lightness level 4. The darker small triangles show the same colors "at their darkest possible level." Some colorant deterioration is evident (Grégoire ca. 1820b).

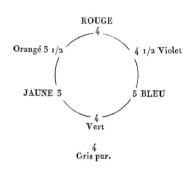

Figure 4.23.
Lightness levels of the principal colors of the hue circle in figure 4.21 (Grégoire ca. 1820b.

color order system in which he advertised the 1,351-sample collection as well as a collection of 343 samples. No copies of these atlases appear to have survived. At about the same time, Grégoire published an even smaller collection containing the concept of his color system (figure 4.21) on one page and 144 painted samples (figure 4.22) on three pages.

Grégoire's system is conventionally based on the chromatic primaries yellow, red, and blue, "from which all other colors are composed; this system is well described in many works on chemistry, painting and dyeing. However in physics, primarily in optics, the system of seven primitive colors remains dominant" (Grégoire 1820a, p. 1, trans. by R.G.K.).

It is evident from his discussion of the problems with gamboge, carmine, and Prussian blue, and from his remarks on loss of intensity and lightness when mixing them, that Grégoire had extensive colorant mixing experience. He described the conceptual idea that a primary color can be reduced to gray by adding equal amounts of the other two. However, from practical experience, he knew that perceptual continuity required much adjusting of colorant concentrations. Grégoire assigned gray scale value 3 to his yellow primary colorant, the value 4 to red, and 5 to blue (figure 4.23). To avoid the effects of desaturation of his primaries in binary mixtures, he used additional colorants for intermediate hues.

Grégoire's system has a hue circle consisting of 24 hues, with red at the 12 o'clock position (figure 4.23). The small atlas shows only 12 hue samples, beginning with yellow at the 8 o'clock position. He used perceptual halving to obtain intermediate grades. In concept, all steps within an attribute are perceptually equidistant. A six-step (five grades plus white and black) gray scale provides a lightness standard. At each lightness level, hues are desaturated in eight relative chroma steps (full color at grade 1 and gray at grade 9). In the small atlas (figures 4.21 and 4.22), only four of the nine grades are shown. Every hue is shown "at middle lightness [4] ... as well as in the deepest possible tone of the same color" (p. 29).

In concept, the system is cylindrical, with a hue circle, planes of constant lightness and cylindrical surfaces of constant relative chroma (see figure 4.24). Each color is identified with three numbers, the first indicating hue (*teinte*, from 1 to 24), the second saturation (*ton*, on a scale from 1 to 9), and the third lightness (*nuance*, gray scale from 1 to 7). Grégoire described all three atlases as having the same structure and varying only in number of samples. However, the theoretical number of color chips resulting from his cylindrical arrangement is 962, and it is not evident how the 1351 samples of the large edition were arranged. He stated: "[T]he number of colors might be multiplied to infinity; but with colors it is as with musical tones; in music even a quarter tone can barely be distinguished" (p. 31).

Grégoire described application of his system to embroidery, general commerce and manufacture, painting, natural history, chemistry, and pharmacology and as a general means for easily producing all colors. Somewhat similar to Mayer and Lambert (see entries in this chapter), he also introduced a calculation system. The table in figure 4.25 shows the content of (normalized) primary colors yellow, red and blue in the 24 hues of the hue circle. Orange (#5) consists of four parts each of yellow (#1) and red (#9), and so forth. In this manner and by including grays, the perceptual color content of any color can be (in theory) calculated. The table includes a hue-naming proposal using primary and secondary hue names and combinations.

Grégoire's system represents a considerable advance in color order, anticipating the system of Munsell (see entry in chapter 5) in the placing of colors of equal perceived lightness on its constant lightness planes. Despite the fact that it was published in Paris, Grégoire's system did not leave any strong marks. Very few copies are extant. Somewhat unfortunately, his system soon was overshadowed by Chevreul's work, published 29 years later (see entry in this chapter).

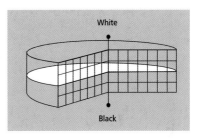

Figure 4.24.
Conceptual sketch of Grégoire's implied color order system.

NOTE des couleurs brillantes N.º 1.				NOMENCLATURE pour les sciences naturelles.
Série	Jaune.	Rouge.	Bleu.	
1.re	8.	0.	0.	JAUNE.
3.e	6.	2.	0.	Jaune-orangé.
5.e	4.	4.	0.	Orangé.
7.e	2.	6.	0.	Rouge-orangé.
9.e	0.	8.	0.	ROUGE.
11.e	0.	6.	2.	Rouge-violet.
13.e	0.	4.	4.	Violet.
15.e	0.	2.	6.	Bleu-violet.
17.e	0.	0.	8.	BLEU.
19.e	2.	0.	6.	Bleu-vert.
21.e	4.	0.	4.	Vert.
23.e	6.	0.	2.	Jaune-vert.

Figure 4.25.
Table of the composition of the 12 hues in figure 4.11 in terms of primaries and proposed nomenclature for these hues "for the use of the natural sciences" (Grégoire ca. 1820a).

MICHEL-EUGÈNE CHEVREUL   1839

M.-E. Chevreul, De la loi du contraste simultané des couleurs et de l'assortiment des objets colorés, 1839

Michel-Eugène Chevreul (1786–1889), raised during the French revolution, studied chemistry in Paris and became an assistant professor at the Collège de France at age 20. As a chemist, he is best known for pioneering work in the composition of animal fats. In 1824, when Chevreul was 38, Charles X named him director of the *Atelier des Teintures* (Dyeing department) at the *Manufacture Royale des Gobelins*, the fabled producer of tapestries. Here he became involved in textile dyeing problems and color issues, while also lecturing in chemistry at the *Muséum d'Histoire Naturelle* (Natural history museum).

Customer complaints about the quality of black dyeings produced in his department led Chevreul to study simultaneous color contrast: the effects of colored fields on neighboring colored fields. The importance of producing tapestries of high taste and refinement also created interest in laws of color harmony. In 1839, he published his findings on simultaneous color contrast and harmonic color composition. Translated into German and English, and influential in arts and crafts circles, the book also described his novel color order system.

The base plane of Chevreul's system consists of a 72-hue color circle based on the primary colors yellow, red, and blue, separated by equal segments. With the help of a colleague, the physicist Antoine Henri Becquerel, he fixed the positions of most of the hues relative to the spectrum, using Fraunhofer lines as reference. Twenty tint/shade grades of the corresponding hue are located on radial lines from the white center, ending in black on the periphery of the circle (figure 4.26). He considered locating the full color of each hue in this ladder at its appropriate level of lightness:

> In each of the scales . . . there is one tone which, when pure, represents in its purity the colour of the scale to which it belongs: therefore I name it the normal tone of this scale. . . . If the tone 15 of the Red scale is the normal tone, the normal tone of the yellow scale will be a lower number, while the normal tone of the Blue scale will be of a higher number. This depends upon the unequal degree of brilliancy and luminousness of the colours. (Chevreul 1987, p. 71)

However, later Chevreul placed all full colors on the same, middle grade, a fact that confused readers of different book editions. Colors toward the center from the "normal tone" are mixed with white, and those toward the periphery are mixed with black in appropriate amounts to make the constant-hue ladder visually equidistant.

Chevreul realized that he had not represented many other colors of a given hue in this plane (the plane in essence is a

flat representation of the surface of a color solid). To represent the missing interior colors of an implicit solid, he chose to erect a hemisphere above the plane (figure 4.26). The surface of the hemisphere is black and a 20-step gray scale forms the line between white at the center of the plane and the top of the hemisphere (figure 4.27).

Chevreul then proposed to mix each of the colors of the base plane in 10 steps with increasing amounts of black, but commented: "It is understood that these proportions relate to the effect of the mixtures upon the eye and not to material quantities of the red and black substances" (p. 72). In this manner, as the angle of the radial lines increases toward 90º, the colors become increasingly blackish.

Such a system would have large visual steps between the ninth line of colors, nearing the vertical center line, and the gray scale of the center line, particularly near the core of the hemisphere. The large steps can be avoided, if one assumes that Chevreul had in mind mixtures of the base plane colors (ordered, as originally planned by lightness) with the appropriate gray rather than black so that a smooth transition to the gray scale would result (Schwarz 1997). Calculations

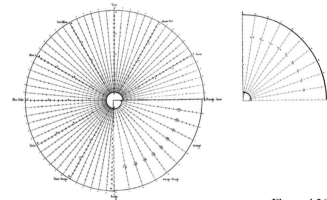

Figure 4.26.

Conceptual diagram of the 72-hue circle with 20 grades each. White is located at the center and black on the periphery. The quarter from 3 o'clock to 6 o'clock (a) is covered by a flap that can be raised and that illustrates the form and structure of the hemisphere (b) (Chevreul 1839).

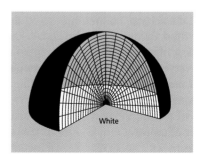

Figure 4.27.

Sketch of Chevreul's hemispherical system.

show that explicit redundancy of colors occurs in neither case (even though differences would be far from equidistant and some samples difficult to distinguish from neighbors), thus negating later criticism of Ostwald (1921) and others in this respect.

Mixtures with gray would have the advantage that all colors in a hemispheric layer would have similar lightness throughout, a result that Chevreul may have intended. But his system was never fully colored. Twelve constant-hue tint/shade ladders representing the key colors of the base plane and ten 72-hue circles, beginning with full colors and continuing with increasing amounts of black were produced first by hand-painting and, in 1864, also using a paper printing technique called *chromocalcographie* (figures 4.28 and 4.29). Reflectance data of color areas on copies of the original charts have been measured; the results indicate the kind of irregularities that are to be expected in such a system (Viénot and Chiron 2001).

Chevreul's color hemisphere created much discussion but was soon replaced by the new psychophysical systems of Hermann von Helmholtz and James Clerk Maxwell (see entries in chapter 6) and the psychological system of Ewald Hering (chapter 5).

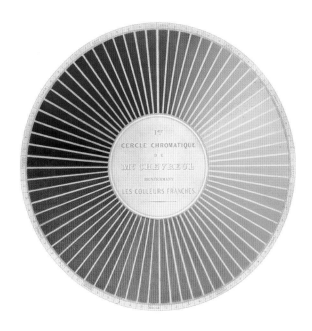

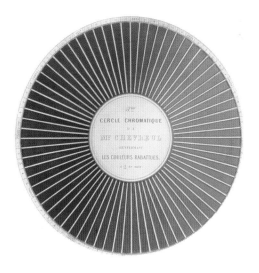

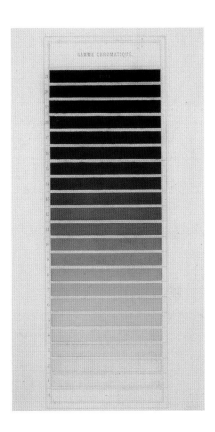

Figure 4.28.
Chevreul's (1864) tint/shade scale in 20 grades (one of a series of 12).

Figure 4.29.
Full color circle (circle 1, top) and circles 3 (middle) and 10 (bottom) shaded progressively toward black (from a series of 10) (Chevreul 1861).

OTTO RADDE   1878

O. Radde, Radde's internationale Farben-Skala, 1878

Little is known of German **Otto Radde** except that he was a publisher in Hamburg distributing his "international color scale" for six marks. The work was printed by the *Sociéte Sténochromique* in Paris. Chevreul's influence on this effort is undeniable. The system consists of, using Chevreul's term, 42 *Gammen* (scales). In each case, the most saturated color (*Cardinalton*) was placed in the scale according to its lightness (as Chevreul originally described it; see entry in this chapter). The color was varied in the tint direction toward white and in the shade direction toward black, with a total number of 21 grades (figure 4.30).

Thirty of the scales are based on highly saturated; 10 are based on considerably desaturated hues. There is also a brown and a gray scale. The system has a total of 882 color samples measuring 8 × 60 mm. The hues of the system are shown in figure 4.30, with the implicit hue circle shown in figure 4.31. Individual color samples are identified with a hue number (1–42) and a tint/shade letter (a–v). Radde's implicit double-cone solid (figure 4.32) demonstrates the gaps in his system.

The process used to print Radde's scales is not known with certainty but believed to involve oil-based printing inks. The scales are printed on stiff paper with a shiny surface turned slightly yellowish. Radde's scales were produced in the form of 15 charts, one with a hue scale and 14 with three tint/shade scales each. It was also produced in a pocket edition. In Europe, the scales quickly succeeded as a means of color communication in scientific and trade applications. In 1903, the spectral properties of the color samples were measured (Topolansky 1903) to safeguard the scientific results based on the *Farben-Skala* because it was no longer available.

CHARLES HENRY   1888

C. Henry, Cercle chromatique présentent tout les compléments et toutes les harmonies de couleurs, avec une introduction sur la théorie générale du contraste, du rythme et de la mesure, 1888

Charles Henry (1859–1926) was a librarian at Sorbonne University in Paris, where he later assumed the position of director of the Laboratoire de Physiologie des Sensations (Laboratory for the physiology of sensations). In 1886, he met George Seurat, with whom he maintained a friendship until the painter's premature death in 1891. Through Seurat, he came to know several postimpressionist painters and became interested in harmony of color and form.

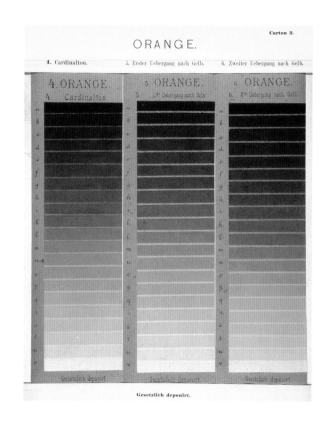

Figure 4.30.
Atlas page from Radde's internationale Farben-Skala *(1878).*

Figure 4.31.
Radde's (1878) hues in linear form.

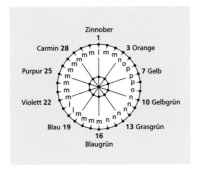

Figure 4.32.
*Schematic drawing of Radde's implicit hue
circle with major colors identified by name
(clockwise beginning at 1): cinnabar, orange,
yellow, yellow-green, grass green, blue-green,
blue, violet, purple, carmine.*

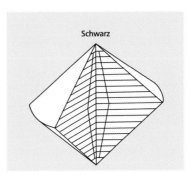

Figure 4.33.
*Schematic drawing of the double cone
solid implied in Radde's system.*

In 1885, Henry published his ideas on the possibility of scientific aesthetics, *Introduction à une esthétique scientifique* (Introduction to a scientific aesthetic), an article widely read in arts and aesthetics circles. He attempted to form a general aesthetic theory from the two psychophysical ideas of arousal (for which he created the term *dynamogénie*) and inhibition. He envisaged this theory to be universally valid, including in all the arts (Homer 1964). In 1888, Henry published *Cercle chromatique présentent tout les compléments et toutes les harmonies de couleurs, avec une introduction sur la théorie générale du contraste, du rythme et de la mesure* (Chromatic circle showing all the complements and harmonies of colors, with an introduction to the general theory of contrast, rhythm, and measure). That year, he also published *Rapporteur esthétique, permettant l'étude et la rectification de toutes formes, avec une introduction sur les applications à l'art industriel, à l'histoire de l'art, à l'interprétation de la méthode graphique* (Aesthetic protractor, making possible the study and the correction of all forms, with an introduction to applications in industrial arts, in the history

of art, and in interpretation of the graphical method). Both works were considerably important in the development of postimpressionism.

Henry's color circle is continuous, in the general form borrowed from Chevreul (see entry in this chapter) but based on the spectrum. White is at the center and black on the periphery (figure 4.34). Its diameter is a surprising 40 cm, and the circle is an accomplished example of the printer's art in the late nineteenth century. Red, the most arousing color, is placed on top (following Goethe [see entry in chapter 3] and unlike in Chevreul's work). Henry locates the most intense (spectral) colors halfway between center and periphery.

Henry claimed high scientific accuracy for his color circle. It is based on four pigment primaries, red, reddish blue, greenish blue, and yellow (figure 4.35). For lights, he accepted the three primaries described by Maxwell and

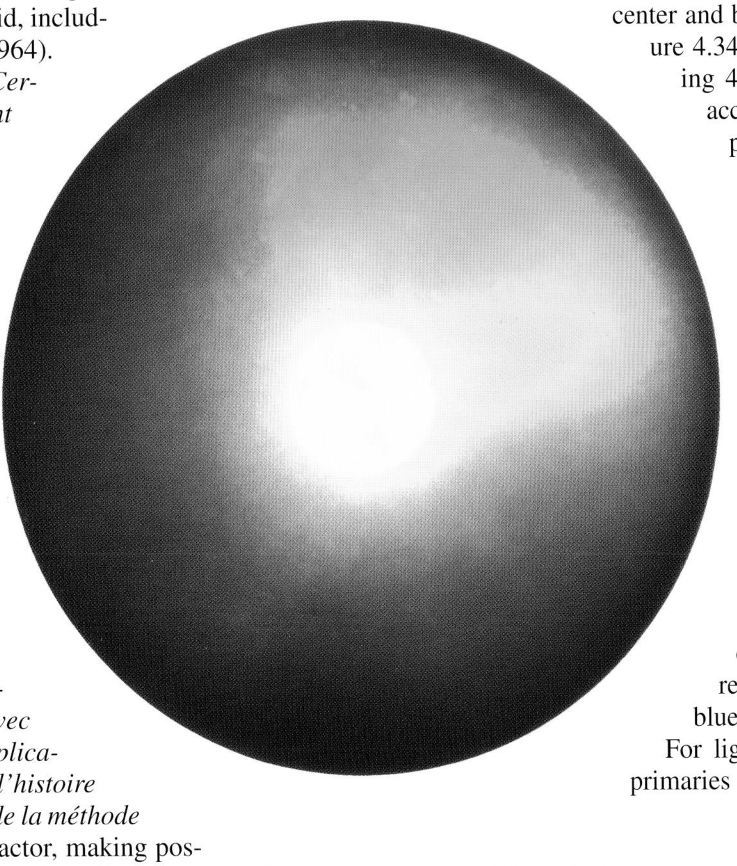

Figure 4.34.
The cercle chromatique *(Henry 1888a).*

Helmholtz, attributing the differences in wavelengths of their respective primary choice to differences in their personal arousal and inhibition processes.

Henry reported that it took six plates to print the color circle, and he described the printer's difficulties in exactly obtaining the desired results. The then young painter Paul Signac described himself as an obedient collaborator of Henry around 1890, working with the latter's color circle and *Rapporteur esthétique*, analyzing or calculating lengths, rhythms, hues and harmonies.

The aesthetic protractor (figure 4.36) allows determination of color harmonies on the color circle as well as rhythms of points, lines, and colors. Seurat may have used it in the design and coloration of some of his later paintings. Despite the existence of several three-dimensional systems, Henry decided to remain in two dimensions, and thus his *cercle* represents one of the last proposals of this kind.

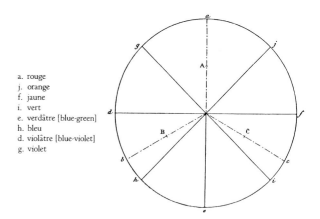

a. rouge
j. orange
f. jaune
i. vert
e. verdâtre [blue-green]
h. bleu
d. violâtre [blue-violet]
g. violet

Figure 4.35.

Diagram of the chromatic circle (Henry 1888a).

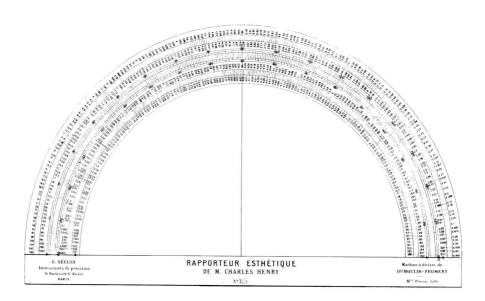

Figure 4.36.

Henry's rapporteur esthétique *(aesthetic protractor) (Henry 1888c).*

Note

1. Lichtenberg, in his 1775 version of Mayer's 1758 essay, defined the relative amounts of the primary colorants g, r, b (for yellow, red, and blue) as factors identified with capital letters. Thus, a given color x can be defined as $x = Gg + Rr + Bb$. He also defined weights γ, ρ, and β in the same way used in the twentieth century by the International Commission on Illumination (CIE) to define chromaticity coordinates, for example, $\beta = B/(G + R + B)$. Further, he expressed the third dimension of the color solid as *claritas* (brightness) but used change in thickness of colored glass as an example of its operation.

References

Algoud, H. 1908. *Gaspard Grégoire et ses velours d'art*,
Paris: Société Française d'Imprimerie et de Librairie.

Anonymous. 1794. *Wiener Farbenkabinet, oder vollständiges Musterbuch aller Natur- Grund- und Zusammensetzungsfarben, wie solche seit Erfindung der Malerei bis auf gegenwärtige Zeiten gesehen worden, mit fünftausend nach der Natur gemalten Abbildungen und der Bestimmung des Namens einer jeden Farbe, darin eine ausführliche Beschreibung aller Farbengeheimnisse, in Seide, Baum- und Schafwolle, Lein-, Leder, Rauch- und Pelzwaren, Papier, Holz und Bein, usw. schön und dauerhaft zu färben. Herausgegeben zum Gebrauche aller Naturforscher, Eltern und Erzieher, Maler Färber, Drucker, Fabrikanten, Künstler und Handwerker und überhaupt aller Menschen die sich mit Farben beschäftigen*, 2 Bände (Vienna color cabinet, or complete book of samples of all natural, basic and composed colors, such as they have been seen from the invention of painting to the present, with 5000 samples painted after nature and the determination of their names, wherein there is a detailed description of all color secrets, how to dye beautifully and lastingly on silk, cotton, wool, linen, leather, fur, paper, wood, and bones etc. Published for use by natural scientists, parents and teachers, painters, dyers, printers, manufacturers, artists, and craftspeople and in general all people that concern themselves with color, 2 vols.), Wien und Prag: Schönfeld.

Chevreul, M.-E. 1839. *De la loi du contraste simultané des couleurs et de l'assortiment des objets colorés*, Paris: Pitois-Levrault. English translation: *The principles of harmony and contrast of colours and their applications to the arts*, London: Longman, Brown, Green, and Longmans, 1854.

Chevreul, M.-E. 1861. *Exposé d'un moyen de definir et de nommer les couleurs*, Paris: Didot.

Chevreul, M.-E. 1864. *Des couleurs et leurs applications aux arts industriels*, Paris: Baillière.

Chevreul, M.-E. 1987. *The principles of harmony and contrast of colours and their applications to the arts*, Birren F., ed., West Chester, PA: Schiffer.

Fiorentini, A., and **B. B. Lee.** 2000. Tobias Mayer's On the relationship between colors, *Color Research and Application* 25:66–74.

Grégoire, G. 1810. *Table des couleurs*, Paris.

Grégoire, G. ca. 1820a. *Théorie des couleurs, contenant explication de la table des couleurs*, Paris: Brunot-Labbe.

Grégoire, G. ca. 1820b. *Table des couleurs sur trois feuilles, précédée d'une planche indicative des couleurs*, Paris: Brunot-Labbe.

Hay, D. R. 1845. *A nomenclature of colours applicable to the arts and natural sciences, to manufacturers, and other purposes of general utility*, Edinburgh: Blackwood.

Henry, C. 1888a. *Cercle chromatique présentent tout les compléments et toutes les harmonies de couleurs, avec une introduction sur la théorie générale du contraste, du rythme et de la mesure*, Paris: Verdin.

Henry, C. 1888b. Cercle chromatique et sensation de couleur, *La Revue Indépendante*, May, 238–289. Henry, C. 1888c. *Rapporteur esthétique, permettant l'étude et la rectification de toutes formes, avec une introduction sur les applications à l'art industriel, à l'histoire de l'art, à l'interprétation de la méthode graphique*, Paris: Seguin.

Homer, W. I. 1964. *Seurat and the science of painting*, Cambridge, MA: MIT Press.

Klotz, M. 1816. *Gründliche Farbenlehre*, München: Lindauer.

Lambert, J. H. 1772. *Beschreibung einer mit dem Calauischen Wachse ausgemalten Farbpyramide wo die Mischung jeder Farbe aus Weiss und drey Grundfarben angeordnet, dargelegt und derselben Berechnung und vielfacher Gebrauch gewiesen wird*, Berlin: Haude und Spener.

Lang, H. 1980. Tobias Mayer's Abhandlung über die Verwandtschaft der Farben, *Die Farbe* 28:1–34.

Matile, H. 1979. *Die Farbenlehre Phillip Otto Runges*, 2nd ed., Munich: Mäander.

Mayer, T. 1758. De affinitate colorum commentatio, in *Opera inedita Tobiae Mayeri*, Lichtenberg, G. C., ed., Göttingen, 1775.

Ostwald, W. 1921. *Mathetische Farbenlehre*, Leipzig: Unesma.

Pfannenschmid, A. L., and **E. R. Schulz.** 1781. *Versuch einer Anleitung zum Mischen aller Farben [aus] blau, gelb und roth, nach beiliegendem Triangel*, Hannover.

Prange, C. F. 1782. *Farbenlexicon, worinn die möglichsten Farben der Natur nicht nur nach ihren Eigenschaften, Benennung, Verhältnissen und Zusammensetzungen, sondern auch durch wirkliche Ausmahlung enthalten sind. Zum Gebrauch für Naturforscher, Maler, Fabrikanten, Künstler, und übrigen Handwerker, welche mit Farbe umgehen* (Color lexicon, in which are contained all possible colors of nature, not only according to their properties, naming, relationship and composition, but also executed in real paint. For the use of natural scientists, painters, manufacturers, artists, and other trades people concerned with color), Halle, Germany.

Radde, O. 1878. *Radde's internationale Farben-Skala*, Hamburg: Radde.

Runge, P. O. 1810. *Farben-Kugel, oder Construction des Verhältnisses aller Mischungen der Farben zu einander, und ihrer vollständigen Affinität, mit angehängtem Versuch einer Ableitung der Harmonie in den Zusammenstellungen der Farben*, Hamburg: Perthes.

Schwarz, A. 1997. Michel Eugène Chevreuls chromatisch hemisphärische Konstruktion von 1839, Systematik mit Tücken, *Die Farbe* 43:205–220.

Topolansky, M. 1903. Bestimmung der Farben der Raddeschen Internationalen Farbenskala. *Sitzungsberichte der Mathematisch-naturwissenschaftlichen Klasse der Kaiserlichen Akademie der Wissenschaften, Wien*, 112, Abt. IIa, 67–81.

Viénot, F., and **A. Chiron.** 2001. Michel-Eugène Chevreul and his color classification system, *Color Research and Application* 26 (suppl. vol.):S20–S24.

Werner, A. G. 1814. *Werner's nomenclature of colours*, London: Blackwell and Caddell.

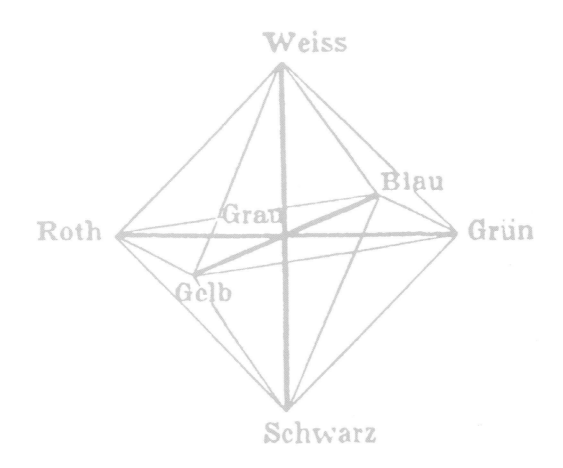

A description of purely empirical psychological color order systems, with a basis of more or less detailed psychological scaling.

As mentioned in chapter 1, colors are psychological experiences (whose nature is not yet known). Those experiences lack a close relationship with color stimuli, so color order systems that are based on simple mathematical manipulation of cone absorption or color-matching function data cannot be expected to agree with average observer judgment data. As distinct from color stimulus or colorant order systems, perceptual color order systems must be based on judgments of psychologically significant distances between color percepts.

In Germany in the second half of the nineteenth century, that was the topic of extended academic debate between two schools of thought: The Helmholtz school explained color experiences in terms of the response of cones in the retina to color stimuli, while the Hering school viewed colors as entities that can and should be treated independent of wavelengths and laws of light or colorant mixture. In Ewald Hering's view (see entry in this chapter), the Newtonian attributes hue, intenseness, and luminosity had been tainted by association with physical measurements. Hering proposed an independent, purely psychological system, thus generating a persisting dichotomy.

When attempting to arrange a large collection of color samples in a systematic manner, the most important aspect is hue. All psychological order systems agree on representing the hue attribute as a closed circle. Where they disagree is the meaning of distance segments along the circle. When grouping samples that have the same hue, we find that many variations of color are possible within a given hue. In the colloquial language of one system, these variations are describable in terms of lightness and intensity of chromatic coloration (chroma), as shown in figure 5.1. Typical terms to describe perceptual differences in this diagram are *light* and *dark*, *bright* and *dull*, *deep* and *pale*, *colorful* and *grayish*. From these categorical terms, one can derive attributes by which one can order colors.

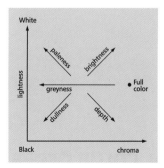

Figure 5.1.
Schematic representation of properties ascribed to colors depending on lightness and chroma.

The cognitive formation of categories based on sensory information is a complex issue that still lacks a clear basis (see, e.g., Harnad 2003). However, it is already quite obvious that, on the simplest level, our sensory apparatus influences category formation. On the most primitive level, ingested liquids and solids taste either sweet, sour, salty, bitter, or "umami" (typically described as savory, e.g., the taste of shiitake mushrooms). From these categories, descriptions of the taste of individual food items, say, meats, fruit, wines, and cheeses, derive in endless culturally specific ways. In the case of colors, it is very likely that the psychological primaries are, as Hering indicated, black, white, yellow, red, blue, and green, and these can be used to at least partially describe any color experience. These two perspectives of color categorization (Hering's and that of the three Newtonian attributes) have proved most influential.

There is general agreement that three attributes are sufficient to fully define color percepts in any given viewing situation but not across viewing situations. As described in chapter 1, the relationship between stimuli and percepts usually changes as a function of surround and lighting quality. As Manfred Richter (1967) pointed out, more than three perceptual attributes in a given system rarely can be made to coincide: "[A] color-order system is a selection of discrete color samples. The samples can be selected and placed according to [for constant-hue samples] two of the attributes; then one can determine if by chance one or more of the samples coincide on an integer basis with additional attributes" (p. 128).

Another issue to be addressed is the matter of absolute and relative attributes. It is possible to scale lightness and chromatic intensity in absolute or relative terms. In the case of lightness, absolute scaling is scaling in terms of a perceptually equidistant gray scale between white and black. Relative scaling begins by taking the lightness of the most intense perceived color of a given hue as having lightness 1 (schematically). Then one scales the lightness of all other samples of the same hue accordingly.

Similarly, chromatic intensity can be scaled absolutely on an interval scale by declaring a given perceptual increment as having the value of 1 and judging how many increments exist between the achromatic color and the color of highest intensity. Chromatic intensity also can be scaled absolutely on a ratio scale.

Relative chromatic intensity is obtained by uniformly assigning the value of 1 to the distances between achromatic color and color of highest intensity, regardless of hue. It is evident that the resulting scales differ significantly. As a result of relative scaling, it is possible that given color samples can coincide according to four different attributes. When comparing absolute and relative systems, any coincidence of points in space is accidental.

Psychological scaling of attributes is a difficult task that requires understanding the attribute concept. Assessments vary by observer because of considerable variability in what's considered normal color vision, as exemplified by the wide hue ranges of color stimuli reliably selected by individual observers as representing the unique hues (see, e.g., Kuehni 2004). In addition, a growing school of thought believes that individual judgments of sensory experiences are to a significant degree determined by the individual's past experiences (see, e.g., Purves and Lotto 2002).

The idea behind experimental psychology, established in the mid-nineteenth century, was to determine the relationship between stimuli and experiences. Psychophysicists began to use objective psychophysical systems resulting from physical specification of stimuli and determination of average cone-sensitivity data to express individual and average perceptual data. As a result, systems that subjectively describe color experiences and their relationships increasingly have been related to systems that objectively describe stimuli. Such systems are described in chapters 6 and 7. This chapter describes purely psychological systems.

The psychological attributes other than hue can be represented in a schematic diagram based on perceptual lightness and a loosely defined attribute called chromatic intensity. In this diagram, the border of optimal object color stimuli, beyond which there are no object color experiences, is shown in the form of a curved outline (see figures 5.2–5.9).

Psychological attributes

Hue

Hue is often considered to be the most important aspect of a color. Color naming in most cultures appears to be based fundamentally on four principle hues, often referred to as the elementary or Hering hues: yellow, red, blue, and green. Perceptual distance between yellow and red allowed formation of the category of orange; perceptual distance between red and blue formed purple and violet. In the yellow-green-blue region, no intermediate categories have established themselves early and strongly enough to result in simple, generally understood hue terms. Hues between yellow and green are known as yellow-green; hues between blue and green, blue-green.

Since the early eighteenth century, the closed perceptual sequence of all existing hues has resulted in hue sequences being represented in the form of circles (see chapter 3). As mentioned above, each hue has many colors differing from each other in terms of lightness and chromatic intensity. As a result, hue is an attribute applicable to a class of colors. A gray scale from white to black is always placed perpendicular to the hue circle and in the center of it. Differing colors

of constant hue can be regarded as variations of the color of highest intensity (in Hering's term, "full color"), black, and white. Comparable planes for different hues form a solid around the central axis (figure 5.2).

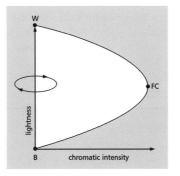

Figure 5.2.
Plane of colors of constant hue. Rotating the plane around its vertical axis represents changes in hue and results in the color solid. In figures 5.2–5.9, FC stands for full color, the color of highest intensity.

By the mid-twentieth century, the idea that constant dominant wavelength represents perceptually constant hue generally had been abandoned based on experimental evidence. Only the DIN 6164 system (see Manfred Richter and DIN 6164, chapter 7), for simplification purposes, is organized according to this idea. In particular, darker colors of a given constant dominant wavelength tend to perceptually differ in hue from their lighter versions. An example is the case of yellow, where darker colors generally appear greener than the lighter ones. In a psychophysical system, this tendency leads to curved surfaces of constant hue, which makes perceptual systems in psychophysical systems mathematically challenging to express.

In addition to the large group of chromatically different colors is a much smaller group of hueless, or achromatic, colors: those of the gray scale, bounded by white and black.

Lightness

In a given set of conditions, color perceptions, always based on light stimuli, are a function of the stimuli's intensity. The magnitude of most stimuli affects not only perceived lightness but also perceived chromatic intensity. For achromatic colors, which can perceptually vary only according to lightness, the easiest way to study lightness is experimentally. Measurements of light intensity (photometry) began to be made in the seventeenth century and were soon used in attempts to establish perceptually uniform lightness scales by allowing objective specification of the stimuli.

Psychological lightness scales can be prepared in several ways and the results differ. It was discovered early that the relationship between psychophysical and perceptual lightness scales is complex. In fact, it is so complex that a complete theory of how our visual system assigns lightness percepts to stimuli does not yet exist (or is in its infancy; see chapter 1).

For psychophysical lightness, a specific kind of measurement (flicker photometry) was needed to obtain additivity within the scale. This method does not generally agree with normal viewing situations. In addition, it was found that the psychophysical lightness scale is not applicable to chromatic colors and that it needs to be corrected to agree with psychological data on lightness of all colors (Helmholtz-Kohlrausch effect; see glossary).

Absolute lightness

Absolute lightness is expressed as a nonlinear mathematical relationship between the results of a method known as psychometric flicker lightness (International Commission on Illumination [CIE] tristimulus value *Y*, luminous reflectance; see CIE X, Y, Z Color Stimulus Space, chapter 6) and a perceptual lightness scale. Different relationships apply in different experimental conditions, but these differences are generally overlooked in color order systems in favor of one particular relationship between stimulus and perception (e.g., that between reflectance and Munsell value). The only color order system that includes the Helmholtz-Kohlrausch effect is the Optical Society of America Uniform Color Scales (OSA-UCS; see OSA Uniform Color Scales, chapter 7). In figure 5.3, lines of constant absolute lightness are drawn schematically into the chromatic intensity/lightness diagram.

Relative lightness

The concept of "natural equality" between colors of highest chromatic intensity and different hues has a long history in color order, as chapters 2–4 indicate. Although full-color blue is darker than full-color yellow, they can be considered psychologically equivalent. This equivalence can be achieved by expressing their lightness in relative terms by assigning the same lightness value to both full colors. Corresponding relative lightness is then assigned to all other members in the same hue class. A simple perceptual schematic diagram of the kind in figure 5.1 cannot show this, because the diagram is different for each hue. Relative lightness is shown schematically for a given full color in figure 5.4.

Measures of chromatic intensity

Saturation

Saturation is one of the more complex psychological-attribute concepts. Its definition has changed several times in the last 150 years. In a categorical sense, saturation is colloquially defined as the degree to which a chromatic color differs from an achromatic one, in the case of unrelated colors (lights) from "white" light. Saturation has also been colloquially defined as chroma (see below) divided by lightness (Berns 2000).

Saturation's perceptual assessment is particularly difficult. Figure 5.5 schematically illustrates lines of constant saturation in our lightness/chromatic intensity diagram. Lines of increasing saturation indicate the distance from a gray of the same lightness, a distance that increases in size as the color becomes lighter and more chromatically intense. A definition of psychophysical metric saturation is found in the DIN 6164 system (see Manfred Richter and DIN 6164, chapter 7).

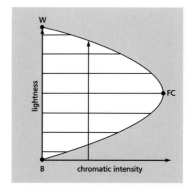

Figure 5.3.
Horizontal lines represent constant absolute lightness, increasing in the direction of the arrow, from bottom to top.

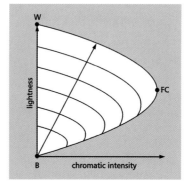

Figure 5.4.
Colors of equal constant relative lightness fall on curved lines the form of which is different for each hue. This property increases in the direction of the arrow.

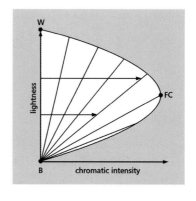

Figure 5.5.
Colors of equal saturation fall on angled lines.

Chroma

Chroma is the attribute of color used to indicate the degree of departure of the color from a gray of the same lightness. It is an absolute measurement of the perceptual distance between a chromatic color and a gray of the same perceptual lightness. It is based on the chroma distance of a given perceptual unit. As mentioned above, the chroma values of optimal object colors of different hues are not identical.

Metric chroma of a chromatic stimulus (the psychophysical version of chroma, e.g., in the CIELAB formula; see CIELAB, CIELUV, chapter 7) is calculated as the length of the line between the metric chroma value of the chromatic stimulus and the zero point (gray with the same luminous reflectance value). Figure 5.6 shows that lines of constant chroma run parallel to the vertical axis of the diagram. The concept of chroma was introduced and first experimentally investigated by Albert Henry Munsell (see entry in this chapter).

Chromatic content

According to Hering, every color can be expressed as the sum of full color content v plus black content b plus white content w, with the sum always 100. Two of these values are sufficient to define the color. Chromatic content can be determined by perceptual judgment or, as Wilhelm Ostwald did (see entry in chapter 10), by calculation from normalized additive mixture components. Chromatic content is a relative measurement, meaning that the chromatic content of optimal object colors of any hue equals 100 (figure 5.7).

Mixed measures of relative lightness and chromatic intensity

Blackness content

In the Hering system, the remaining two attributes of the equation $v + b + w = 100$ indicate that whiteness and blackness content are also relative measures, with maximum values of 100. Lines of constant blackness run parallel to the line between white and full color in our diagram (figure 5.8).

Whiteness content

Whiteness content expresses the degree of similarity to white. Lines of constant whiteness run parallel to the line between black and full color (figure 5.9). Of the three Hering attributes, degree of whiteness is said to be the most difficult to determine. According to Hering's definition, the combination of blackness and whiteness content also defines a color's relative chromatic intensity.

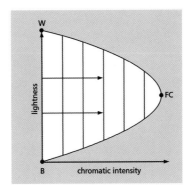

Figure 5.6.
Colors of constant chroma fall on vertical lines.

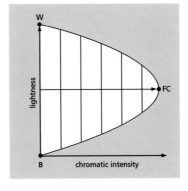

Figure 5.7.
Colors of constant relative chromaticness also fall on vertical lines. All hues have identical maximal values.

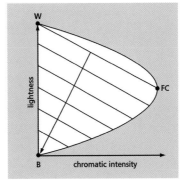

Figure 5.8.
Colors of constant blackness fall on angled lines. Blackness increases toward the location of black.

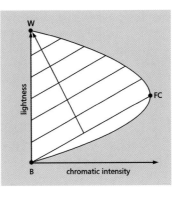

Figure 5.9.
Colors of constant whiteness fall on angled lines. Whiteness increases toward the location of white.

Perceptual equidistance

For attributes to have (linearly) quantitative perceptual meaning, they must be scaled and expressed in uniform distances of some kind. Intersection points of lines of all three attributes can then be defined as aim colors for populating the order system with samples. The necessary absolute or relative scaling must be done perceptually. Historically, such scaling has often involved a single observer with results limited to that observer.

In the case of more than one observer, the results have been averaged to obtain average scaling data. Averaging of perceptual results requires psychophysical support to define the color stimulus that is the average of the individual perceptual results. Individual observers differ to a smaller or greater degree from the average.

With the help of psychophysical models of perceptual data, such data can be extrapolated beyond the existence of real object color samples all the way to the optimal object color limit. This is known as colorimetric extrapolation. On the other hand, colorimetric interpolation can be used to subdivide the perceptually determined scaling points.[1]

The attributes mentioned above and related scales do lend themselves only to uniform perceptual distance along the attribute lines, not to uniform distance throughout the whole of optimal object color space. In most realized systems, the scaling is uniform only within an attribute and differs between, or even within, attributes. And in the case of relative scaling, perceptual equidistance is not obtained even within an attribute because separating a perceptual distance into, say, 10 equal parts produces different results from scaling the same distance according to a defined unit perceptual distance. Both scales can be termed perceptually uniform (in some fashion), but they are clearly not identical.

Perceptual color scaling has been discussed at some length in chapter 1 (see section titled "Color spaces: dimensions, attributes, and scales"). Here we cover additional related issues.

Threshold differences

The smallest unit of perceptual difference is the threshold difference. It is represented by the minimal change (on a continuous scale) in attribute magnitude that is needed for an observer to detect a perceptual difference between two colored fields. The distances are so small that psychophysical support is typically required to detect them, so most threshold experiments involve visual colorimeters. Many color scientists regard threshold experiments as the only meaningful way to assess color difference. That is because such experiments involve only the question of whether a difference is perceived, not how big the difference is, which is

thought to involve subconscious or conscious strategies of individual observers.

There are several classical methods for determining threshold differences, described in the nineteenth century by Gustav Theodor Fechner (1860). When the optimal object color space is divided by threshold differences, millions of distinguishable colors are the result (see chapter 1). The results in each test likely differ depending on the color that is considered to be the reference from which the threshold distance is determined, as well as the surround color.

Differences between the Hering fundamental colors

Hering's chromatic fundamental colors cannot be shown in the form of object colors, while white and black can. The true nature of Hering's fundamental colors is unknown. As mentioned in chapter 1, many experiments have shown that individuals can reliably select samples representing unique hues but that the selections differ considerably from individual to individual. What this means in terms of perceptual distances between the samples involved has not yet been investigated.

Proportional scales

Proportional scales are relative scales where a large distance, say, from white to black or from the full color to the achromatic color of the same lightness, is arbitrarily divided into a given number of perceptually equal steps. Depending on the number of steps, this can be done in two methods. The first method is consecutive halving of the distances. The second begins by having an observer simultaneously assess the equidistance of samples of all steps and make suggestions for adjustments; the resulting samples are newly assessed until relative equidistance is achieved.

Distance scales

These are absolute scales that are established in a certain direction against a reference pair that displays the selected unit distance. The reference pair is often a gray-scale pair. The implicit assumption is that the observer can judge the magnitude of hue, chromatic intensity, or mixed differences against a reference lightness difference. As Jäkel-Hartenstein (1964) commented: "There is the difficulty that the attributes lightness, chromatic intensity, and hue are of such different quality that there is a degree of freedom in the choice of perceptually equal differences" (p. 204, translated by R.G.K.). In other words, individuals will disagree more or less on the magnitude of perceptual differences depending on the nature of the differences. So they also will disagree about the level of uniformity of average scales.

Kinds of perceptual uniformity

For the purposes of this discussion, perceptual uniformity can be defined in several different ways.

Absolute and relative unidirectional uniformity

Here there is perceptual uniformity within a given attribute, either as measured absolutely or relatively. In the case of absolute unidirectional uniformity, all individual scales of a given attribute, say, chroma, are scaled by comparison to a single reference pair. For the other two attributes, the reference pair difference is of different magnitude (this applies, e.g., to the Munsell system). In case of relative unidirectional uniformity, there is uniformity only within a single scale of an attribute. Other scales of the same attribute are of the same relative but different absolute magnitude.

An example of the difference between absolute and relative unidirectional uniformity is given by the hue differences between samples that the observer considers to represent

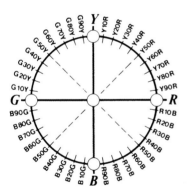

Figure 5.10.
Hue circle of Natural Color System with hue grades changing at identical angular increments (Scandinavian Colour Institute 1988). Reprinted with permission.

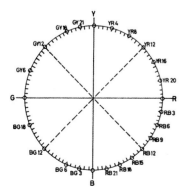

Figure 5.11.
Hesselgren's (1953) Hering-based hue circle. Distances between grades are perceptually equidistant.

unique hues. In Hering's system, the four unique hues are placed on the Cartesian axes of the polar diagram (see figures 5.10 and 5.11). In the Natural Color System (see entry in this chapter), according to Hering's proposal, the hue distances between the unique hues are relatively scaled into 10 equal steps (figure 5.10). Sven Hesselgren (see entry in this chapter) scaled the same distances in an absolute manner, with the resulting scales much different (figure 5.11).

Absolute uniformity within three attributes

In the case of absolute scaling, the reference pair is identical for all three attributes. Thus, units of hue, lightness, and chromatic intensity are the same in perceptual magnitude. Absolute uniformity is possible only in systems conceived accordingly.

Isotropy

Here the attempt is to have the reference distance apply to any color within the system so that there is absolute uniformity in any direction from any color sample. As discussed in chapter 1, this is not possible for three reasons:

1. The resulting spheres of unit difference around any color, selected as standard, cannot be packed solidly without overlap or spaces between them.

2. The hue superimportance effect (see chapter 1) describes the fact that unit hue differences require smaller changes in cone activation than do unit chromatic intensity differences. As a result, the unit difference contours in a perceptual Euclidean geometric space as well as the related psychophysical color space are elongated (and distances thereby not uniform).

3. The magnitude of perceptual hue distances between neighboring colors of the same absolute chromatic intensity is not in the relationship implied by the polar coordinate diagram of constant lightness. In the geometric diagram, the ratio of hue differences between metric chroma 10 and chroma 100 is 1:10 (figure 5.12). The scaling implied in the most accurate small color-difference formulas gives a ratio of only approximately 1:4, again indicating failure of Euclidean geometry. This does not mean that an isotropic space cannot be expressed in a space of another kind of geometry, but rather that the practical problems of comprehensibility have precluded serious research in this regard.

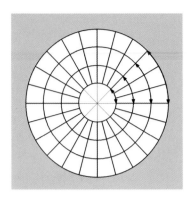

Figure 5.12.

Constant lightness plane of the color solid, with circles of equal chroma and lines of constant hue. The changes in geometric distance are not in agreement with perceptual distances at different points along the radial lines

These difficulties were understood in principle in the 1950s by Deane Brewster Judd (see entry in chapter 7), David Lewis MacAdam (see entry in chapter 6), and others. Rita Halsey (1954) stated the situation succinctly as follows:

> a) [I]ndividual differences in basic visual functions or in experience with color problems make scales based on a few observers inapplicable to others; b) different observing conditions yield markedly different functional relationships; c) color space cannot be described in terms of Euclidean geometry; d) scales derived from liminal [threshold] data and scales for large color differences are incompatible. (p. 199)

Regarding point (a): "[I]t is not only individual differences in basic visual functions but perhaps even more differences in the individual personal visual history that will produce individual differences in perception" (p. 199).

When members of an OSA committee attempted to develop an isotropic system in the 1950s (see OSA-Uniform Color Scales, chapter 7), they used a geometric structure closest to a sphere that allowed for solid packing of the space. Its unit is the cubo-octahedron, with equidistance in 12 directions from a central point. As a result, the committee did not use polar-coordinate attributes hue and chroma, but scaled the constant lightness plane in triangles. But this did not obviate hue superimportance, and, as a result, this Euclidean system is not uniform along the 12 directions.

Aesthetic uniformity

Aesthetic uniformity is a concept introduced by Antal Nemcsics (see entry in this chapter) in connection with the development of the Coloroid system. An explicit cognitive theory for this concept is lacking. The concept is considered applicable to perceptual differences at the suprathreshold level. According to Nemcsics and Béres (1985–1986), "It has been determined that in various places of color space the distances that conform to harmonic thresholds correspond to different numbers of perceptual threshold steps" (p. 328).

Aesthetic uniformity, therefore, appears to be a variant of perceptual uniformity. As the name "aesthetic uniformity" implies, the claim is made that colors can be related in terms of harmonic thresholds. Use of aesthetically uniform colors presumably results in harmonic coloration schemes. However, the results of many other investigations have not indicated the existence of universal principles of harmony.

Although color order systems described in this chapter are psychological, there are often significant differences depending on the adherence to the Hering or the Newton/Helmholtz models. Some systems represent attempts to bridge the gap between the two paradigms. Several authors, including Hering (considered to be the father of psychological color order systems), have not developed their systems beyond purely conceptual ideas. Others have populated their color spaces with samples. In these systems, significant differences also depend on the kind of scaling employed, the size of the scale's steps, the number and quality of color samples, and the number of observers on basis of whose data the scales were developed.

The work of Hering (see entry in this chapter) represents a significant paradigm shift in color order. Before Hering, the psychological attributes of hue, lightness, and chromatic intensity had established themselves among the theoreticians, who were clearly influenced by Isaac Newton's position as a leading scientist and the interest in objective measurability. They also were aided by the apparent promise of a direct relationship between psychological attributes and physical descriptions of stimuli.

Hering recognized that perceptual color order was a world unto itself, complexly related to physical stimulus description. As shown in preceding chapters, hue order, essentially according to the spectrum plus the purple colors, had been well established in the late seventeenth century. At the beginning of the nineteenth century, German painter Matthias Klotz and French textile artist Gaspard Grégoire separately developed implicit color spaces based on hue, absolute lightness, and relative chromatic intensity as independent attributes (see entries in chapter 4). Hering's idea of six fundamental perceptual entities was an entirely new concept. Hering commented on colors:

> What we want is to classify the great multiplicity of colors to get a systematic perspective of them, and designations for them such that the reader is given a comprehensible expression as precise as possible for every color, so that he can mentally reproduce any color with some exactness. To do this we must first disregard altogether the causes and conditions of their arousal. For a systematic grouping of colors the only thing that matters is *color* itself. Neither the qualitative (frequency) nor quantitative (amplitude) physical properties of the radiation are relevant. (Hurvich and Jameson 1964, p. 25)[2]

Hering also delivered an order scheme that he called the *natural color system*. It explains how his six fundamental color entities interact to generate all possible color experiences. However, his proposal was conceptual only without an attempt to color the implicit solid. His ideas were fodder in the growing international efforts of experimental psychology to understand the human senses and emotions, and they helped fuel speculation about the form of the *psychological color solid*. Among Hering's adherents were Alois Höfler, Hermann Ebbinghaus, Edward Bradford Titchener, Franz Hillebrand, Hans Podestà, Adolphe Bernays, and Edwin Garrigues Boring, all with entries in this chapter.

What unites their system proposals is that they considered the form of the psychological color solid in a conceptual, qualitative form, without considering its interior order in detail. For some, the most important concern was the placement of the fundamental colors in the solid. That placement depends on assumptions concerning lightness and chromatic intensity of the fundamentals. But several other Hering adherents regarded fundamental colors as turning points in the solid, replacing the hue circle with right-angled forms, and ending up in regular or irregular octahedrons. Some of the authors combined the paradigms of Hermann von Helmholtz (see entry in chapter 6) and Hering by placing the fundamental colors in the solid in terms of perceptual lightness and chromatic intensity, but with only qualitative definition of these concepts.

In the 1960s, based on the results of their psychological investigations, American psychologists Leo Hurvich and Dorothea Jameson (see entry in this chapter) became strong supporters of Hering's ideas and initiated their international revival.

The first to attempt to populate a Hering-based psychological color solid with samples was the Swede Tryggve Johansson, even though his attempt did not proceed beyond a largely conceptual level. Johansson defined several psychological attributes in addition to hue, saturation, and (absolute) lightness, such as color depth and cleanness. He used these attributes with the fundamental colors to structure color space differently.

Johansson's compatriot Hesselgren colored one of these models using the attributes hue, lightness, and chromatic intensity. In this effort, he discovered the differences between scaling of perceptual differences and scaling in increments of Hering fundamental perceptions, as can be seen in his hue circle that represents a compromise between the two ideas (see figures 5.11 and 5.38).

The Scandinavian Color Institute decided to return to Hering's original thinking and produced the Natural Color System color solid and atlas, based on the idea of the six fundamental colors. Limited data of average results from multiple observers were used as a basis for the scaling. The color samples were defined in the colorimetric system. This system was also used for data inter- and extrapolation. About 2,000 aim colors were specified for the atlas and some 1,750 samples prepared. The use of relative scales for hue, whiteness, and blackness assures the automatic definition of the chromatic intensity attribute (named chromaticness), so that all samples are simultaneously related according to four attributes. Implicit in these attributes is that absolute lightness is not an attribute of interest.

Swedes produced two more systems based on Hering-type psychological attributes. The dyer Sven A. Barding organized his atlas for the textile industry according to attributes hue, grayness, and color degree, without clearly defining the latter two. Similarly, Perry Marthin's International Colour Data (ICD) system employs the attributes hue, saturation, and darkness degree and, as such, has considerable resemblance to the German DIN 6164 system (see Manfred Richter and DIN 6164, chapter 7). Unlike the latter system, ICD does not have samples regularly distributed on its parameter intersection points.

In Switzerland Aemilius Müller found himself dissatisfied with the scaling of the Hering attributes in Ostwald's order system (see entry in chapter 10). Based on his own observations and without any colorimetric support, Müller issued various versions of double-cone color space atlases in which sample colors have been selected to represent scales of improved relative perceptual uniformity.

On the other side of the trench, the first authors to attempt population of psychological color space according to Newton-like attributes were Grégoire and Klotz (see entries in chapter 4). They were followed at the turn of the twentieth century by American artist and educator Munsell, who initially believed the shape of the psychological, uniform color space to be a sphere. Munsell settled on five basic hues, yellow, green, blue, red, and purple and on the attributes absolute perceptually uniform hue, absolute perceptual lightness, and absolute perceptual chromatic intensity, named chroma.

When beginning to populate the solid, Munsell discovered that different colorants (and thereby different hue experiences) vary in terms of maximum chroma. As a result, he gave up the sphere form and produced an irregularly spaced Euclidean solid that he termed the color tree. The Euclidean format was possible because the size of the unit differences in the three attributes differs.

In a time that stressed the physical sciences over psychology, the presence of a degree of agreement between Munsell's attributes and psychophysical system attributes (derived from colorimetric systems) soon gained the Munsell system wide recognition as a rational interpretation of psychological color order.

The Munsell system was "cleaned up" in the 1940s by a committee of the OSA, resulting in the extrapolated Munsell Renotations (see entry in chapter 7), which continue to be the aim color specifications for the system. Since the Renotations, the Munsell system has been considered complete.

The planned successor system, OSA-UCS, was also derived from psychological data. These data related to general perceptual color differences, not to attribute differences. The system failed in public appeal because it was not isotropic, it has a relatively small number of samples, and the samples in the "atlas" are not intuitively arranged.

In Eastern Europe, there was also color order activity. An example of a psychological color order system with considerable similarity to Munsell's came from Russians T. M. Kaptel'ceva, T. Liu Fa-Čun, and S. P. Kričko from the early 1980s. And the Coloroid system of the Hungarian Nemcsics is suggestive of Ostwald's and Munsell's systems. Its claim of aesthetic uniformity lacks scientific replication elsewhere.

EWALD HERING   1878

E. Hering, *Zur Lehre vom Lichtsinne*, 1878

German physiologist **Karl Ewald Konstantin Hering** (1834–1918) studied in Leipzig under Fechner and Ernst Heinrich Weber. Most of his work was in physiology: glands, breathing, temperature sense, and vision. He was chair of physiology at Prague University and later in Leipzig. He invented the *Hering illusion*, two parallel lines that appear to be bent.

During much of his life, Hering was engaged in an acrimonious scientific debate with Helmholtz about the nature of vision in general, and color vision in particular. Helmholtz and his allies worked on the experimental foundations of trichromatic vision, which came to be known as the Young-Helmholtz theory of color vision. It posits three fundamental processes, later identified as sensitivities of three different cone types in the retina.

Hering, on the other hand, looked at color from a phenomenological point of view. That there is no fundamental disagreement between the Young-Helmholtz and the Hering theories had already been pointed out by the physician Hermann Aubert in 1876. The physiologists Franciscus Cornelius Donders and Johannes von Kries proposed a combination of the two theories into a "zone" theory. Helmholtz himself suggested how the two theories could be combined (see Ludwig Pilgrim, chapter 6).

Hering sought to support his insights with physiological evidence but was not able to do so. Only in the 1960s, with the identification of retinal and brain cells that have kinds

of opponent character, was some support provided for his findings. However, at present, there is no valid theory concerning the neurophysiological basis of Hering's *Urfarben* (fundamental colors).

Hering posited the existence of four fundamental chromatic color perceptions (*Urfarben*): yellow, red, blue, and green. Among the reasons was the observation that a color perception cannot simultaneously be yellowish and bluish, or reddish and greenish. He called these pairs opponent color pairs. Simultaneous contrast effects, the result of the perceptual effect of the two adjacent colored fields on each other, also indicate that color appearance of such fields changes in the direction of the four simple hues.

Urfarben are not to be confused with colorants but are products of the normal human color-vision apparatus. By gradual mixtures of two adjacent ones, all intermediate hues in a hue circle are created (figures 5.13 and 5.14). Hering wrote:

> [T]here are four outstanding loci in the series of hues that make up the closed circle: first, the locus of the yellow that shows no remaining trace of redness and yet reveals no trace of greenness; second the locus of blue for which the same is true [Urgelb and Urblau]. . . . Likewise we can name, third, the red, and, fourth, the green that are neither bluish nor yellowish [Urrot and Urgrün]. (Hurvich and Jameson 1964, p. 42)

Hering took the fundamental chromatic perceptions to be identically saturated. The hues of the *Urfarben* are not tinged with any other hues, so the unique red hue is a red appearing neither yellowish nor bluish but just red, and comparably for the other three. His hue circle is graded according to perceived composition from unique hues, as figures 5.13 and 5.14 indicate. The steps between them are not of perceptually equal magnitude. He used the term *Vollfarbe* (full color) for any color perception composed of one or two chromatic *Urfarben* alone (presumably experienced in the form of spectral hues).

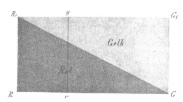

Figure 5.13.
Bipolar diagram for the fundamental colors red and yellow (Hering 1905–1911).

Full colors can, in Hering's terminology, be veiled, that is, partly hidden by black and/or white. For this purpose, he defined the third pair of *Urfarben* as black and white. In graded mixtures, they form a gray scale. He placed all possible color perceptions of a given hue into veiling triangles (figure 5.15). In figure 5.15, the color perception γ consists of the sum (equals 100) of one or two fundamental chromatic color perceptions as well as black and white perceptions. The *Reinheit* (purity) of such a perception is expressed $r/(r$

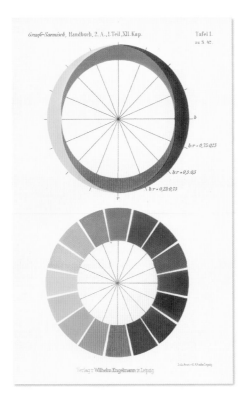

Figure 5.14.
Hering's hue circle with the our fundamental chromatic colors in bipolar form (top) and the resulting hues (bottom) (Hering 1905–1911).

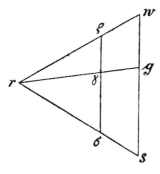

Figure 5.15.
Triangle schematically representing the results of veiling full color r with white w and black s (Hering 1905–1911).

$+ w + s$), where r is the chromatic content and w and s are the white and black content, respectively.

When arranging all possible hue triangles in sequential order along their common gray scale, a double-cone solid results. Hering did not draw such a figure. He was well aware that his chromatic *Urfarben*, and thereby their mixtures, vary distinctly in lightness. In 1889, he commented on this subject in introductory comments to a paper by his student Franz Hillebrand (see note 2). In *Grundzüge der Lehre vom Lichtsinn* (1905–1911), he commented:

> If one has a primary blue that is as clear as possible and finds a primary yellow that one cannot say is either lighter or darker than the blue, then anyone with good color vision who has even a little practice in color analysis will also observe that the yellow is less clear than the blue or that it is more or less grayish or blackish. On the other hand, if he has next to the clearest possible yellow a blue that does not look decidedly darker than the yellow, then he will see that the blue is whitish. . . . Moreover, I find a good primary red, which is the clearest possible, lighter than the clearest primary green available. (p. 61)

Hering's realization of the different lightness of his chromatic fundamental colors can be taken to imply a tilted double-cone color solid.

Hering named his system *Natürliches Farbsystem* (Natural color system), in comparison to what he considered the unnatural trichromatic system of Helmholtz. He did not attempt to create an atlas representing it. He saw it as a tool to mentally order color perceptions according to their similarities and differences in opponent color content. He also hoped to fill a practical void with it, to help people acquire the fundamentals of a common understanding for systematic color analysis.

Hering's system was accepted widely by psychologists but suffered at the time from lack of physiological support. The Young-Helmholtz theory and its technical version, the CIE colorimetric system (see CIE X, Y, Z Color Stimulus Space, chapter 6), became the leading paradigm in the first half of the twentieth century. In the 1960s, American psychologists Jameson and Hurvich (see entry in this chapter) used hue-cancellation experiments to provide support for the Hering theory. These results and the discovery of cells in the retina and the brain with opponent color character (if not exactly in the Hering sense) appeared to create psychological and neurophysiological support for Hering's system, but to date, there is no generally accepted neurophysiological mechanism for unique hues.

ALOIS HÖFLER 1897, 1911

Höfler, *Psychologie*, 1897
Zwei Modelle schematischer Farbenkörper und die vermutliche Gestalt des psychologischen Farbenkörpers, 1911

Austrian philosopher, psychologist, and educator **Alois Höfler** (1834-1918) wrote textbooks on logic and psychology. His psychology textbook of 1897 contains the first psychological color solid in form of a conceptual double-square pyramid (octahedron) (figure 5.16). Hering-type opponent colors are placed at the ends of the horizontal axes, and white and black are placed at the end of the vertical axis, with middle gray in the center of the solid.[3]

In 1911, Höfler presented what he viewed as an improved model based on Ebbinghaus's tilted octahedron (see entry in this chapter). Höfler's version was supported by perceptual data. He saw the perceptual distance between central gray and primary green as smaller than that between gray and red (figure 5.17). He also believed that the perceptual distances between gray and black and gray and white were much larger than those between gray and the chromatic primaries, thus elongating the tilted irregular double pyramid.

In the same article, Höfler offered color illustrations of two conceptual color solids: an octahedron and a double tetrahedron (figure 5.18). In the former, a tetrahedron is removed for an interior view of the octahedron. The tetrahedral space is based on the subtractive primaries yellow, red, and blue. With no quantitative data on perceptual distances at his disposal, Höfler concluded that his original octahedron was the correct conceptual form of the psychological color solid.

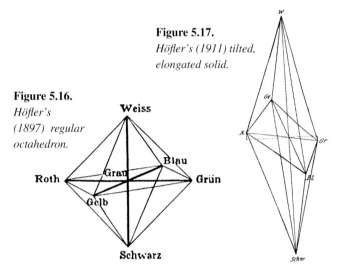

Figure 5.17.
Höfler's (1911) tilted, elongated solid.

Figure 5.16.
Höfler's (1897) regular octahedron.

HERMANN EBBINGHAUS   1897

H. Ebbinghaus, *Grundzüge der Psychologie*, 1897

Hermann Ebbinghaus (1850–1909) was a German psychologist and philosopher. His most important work was on the workings of memory. He also composed a sentence-completion test for juveniles still in use today. And he wrote a psychology textbook that saw three editions.

The textbook included the topic of visual perception. In it, Ebbinghaus proposed a tilted version of Höfler's double pyramid (an irregular octahedron) as a conceptual color solid for what he estimated to be approximately one million distinguishable colors (figure 5.19). The tilt gives geometric meaning to perceived lightness by placing the four full colors at different height. The Hering-type fundamental opponent colors are located at the six corners, rounded to demonstrate the continuous nature of color experiences.

Colors are described as having three, and only three, independent properties: a certain hue, a degree of saturation, and a degree of lightness. With regard to these three properties, every color is related to other colors in a relationship of larger or smaller similarity. Ebbinghaus considered color experiences to be simple, regardless of the colors involved. He distinguished this from the physical complexities of stimuli. His solid was not to be seen as a mixture diagram, but as a geometric model of all possible color experiences that shows their psychological relationships in terms of the three properties.[4]

Ebbinghaus's solid is an attempted compromise between Hering's four basic chromatic perceptions and the Newtonian three attributes. Such a representation required a tilted central plane. Like Höfler, he considered the four basic chromatic sensations to have higher saturation than any of the intermediate full colors.

Figure 5.19.
Ebbinghaus's (1897) tilted color octahedron.

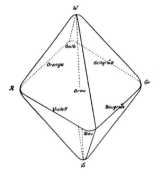

Figure 5.18.
Colored illustration of two color solids (Höfler 1911). On the left is an illustration of Höfler's (1897) octahedron (figure 5.16). A view of the interior is made possible by the removal of a tetrahedron. The model on the right is in the form of a double tetrahedron. Above, the central triangle is illustrated with continuously varying colors; below, the three subtractive primary colors on which the solid is based are shown together with their intermediate mixtures.

EDWARD BRADFORD TITCHENER   1901

E. B. Titchener,
Experimental psychology: a manual of laboratory practice, 1901
A demonstrational color pyramid, 1909

The English psychologist **Edward Bradford Titchener** (1867–1927) studied in Oxford and under Wilhelm Wundt in Leipzig. In 1892, he moved to Cornell University in the United States, where he spent the rest of his life. He was one of the pioneers in bringing German experimental psychology into the United States and changing what was known as mental philosophy into psychology. He was the leader of the structuralist school using the method of introspection.

The first volume of his vast, four-volume textbook *Experimental Psychology* contains a discussion of color perception where Titchener used Ebbinghaus's irregular octahedron as a conceptual solid representing the order of color experiences (figure 5.20).

In 1909, Titchener published a paper detailing the construction of a model of his double pyramid (figure 5.21) prepared at Cornell for instruction. (Nine years earlier in Germany, cardboard models of tilted octahedral already had been used for the same purpose.)

In his paper, Titchener described the use of 14 hand-ground pigments applied in linseed oil to color. The central gray scale consists of white, black, and 23 gray grades, and there are 15 grades between pairs of neighboring basic colors. The total number of different colors on the surface is 750. The well-known nature painter L. A. Fuertes was employed to color the perceptually spaced fields. The completed copy was at Cornell University for many years, but appears to be lost.

FRANZ HILLEBRAND   1929

F. Hillebrand, *Lehre von den Gesichtsempfindungen*, 1929

Franz Hillebrand (1863–1926) was a German experimental psychologist and a student of Hering (see entry in this chapter). He became one of Hering's chief academic defenders. With his article "Über die specifische Helligkeit der Farben" (On specific lightness of colors), Hillebrand contributed important findings to Hering's later view on color order, as shown in Hering's introduction to Hillebrand's article (see note 3).

Hillebrand's textbook, *Lehre von den Gesichtsempfindungen* (Treatise on the light sense) was published by his wife after his death. This work contains a figure of a color cylinder (figure 5.22) that represents an individualistic and somewhat vague solution, due to the fact that Hillebrand doubted the possibility of graphically representing the shape of the psychological color solid.

Saturated colors are located on the edge of disks with the fundamental colors placed on the horizontal axes at right angles. Chromatic intensity diminishes toward the central achromatic color (white on top, white and black in the middle, and black on the bottom). A series of such disks forms a cylinder. A similar proposal, based on disk mixture, had been offered in 1879 by Ogden Nicholas Rood (see entry in chapter 10). Hillebrand criticized his own system for not representing each color sensation as a point in the cylinder.

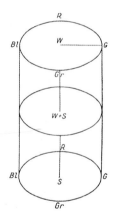

Figure 5.22.
Hillebrand's (1929) color cylinder.

HANS PODESTÀ   1930

H. Podestà, *Beiträge zur Systematik der Farbempfindungen*, 1930
Der ordnungswissenschaftliche Aufbau des Farbenkörpers, 1941

Hans Podestà (1872–1953) was an ophthalmologist and assumed high ranks in the German governmental as well as the navy medical hierarchies. He is the author of volume 4 of Ostwald's 1922 *Farbenlehre* (see entry in chapter 10). He is also the author of an early color-vision test chart (in 1916) widely used in the German armed forces.

In 1930, Podestà developed a purely psychological color order system:

> The mathematical structure of the complete continuum of colors is developed on basis of the relevant laws into an ideal color solid. In this spatial scheme it demonstrates

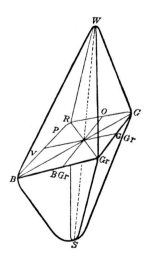

Figure 5.20.
Titchener's (1901) tilted double pyramid.

Figure 5.21.
Model of the tilted pyramid, constructed according to the instructions in the related publication (Titchener 1909).

symbolically the diverse relations in several directions between the color perceptions, after internal conversion of the external color stimuli, that is, without consideration of physical, chemical, or neurophysiological relationships and interdependencies. (p. III, trans. by R.G.K.)

In 1941, Podestà described his contribution a second time specifically for ophthalmologists.

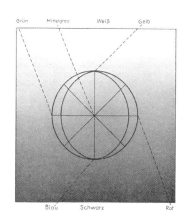

Figure 5.23.

Determination of the brightness of Hering's fundamental colors and middle gray in a field with continuous variation from white to black (Podestà 1930).

Unlike Ebbinghaus's tilted double pyramid (see entry in this chapter), but like August Kirschmann's tilted double cone (see entry in chapter 10), all of Podestà's full colors have identical saturation (or purity; Podestà did not distinguish between the two terms and always used them together). Figure 5.23 illustrates the locations of the psychological full colors, as well as white, black, and middle gray on a plane, continuously varying in lightness. Full red and green have the same lightness level as middle gray. In figure 5.24, full colors blue and yellow are both located eight grades from the extremes on the central 32-grade gray scale. Figure 5.25 shows the tilted double-cone color solid, which Podestà called the "ideal model of the color solid." Podestà's geometric representations are not based on psychological measurements, but conceptual in nature.

ADOLPHE BERNAYS   1937

A. Bernays, Versuch einer neuen Farbenordnung, 1937

In 1937, Swiss psychologist **Adolphe Bernays** published the article "Versuch einer neuen Farbenordnung" (Attempt at a new color order) in which he described a conceptual color space in the form of an irregularly shaped tilted octahedron (figure 5.26). Bernays used the term *Buntheit* (chromaticness) in place of saturation, a term still used today. He found the chromaticness of the four Hering chromatic fundamental colors to be different and, in his model, placed them at different distances from the gray axis. Yellow and red have considerably higher chromaticness than green and blue.

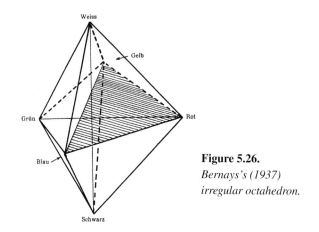

Figure 5.26.

Bernays's (1937) irregular octahedron.

The central trapezoid is tilted to indicate the different lightness of the fundamental colors. This results in the axes connecting yellow and blue and the gray-scale axis intersecting red and green at different locations. The hatched triangle indicates the colors obtainable in the central play using the artist's primaries yellow, red, and blue.

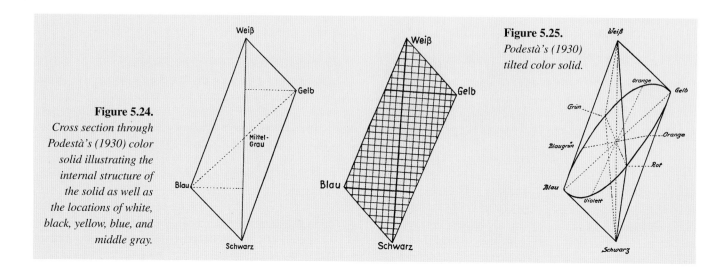

Figure 5.24.

Cross section through Podestà's (1930) color solid illustrating the internal structure of the solid as well as the locations of white, black, yellow, blue, and middle gray.

Figure 5.25.

Podestà's (1930) tilted color solid.

BORING, LANGFELD, and WELD   1948

E. G. Boring, H. S. Langfeld, and H. P. Weld,
Foundations of psychology, 1948
E. G. Boring, *A color solid in four dimensions*, 1951

Edwin Garrigues Boring was a psychologist and historian at Harvard University, **Herbert Sidney Langfeld** a psychologist at Princeton University, and **Harry Porter Weld** a psychologist at Cornell University when they jointly published the widely used textbook *Foundations of Psychology*. Boring is the author of well-known books on the history of experimental psychology.

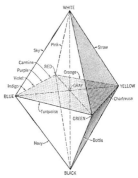

Figure 5.30.
View of the double pyramid with unique and some complex colors identified (Boring et al. 1948).

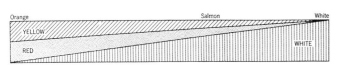

Figure 5.27.
Representation of a triplex series of colors with varying amounts of yellow, red, and white (Boring et al. 1948).

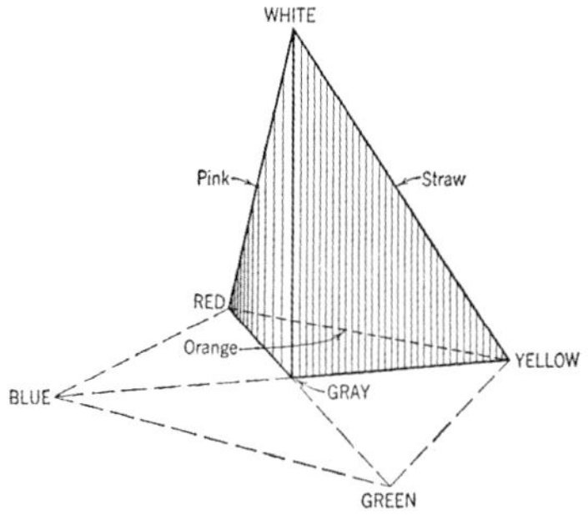

Figure 5.31.
Tetrahedron from the double pyramid of figure 5.30. All colors in this tetrahedron are specifiable as the sum of red, yellow, white, and gray (Boring et al. 1948).

Figure 5.28.
Schematic arrangement of the five unique colors red, yellow, green, blue, and gray in the central vertical plane of the double pyramid. Examples of duplex colors are located along the straight lines connecting the unique colors (Boring et al. 1948).

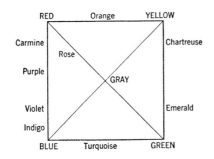

Influenced by work of the psychologist F. L. Dimmick (1929), the authors proposed seven fundamental colors experiences: To the six Hering colors (see entry in this chapter), they added middle gray. In their chapter on color perception, they distinguished between duplex, triplex, and quadruplex colors. Duplex color series involve mixtures of two fundamentals. Triplex colors, of which there are 12 series, involve the six Hering fundamentals and gray, for example, G-B-Wh, G-B-Bk, and G-B-Gy (where G is green, B is blue, Wh is white, Bk is black, and Gy is gray). An example is graphically illustrated in figure 5.27. Quadruplex colors form the majority of colors, and there are eight combinations, for example, Y-G-Bk-Gy, or B-R-Wh-Gy.

The chromatic plane of the pyramid is shown in figure 5.28, a rotated vertical section in figure 5.29, and the complete pyramid in figure 5.30. It is formed by eight tetrahedra representing quadruplex colors (figure 5.31). The tetrahedron of figure 5.31 is expressed by the equation C = red + yellow + white + gray, where C is any color in the tetrahedron. In a later edition, the pyramid was rotated so that the blackness–whiteness axis is horizontal.

The authors compare this "new system" to the "older system" of arrangement in the Munsell attributes (see Albert Henry Munsell, this chapter). The proposal, though studied by countless psychology students, has left no noticeable impact.

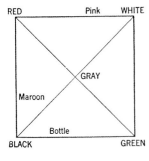

Figure 5.29.
Schematic arrangement of five unique colors in one of the two (rotated) vertical planes of the double pyramid with unique and duplex colors (Boring et al. 1948).

LEO M. HURVICH and DOROTHEA JAMESON   1956

L. M. Hurvich and D. Jameson, *Some quantitative aspects of an opponent-colors theory. IV. A psychological color specification system*, 1956
L. M. Hurvich, *Color vision*, 1981

The American experimental psychologists **Leo M. Hurvich** (1910–) and **Dorothea Jameson** (1920–1998) are known for reintroducing the work of Hering to American psychology. It had largely been forgotten as a result of the rise of the Young-Helmholtz theory and its technical implementation in the CIE colorimetric system (see CIE X, Y, Z Color Stimulus Space, chapter 6). In 1964, Hurvich and Jameson published an English translation of Hering's *Grundzüge der Lehre vom Lichtsinn* (1905–1911) as *Outlines of a theory of the light sense*.

In a series of psychological investigations, the two psychologists attempted to erect a quantitative foundation for Hering's theory of fundamental colors by empirically determining "chromatic response functions." The technique they employed is known as the hue-cancellation method. First, observers determined the spectral lights that they saw as having the four unique hues (found to vary significantly among observers). Then, the same observers determined how much light resulting in unique yellow hue needed to be added to a standard amount of light of unique blue hue to meet the criterion "neutral as between blueness and yellowness." Observers did the same for red and green to obtain "neutral as between redness and greenness."

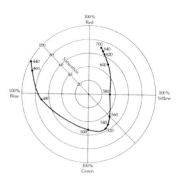

Figure 5.33.

Spectral trace plotted in the polar coordinate hue-saturation diagram, single observer (Hurvich 1981). Reprinted with permission.

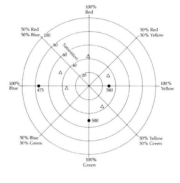

Figure 5.34.

Hue-saturation diagram with the locations of three spectral (dots with wavelength) and five broadband light stimuli ("blue, green, yellow, yellowish red, and purple," triangles) for a single observer (Hurvich 1981). Reprinted with permission.

In most cases, the mixtures were not achromatic because the selections of stimuli of opposing unique hues usually were not complementary. In a similar manner, perceptual hue neutrality was obtained for spectral lights at every 10 nm from 420 to 700 nm, resulting in the spectral chromatic response functions (figure 5.32). This figure contains the chromatic response functions with the achromatic response function (the luminosity function) to provide the basis for a three-dimensional perceptual color order system for lights.

From these curves, Hurvich and Jameson calculated the total chromatic response at any given wavelength, and the ratio of chromatic to achromatic response. This allowed them to plot the spectral trace in a hue/saturation diagram (figure 5.33). The brightness dimension can be erected perpendicularly over the center of this plane. The perceptual color of all real lights must fall on or within the spectral outline and the straight line connecting the ends of the spectrum, such as shown in figure 5.34.

The validity of Hurvich and Jameson's perceptual color order system depends, as David Krantz pointed out (1974), on the applicability of the implied linearity, additivity, and proportionality laws. Direct experimental tests of these laws by Krantz and colleagues (Larimer et al. 1974–1975) have indicated certain deficiencies in this respect. The hue-saturation-brightness (HSB) system of Hurvich and Jameson has found only indirect application in the technological HSB system used in computer display systems (see entry in chapter 9).

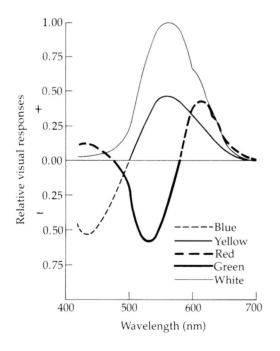

Figure 5.32.

Chromatic and achromatic response functions for a single observer (Hurvich 1981). Reprinted with permission.

TRYGGVE JOHANSSON   1937

T. Johansson, *Färg, Den allmänna färglären grunder*, 1937

As a result of a visit by Hesselgren (see entry in this chapter) in the early 1930s, Swedish physicist **Tryggve Johansson** (1905–1960) gave up his academic career to devote himself to the study of color phenomena.[5] Hoping to find laws of color harmony, he deeply concerned himself with perceptual color order. He was influenced by Hering's natural color system (see Ewald Hering, this chapter), which intrigued him, and by the wide acceptance of the Newtonian attributes.

Johansson decided to have one dimension of the color solid express lightness. Beginning in 1937, he published three perceptual color order systems based on three sets of attributes. He was instrumental in founding the Swedish Color Research Institute in 1946. With Hesselgren, he initiated a research effort that culminated in the Natural Color System (1964; see entry in this chapter).

In Johansson's view, a color order system for the purpose of aesthetic studies can only be based upon perceptual data:

> A colour-system, that should be a basis for studies in colour aesthetics, must only use attributes of visual colour perception, e.g., such attributes of a colour which could be found and evaluated only with the help of the colour sense of human beings. All properties which are related to the material of colour stimulus must be excluded from these studies. (Hård 1965, p. 288)

Hering had not concern himself with a system based on perceptually equal differences. Instead, he placed his six fundamental colors in certain geometric positions in a double cone and perceptually scaled the distances between them. Johansson followed Hering in regard to the four chromatic primary colors, but placed these on their respective lightness levels. Figure 5.35 illustrates his hue circle as well as the gray scale. Figure 5.36 shows his three perceptual color solids based on five attributes: hue (in all three models), color strength (approximately chroma, left and right model), saturation (middle model), lightness (left and middle model), and a new attribute of his invention, cleanness (right model).

In all three models, Johansson emphasized "oversaturated" colors (shown with hatched lines in the cross sections). The constant-hue plane of primary red from the left model is shown in figure 5.37. Johansson's middle model of figure 5.36 was later used as the basis for Hesselgren's color atlas (see entry in this chapter).

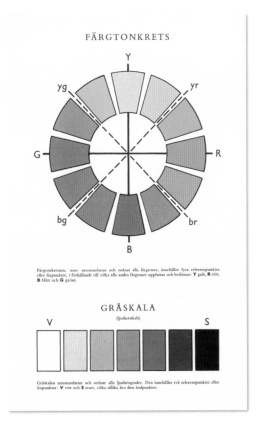

Figure 5.35.
Color illustration of Johansson's hue circle and gray scale (1952 edition).

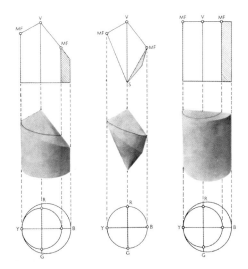

Figure 5.36.
Johansson's three versions of the color solid based on five attributes. Left: hue, color strength, and lightness. Center: hue, saturation, and lightness. Right: hue, colour strength, and cleanness. Hatched areas in the cross sections represent oversaturated colors (1952 edition).

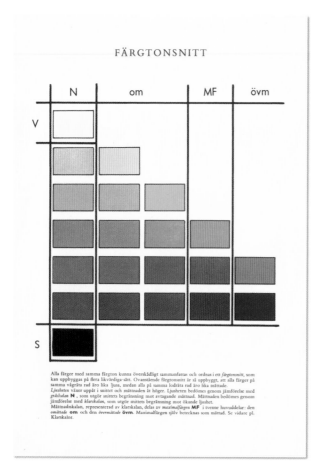

Figure 5.37.
Color illustration of the red constant-hue plane of the left solid in figure 5.36.

The hue circle of the atlas is based on the four Hering fundamental chromatic colors with unique hues that geometrically divide the circle of 24 hues into four equal quadrants (figure 5.38). Unlike Hering's model, the distances between the hues were scaled only perceptual equidistant. As a result, most grades are between red and blue, and there are more grades between yellow and red than between blue and green or green and yellow. Perceptual equidistance is different from equal apparent changes in Hering fundamentals.

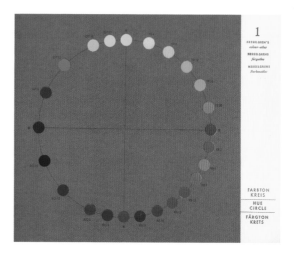

Figure 5.38.
Hesselgren's 24-grade hue circle with an uneven number of steps between the fundamental hues yellow, red, blue, and green (Hesselgren 1953).

SVEN HESSELGREN   1953

S. Hesselgren, *Hesselgrens Färgatlas*, 1953
Subjective colour standardization, 1954

Swedish architect **Sven Hesselgren** (1907–1993) became involved in the subject of architectural color choice as a student. After meeting Johansson (see entry in this chapter) in the early 1930s, the two studied Hering's system (see entry in this chapter) and began to collaborate on developing a color order system. In 1953, Hesselgren published his realization of Johansson's hue, saturation, lightness color solid in the form of an atlas to offer architects and designers a useful tool for choosing and specifying colors. The following year, in a brief publication, he described the perceptual scaling data that are the basis of his specific implementation of the hue, saturation, lightness model.

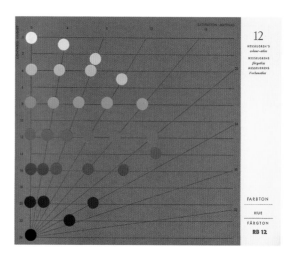

Figure 5.39.
Page of hue RB12 of the atlas. Lines of constant saturation radiate from the black point and lines of constant lightness are horizontal (Hesselgren 1953).

The result is a compromise between Hering's and a perceptually uniform hue circle. In the atlas, samples are arranged on constant-hue pages with colors of constant saturation falling on lines radiating from the black point and those of constant lightness on horizontal lines (figure 5.39). Additional attributes – clarity and intensity – are illustrated by vertical and curved lines, respectively (figure 5.40). Hesselgren did not provide an illustration of the implicit color solid, but it corresponds to Johansson's middle model of figure 5.36.

Hesselgren's atlas contains a total of 507 circular painted color samples with a diameter of 9 mm. He wrote that another 2,000 samples can easily be interpolated among them. He supported practical use of the system with his publication of the detailed pigment formulations of the atlas samples.

In 1984, he offered the following comment about his atlas: "It reflected first of all the six primary colours yellow, red, blue, green, white, and black, but also reflected Munsell's ideas of lightness and something that Johansson and I called 'saturation.' This was a big mistake" (p. 224). In the 1960s, Hesselgren became one of the chief supporters of the development of the Natural Color System (see entry in this chapter).

NATURAL COLOR SYSTEM 1978

Scandinavian Colour Institute, *Natural color system*, 1978

The **Swedish Colour Centre Foundation**, faced in the 1960s with either updating Hesselgren's color atlas of 1952 or replacing it, chose to support the development of a system entirely based on the ideas of Hering (see entry in this chapter). According to Hering: "For a systematic grouping of colors the only thing that matters is *color* itself. Neither the qualitative (frequency) nor the quantitative (amplitude) physical properties of the radiations are relevant" (Hurvich and Jameson 1964, p. 25).

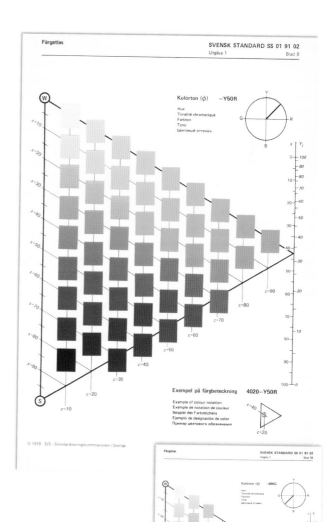

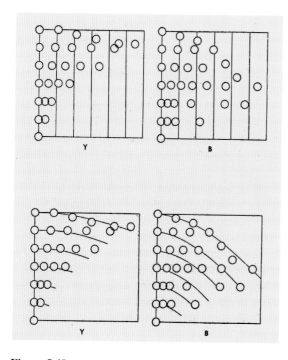

Figure 5.40.

Schematic representation of lines of constant intensity for the fundamental hues yellow Y and blue B (top), parallel to the gray axis, and lines of constant clarity for the same hue pages (bottom) (Hesselgren 1953).

Figure 5.41.

Examples of two constant-hue triangles in the Natural Color System (Scandinavian Colour Institute 1978). Reprinted with permission.

As a result, researchers performed experiments in which observers estimated the content of Hering's full colors, whiteness, and blackness in various color samples, some with the aid of six reference samples and some without. Good rank-order correlation between the two sets of data was reported, so research proceeded toward developing a complete system based on average introspective judgments of color content.

Fundamentally, the Natural Color System (NCS) is an empirical system according to which the Heringian color content of any object color in any given situation can be estimated. Any color perception can be expressed by the person experiencing it as the relative content of one or two fundamental colors having unique hues, as well as blackness and whiteness. The numbers must always add up to 100.

However, it also was considered necessary to express the system in the form of an atlas representing a specific set of conditions. For this purpose, researchers obtained average judgments from some 50 observers involving some 200 painted color samples. They colorimetrically quantified samples and compared perceptual and psychophysical data.

Regularities discovered in the data have been used to interpolate and extrapolate aim colors for an atlas. The aim colors are defined in terms of the CIE 2° standard observer (see glossary) and CIE standard illuminant C (daylight) (Hård 1965; Hård and Sivik 1981, 1996). The first edition of the atlas,

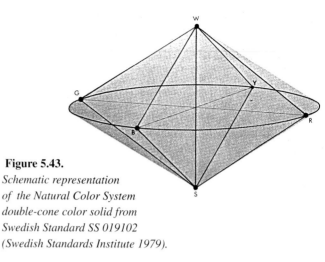

Figure 5.43.
Schematic representation of the Natural Color System double-cone color solid from Swedish Standard SS 019102 (Swedish Standards Institute 1979).

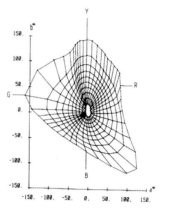

Figure 5.44.
Plot of lines of constant hue and constant chromaticness of the Natural Color System in the CIELAB a*, b* diagram (Derefeldt and Hedin 1987). Reprinted with permission.

published in 1978, consisted of 1,412 samples and was made a Swedish standard (Swedish Standards Institute 1979). The second edition, published in 1995, consists of 1,750 color chips arranged in triangular format on 40 pages (figure 5.41). Most triangles are incomplete because of lack of suitable pigments to express certain colors that can be experienced. The hue circle of the system is shown in figure 5.42. The solid formed by the system is a double cone with all full colors located on the central plane (figure 5.43). A 10-step gray scale forms the common center of the double cone, its grades differing in equal perceived changes in blackness and whiteness. The best fit between perceptual scale and colorimetric data of the samples was obtained with a hyperbolic psychophysical formula. The gray scale is not representative of the vertical dimension because all full colors are located on the same horizontal plane.

Full specification of a color involves blackness, chromaticness, and hue. For example, 4030R70B identifies a color with blackness $s = 40$, chromaticness $c = 30$, and a hue consisting of 30 parts (100-70) unique red and 70 parts perceptual unique blue. Lines of constant blackness and chromaticness are drawn into the triangular diagrams (see figure 5.42). Lines of constant hue (radial) and constant chromat-

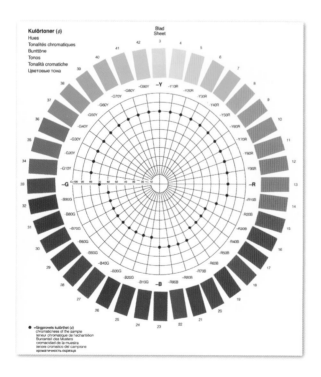

Figure 5.42.
Illustration of the hue circle of the Natural Color System from the Color Atlas 2 (Scandinavian Colour Institute 1988). Reprinted with permission.

icness (circular) plotted in the CIELAB *a**, *b** diagram (see CIELAB, CIELUV, chapter 7) are found to be very irregular (figure 5.44), indicating that the atlas samples refer to an ordering system much different from one based on judgments of perceptual equality.

NCS is the property of and marketed by The Scandinavian Colour Institute AB and enjoys wide popularity, particularly as a reference system for architects, designers, and paint manufacturers. Its claim to naturalness is impaired by Hering's arbitrary decision to define the chromaticness of all full colors as identical. The claim is also impaired by lack of evidence that judgments of whiteness and blackness content are more natural than those of lightness and chroma. The structure of NCS implies a different set of facts and cultural constructs from those implemented in Hering's time by Munsell (see entry in this chapter).

The hue circle is divided into 48 steps, attempted to be perceptually uniform, grades numbered from 02 to 96 (figure 5.45). Each hue is shown in six degrees of saturation, and each degree, in six levels of grayness. Saturation degrees and grayness levels are designated by numbers between 20 and 70. Individual samples are designated with a six-digit number, representing hue, saturation degree, and grayness. For example, color 243070 is a red with hue 24, grayness 30 (relatively light), and saturation degree 70 (saturated).

Two representative pages of the atlas are reproduced in figures 5.46 and 5.47. The atlas contains neither colorant formulations nor colorimetric data, a fact that diminishes its value as a standard. The user's guide mentions that the atlas is to be viewed in "daylight from overcast sky, North-facing window," followed by "Comparison can also be made in artificial light. Key is that the illumination is representative of the conditions in which the material is to be used."

SVEN A. BARDING   1956

S. A. Barding, *Nordisk Textil Unions standard farvekort*, 1956

Sven A. Barding was a Scandinavian textile technologist who produced and published a standard textile color atlas (1956) for the Nordic Textile Union, an association of Scandinavian textile producers. Its 1,687 samples have been dyed on wool felt and measure 15 × 20 mm. The atlas represents a color order system on basis of hue, grayness, and color degree (saturation). Grayness is related in an undefined manner to lightness and appears to correspond approximately to the cleanness attribute of Johansson and Hesselgren (see entries in this chapter), or to DIN 6164 darkness (see Manfred Richter and DIN 6164, chapter 7). The perceptual scaling on which the atlas is based is unknown.

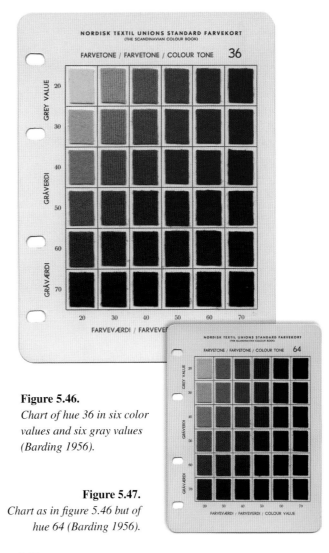

Figure 5.46.

Chart of hue 36 in six color values and six gray values (Barding 1956).

Figure 5.47.

Chart as in figure 5.46 but of hue 64 (Barding 1956).

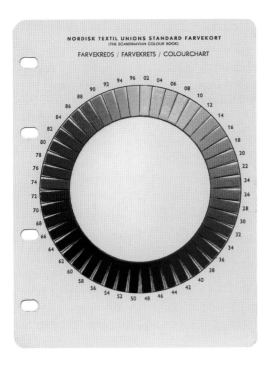

Figure 5.45.

Hue circle of the Standard Farvekort *(Barding 1956).*

PERRY MARTHIN   1974

P. Marthin, *Praktisk färglära med övningar*, 1974

Perry Marthin was a Swedish interior and exterior architectural designer and a devotee of Heinrich Frieling (see entry in chapter 11). In 1974, Marthin published the International Colour Data (ICD) system, whose purpose was to aid in visual color analysis in modern building design. He believed it to be the basis of objective color design.

Marthin was well acquainted with the works of his fellow Swedes Johansson and Hesselgren (see entries in this chapter) but did not find their systems well suited for his purposes. In the early 1970s, Richter, the developer of DIN 6164 (see entry in chapter 7), advised him, which accounts for similarities between the two systems: The three system attributes are hue (*Färgton*), saturation (*Mättnad*), and darkness degree (*Mörkhet*). However, ICD's attributes are defined slightly differently than those of DIN 6164, and ICD is not based on the colorimetric system.

The hue circle (*Färgtoncirkel*) is based on the pigment primaries yellow, red, and blue (*Huvudfärger*), located at positions 10, 20, and 30 (figure 5.48). The intermediate colors (*Mittfärger*) orange, violet, and green are located in positions 15, 25, and 35. Between primaries and intermediate colors are four additional grades, resulting in a 30-hue circle.

The constant-hue diagram (figure 5.49) is a 10 × 10 square of an organization similar to that of the system of Andreas Kornerup and Johan Henrik Wanscher (see entry in chapter 10), with darkness (as illustrated in the gray scale, figure 5.48) on the abscissa, and saturation on the ordinate. The directly corresponding color solid is a horizontal cylinder. Figure 5.50 illustrates the (varying) constant-hue diagrams of six hues. The color identification system is designed as shown in figure 5.51, with the first two digits indicating hue, the third saturation, and the fourth darkness degree.

ICD is a color order system for design work. A corresponding atlas has not been developed. ICD contains 30 hue samples, 10 gray scale samples, and 7 samples each for the six major constant-hue planes, but no clear details about how the three attribute scales are defined.

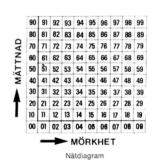

Figure 5.49.
Schematic representation of the arrangement of a constant-hue plane (Marthin 1974a)

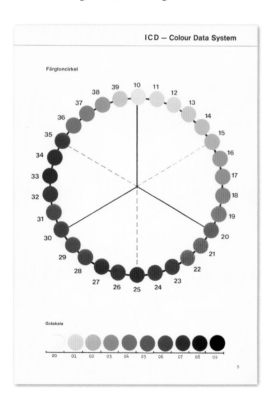

Figure 5.48.
Hue circle and gray axis of the International Colour Data system (Marthin 1974a)

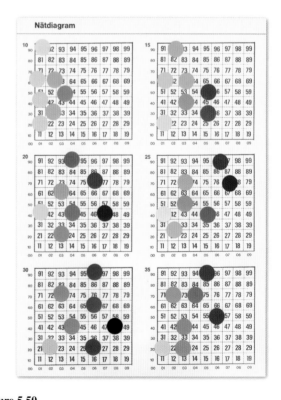

Figure 5.50.
The number of color samples on the six constant-hue planes of the International Colour Data system is not uniform (Marthin 1974a). As a result, there are no analogous color series in the system.

Figure 5.51.
*Color identification system in International
Colour Data (Marthin 1976).*

ADELBERT AMES   circa 1910

A. Ames, Jr., *Systems of color standards,* 1921

Adelbert Ames, Jr. (1880–1955) was an American psychologist and ophthalmologist, and founder of the Dartmouth Eye Institute (1937) at Dartmouth College in New Hampshire (USA). He also taught at Harvard and Princeton, and invented the famous visual illusion named the Ames Room (1946), and a viewing device allowing two-dimensional pictures to be seen as three-dimensional.

Ames was also a painter and in his younger years was interested in color order systems that might aid painters. In a 1921 paper, he described a color order system that he and his sister Blanche (Mrs. Oakes Ames) developed (ca. 1910) in three copies ("[B]efore Ridgway's second system was published and before Munsell completed his work" [Ames 1921, p. 168]). The system was never published and is not extant. But it is of interest because of its development time frame and its relationship to the systems of Robert Ridgway (see entry in chapter 10) and Munsell (see entry in this chapter).

The Ames system was a collection of "about thirty-three hundred different cards" (p. 168). The arrangement is cylindrical, following in most respects that of Munsell, except that chroma is relative, with an equal number of grades for each hue. There were 27 hues, 15 value grades, and 17 relative chroma grades (figure 5.52).

Ames stated five requisites for a useful color order system:

1. Arrangement: He believed that Munsell's arrangement "cannot be improved upon."

2. Notation: He believed that, here, too, Munsell's method was unsurpassed.

3. Number of cards: In this respect Ames found his system, with approximately 3,300 samples, superior to Ridgway's and Munsell's.

4. Spacing: Ames used the Maxwell disk method to assure that his lightness scale followed the Weber-Fechner law (see glossary). But "spacing in hue and chroma was done only by eye" (p. 169).

5. Standardization: Here Ames (in 1921) called for the most accurate measurement possible. Hue, value, and chroma should be calculated in terms of wavelength, intensity, and saturation. The measurement data should be obtained from the disk apparatus by means of a colorimeter with a standard light source in a standard lighting setup. Cards should then be produced to match the resulting data. Ames proposes that this work is done "by a government bureau, as the Bureau of Standards, and the colorimeter kept by them as a standard. A set of measurements of every card in the standard should accompany the standard" (p. 166). Ames was proposing a standard for color atlases at a level rarely achieved even today.

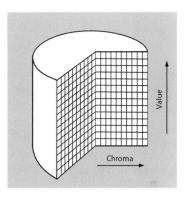

Figure 5.52.
Conceptual sketch of the Ames system.

It is interesting to note that in 1921, Ames called Ridgway's system "probably the most used to-day, and on the whole is believed to be the best system" (p. 166). Munsell's system (in the form of the 1915 *Atlas*) is considered perfect in terms of arrangement and notation, but woefully lacking in number of color chips. In regard to spacing, Ames made note of the different perceptual units for the hue, value, and chroma scales and the fact that the value scale is not in accord with the natural system expressed by the Weber-Fechner law. Lastly, Ames saw Munsell's system as lacking standardization because the inventor had not provided a detailed account of the perceptual data behind his color samples.

Ames's system of 3,300 cards is a remarkable personal effort by Ames and his sister. In 1921, he called for a collaborative effort "by various laboratories and as soon as possible, in order that the great need for a proper color standard can be met" (p. 170).

ALBERT HENRY MUNSELL 1905

A.H. Munsell, *A color notation*, 1905
Munsell color system, Atlas of the color-solid, 1907

American art educator and artist **Albert Henry Munsell** (1858–1918) became interested in color science after reading Ogden Nicholas Rood's book *Modern Chromatics* (see entry in chapter 10). Inspired by Rood, in 1879 Munsell painted pigments on a triangular pyramid and "twirled" the result on a string, thus obtaining optical mixtures similar to a spinning disk. After studying art in the early 1880s in Paris, he was appointed lecturer at the Normal Art School in Boston, where his interest turned to color education. He saw a need for a didactical tool to teach relations between colors and to develop rules of color harmony.

By end of 1898, Munsell had constructed a color sphere whose surface was painted with "balanced colors," that is, colors that resulted in a neutral gray appearance when the sphere was spun. In 1900, he obtained a patent for the sphere (figure 5.53) and began to consider coloring its interior. Munsell concluded that Hering's system (see entry in this chapter) could not be correct, He was intrigued with its use of the decimal system. As a result. he selected five (rather than Hering's four) primary colors, yellow, red, purple, blue, and green for the hue scale (figure 5.54). Also in 1900, Munsell realized that the chromatic powers of pigments differed, so a sphere could not represent the surface of a complete color solid. He studied the works of Johann Heinrich Lambert, Ignaz Schiffermüller, and Phillip Otto Runge (see entries in chapter 4), thus gaining a view of historical color order efforts.

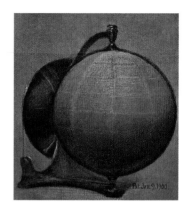

Figure 5.53.
Artist's rendition of Munsell's (1913) balanced color sphere, patented in 1900.

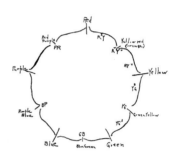

Figure 5.54.
Traced sketch of Munsell's division of the hue circle into five primary hues (April 5, 1900; Munsell 1918, vol. 1, p. 20).

Figure 5.55.
Traced sketch by Munsell of the "color tree" with irregularly shaped vertical constant-hue leaves, with a central gray axis (March 20, 1902; Munsell 1918, vol. 1, p. 97).

Figure 5.56.
Image from a 1906 advertisement for Munsell's No. 2 standard watercolor box, with the five primary middle colors, gray, black, and the maxima of red, yellow, and blue; price $0.50.

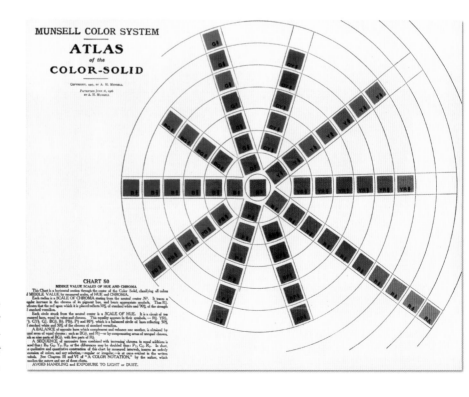

Figure 5.57.
Image of chart 50 (value 5) of the 1907 Munsell Atlas of the Color-Solid, with 10 constant-hue colors at value 5, differing in chroma.

Munsell agonized over the lightness scale that he called "value" scale: Should its scaling be logarithmic or square root in nature? His answer was based on experimental data. Initially, he called the third attribute "energy or purity" and also "strength" or "intensity." In 1901, he began to call it "chroma" (Munsell 1918, p. 80). Inspired by Rood, Munsell thought in terms of constant value slices through his color solid, but in 1902, he sketched a "color tree" with constant-hue slices (figure 5.55) that eventually became the model for the Munsell color atlas.

Munsell's original concept for hue arrangement was in terms of compensatory colors, but by 1904, he had changed to one of perceptual uniformity. The following year, he published the first edition of *A Color Notation*, a description of the system he was developing, which continues to be in print. Munsell also invented a visual photometer and sold color spheres of various sizes for educational purposes, as well as sets of watercolors and oil crayons in standard Munsell colors (figure 5.56).

Munsell began to assemble color chips for horizontal slices through the solid, and in 1907, he published the first *Atlas of the Color-Solid* with eight charts (see, e.g., figure 5.57). Of the eight charts, two are of a general nature, and six are constant value charts, at values 30 to 70. The value 30 charts is shown on both white and black paper to illustrate the change in appearance of the color chips as a function of the surround. A 1915 edition of the *Atlas* grew to 15 charts (three general, five constant hue, and seven constant value).

The Munsell Color Company was formed in 1918, shortly before Munsell's death. It continued, with help of the U.S. National Bureau of Standards, to develop uniform color scales along the three attributes. In 1929, the company issued the first version of an even further enlarged atlas, named the *Munsell Book of Color* (Munsell Color Company 1929).

Among the pioneers of color order who worked at one time for the Munsell Color Company or were closely involved with Munsell's system through the Bureau of Standards were Judd and Dorothy Nickerson (see entries in chapter 7). In the 1940s, the OSA installed a subcommittee to investigate the spacing of colors of the *Munsell Book of Color*. The investigation resulted in the Munsell Renotations (see entry in chapter 7), the redefinition of many color chips.

Munsell advanced color order in several respects. He clearly defined the vertical dimension of the system in terms of the value attribute. As a result, colors of highest chroma were not placed on the same horizontal plane, as had earlier inventors (except for Grégoire and Klotz; see entries in chapter 4). Munsell also defined chroma as an attribute in which colors of constant hue and value can vary. Its open-ended scale brought an end to the idea that a perceptually uniform color solid can fit into a simple geometric solid.

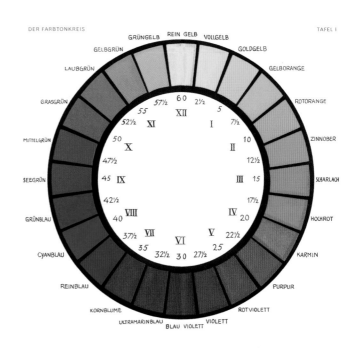

Figure 5.58.
Sixty-grade hue circle reduced to 24 grades (Müller 1948/1959).

AEMILIUS MÜLLER   1945–1973

Aemilius Müller, *Schweizer Farbenatlas 1210*, 1945
Dreifarbenwürfel 1000, 1951
Mobiler Farbtonkörper 743, 1953
Farbbestimmer CUC 12000, 1958
Swiss Color Atlas 2541, 1962
Ästhetik der Farbe in natürlichen Harmonien, 1973

Aemilius Müller (1901–1989) studied economics and worked in advertising and journalism in Switzerland. His hobbies were drawing, painting, and sculpture. In 1941, he came across remaindered copies of Ostwald's *Farbenatlas* (see entry in chapter 10), becoming immediately fascinated with color order and harmony. He decided to devote the rest of his life to expanding upon and popularizing Ostwald's work.

Müller decided to improve the perceptual uniformity of Ostwald's hue circle and to simplify naming conventions. From 1946 to almost the end of his life, he produced some two dozen different kinds of color atlases (only a few key publications are listed above). He produced nearly all of them by hand-coloring paper with dyestuffs. Using dyes allowed him to achieve higher chroma levels than with pigments. He manufactured (in its true sense) hundreds of copies of the atlases after he learned that printing would triple production cost and limit saturation.

Most of Müller's atlases and other works are based on Ostwald's double-cone color order system. His *Swiss Color Atlas* (1945) contained 1,210 samples in 24 constant-hue

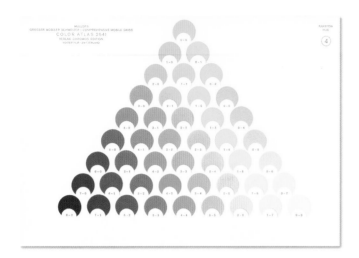

Figure 5.59.
Hue 4 from the Swiss Colour Atlas
2,541, 2nd ed. (Müller 1962/1964–
1965).

Figure 5.60.
Hue 55 from the Schulfarbenatlas
with 186 color samples
(Müller 1955).

triangles (figure 5.58). The 1962 atlas contained 2,541 samples on 60 constant-hue triangles (figure 5.59). It originally was produced at the behest of the Bally of Switzerland shoe factories. The Mobile color solid 743 contains 30 constant-hue triangles and was later expanded to the 1,093-sample version. This version was issued as the standard color atlas with dye recipes for nylon fabric by the German dye manufacturer Hoechst AG.

In 1958, Müller issued *Farbbestimmer CUC 12000* containing 60 hues each with 16 different tints (in direction of white). A 17-grade scale of transparent gray filters allowed the generation of more than 12,000 different color perceptions. Earlier, he offered a color order system in form of a cube with 1,000 samples on 10 charts (Müller 1951). Eight corners of the cube were occupied by the three subtractive primaries yellow, magenta, and cyan, their 1:1 mixtures, and white and black. He also hand-colored these samples with dyes. Color education was of particular interest to Müller. In 1946, he produced a school color atlas with 228 fixed samples. The second edition, issued in 1955, contained 186 detachable samples (figure 5.60) and instructions for practical experimentation. In 1963, 5,000 copies of a large wall chart were produced for distribution to schools in Switzerland. In 1948, he published *Die Moderne Farbenharmonielehre* (Modern color harmony) with 184 samples.

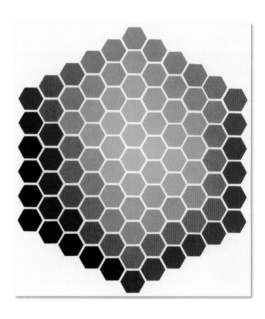

Figure 5.61.
Table 140 of Müller's (1973) Aesthetik der Farbe

In Müller's advancing years, his interest in color harmony grew, and in 1973, he issued *Ästhetik der Farbe in natürlichen Harmonien* (Aesthetics of color in natural harmonies), consisting of 200 charts (figure 5.61) that contained more than 90 samples each. It was made available in three different formats, the largest in poster form.[6]

Müller used only his eyes, not colorimetry, in developing his color samples. The samples are the result of a remarkable personal effort in helping to keep Ostwald's color legacy alive and to teach the intricacies of color order and harmony.

T. M. KAPTEL'CEVA, L. LJU FA-ČUN, and S. P. KRIČKO   1981

T. M. Kaptel'ceva, T. Liu Fa-Čun, and S. P. Kričko, *Katalog cvetov*, 1981

The *Katalog cvetov* (Color catalog) was the cooperative effort of three Soviet Russian institutions: the Municipal Executive Committee of Moscow, the Architectural Planning Committee of Moscow, and the Moscow Scientific Research and Planning Institute for Normalized and Experimental Projects. The color catalog was to be a tool in a general concept of color standardization in the architectural design of apartment buildings.

In *Katalog cvetov*, color order is based on the attributes hue, lightness, and chroma. The hue circle contains ten spectral and one purple hue, identified by wavelengths. The hues are red (K), purple (P), violet (F), blue I (S), blue II (G),[7] green blue (ZG), green (ZL), yellow (Z), yellowish orange (ŽL), and orange (OR) (figure 5.62). Each of these has an associated constant-hue plane with all colors having nearly identical dominant wavelengths. Tonal colors were produced by adding white and/or black.

Exact details concerning the composition of the samples in the system are not provided. Perceptual uniformity, to the extent possible, has been attempted throughout the system (figure 5.63). The lightness scale ranges from 0 to 100 and samples are displayed in steps of 10. Chroma has a maximum value of 90 and is displayed in steps of 10. The result is a cylindrical system of the Munsell type (figure 5.64).

Each color sample is identified by the first letter of the hue name and a two-digit number, the first identifying lightness and the second chroma. The gray scale is shown in 17 grades. In total, the *Katalog cvetov* contains 370 painted samples each measuring 11 × 12 mm.

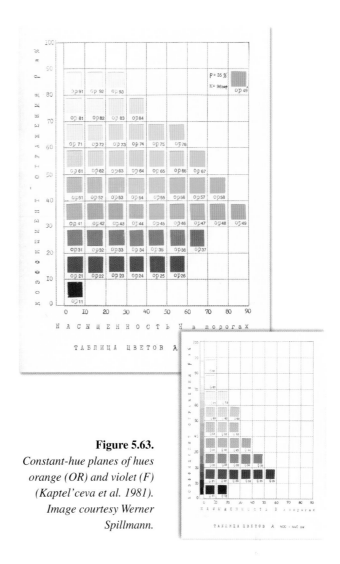

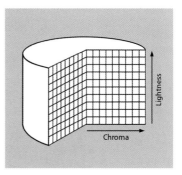

Figure 5.63.
Constant-hue planes of hues orange (OR) and violet (F) (Kaptel'ceva et al. 1981). Image courtesy Werner Spillmann.

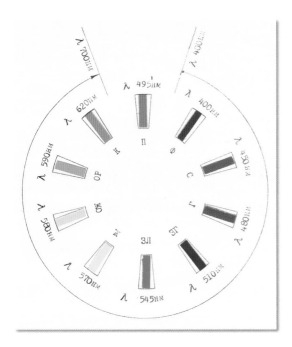

Figure 5.62.
Hue circle with corresponding dominant and complementary wavelengths, respectively (Kaptel'ceva et al. 1981). Image courtesy Werner Spillmann.

Figure 5.64.
Schematic representation of the color solid of Katalog cvetov.

ANTAL NEMCSICS   1982/1988

A. Nemcsics, *Coloroid colour atlas*, 1982, 1988

The Coloroid system was developed by the Hungarian ar-chitectural color designer **Antal Nemcsics** (1927 -) between 1960 and 1980. Since 1982, it has been a Hungarian stan-dard (MI 17063-81, now MSZ 7300). Reportedly more than 70,000 observers participated in the aesthetic judgments preceding its development. Coloroid is based on the premise that a particular arrangement of samples in a color solid can result in rules that result in harmonious color combinations.

Coloroid colors are defined by three attributes: hue (A), sat-uration (T), and luminosity (V). Hue is defined by the hue angle in the CIE 1931 chromaticity diagram (figure 5.65), similar to DIN 6164 (see Manfred Richter and DIN 6164, chapter 7). Hue designations range from 10 to 76 (figure 5.66). Colors of constant saturation are located on cylindri-cal surfaces in the Coloroid color solid (figure 5.67). Colors of maximum saturation (full colors) are called limit colors (H) and have a value of T = 100 regardless of the hue. The lightness scale found to best agree with "aesthetic judg-ments" is well represented by a square-root compression of luminous reflectance Y. The compression results in compar-atively more light color samples than in the Munsell system (useful in architectural design).

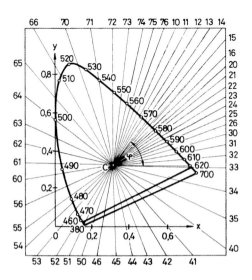

Figure 5.65.

Identification of the Coloroid hues in the CIE chromaticity diagram (Nemcsics 1993). Reprinted with permission.

The Coloroid atlas is considered "aesthetically uniform." Individual colors are defined as the sum of full color perception *p*, blackness *s*, and whiteness *w*, where the sum always is 1. Formulas connecting these parameters with colorimetric values have been established. Nemcsics also de-veloped rules of harmony based on the system.

Figure 5.65 indicates a (aes-thetically demanded) rich-ness of hue steps in the greenish yellow to yel-lowish red region (hues 75–30). This is mirrored to a lesser extent in the blue region (hues 50–55). Hues are sparser in the remaining regions. The first *Coloroid Co-lour Atlas*, containing some 3,000 painted sam-ples on 48 constant-hue pages (figure 5.68), was is-sued in only a few copies in 1982. Since 1988, the *Coloroid Color Atlas* has been available with 1,647 printed samples measur-ing 28 × 15 mm in 48 Coloroid hues

(figure 5.69). A corresponding software package, *Color-oid Professional*, converts stimulus data into designations in several systems. Its *Coloroid Harmony Wizard* al-lows generation of harmonious color de-sign schemes according to the system.

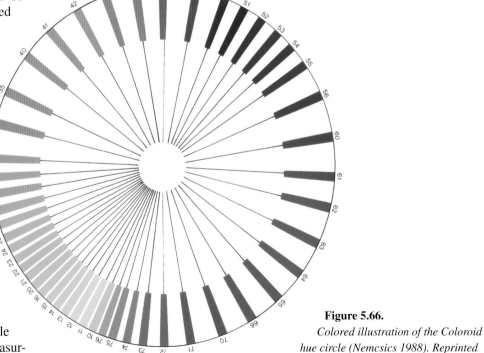

Figure 5.66.

Colored illustration of the Coloroid hue circle (Nemcsics 1988). Reprinted with permission.

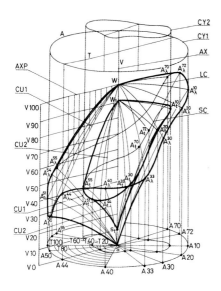

Figure 5.67.
Sketch of the cylindrical nature of the Coloroid system. Colors of constant Coloroid luminosity are located on horizontal planes; the central vertical axis runs from white to black. Full colors are designated by the hue indicators Axx and fall on a curved line. Color samples on the hue pages (see figures 5.65 and 5.66) fill the irregular solid (Nemcsics 1993). Reprinted with permission.

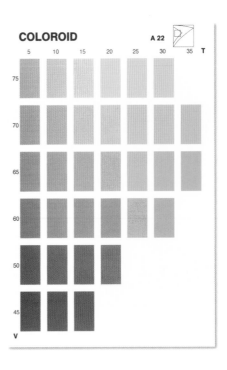

Figure 5.69.
Page with samples of hue A22 from the first official version of the Coloroid system (Nemcsics 1988). Reprinted with permission.

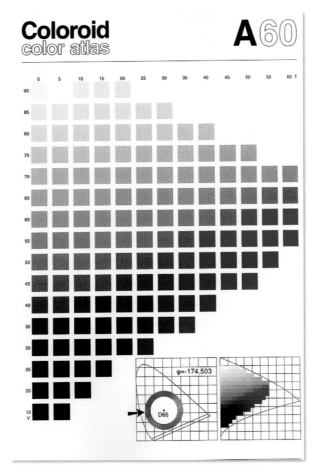

Figure 5.68.
Page with samples of hue A60 from the 1982 version of the Coloroid system (Nemcsics 1988). Reprinted with permission.

Notes

1. Such interpolation is meaningful only to a limited degree because, as mentioned in chapter 1, important visual parameters change as a function of the size of the unit difference. What applies at the level of thresholds may not apply any longer at the level of Munsell system–sized differences.

2. In his concluding work *Grundzüge der Lehre vom Lichtsinn* (1905–1911), Hering stated: "[I]t was assumed that chromatic colors, whatever their hue, would appear equally bright for exactly the same degree of veiling and the same kind of black-white components. However, I soon altered this opinion, as I have already had the opportunity to report in the year 1886. . . . Franz Hillebrand, who later worked with me on this problem, has designated as specific brightness what I have called here the intrinsic brightness and intrinsic darkness of color hues" (Hurvich and Jameson 1964, pp. 64–65).

3. Five years later, Ebbinghaus published his tilted, rounded octahedron model (see entry in this chapter) as an improvement on Höfler.

4. A slightly different version of Ebbinghaus's solid was published by the American psychologist Mary W. Calkins in the second edition of her psychology textbook (Calkins 1905)

5. As reported in Hesselgren (1984).

6. The work of Müller has been described in detail by Werner Spillmann (1984, 1989).

7. The Russian language has two basic color terms for blue, *sinij* and *goluboy*, the former having a focal color of (comparatively) slightly reddish, darker blue, the latter one of a slightly greenish, lighter blue.

References

Ames, A., Jr. 1921. Systems of color standards, *Journal of the Optical Society of America* 5:160 -170.

Barding, S. A. 1956. *Nordisk Textil Unions standard farvekort*, Copenhagen: Barding.

Bernays, A. 1937. Versuch einer neuen Farbenordnung, *Vierteljahresschrift der Naturforschenden Gesellschaft in Zürich* 82:161-196.

Berns, R. S. 2000. *Billmeyer and Saltzman's principles of color technology*, New York: Wiley.

Boring, E. G., H. S. Langfeld, and **H. P. Weld.** 1948. *Foundations of psychology*, New York: Wiley.

Calkins, M. W. 1905. *Introduction to psychology*, 2nd ed., New York: Macmillan.

Derefeldt, G., and **C. E. Hedin.** 1987. A color atlas for graphical displays, in *Work with display units 86*, Knave, B., Wiedebäck, P. G., eds., Amsterdam: North-Holland.

Dimmick, F. L. 1929. A reinterpretation of the color-pyramid, *American Journal of Psychology* 40:83 - 90.

Ebbinghaus, H. 1897. *Grundzüge der Psychologie*, vol. 1, Leipzig: Veit.

Fechner, G. T. 1860. *Elemente der Psychophysik*, Leipzig: Breitkopf und Härtel.

Halsey, R. 1954. A comparison of three methods of scaling, *Journal of the Optical Society of America* 44:199 -206.

Hård, A. 1965. Philosophy of the Hering-Johansson Natural Colour System, *Die Farbe* 15:287-295.

Hård, A., and **L. Sivik.** 1981. NCS-Natural Color System: a Swedish standard for color notation, *Color Research and Application* 6:129 -138.

Hård, A., and **L. Sivik.** 1996. NCS, Natural Color System – from concept to research and applications, pts. I and II. *Color Research and Applications* 21:180 -220.

Harnad, S. 2003. Categorical perception, in *Encyclopedia of cognitive science*, Polk, T. A., Seyfert, C. M., eds., New York: Macmillan.

Hering, E. 1878. *Zur Lehre vom Lichtsinne*, Vienna: Gerolds Sohn.

Hering, E. 1905-1911. *Grundzügeder Lehre vom Lichtsinn*, Berlin: Springer (for English trans., see Hurvich and Jameson [1964]).

Hesselgren, S. 1953. *Hesselgrens Färgatlas*, Stockholm: Palmer.

Hesselgren, S. 1954. *Subjective colour standardization*, Stockholm: Almquist and Wiksell.

Hesselgren, S. 1984. Why colour order systems? *Color Research and Application* 9:220 -228.

Hillebrand, F. 1889. Über die specifische Helligkeit der Farben (mit Vorbemerkungen von E. Hering), *Sitzungsberichte der kaiserlichen Akademie der Wissenschaften, Wien Mathematisch-Naturwissenschaftliche Klasse* 98(Abt. 3):70 -122.

Hillebrand, F. 1929. *Lehre von den Gesichtsempfindungen*, posthumously published by Dr. Franziska Hillebrand, Wien: Springer.

Höfler, A. 1897. *Psychologie*, Vienna: Tempsky.

Höfler, A. 1911. Zwei Modelle schematischer Farbenkörper und die vermutliche Gestalt des Psychologischen Farbenkörpers, *Zeitschrift für Psychologie* 58:356 -371.

Hungarian Standards Organization (Magyar Szabványügyi Hivatal). 1982. *MI 17063-81, A Szinoid-Színjellemezök Meghatározása*, Budapest.

Hurvich, L. M. 1981. *Color vision*, Sunderland, MA: Sinauer.

Hurvich, L. M., and D. Jameson. 1956. Some quantitative aspects of an opponent-colors theory. IV. A psychological color specification system, *Journal of the Optical Society of America* 46:416-421.

Hurvich, L. M., and D. Jameson (trans. and eds.). 1964. *E. Hering's Outlines of a theory of the light sense*, Cambridge, MA: Harvard University Press.

Jäkel-Hartenstein, B. 1964. Empfindungsgemässe Farbkörper, *Die Farbe* 13:201-207.

Johansson, T. 1952. *Färg, Den allmänna färglären grunder*, 2nd ed., Stockholm: Esselte; 1st ed., Stockholm: Lindfors, 1937.

Kaptel'ceva, T. M., T. Liu Fa-Čun, and S. P. Kričko. 1981. *Katalog cvetov*, Moskau: Moskauer wissenschaftliches Planungs- und Forschungsinstitut.

Krantz, D. H. 1974. Measurement theory and qualitative laws in psychophysics, in Measurement, psychophysics, and neural information processing, Krantz, D. H, Luce, R. D., Atkinson R. C, and Suppes P. eds. vol. 2, pp. 160 -197, San Francisco: Freeman.

Kuehni, R. G. 2004. Variability in unique hue selection, *Color Research and Application* 29:158-162.

Larimer, J., D. H. Krantz, and C. M. Cicerone. 1974-1975. Opponent process additivity. I. Red/green equilibria, *Vision Research* 14(1974):1127–1140; II. Yellow/blue equilibria and nonlinear models, *Vision Research* 15(1975):723-731.

Marthin, P. 1974a. *Praktisk färglära med övningar*, Stockholm: Ltsförlag.

Marthin, P. 1974b. *Färgungskap*, Stockholm: Ltsförlag.

Marthin, P. 1976. *Ljus, färg & funktion*, Stockholm: Ltsförlag.

Müller, A. 1945. *Schweizer Farbenatlas 1210*, Winterthur, Switzerland: Chromos Verlag.

Müller, A. 1948. *Die moderne Farbenharmonielehre*, Winterthur, Switzerland: Chromos Verlag; 2nd ed., 1959.

Müller, A. 1951. *Dreifarbenwürfel 1000*, Winterthur, Switzerland: Chromos Verlag.

Müller, A. 1953. *Mobiler Farbtonkörper 743*, Winterthur, Switzerland: Chromos Verlag.

Müller, A. 1955. *Schweizer Schulfarbenatlas*, Winterthur, Switzerland: Chromos Verlag.

Müller, A. 1958. *Farbbestimmer CUC 12000*, Winterthur, Switzerland: Chromos Verlag.

Müller, A. 1962. *Swiss Color Atlas 2541*, Winterthur, Switzerland: Chromos Verlag; 2nd ed. 1964 -1965.

Müller, A. 1973. *Ästhetik der Farbe in natürlichen Harmonien*, Winterthur, Switzerland: Chromos Verlag.

Munsell, A. H. 1907. *Atlas of the color-solid*, Malden, MA: Wadsworth-Holland.

Munsell, A. H. 1913. *A color notation*, 3rd ed., Boston: Ellis.

Munsell, A. H. 1915. *Atlas of the Munsell color system*, Malden, MA: Wadsworth-Holland.

Munsell, A. H. 1918. Color diary 1899 -1918, unpublished typed diary, copy in possession of R.G.K.

Munsell Color Company. 1929. *Munsell book of color*, Baltimore, MD: Munsell Color Company.

Nemcsics, A. 1988. *Coloroid colour atlas*, Budapest: Innofinance.

Nemcsics, A. 1993. *Farbenlehre und Farbendynamik*, Göttingen: Muster-Schmidt.

Nemcsics, A., and E. Béres. 1985 -1986. Der Farbenraum des Coloroid-Farbensystems, *Die Farbe* 32/33:327-345.

Ostwald, W. 1918. *Die Farbenlehre, II. Buch, Physikalische Farbenlehre*, Leipzig: Unesma.

Podestà, H. 1930. Beiträge zur Systematik der Farbempfindungen, in *Psychologische Optik*, Krüger, F. ed., vol. 4, pp. 1 -92, München: Beck.

Podestà, H. 1941. *Der ordnungswissenschaftliche Aufbau des Farbenkörpers* (Bücherei des Augenarztes, vol. 9), Stuttgart: Enke.

Purves, D., and R. B. Lotto. 2002. *Why we see what we do*, Sunderland, MA: Sinauer.

Richter, M. 1967. Gedanken über Farbsysteme, *Die Farbe* 16:121-130.

Spillmann, W. 1984. Ein Leben für die Farbe – das Werk von Dr. Aemilius Müller, Winterthur, *applica* 24, Dec., pp. 7-17.

Spillmann, W. 2001. Tausende von handgefärbten Farbmustern, *applica* 21-22, Nov., pp. 4-13.

Swedish Standards Institute. 1979. *Swedish natural color system*, Swedish Standards SS 019100, SS 019102, and SS 019103, Stockholm: Swedish Standards Institute.

Titchener, E. B. 1901-1905. *Experimental psychology*, New York: Macmillan.

Titchener, E. B. 1909. A demonstrational color-pyramid, *American Journal of Psychology* 20:15-21.

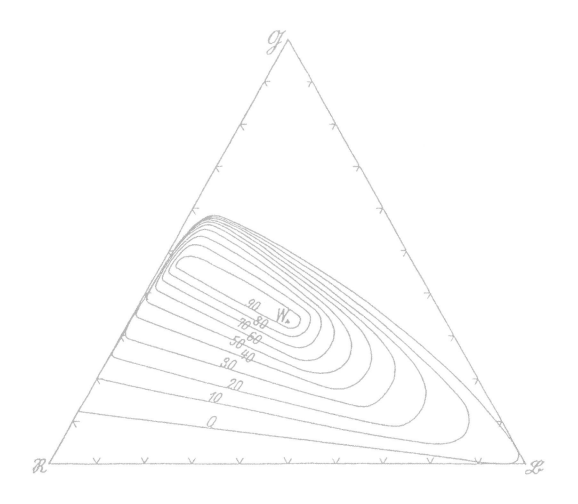

A description of scientifically and technically important systems derived from psychophysical or neurobiological descriptions of color stimuli.

The Greek atomist philosophers believed that material images (*eidola*) separate themselves from objects and penetrate the eyes to cause vision. Aristotle could not accept the material nature of *eidola* and looked for a medium between object and eye. He found it in the diaphanous or transparent (later to become the ether). He thought light to be a certain aspect of the transparent. Colors on the surface of objects can set the transparent in motion and thereby transmit their quality to the eye. Aristotle's theory was defended in the late Middle Ages by Mideastern philosophers Avicenna and Averroës, and, largely via southern Spain, this theory established itself in European thought.

Absent any obvious indications to the contrary, people with color vision take for granted an immediate relationship between color stimulus and perceptual result. Thus developed the idea to define color experiences by quantitative description of the stimuli. A given stimulus is generally assumed to result in the same experience (within a small error) for all color-normal observers. By the later nineteenth century, the idea that color experiences can be specified with color stimuli was in some circles firmly established. Given what we now know about colors, this is not valid. Similarly, it was (and is) assumed that a given vibratory frequency results for all hearers in the same sound experience. There is less certain popular belief in this regard in respect to taste and smell.

First steps on the path to psychophysical color diagrams were paved by Isaac Newton. His spectral hue circle was designed by 1694 but not published in *Opticks* until 1704, the year after his nemesis Robert Hooke passed away. Newton did not explicitly explain his motivation for a circular format, but it can be found in its purpose as a quantitative mixture diagram of spectral colors. The pan-harmonist Newton used musical ratios to place his seven primary colors along the circumference. He knew about colors not found in the spectrum but did not leave room for them in his diagram. The idea of specifying colors by specifying stimuli as filtered by the sensor functions came into high gear in the middle of the nineteenth century. The International Commission on Illumination (CIE) colorimetric system of 1932 (see CIE X, Y, Z Color Stimulus Space, this chapter) represents a high point in this process.

A colorimetric system has the following properties:

1. It provides precise specification of object and light stimuli based on precise measurements of, respectively, reflectance functions of objects and spectral power distributions of lights.

2. It permits arranging the stimuli in a three-dimensional system after reducing the stimuli to three dimensions.

Reduction occurs by "filtering" stimuli through color-matching functions representing the cone functions of a presumed average observer. The arrangement of the stimuli in the space corresponds on an ordinal basis with the perceptual arrangement as viewed against a simple achromatic surround.

3. The light mixtures (additive mixture) that the standard observer sees as identical can be accurately predicted in the system (they fall on identical locations in the system). However, the system performs unsatisfactorily regarding more complex perceptual results such as the prediction of the appearance of stimuli in simple and, even more so, in complex surrounds, and to predict average perceived differences between two stimuli. Despite shortcomings, psychophysical systems have been found technically useful to demonstrate relationships between stimuli, results of metamerism, comparison of gamuts of light and colorant mixture, and other purposes.

From Newton to Schrödinger

To begin this chapter with Newton is to reach quite far back in history, but it is with Newton that the idea of association between color perceptions and specific physical stimuli began. Newton, as to some extent his predecessor Francesco Maria Grimaldi (who, already in 1665, discussed the composition of white light), showed that hues (in general) can be associated with specific components of refracted sunlight. He found these associations to be stable and repeatable for himself. Because he had no means to measure the wavelengths of lights, he used color names to identify seven salient hues and thereby the (approximate) corresponding positions in the spectrum.

By clever experimentation, Newton found that mixtures of neighboring spectral lights were seen as hues judged perceptually intermediate between those of the generating lights. The farther apart the lights were along the spectrum, generally the more dramatic the hue changes were: Green-appearing spectral light mixed with red-appearing spectral light resulted in yellow-appearing light. Certain combinations, when mixed in appropriate ratios, resulted in light without hue (see chapter 1).

Newton invented his color circle to express these relationships in a semiquantitative manner. He realized that different spectral lights had different powers to influence the appearance of mixtures; he expressed these powers in the diagram with differently sized small circles. He invented a geometric method to determine the perceptual results of stimulus mixtures. Thus, in a manner of speaking, he initiated psychophysics. Given the technology he had available, his spectra were relatively crude. And he was not concerned with whether the wavelengths he described as green were identical to those described by another person as green.

Not given to speculation, Newton did not address the question of how the visual sense operates. It was left to the English glass merchant George Palmer to propose, in 1777, the existence of three kinds of "fibres" in the eye, sensitive to different rays of light. At the end of that century, Thomas Young addressed the question of the nature of light and experimentally concluded that light propagated in the form of waves. He also proposed three kinds of retinal mechanisms. In support of his theory, he calculated the wavelengths of various color stimuli, thus initiating a quantitative relationship between color names and light properties.

The quantification of the wavelength scale of the spectrum posed some technical challenges. In 1802, William Hyde Wollaston used dark lines in the spectrum to separate the visible spectrum into four parts. Some 10 years later, German lens maker Joseph Fraunhofer investigated these lines in much more detail, making them convenient markers. Young's undulatory theory of light, resulting in the definition of wavelengths, faced much opposition because the great Newton had rejected it. James Clerk Maxwell, in his early work in the middle of the nineteenth century, still used an arbitrary spectral division scale.

Physicists Maxwell and Hermann von Helmholtz and mathematician Günter Grassmann were primarily interested in clarifying the fundamental relationship between stimuli and color perceptions. Maxwell and Helmholtz attempted to determine if Palmer and Young's hypothesis of three chromatic mechanisms in the retina was true. Not only were they able to prove its validity, but they also experimentally measured more and more accurate estimates of the sensitivities of the three retinal processes.[1] They soon discovered a noticeable individual variation in these sensitivities and that some people with impaired color vision were apparently missing one or more of the sensor types.

The meme of a close relationship between color perceptions and stimuli for observers with normal color vision was already planted in the human mind and was only strengthened by the findings of Maxwell and Helmholtz and their associates. Known since antiquity, effects, such as simultaneous contrast or inconstancy of the appearance of colored objects as a function of illuminating light, were nonetheless considered aberrations of the visual sense.

Maxwell's findings were at first based on disk mixture (see glossary) and thus relate to primaries derived from object colors. Later he also experimented with spectral lights and created a triangular basis plane with red, green, and reddish blue spectral primaries (see figure 6.5). He used Young's equilateral triangle (itself most likely based on Mayer's) to show his results, and the Maxwell triangle, as it came to be known, developed into the standard format for representing color mixture data. In 1867, Helmholtz used a Maxwell triangle to show an estimated spectral trace in a triangle form based on derived color fundamentals (now taken to mean cone functions).

Without detailed experimental data, Helmholtz sketched a conceptual three-dimensional solid of spectral and purple light mixtures in the form of a dented cone filled with points representing mixtures of these lights (figures 6.8 and 6.9 in his entry). The cone shows that at near zero intensity, the lights are very faint. As their intensity increases, so does saturation and the perceptual saturation steps between the central white and the unmixed spectral color.

Helmholtz did not describe in detail the nature of the central axis, or identify the intensity level of spectral and extraspectral lights at the basis plane of his cone. What is known is that as the intensity increases, the hues of most lights change, and eventually, at very high intensity, fade. Given the then-available knowledge, Helmholtz's assistant Artur König optimistically predicted that a formula for quantitative description of the perceptual spacing of this kind of space would soon be available.

Grassmann, by taking the mixture of spectral lights to be additive, was able to state three laws and describe Newton's experimental findings of the result of mixing spectral lights in fully quantitative form. A colorimetric system based on these developments required the ability to accurately spectrally measure lights and objects and standardize a set of functions representing the spectral sensitivities of the human observer. In the first quarter of the twentieth century, such measurements became routine. These developments helped solidify the idea of a close relationship between measured stimuli and perceptual experience despite opposition from some psychologists.

In the first edition of Munsell color sample charts of 1907 (see Albert Henry Munsell, chapter 5), samples of one constant value (lightness) were shown against a paper-white and a black background to demonstrate the large effect of the background on perceived colors. This demonstration was dropped from the second, enlarged edition of 1915. Psychophysicists comparing stimuli and perceptions to find and state the connection between them began to assign numbers to perceptual scale values. They did so despite Johannes von Kries's charge that assigning numbers to perceptions did not indicate mathematical meaning for such symbols.

In the early 1920s, Austrian physicist Erwin Schrödinger helped to solidify the mathematical foundation of color stimulus science. His "spectrum bag" illustrated more accurately the form of Helmholtz's spectral light solid and the spectral lights' position in it. For object colors, he devised a sphere octant model, similar to Christian Doppler's (see entry in chapter 10).

In addition, following Wilhelm Ostwald (see entry in chapter 10), Schrödinger further clarified the concept of optimal object colors, the highest saturated object colors possible at a given level of luminous reflectance. He also attempted to fulfill König's promise of formulating the space's mathematics in terms of perceptual uniformity (at the level of just noticeable differences) by developing what is known as a *line element*. Such line elements lost their importance, but not until the later twentieth century, in light of their practical failures to predict average perceptual differences between stimuli.

Development of the colorimetric system

The concept of a colorimetric system was well established before the CIE promulgated a quantitative system. By the turn of the twentieth century, German educator Ludwig Pilgrim had developed a quantitative system of light mixture based on the best available color fundamentals data. He demonstrated Helmholtz's proposal of a stimulus-based interpretation of Ewald Hering's opponent system (see entry in chapter 5).

But Pilgrim had noted some difficulties. For purposes of color perception, he found the spectral power distribution of a light not to be a fully independent variable. The concept of adaptation (see chapter 1) applies not only to intensity but also to hue and saturation perception, if perhaps in an individual manner. As a result of adaptation, lights of a wide degree of chromaticity can be seen as "white," and many objects change their appearance in such lights much less than pure colorimetry predicts. For this reason, spectrally neutral, equal-energy light sources already were often used for calculations early on.

In the early twentieth century, spectral power or reflectance measurement capability was more widely available, and the König measurements of fundamental sensitivities were generally accepted as valid, having been confirmed more or less closely by other investigators. Thus, Robert Luther in 1927, Nikolaus Nyberg in 1928, and Sigfried Rösch in 1928 were able to calculate different versions of optimal object color stimulus solids and to determine where real color stimuli are placed in these solids. Luther's opponent color stimulus formulation modified Pilgrim's implementation of Helmholtz's earlier one.

Proposals like these and others made international standardization of a colorimetric system desirable. In 1924, the CIE established the photopic standard observer, defining brightness and lightness perception in daylight. In 1931, it established the spectral sensitivity data of the 2° visual field standard observer (see glossary entry CIE color-matching functions), based on recent new measurements. It also defined the spectral powers of three standard light sources as well as a method for spectral reflectance measurement.

Thus, a technological system for accurately defining color stimuli in three dimensions, related to average human cone sensitivities, was in place and soon in use. For example, chips of the Munsell Color Company's 1929 enlarged *Munsell Book of Color* were measured and CIE tristimulus values calculated at the (U.S.) National Bureau of Standards, thus providing their objective specification in terms of three numbers. The stimulus space implicit in the CIE colorimetric system, the X, Y, Z space, was found to be impractical for illustrating chromatic color relationships, and a rectangular (rather than the more cumbersome Young equilateral triangle) chromaticity diagram (x, y) was also defined. With the luminance value Y, this transformation provides a second kind of quantitative information about relations of light stimuli.

In 1935, David Lewis MacAdam recalculated the Rösch optimal color solid using the CIE colorimetric system for both standard daylight C and tungsten light A, raising them over the chromaticity diagram. The significantly different form of the daylight and tungsten light solids is an indication of the discrepancy between color perception and stimulus specification: Chromatic adaptation changes the perceived colors of objects seen in tungsten light compared to those seen in daylight in a manner much different from the colorimetric results.

Also in 1935, Hans Neugebauer, a student of Luther, calculated object color solids based on idealized and real chromatic printing ink primaries in his effort to develop a theory of halftone printing.

Since their introduction, the CIE standards have seen growing use for color stimulus specification in research, technology, and manufacture. It was perhaps inevitable that the system for standardized, fine-grained specification capability, presumed to relate closely to the human color-vision system, would also be used (with CIE's blessing) to describe average color perception data. Metric hue, saturation, and lightness were defined colorimetrically, followed by metric color difference and, more recently, metric color appearance.

Many colorimetrically based color space, difference, and appearance formulas have been developed over the last century, but their usefulness continues to be limited in many ways, and the causes of the limitations are mostly unknown. The human color-vision system appears to operate individually according to different and probably much more complex principles than such formulas have defined. But this fact does not detract from their useful application in specifying colorimetric stimulus or calculating difference and appearance values for the standard observer.

Color stimuli arriving at the eye are spectral. For example, when measured at every 10 nm throughout the visible range, a stimulus can be considered 31-dimensional. "Filtering" stimuli with cone absorptions reduces the information from

31 dimensions to three, making possible its representation in a Euclidean space (as discussed in chapter 1). A basic phenomenon of such dimension reduction is what Ostwald (see entry in chapter 10) called "metamerism" in connection with color vision.

Technologically, metamerism is both a boon and a problem. On the one hand, it makes possible color television, color monitors, and color printing with three inks. On the other hand, although metameric formulations against a reference reflectance function can result in color matches when they are viewed in, say, daylight, they can fail to even come close to matching when viewed in another kind of light, such as fluorescent.

But colorists usually need to find formulations that match the reference sample under several light sources. For reasons such as apparent differences in individuals' sense of color perception, observers rarely agree on the quality of a color match in different lights.

In 1953, Günter Wyszecki postulated that all spectral power functions resulting for, say, a CIE standard observer in the same tristimulus values had a common component—the fundamental—and variable components he called metameric blacks. In the 1980s, psychologist Jozef B. Cohen developed a mathematical procedure to separate fundamental from metameric black components, and he defined a fundamental color space housing all fundamentals.

In principle, such a space contains the same information as the CIE tristimulus space, and its form depends only on the choice of axes. Therefore, Cohen's is simply one of many possible forms of linearly related color-stimulus spaces, all containing the same information. However, his has the distinction of being orthonormal (see glossary).

Helping to solve the practical problems of matching the appearance of reference samples with different colorants and on different substrates became a major application of colorimetry beginning in the mid-1950s. The relative success of this technology strengthened the idea of the close relationship between stimulus and resulting color experience. In computer-assisted color matching, the key property is light absorption, not reflection, by colorants. As a result, absorption stimulus spaces (see Absorption Space, this chapter) were developed that use the colorimetric framework to help computer colorant formulations to match the tristimulus values of specific reference samples. The spaces no longer have a direct link with color perception.

CIE's choice to make one of the three sensitivity functions coincide with the luminance function causes the other two functions to represent the implied chromatic content of a color stimulus. This makes two of the three functions different from cone functions (the S cone function is considered identical to the CIE color-matching function $\bar{z}$. Cone sensitivity functions (see figure 1.1) are considered linearly related to color-matching functions, and in recent years, their form (for a standard observer) has been elaborated in finer and finer detail.

Cone-function–based systems

Color-vision scientists want to work with cone functions because they have a physiological basis. Cone functions are not influenced by additional ideas or information about color vision. An optimal object color solid can also be calculated in the cone color space (see figure 6.38). But not as in CIE tristimulus space, none of the solid's dimensions has any implied direct relationship to color perception attributes.

In the second half of the twentieth century, signals generated in the retina were followed along the visual path into the brain. In a major "switching station" on this path, the lateral geniculate nucleus (LGN), neuroscientists in the 1960s identified cells that have a kind of opponent color function. Closer analysis of the workings of these cells in the LGN of macaque monkeys (whose color-vision system is, in its neurophysiology, believed to closely resemble that of humans) has indicated the following inputs into two types of opponent cells: $L - M$ (or the opposite) and $(L + M) - S$ (or the opposite) (see figure 1.1 for the spectral sensitivity functions of the three cone types L, M, and S).

However, the implied perceptual color experiences associated with such cell wiring do not relate simply to experiences of unique hues. The resulting DKL space (see DKL Color Space, this chapter) is widely used in color-vision research to display experimental results, usually with implicit or explicit connection to color experiences.

The ability to accurately measure and describe color stimuli at an early point in our color-vision system has been widely interpreted as an ability to predict color perceptions, despite much evidence to the contrary. The fundamental weakness of this idea has become more widely understood only since the 1990s. To establish correlations of moderate technical and commercial usability between stimuli and perceptions has been possible only for very specific conditions of viewing and surround. Much more detailed knowledge of the functioning of and variability in the human color-vision apparatus is necessary to develop a useful model of how color stimuli (or visual stimuli in general) are related to visual experiences.

ISAAC NEWTON   1704

I. Newton, Opticks, 1704

Isaac Newton (1642-1727), celebrated English mathematician, physicist, alchemist, and discoverer of the law of universal gravitation, experimentally determined the composition of white light to be a mixture of spectral lights of different wavelengths. When these are viewed individually, they create various hue experiences, the spectral colors. In *Opticks*, his mature reflections on colors, Newton introduced a circular diagram of spectral colors (figure 11 of plate 2, part II, book 1; see figure 6.1). He described it (in part) as follows:

> With the Center O and Radius OD describe a Circle ADF and distinguish its circumference into seven parts . . . proportional to the seven musical Tones or Intervals of the eight Sounds, contained in an Eight. . . . Let the first part DE represent a red Colour, the second EF orange, the third FG yellow, the fourth GH green, the fifth AB blue, the sixth BC indico, and the seventh CD violet, And conceive that these are all the Colours of uncompounded Light gradually passing into one another, as they do when made by Prisms. . . . Let p be the center of gravity of the Arch DE [comparably for q, r, s, t, v, x] and about those centers of gravity let Circles proportional to the number of rays of each Colour in the given mixture be described. . . . Find the common center of gravity of all those Circles p, q, r, s, t, v, x. Let that center be Z; and from the center of the Circle ADF, through Z to the circumference, drawing the right line OY, the place of the point Y in the circumference shall shew the Colour arising from the composition of all the Colours in the given mixture, and the line OZ shall be proportional to the fullness or intenseness of the Colour, that is, to its distance from whiteness. As if Y fall in the middle between F and G, the compounded Colour shall be the best yellow; if Y verge from the middle toward F or G, the compounded Colour shall accordingly be a yellow, verging toward orange or green. If Z fall upon the circumference the Colour shall be intense and florid in the highest degree; if it fall in the mid way between the circumference and center it shall be but half so intense, that is, it shall be such a Colour as would be made by diluting the intensest yellow with an equal quantity of whiteness; and if it fall upon the center O, the Colour shall have lost all its intenseness and become a white. . . . [I]f the point Z fall in or near the line OD, the main ingredient being the red and violet, the Colour compounded shall not be any of the prismatic Colours, but a purple, inclining to red or violet. (Newton 1704, pp. 114–116)

Newton's figure represents an incomplete hue circle: It lacks the extraspectral red and purple colors. It is also the first representation of an additive stimulus mixture diagram, as demonstrated with a specific example:

[S]uppose a Colour is compounded of these homogeneal Colours, of violet 1 part, of indico 1 part, of blue 2 parts, of green 3 parts, of yellow 5 parts, of orange 6 parts, and of red 10 parts. Proportional to these parts I describe the Circles x, v, t, s, r, q, p respectively, that is, so that if the Circle x be 1, the Circle s 3. . . . Then I find Z, the common center of gravity of these Circles, and through Z drawing the line OY the point Y falls upon the circumference between E and F, . . . and thence I conclude, that the Colour compounded of these ingredients will be an orange, verging a little more to red than to yellow. Also I find that OZ is a little less than one half of OY, and thence I conclude, that this orange hath a little less than half the fullness or intenseness of an uncompounded orange . . . this proportion being not of the quantities of mixed orange and white powders, but of the quantities of the lights reflected from them. (pp. 116–117)

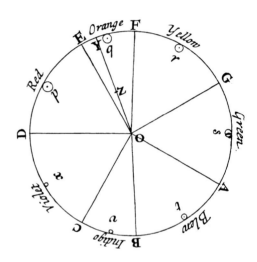

Figure 6.1.

Newton's spectral color circle and color mixture diagram (Newton 1704).

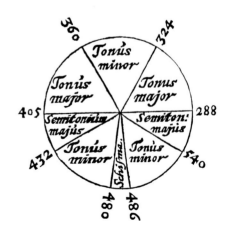

Figure 6.2.

Descartes's (1650) circle of major and minor tones from Compendium musicae.

Newton's diagram combines Pythagorean beliefs in a universal harmony with aspects of a quantitative mixture diagram. Its circular form may have been influenced by an illustration of the relationship between musical ratios in the form of a circle (figure 6.2) by the French philosopher and natural scientist René Descartes. Parallel to the musical scale, Newton used the classical number of seven for the basic colors of the spectrum (the Aristotelian seven colors, however, include white and black).

Newton was intrigued by the possibility of parallels between musical harmony and color harmony. He pointed out the common perceived redness of shortest and longest wave colors of the spectrum and compared it with the similarity of the tones at the beginning and end of an octave. The quantitative aspects are represented by the common white center, the result of mixing two or more spectral colors or all spectral colors together.

Newton also described for the first time (if not in the same place in the book) three kinds of attributes for any color experience: color (hue), luminosity (brightness), and intenseness (saturation), still in general use. He recognized seven primary hues and an infinite number of intermediate hues. As demonstrated in figure 6.1, intenseness is related to the relative amounts of spectral and white light in a mixture. He described lightness in connection with colorants as follows:

> Now considering that these grey and dun Colours may also be produced by mixing whites and blacks, and by consequence differ from perfect whites not in Species of Colours but only in degree of luminousness, it is manifest that there is nothing more requisite to make them perfectly white than to increase their Light sufficiently (p. 112).

Newton's is a prototypical version of several more quantitative diagrams of color stimulus mixture that followed. His work, as is well known, represented a sea change in thinking about colors, opening furious and extended discussions settled only some 200 years later.

THOMAS YOUNG 1807

T. Young, A course of lectures on natural philosophy and the mechanical arts, 1807

English physician and physicist **Thomas Young** (1773-1829) was of wide-ranging education. He was elected to the Royal Society at age 21 and became its foreign secretary at 29, a position he held for life. He is known for developing the wave theory of light, unpopular at the time because of Newton's earlier objections to it. Among other things, he determined the wavelengths of spectral lights from diffraction, experimented on elasticity of materials (Young's modulus), and helped to decipher Egyptian hieroglyphic writing.

In 1795-1796, Young studied in Göttingen, Germany, where he heard lectures by the physicist Georg Christoph Lichtenberg, who had published Mayer's unpublished works after Mayer's death in 1775, including his paper on color order (see Tobias Mayer, chapter 4). Lichtenberg lectured on Mayer's system, based on three primary colors (including the triangular form of the central plane of the color solid) (Mollon 2003).

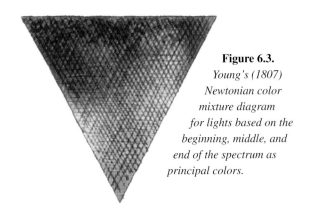

Figure 6.3.
Young's (1807) Newtonian color mixture diagram for lights based on the beginning, middle, and end of the spectrum as principal colors.

This early contact with Mayer's work is likely to have inspired Young's interest in color and the nature of light. In the Bakerian Lecture "On the theory of light and colours," given in November of 1801, Young presented arguments in favor of the undulatory (wave) theory of light. In this connection, he suggested that there are three kinds of particles in the retina responsible for color vision:

> Now, as it is almost impossible to conceive each sensitive point of the retina to contain an infinite number of particles, each capable of vibrating in perfect unison with every possible undulation, it becomes necessary to suppose the number limited, for instance to the three principal colors, red, yellow, and blue, of which the undulations are related in magnitude nearly as the numbers 8, 7, and 6; and that each of the particles is capable of being put in vibration less or more forcibly, by undulations differing less or more from a perfect unison. (Young 1802a, pp. 20-21)

Colours.	Length of an Undulation in parts of an Inch, in Air.	Number of Undulations in an Inch.	Number of Undulations in a Second.
Extreme –	.0000266	37640	463 millions of millions
Red – –	.0000256	39180	482
Intermediate	.0000246	40720	501
Orange – –	.0000240	41610	512
Intermediate	.0000235	42510	523
Yellow –	.0000227	44000	542
Intermediate	.0000219	45600	561 ($= 2^{48}$ nearly)
Green – –	.0000211	47460	584
Intermediate	.0000203	49320	607
Blue – –	.0000196	51110	629
Intermediate	.0000189	52910	652
Indigo – –	.0000185	54070	665
Intermediate	.0000181	55240	680
Violet – –	.0000174	57490	707
Extreme – –	.0000167	59750	735

Figure 6.4.

Table of wavelengths of spectral colors and their intermediates and the extremes of the spectrum as calculated by Young (1802a).

In a second Bakerian lecture in 1802, Young redefined the three principal colors based on Wollaston's reinterpretation of the colors of the spectrum:

> In consequence of Dr. Wollaston's correction of the description of the prismatic spectrum, compared with these observations, it becomes necessary to modify the supposition that I advanced in the last Bakerian lecture, respecting the proportions of the sympathetic fibres of the retina; substituting red, green, and violet, for red, yellow, and blue, and the numbers 7, 6, and 5 for 8, 7, and 6. (Young 1802b, p. 395)

In *A Course of Lectures on Natural Philosophy and the Mechanical Arts* of 1807, Young illustrated a Newtonian color mixture diagram in form of an equilateral triangle based on the new principal colors (figure 6.3, originally executed in colored pencil). Here he commented as follows:

> [It is] certain that the perfect sensations of yellow and blue are produced respectively, by mixtures of red and green and violet light, and there is reason to suspect that those sensations are always compounded of the separate sensations combined . . . and we may consider white light as composed of a mixture of red, green, and violet only . . . (p. 439)

Though Newton had provided hints about the difference between light mixture and colorant mixture, Young did not make such a distinction. The center of his triangle is gray, and he observed that when viewing balanced mixtures of colorants in strong light, they appeared white, if gray at nor-

mal light levels. He also mentioned disk mixture as a technique for "combining sensations of various kinds of light." He posited seven primitive "colour distinctions" arising from the three primitive sensations: red, yellow, green, blue, violet, and crimson, and white.

Young described his triangle as follows:

> A triangular figure, exhibiting in theory all possible shades of colours. The red, the green, and the violet, are single at their respective angles, and are gradually shaded off towards the opposite sides: a little yellow and blue only are added in their places, in order to supply the want of brilliancy in the colours which ought to compose them. The center is grey, and the lights of any two colours, which are found at equal distances on opposite sides of it, would always very nearly make up together white light, as yellow and violet, greenish blue and red, or blue and orange. (p. 786)

Making an informed assumption about the speed of light (500 billion feet in 8 1/8 minutes), Young also calculated the wavelengths of spectral colors (figure 6.4).

The form of an equilateral triangle for a color mixture diagram was also later employed by Maxwell and Helmholtz (see entries in this chapter). In 1777, Palmer anticipated Young's idea of three different kinds of "fibres" in the retina.

JAMES CLERK MAXWELL 1857, 1860

J. C. Maxwell, Experiments on colour, 1857
On the theory of compound colors, 1860

Scottish physicist **James Clerk Maxwell** (1831-1879) was the first professor of experimental physics at Cambridge University and is primarily known for his work on electromagnetism. He showed light to be a form of electromagnetic radiation. A student of James David Forbes (see entry in chapter 11), he continued working on color problems using disk mixture after moving to Cambridge. In a 1857 paper, "Experiments on colour, as perceived by the eye, with remarks on colour-blindness," he described the results of disk mixture with varying ratios of disks colored with the three primary colorants vermilion, ultramarine, and emerald green on Young's quantitative triangular force diagram. It came to be known as the Maxwell triangle (figure 6.5).

The diagram shows the results of matching colors of 14 colorants with quantitative disk mixture using the above-mentioned three primary colorants as stimuli. He expressed the results as an angle and a saturation coefficient. In doing so, he found that the pigments in the diagram were arranged in prismatic order. Most of the individual pigments have higher saturation than the three primaries used by Maxwell; only their hue, not their saturation, can be matched by disk mixture of his three primaries.

In an 1860 two-part paper, Maxwell proposed a mathematical theory of Newton's color diagram as well as a description of color stimulus mixture represented by vectors. He built a visual spectrometer to produce quantitative mixtures of spectral lights. It allowed him to match the appearance of sunlight on white paper with a mixture of three primary spectral lights. Selecting as primaries the wavelengths 24, 44, and 68 (in his nomenclature), he and observer K (his wife) matched the standard "white" light with combinations of two primaries and a third light of various other wavelengths.

From the results, Maxwell located the spectral colors of 16 wavelengths in the triangular diagram based on his spectral primary lights (figure 6.6). He found that endpoints of most spectral color vectors are outside the triangle (indicated by the dashed lines), showing again that it is impossible to match in saturation all spectral colors with any three of them. He presented his results in the form of spectral functions that show the relative amounts of primary lights required to match the hue at each wavelength (figure 6.7). Although generally similar for the two observers, the results show some individual differences:

> [Figure 6.7 is] intended to indicate the intensities of the three standard colours at different points of the spectrum. The curve marked (R) indicates the intensity of the red or (24), (G) that of green or (44), and (B) that of blue or (68). The curve marked (S) has its ordinates equal to the sum of the ordinates of the other three curves. (Maxwell 1860, pp. 75–76)

The color mixture diagram of figure 6.6 represents a fully quantitative version (with respect to the equipment and the wavelengths of the spectral lights used as primaries) of Newton's color diagram of spectral lights mixture. As in Newton's diagram, brightness is not considered. Pilgrim (see entry in this chapter) recalculated Maxwell's diagram based on the König functions. These developments are direct antecedents of the CIE colorimetric system in use today.

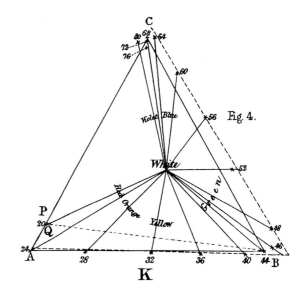

Figure 6.6. *Maxwell triangle based on results of mixtures of spectral colors. The primaries are the wavelengths identified by numbers 24, 44, and 68, observer K (Maxwell 1860).*

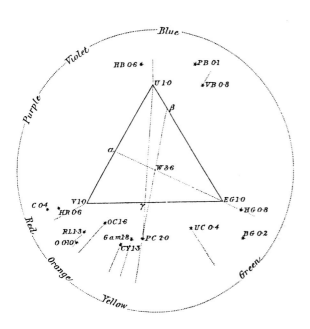

Figure 6.5.
Disk mixture diagram based on three arbitrary pigment standards (as painted on paper) embedded in a spectral circle, with quantitatively determined locations of other pigments relative to the primaries (Maxwell 1857).

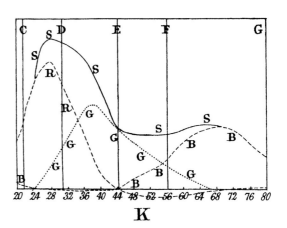

Figure 6.7.
Relative amounts of the three primary lights R, G, and B required by observer K to match other wavelengths and the curve S representing the sum of all three lights. The vertical lines with letters represent Fraunhofer lines of the spectrum (Maxwell 1860).

HERMANN VON HELMHOLTZ   1856–1866

H. L. F. von Helmholtz, *Handbuch der physiologischen Optik*, 1856–1866

Hermann Ludwig Ferdinand von Helmholtz (1821-1894), a German physicist, physiologist, mathematician, philosopher, and inventor, formulated the first law of thermodynamics. His inventions include the ophthalmoscope, an instrument to view the interior of the eye. He was the first to measure the speed of propagation of nervous excitations and wrote the two-volume *Handbuch der physiologischen Optik* (Treatise on physiological optics).

With Maxwell, Helmholtz demonstrated the validity of Palmer's and Young's sketched trichromatic theory, which came to be known as the Young-Helmholtz theory of color vision. Maxwell and Helmholtz closely followed each other's work on color and were in regular contact. Helmholtz made mixtures of spectral lights at the same time that Maxwell made disk mixtures, but Helmholtz had somewhat different results due to initial difficulties with his equipment. The discrepancy was conceptually resolved by Grassmann (see entry in this chapter) and later by Helmholtz with improved equipment.

Helmholtz connected perceptual attributes to spectral and intensity variations of lights, and he demonstrated optimal primary colors:

> Every difference of impression made by light . . . may be regarded as a function of three independent variables: and the three variables which have been chosen thus far were (1) the luminosity, (2) the hue, and (3) the saturation. . . . However, instead of these variables, three others may also be employed; and in fact this is what it amounts to, when all colors are regarded as being mixtures of variable amounts of *three so-called fundamental colors*, which are generally taken to be *red, yellow* and *blue*. . . . Least suited for this purpose are red, yellow and blue. . . . It would be rather better to take *violet, green* and *red* for fundamental colors. (Southhall 1924, p. 141, emphasis original)

> [S]uppose at first we leave out of account differences of luminosity; then there will still be two variables left on which the quality of the color depends, namely the hue and the proportion between colored light and white light. All the various colors may be represented, therefore, according to their two dimensions, by points lying in a plane, just like the values of any other function of two variables. The saturated colors constitute a closed series. . . . The transitions from white to any saturated color at a point on the circumference of the circle lie along the radius drawn to this point, so that the paler shades of this hue are nearer the center, and the more saturated ones nearer the circumference. Thus we get a *color chart* containing all possible kinds of colors of the same luminosity arranged according to their continuous transitions [figure 6.8]. When the different degrees of luminosity are also to be

taken into account, the third dimension of space has to be utilized, as was done by Lambert. The darkest colors for which the number of discriminable shades continually gets less and less may be made to culminate in a point corresponding to black. In this way we get a *color pyramid* or color cone [figure 6.9]. (pp. 132–133, emphasis original)

In figure 6.8, Helmholtz showed that mixtures between spectral violet and red (the purple colors) must lie on a straight line (also see Wilhelm von Bezold, chapter 10). Figure 6.9 conceptually represents the color cone whose surface contains color perceptions from lights at different intensities, beginning with no light (black) at the top.

On object colors, Helmholtz commented as follows:

> [I]n ordinary speech we are wont to describe differences of luminosity as differences of color, however, only in case color is considered as a characteristic of bodies. Absence of light is called *darkness*. But when a body does not reflect any light that falls on it, we say it is *black*. On the other hand, a body that scatters all incident light is called *white*. A body that reflects an equal share of all the incident light without reflecting all of it is called *grey*; and one which reflects light of one color more than that of another is said to be a colored body. Accordingly, in

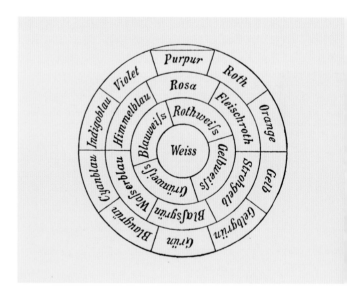

Figure 6.8.

Two-dimensional conceptual representation without consideration of luminosity of spectral and purple lights and their mixtures, with white the result of mixture of all of them. The mixtures of violet and red (nonspectral purple colors) are located on a straight line. The colors on the outer circle, beginning at the 1 o'clock position, are red, orange, yellow, yellow-green, green, blue-green, cyan blue, indigo blue, violet, purple; on the second circle red of meat, straw yellow, pale green, water blue, sky blue, pink; on the inner circle reddish white, yellowish white, greenish white, bluish white (Helmholtz 1856-1866).

this sense *white*, *grey* and *black* are colors also. Saturated colors of low luminosity are said to be "dark," as, for example, dark green, dark blue. . . . [R]ed, yellow and green of low luminosity are called *red-brown*, *brown* and *olive-green*, respectively, whereas exceedingly pale colors of low luminosity have names such as *reddish grey*, *yellow-grey*, *blue-grey*, etc. (p. 130, emphasis original)

The *Handbuch* contains a conceptual psychophysical color stimulus diagram in form of a Maxwell triangle (see James Clerk Maxwell, this chapter). But it is based on assumed color fundamentals (not spectral primaries), and it shows the approximate form of the spectral trace in this diagram. The trace's shape is due to intrinsic differences in spectral color saturation (figure 6.10). This diagram is a conceptual forerunner of the CIE chromaticity diagram (see CIE X, Y, Z Color Stimulus Space, this chapter). An experimentally derived version of the diagram was presented by Helmholtz's assistant and collaborator König in an 1886 paper. Helmholtz showed how it is possible to integrate Hering's theory of opposing fundamental color experiences into the Young-Helmholtz theory as a secondary step (see Ludwig Pilgrim, this chapter).

Another key Helmholtz contribution was to clarify, after 2,000 years or more of confusion, the difference between light mixture and colorant mixture:

For the mixed pigment does not give at all a color that would be the resultant of mixing the two kinds of lights that are reflected separately from each of the ingredients. . . . [W]henever we examine pigments . . . we find . . . in the form of thin sheets they are transparent. . . . Now when light falls on a powder of this kind . . . a certain portion of it will be reflected from the outer layer, but most of it will not be reflected until it has penetrated into the interior to some extent. . . . [B]y far the greatest part of it [is] being reflected from the deeper layers. . . . [Light] which comes from the interior begins to be colored by absorption; the farther the light penetrates, the deeper being the color. . . . Thus, most of the light reflected from a mixture of colored powders is due, not to an addition of both colors, but a subtraction. . . . This is the reason . . . why mixtures of pigments are much darker than the separate ingredients. . . . Vermilion and ultramarine, for instance, make a dark gray with scarcely a trace of violet, although that is the compound color of red and blue light. (pp. 123–124)

With Maxwell, Helmholtz paved the way for colorimetry as it is practiced today. His clarification between object and light colors ended previous confusion over this issue and provided the definition for additive (lights) and subtractive (colorant) color mixture.

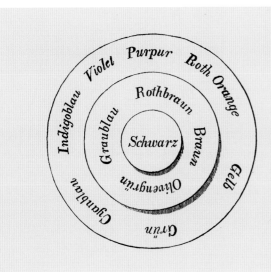

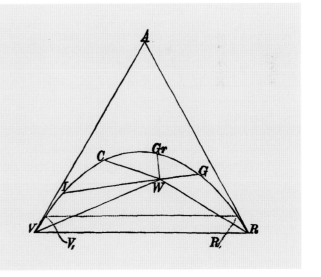

Figure 6.9.
Three-dimensional conceptual representation of the color stimulus solid, with consideration of brightness, in form of a cone with black at the top. The colors of the outer circle, starting at the 1 o'clock position are identical to those of figure 6.8; those of the second circle are red-brown, brown, olive green, gray-blue (Helmholtz 1856-1866).

Figure 6.10.
Qualitative plot of the locus of spectral colors in a Maxwell triangle with red, green, and violet fundamental sensations as primaries. White is located at W. G, Gr, C, and I indicate yellow, green, cyan, and indigo spectral colors (Helmholtz 1866).

HERMANN GÜNTER GRASSMANN   1853

H. G. Grassmann, *Zur Theorie der Farbenmischung*, 1853

Hermann Günter Grassmann (1809-1877) was a German mathematician and Sanskrit scholar who developed fundaments of linear algebra and who translated the *Rig Veda*.[2] He developed an interest in color from a paper by Helmholtz, who, for a time, based on erroneous experiments, disputed Young's theory of three fundamental processes in the eye. In response, Grassmann wrote a paper supporting, based on reasoning alone, the beliefs of Newton and Young and the findings of Maxwell (see entries in this chapter). Helmholtz repeated his experiments with improved equipment and confirmed Grassmann's analysis.

During his analysis, Grassmann developed three rules, now called Grassmann's laws, that are essential to modern colorimetry. They take mixture of color stimuli to be additive, a situation that has been confirmed experimentally under certain limited conditions.

a, but *m* times its length. . . . Having represented in this manner the two colors geometrically, let us construct from these lines the geometrical sum, that is the diagonal of the parallelogram that has the two lines as sides, and assume that this sum or diagonal shall represent the color of the mixture, its direction showing the tint [hue], and its length the intensity.

This done, the tint and intensity of any mixture of colors may be found by simple construction. Thus it is only necessary to determine the lines which represent the tint and intensity of the mixed colors, and then to add these geometrically, that is to compound them as forces, and the geometrical sum (the resultant of the forces) represents the tint and intensity of the mixture. (Grassmann 1853, pp. 261–262, translated by R.G.K.)

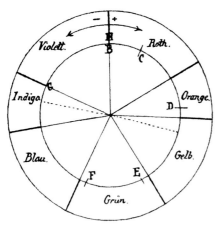

Figure 6.11.
Grassmann's (1853) version of Newton's color circle. Capital letters indicate the location of Fraunhofer lines in the spectrum. No space is provided for purple colors.

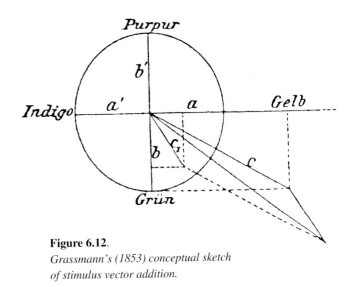

Figure 6.12.
Grassmann's (1853) conceptual sketch of stimulus vector addition.

Grassmann derived his color circle from Newton's. The color segments are identically named and occupy the same circle segments. Like Newton, Grassmann allotted no space to purples. But he attempted to make his circle more quantitative than Newton's by identifying approximate positions of Fraunhofer lines in the diagram (figure 6.11). And, again from reasoning only, he proposed a mathematical method for predicting the result of additive mixture of lights:

> Suppose the intensities of the two colors to be mixed are signified by the length of the lines representing them, so that if a color has the tone *a*, and its intensity is in the same proportion to that of *a* as *m* to 1, then the color may be represented by a line having the same direction as

Figure 6.12 illustrates this process. A color solid represented in this manner is, in modern language, a vector space because each color is represented by a vector arrow originating in the black point. The axis system of such vector spaces can be defined in different ways. The CIE tristimulus space (see CIE X, Y, Z Color Stimulus Space, this chapter) and Cohen's fundamental color space (see entry in this chapter) later implemented an additive Grassmann vector space. Such spaces and distances in the spaces are not linearly related to perceptual distances.

LUDWIG PILGRIM   1901

L. Pilgrim, *Einige Aufgaben der Wellen- und Farbenlehre des Lichts*, 1901

Ludwig Pilgrim was a mathematics professor at the Realanstalt in Cannstadt in southern Germany and a member of a regional mathematical-scientific association. Each year, one of the professors of this school was invited to publish an article as an addendum to the program booklet for the school's year-end celebrations; in 1901, Pilgrim described results of his studies in the optics of light. Despite its modest title, *Einige Aufgaben der Wellen- und Farbenlehre des Lichts* (Some exercises in wave and color theory of light), Pilgrim's article significantly exceeded the normal level of teaching at a *Realanstalt* (a high school specializing in natural science and technology), and it reflects a deep personal interest.

The article consists of five chapters, with the final two discussing the colors of thin layers and the results of polarization, respectively. Of greatest interest is the third chapter, "Interference colors," in which Pilgrim described the spectrum's composition and colors according to several authors. He also discussed results of stimulus mixture according to Newton and Maxwell (see entries in this chapter) and based on König's and Conrad Dieterici's fundamental sensations (König 1886, 1892). And he presented the mathematics of a linear trichromatic system.

Pilgrim listed relative amounts of light of the solar interference spectrum at every 10 nm for the spectral range from 380 to 740 nm, the values for the three normalized König fundamental sensations (each summing to 100), as well as the spectral brightness implicit in cone and rod sensitivity, as he graphically determined it. He also listed the magnitude of the composite spectral vectors, designated as "color intensity." Pilgrim stated:

If the magnitudes *R, G, B* are viewed as projections of lines on the axes of a three-dimensional coordinate system they form a cone surface [figure 6.13] and their end points a doubly curved line named *spectral color line*; the latter can be represented as projected onto a flat surface sectioning equal lengths of the three axes (*R, G, B*), named *color diagram* [figure 6.14]. Orthogonal projection of the *R, G, B* axes onto the color diagram forms angles of 120°. A second plane of projection is vertical to the first and to the projection of the *G* axis. It is to be named *yellow-blue plane*. The line extending from the origin to the plane is the whiteness axis (*w*), the beginning of which, *O*, represents black. . . . A line from *O* onto the yellow-blue plane is designated *u-axis* (positive in direction of *G*) or *carmine-green axis*, and a parallel line through *O* to the intersection of the two planes of projection *v-axis* (positive on the *B* side) or *yellow-blue axis* [figures 6.13 and 6.14]. Between the coordinates of a color point in the (*u, v, w*) and the (*R, G, B*) systems the following relations apply:

$$u = (1/\sqrt{6})\,(-R + 2G - B)$$
$$v = (1/\sqrt{2})\,(-R + B)$$
$$w = (1/\sqrt{3})\,(R + G + B)$$

(Pilgrim 1901, p. 31, trans. by R.G.K.; italics original)

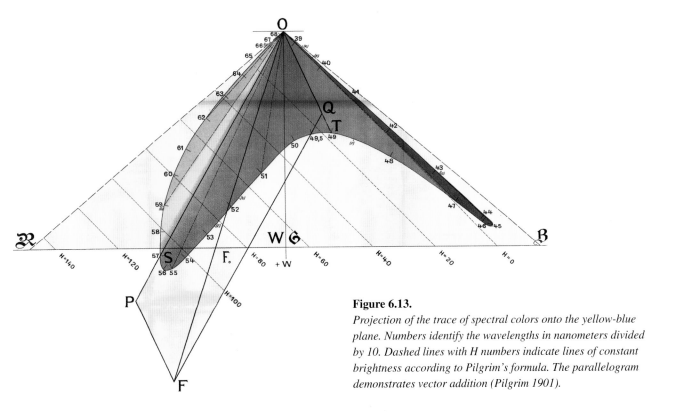

Figure 6.13.
Projection of the trace of spectral colors onto the yellow-blue plane. Numbers identify the wavelengths in nanometers divided by 10. Dashed lines with H numbers indicate lines of constant brightness according to Pilgrim's formula. The parallelogram demonstrates vector addition (Pilgrim 1901).

Figure 6.14.

Projection of the R, G, B space onto the chromatic plane with the spectral trace and the central white point W defined by sunlight. Opponent color axes are defined by u and v. In the u, v diagram the spectral trace is shown as a heavy line. In the original, both figures 6.13 and 6.14 are printed on a large-format (ca. 28 x 37 inches) sheet (Pilgrim 1901).

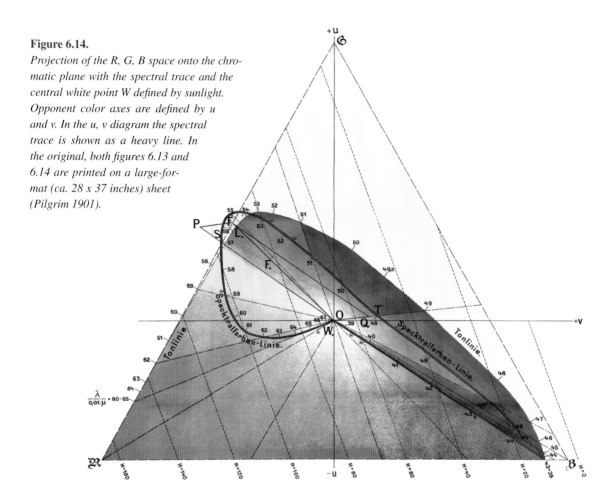

Helmholtz proposed the *u, v, w* system in the second edition of the *Handbuch der physiologischen Optik* (1896) as a psychophysical interpretation of Hering's color theory. Hering assumed that three substances in the eye correspond to the three components *u, v, w*. The decomposition of the substances results in the formation of ±*u*, ±*v*, and ±*w* components, that is, the carmine-green component, the yellow-blue component, and the white-black component:

> The line connecting *O* with a point *S* of the spectral color line represents the corresponding color according to hue and intensity based on equal slit width for all colors (approximately 1). Different points on the line *O S* represent color *S* at different slit widths. To every point of the color diagram corresponds a certain color. The central projection l_O of the spectral line *l* onto the color diagram therefore represents the spectral colors at varying slit widths. Line l_O is to be called (color) *hue line*. The color corresponding to a certain point F_O on the color diagram is obtained by connecting the projection W_O from *O* onto the color diagram with F_O. The section L_O of $W_O F_O$ with hue line l_O produces the hue, the ratio $W_O F_O : W_O L_O$ the saturation. (p. 32)

Under the heading "Color parallelogram" in the same chapter, Pilgrim showed how to obtain the mixture result of two spectral lights of varying slit width *u* and *v* according to Grassmann's vector-addition method. Complementary color *G* of color *F* stands in relationship $G_u : G_v = F_u : F_v$, but their components have opposite signs. Pilgrim showed how to calculate a complementary wavelength from this relationship, and the slit width ratio needed to obtain complete desaturation. Although Pilgrim's trichromatic system differs in several details from the CIE colorimetric system developed in 1932, many basic ideas are the same or similar.

However, Pilgrim used the König color-matching fundamentals (cone sensitivity functions), while the CIE developed its own color-matching functions (see CIE X, Y, Z Color Stimulus Space, this chapter). Pilgrim took the three functions to form an orthogonal space in which he plotted the spectral trace in three dimensions (figure 6.13 is a projection onto a plane). He expressed colors of individual wavelengths as vectors. The resulting figure is similar to the vector representation used some 70 years later by Cohen (see entry in this chapter). Projection onto a plane (see figure 6.14) forms a Maxwell triangle. The central stimulus in this figure is equivalent to Pilgrim's sunlight data.

The Helmholtz *u, v* opponent color system that Pilgrim demonstrated differs in detail from the system of Luther (see entry in this chapter) and from that of Leo Hurvich and Dorothea Jameson (see entry in chapter 5). The red–green axis (dominant wavelength ~503 nm) of those systems reasonably agrees with Helmholtz's. The *v* axis has dominant wavelengths of approximately 488 and 602 nm, indicative of greenish blue and reddish yellow. The white–black axis is the energy sum axis and is not in agreement with the CIE luminance axis.

Pilgrim's chromatic diagram derives from Young's and Helmholtz's. The CIE later decided on an easier-to-calculate method with a diagram of rectangular rather than triangular form. Pilgrim's brightness formula, which he considered to agree with König's and Dieterici's result, is

$$H = 9.6\,R + 1.88\,G + 0.2\,B.$$

In this formula, *H* is largely defined by *R*. Pilgrim did not discuss object color stimuli, but they would fit appropriately into his system. Despite his awareness of many uncertainties and observer variation, Pilgrim calculated with a mathematician's accuracy, and he related color perceptions to specific wavebands. We have not come across any similarly complete earlier or contemporary presentation of a trichromatic system. Pilgrim has not received much credit for his effort, perhaps due to the fact that the paper was published in a program of school festivities, far from the intellectual centers of the time, and thus received little attention.

ERWIN SCHRÖDINGER   1920

E. Schrödinger, Grundlinien einer Theorie der Farbenmetrik im Tagessehen, 1920

Erwin Schrödinger (1887-1961), Austrian theoretical physicist and Nobel Prize winner, mathematically developed wave mechanics; his formulation of the wave equation is known in quantum physics as the Schrödinger equation. In his early years, he also applied his mathematical-analytical skills to color problems. He used Grassmann's laws (see entry in this chapter) and König's (as well as other) fundamental sensitivity data (similar to figure 6.15) as a basis for his efforts.

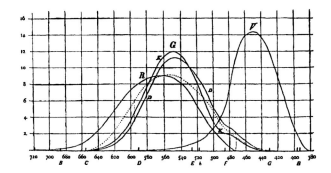

Figure 6.15.
König and Dieterici's experimental fundamental spectral color process functions V, G, and R. Note that individual G functions for König and Dieterici have been plotted. The dashed curve represents a fundamental function of an observer with impaired color vision (König 1886).

Like Pilgrim (see entry in this chapter), Schrödinger showed that all mixtures of spectral lights can be represented in a three-dimensional manifold: "Now, since each light can be regarded with sufficient approximation as a mixture of *n* pure spectral lights, it follows that *the pure spectral colors and the binary mixtures exhaust completely the manifold of colors.* . . . Thereby, the dimension number 3 is finally guaranteed" (Schrödinger 1920, p. 136; trans. by R.G.K.; emphasis original).

Figure 6.16.
Schrödinger's (1920) sketch of the Farbentüte *(spectrum bag), originating at black O. The sectors ROG and VOI are planes. The locus of spectral colors is represented by the curved line.*

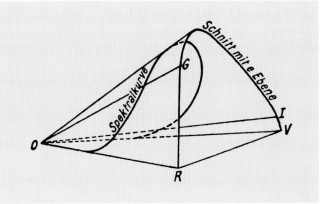

Schrödinger showed that the geometric model of the manifold can be constructed from results of color stimulus mixture experiments, as Maxwell and Helmholtz had envisaged and as Pilgrim had calculated. His conceptual drawing of the model is shown in figure 6.16. He suggested a cone-shaped paper bag, later termed "Schrödinger's spectrum bag," as a visual aid to understanding the space. The chromatic color diagrams of Maxwell, Helmholtz, and Pilgrim are modified views of the (irregular) semicircular frontal plane of the bag.

Another theoretically important Schrödinger contribution was the definition (following Ostwald) of ideal object color reflectance functions representing maximally chromatic colors. They have sharp transitions between 1.0 and 0, and either one or two transitions. Typical examples are shown in figure 6.17.

Reflectance functions of real objects are always curved in various, usually irregular ways that approach the ideal forms. Experimental determination of fundamental color functions of observers and the definition of ideal object reflectance functions allow the calculation of a geometric solid into which all objects with real reflectance functions have to fall. Such solids, in different configurations, were calculated by Luther, Nyberg, Rösch, and MacAdam (see entries in this chapter).

Schrödinger also looked at the metric of color, that is, the change in the form of his stimulus space to create a 1:1 relationship between distances in the new space and those perceived between color experiences from samples. For this purpose, he applied square-root compression to the stimulus space (although he considered it inadequate).

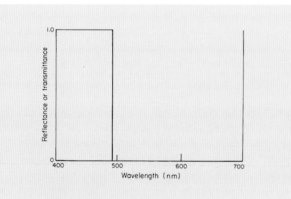

Figure 6.17a.

Reflectance function of a color from the surface of the object color solid, single transition. Mirror image functions are also possible (Schrödinger 1920).

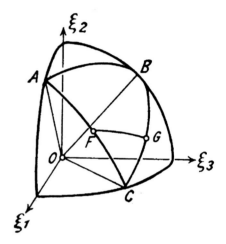

Figure 6.18.

Conceptual sketch by Schrödinger of the result of nonlinear transformation of color stimulus space to a space in form of a sphere octant in which equal distances represent equal perceptual differences (Schrödinger 1920).

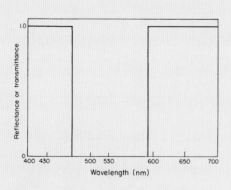

Figure 6.17b.

Reflectance function of a color from the surface of the object color solid, double transition. Reverse functions are also possible (Schrödinger 1920).

The resulting orthogonal space, called ξ (xi) space, forms a sphere octant (figure 6.18) comparable in form to Doppler's (see entry in chapter 10). In this space, shortest distances between colors of equal intensity form geodesics and are calculated from line elements. For example, figure 6.18 shows the geodesic line between the two equally intense colors F and G on the surface of the octant.

With his efforts, Schrödinger put the geometry of color stimulus space on a solid footing. He also attempted a perceptually uniform version of that space. His contribution helped pave the way for further developments, although experimental data at his disposal were limited in quality and quantity to wavelength discrimination data (he had no quantitative perceptual data of color differences).

ROBERT LUTHER 1927

R. Luther, *Aus dem Gebiete der Farbreiz-Metrik*, 1927

Robert Luther (1868 -1945) was a German natural scientist who grew up in Russia (and was distantly related to the religious reformer Martin Luther). In the first decade of the twentieth century, he worked with Ostwald (see entry in chapter 10). Later he headed the photography institute of the Technical University Dresden. His greatest contributions helped elucidate the photographic processes. He developed a surface of optimal object color stimuli.

Luther's stimulus solid is located in a modified tristimulus space where the axes represent what he called the three color moments: $M_{blue-yellow} = 1/3B - 1/3R$, $M_{green-red} = 2/3G - 2/3R$, and $S = R + G + B$, where R, G, and B are tristimulus values related to the red, green, and blue color fundamentals (S is a power moment not related directly to any perceptual property). He also defined luminance as $H = R + G$. For the three fundamentals, Luther used his own data calculated ("not without unavoidable arbitrariness") from various, previously published, experimental data, and an (implied) equal energy light source.

The first of the color moments represents a red–green opponent axis, the second a blue–yellow axis, and the third (following Helmholtz; see entry in this chapter) a summed energy axis. With these moments, Luther hoped to help bridge the gap between the Young-Helmholtz and the Hering color theories.

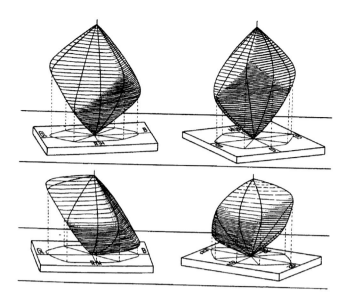

Figure 6.20.

Two views of Luther's (1927) optimal object color solid in the color-moment space $M_{blue-yellow}$, $M_{green-red}$, and S (top) and two views (bottom) of the solid in the $M_{blue-yellow}$, $M_{green-red}$, and H color moment space.

After the 1931 development of the CIE colorimetric system, the moments were redefined in terms of CIE tristimulus values as follows: $M_{1(red-green)} = X - Y$, $M_{2(yellow-blue)} = Y - Z$, $S = X + Y + Z$. The resulting representation became known as the Luther-Nyberg color solid. It is a color stimulus space linearly related to the CIE tristimulus space. Today, the Luther-Nyberg solid is a curiosity, usually no longer included in treatises on color science.

NIKOLAUS D. NYBERG 1928

N.D. Nyberg, *Zum Aufbau des Farbkörpers im Raume aller Lichtempfindungen*, 1928

Nikolaus D. Nyberg was a Russian natural scientist, active in the first half of the twentieth century. At nearly the same time as Luther (see entry in this chapter), he independently developed the mathematics for an object color stimulus solid. Nyberg realized that the shape of the optimal object stimulus solid depends to a considerable extent on the spectral properties of the light source. But he also realized that the appearance of objects is not influenced by the light source to the same degree as stimulus calculations indicate (due to adaptation).

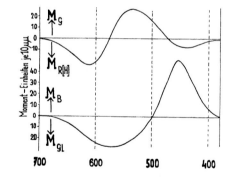

Figure 6.19.
Spectral functions defining the chromatic axes of Luther's (1927) space.

The chromatic moment functions (figure 6.19) are different from Helmholtz's proposal (see Ludwig Pilgrim, this chapter). Optimal object color stimuli are located on the surface of the solid and all other possible object color stimuli, in its interior. Luther created four views of the solid (figure 6.20). The top two are based on the chromatic moment functions and the summed energy function; the bottom two are based on the chromatic moment functions and the luminance function. Intervals in the solids do not agree with perceptually uniform intervals. Elliot Q. Adams (see entry in chapter 7) later defined opponent functions that are nonlinearly related to tristimulus values based on the CIE colorimetric system and that better agree with perceptual data.

Nyberg used König and Dieterici's color fundamentals (König 1892) for the calculations. "For convenience's sake," he used an equal energy light source while indicating: "Any effort to develop a pigment color solid without considering the spectral content of the illumination must be absolutely erroneous, because in this circumstance we cannot say anything about the perception of pigment colors" (Nyberg

1928, p. 410, trans. by R.G.K.). He drew the optimal object stimulus solid in orthogonal *R, G, B* space, looking down along the equal energy axis (figure 6.21). Sections *A* and *B*$_{1-4}$ are at right angle to the achromatic equal energy axis (not shown) and give an impression of the surface form. Contour *V* illustrates the location of Ostwald's *Vollfarben*.

Nyberg's solid is equivalent to Luther's solid, except that he defined his in terms of the *R, G, B* space, not moment axes (see Robert Luther, this chapter). Like Luther's, Nyberg's is a stimulus space, so distances within the solid have no quantitative relationship to those in a perceptually uniform color solid.

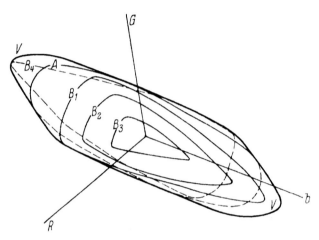

Figure 6.21.

View of the Nyberg (1928) color solid down the equal energy axis in the R, G, B *color stimulus space.*

SIGFRIED RÖSCH   1928
S. Rösch, *Die Kennzeichnung der Farben*, 1928

Sigfried Rösch (1899–1984) was a German mineralogist with an interest in color order. Rösch used red (*R*), green (*G*), and blue (*B*) color fundamental functions based on the data of König (1892) and others to calculate tristimulus values. He raised his optimal object color stimulus solid over a Maxwellian triangular diagram of the kind depicted by Pilgrim (see figure 6.14) representing the fundamental functions. The third dimension in the solid illustrated in figure 6.22 is in lightness units, an approximation of luminous reflectance.

Figure 6.23 is a projection of the solid of figure 6.22 onto the basis triangle. *W* denotes white and falls on the gravimetric center of the triangle, indicating that Rösch used balanced functions and the equal energy illuminant. The interior lines are isoluminance (equal lightness) contours. They show the empirical fact that saturated yellows (halfway between *R* and *G*) have high lightness values, and saturated blues have low ones. Rösch used the term *Farbenreizkörper* (color stimulus solid) for representations of this kind because they provide no information concerning the difference thresholds of the perception.

Rösch's representation contains the same intrinsic information as the earlier ones by Luther and Nyberg and the later one by MacAdam, who used CIE color-matching functions (see entries in this chapter). One system can be converted into the other by linear transformation (assuming identical fundamentals data).

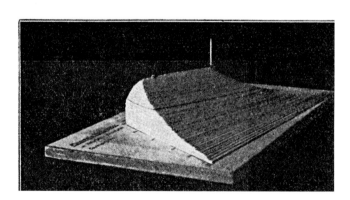

Figure 6.22.

Image of Rösch's (1928) model of the optimal object color solid raised over a Maxwell triangle representing the fundamental sensations R, G, *and* B.

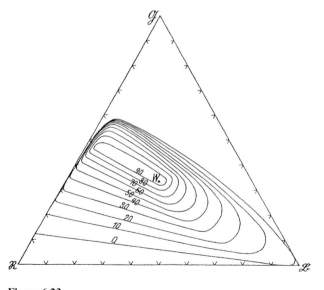

Figure 6.23.

Projection of the solid of figure 6.22 onto the basis triangle. The contours show loci of stimuli of constant lightness value (Rösch 1928).

CIE X, Y, Z COLOR STIMULUS SPACE   1931

Commission Internationale de l'Éclairage, *CIE proceedings,* 1931

In the early 1920s, the **International Commission on Illumination** was formed to develop international standard methods in lighting and associated subjects. Its official name was the French version, with the resulting acronym CIE. Since then, the CIE has been the preeminent technical organization concerned with standardization of color stimulus measurement and technically important aspects of color vision. Over the years, it has issued recommendations on how to physically measure color stimuli, defined standard observers and a colorimetric system, and proposed formulas that relate psychophysical data to various kinds of average perceptual judgments.

Under international pressure to recommend a standardized colorimetric system, in 1931, the CIE proposed a trichromatic system based on standardized measurement of light sources, reflectance properties of objects, and a numerically defined standard observer (CIE color-matching functions). The system is based on the work of Newton, Young, Maxwell, Helmholtz, Schrödinger (see entries in this chapter), and others.

The 1931 standard observer is based on experimental color-matching data of English physicists J. Guild and W. D. Wright. Their experimental data of two groups of observers were averaged and, using Grassmann's laws (see entry in this chapter), linearly transformed to nonreal primary lights **X, Y,** and **Z,** selected so that it is possible to express in the system all spectral lights with positive values. Based on a recommendation by Deane Brewster Judd (see entry in chapter 7), CIE decided that one of the three primaries, the **Y** primary, would agree with the CIE spectral luminance function standardized in 1924. In this manner, implicit brightness or lightness as well as chromatic information is contained in the resulting three CIE tristimulus values *X, Y,* and *Z.*

These values are obtained from integrating the area under the spectral curves. In practice, they are usually calculated by multiplying (for object colors) at each wavelength the spectral reflectance factor (describing the object), with the spectral power of the light source in which the object is viewed and, in turn, with the three color-matching functions $\bar{x}, \bar{y}, \bar{z}$; figure 6.24), and summing the values.

This process is equivalent to determining the relative catch of light quanta by the three cone types in the eyes of the standard observer. The process amounts to a dimension reduction (e.g., from 31, one value every 10 nm from 400 to 700 nm, to 3) of the spectral stimulus functions using the color-matching functions as filters. One of its most important aspects is that it defines metamers for the standard observer; that is, it allows identification of different spectral stimulus functions that cannot be distinguished by the standard observer because they have identical tristimulus values.

The resulting three numbers uniquely describe a particular stimulus (object reflectance and light source spectral power distribution) as "viewed" by the standard observer in unspecified surround conditions. For a given object (or light), these numbers identify a point in the three-dimensional *X, Y, Z* color space whose axes are taken to be orthogonal. However, mathematically speaking, *X, Y, Z* color space is not properly orthogonal; the only truly orthogonal version of a stimulus space based on the CIE standard observer is Cohen's space (see entry in this chapter).

Figure 6.25 shows a view of the solid of optimal object color stimuli in CIE standard daylight D6500 in the CIE tristimulus space. The null point represents an object with zero reflectance (taken to be black) across the spectrum; its

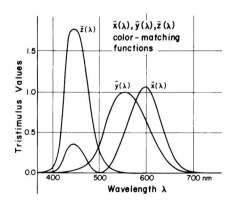

Figure 6.24.

IE 2° standard observer color-matching functions $\bar{x}$-, $\bar{y}$, $\bar{z}$ obtained from linear transformation of average experimental color-matching functions determined by Guild and Wright (Wyszecki and Stiles 1982). Reprinted with permission.

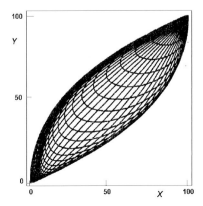

Figure 6.25.

Projection of the optimal object color solid onto the X, Y plane of the CIE tristimulus space: 10° standard observer, illuminant D65 (modified from Koenderink and Kappers 1996). Reprinted with permission.

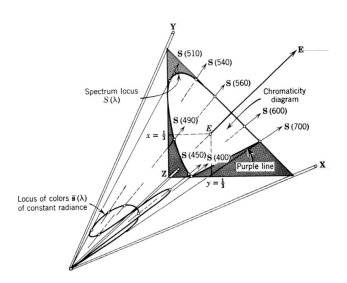

Figure 6.26.

CIE tristimulus space in the form of a right-angled triangular pyramid. The section shows the CIE chromaticity diagram with axes represented by the chromaticity coordinates x and y. E represents the equal energy stimulus. Stimulus vectors S of selected wavelengths are also illustrated. (Wyszecki and Stiles 1982). Reprinted with permission.

opposite represents an object with 100% reflectance (taken to be white). All objects with flat reflectance curves (or metamers for them, grays) fall on the surface of or fill the solid. The points representing all other possible object reflectances fill the solid.

The solid's shape changes as a function of the observer and the light source. Without loss of intrinsic information, the shape of the object color stimulus solid can be changed by mathematical linear transformation of visual functions (see Luther, Nyberg, Rösch, and MacAdam entries in this chapter).

A linear transformation in which the axes are not orthogonal is illustrated in figure 6.26. Here, a plane sectioning the space contains the spectral trace and purple line in the so-called CIE chromaticity diagram. All possible colors fall on or within its horseshoe-shaped outline. The diagram's axes are denoted by *x* and *y*, with *x* calculated by dividing the *X* tristimulus value by the sum of all three tristimulus values (and equivalently for *y*). This diagram is a modified geometrical representation of the Maxwell triangle, but it is based on color-matching functions.

In 1964, a second standard observer was defined that relates to a larger visual field than the 1931 observer; by now, there is an extended series of standard light source data.

The *X, Y, Z* stimulus solid is contained in a stimulus space and does not express anything specific about appearance. The only information related to humans is contained in the color-matching functions, the result of matching spectral stimuli. Examples of discrepancies between colorimetric and perceptual results are the following:

1. Depending on the circumstances of surround and illumination, any object with a flat (or near-flat) reflectance function can be seen as black, any kind of gray, or even white.

2. Stimuli represented by points elsewhere in the solid can have many different chromatic appearances depending on conditions. However, for a given set of conditions, color stimuli are ordered in this space in spectral sequence, with logical continuation of the purple mixtures from stimuli at both ends of the spectrum, as well as by their relative energy level.

The CIE colorimetric system is widely used in technological applications, and depending on the application, the use is more or less successful. Although the system's original application was merely to determine matching color stimuli, its use has expanded. Now there are color-difference and color-appearance models, based on assumption of a simple relationship between stimuli and perceptual responses. But in recent years, it has become increasingly evident that the true relationship is very complex. Modeling of these complex relationships requires much empirical adjustment.

HANS NEUGEBAUER   1935

H. E. J. Neugebauer, *Zur Theorie des Mehrfarbenbuchdruckes,* 1935

German technical scientist **Hans Neugebauer,** a student of Luther (see entry in this chapter), is best known for his equations that predict the result obtained in halftone printing with three chromatic pigments and a black pigment (1935). In his dissertation, he described an object color stimulus solid from pigments used in halftone printing. He published key excerpts from the dissertation in two papers in 1937. Neugebauer later emigrated from Germany to the United States, where he worked for the Xerox Corporation.

Neugebauer's color stimulus solid is based on Luther's version of an opponent optimal object color stimulus solid. Neugebauer described the constants 1/3 and 2/3 used by Luther for weighting the two chromatic axes as "completely arbitrary, replaceable by other constants of choice" (Neugebauer 1937a, p. 45, trans. by R.G.K.). He showed that the color solid based on three idealized halftone printing pigments forms a parallelepiped (see glossary; figure 6.27), where O represents black; R_{12}, R_{13}, and R_{23} represent the locations of the idealized primaries yellow, blue-green, and purple; R_1, R_2, and R_3 represent the locations of their 1:1 mixtures (red, green, and blue); and W represents white, the result of mixture of these three secondaries. For his calculations, he used H. E. Ives's (1915) color-matching functions for white light of 5,000°K.

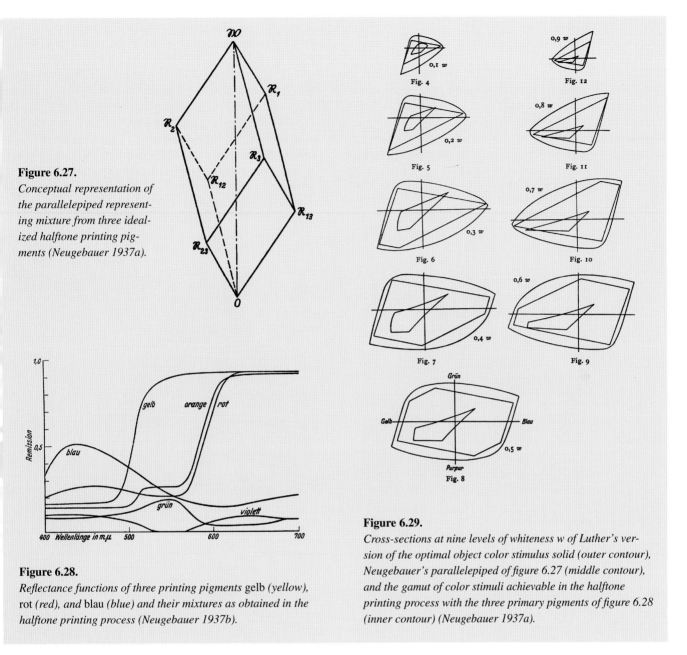

Figure 6.27.

Conceptual representation of the parallelepiped representing mixture from three idealized halftone printing pigments (Neugebauer 1937a).

Figure 6.28.

Reflectance functions of three printing pigments gelb (yellow), rot (red), and blau (blue) and their mixtures as obtained in the halftone printing process (Neugebauer 1937b).

Figure 6.29.

Cross-sections at nine levels of whiteness w of Luther's version of the optimal object color stimulus solid (outer contour), Neugebauer's parallelepiped of figure 6.27 (middle contour), and the gamut of color stimuli achievable in the halftone printing process with the three primary pigments of figure 6.28 (inner contour) (Neugebauer 1937a).

Neugebauer's goal was to determine what part of such a rhombohedron is filled out by the solid formed from actual pigments in halftone printing with some partitive mixture (no overlap of dots) and some subtractive mixture (dots overlapping; see chapter 9). Here, in addition to yellow, Neugebauer used red and blue rather than magenta and cyan.

Figure 6.28 illustrates the reflectance functions of the pigments including their subtractive mixture. Neugebauer illustrated the result in the form of cross sections at constant sums of Luther's solid's R, G, and B values (whiteness w, not identical with CIE luminous reflectance).

Figure 6.29 illustrates nine cross sections of the respective solids at 0.1-unit increments of whiteness w. The outer contour represents Luther's optimal object color stimulus

solid, the middle contour represents the parallelepiped of figure 6.27 based on idealized halftone pigments, and the inner contour represents the results of partially partitive (see glossary), partially subtractive mixture from the pigments of figure 6.28. Due to the subtractive mixture, most lines of the third solid are curved.

Neugebauer's is an effort to demonstrate the relationship between the optimal object color stimulus solid and the solid (three-dimensional gamut) achievable with real pigments in a somewhat simplistic model of halftone printing. In 1943, Nickerson and Newhall built cross sections that were comparable (to a degree) but that represented the solid of the Munsell system samples in the larger perceptual solid of optimal object colors (see Dorothy Nickerson, chapter 7, esp. figure. 7.6).

DAVID LEWIS MACADAM   1935

D. L. MacAdam, *Maximum visual efficiency of colored materials*, 1935

Among the early efforts of the American physicist **David Lewis MacAdam** (1910–1998) was the calculation of object color solids similar to Rösch's of 1928 (see entry in this chapter) but based on the newly developed CIE 2° standard observer and different illuminant functions. MacAdam calculated optimal object color solids for daylight illuminant C and tungsten light illuminant A. Instead of a Maxwell triangle representing the CIE primaries, used by Rösch, MacAdam used the CIE chromaticity diagram as the basis plane and luminous reflectance *Y* for the third dimension.

But like Rösch, MacAdam prepared a physical model. Figure 6.30 is a stereoscopic image of the model for CIE illuminant C. The shape of the surface generated in this manner depends on the illuminant used in the calculation of the tristimulus values. This is evident from figures 6.31 and 6.32, which illustrate projections of isoluminant lines of the solids for the respective illuminants onto the chromaticity diagram. The figures, representing stimuli absorbed by the cones, indicate large differences in the shape of the two solids. However, natural adaptation of an observer to the light from the two different sources results in much reduced perceptual effects. Distances within the spaces are not quantitatively related to perceptual distances.

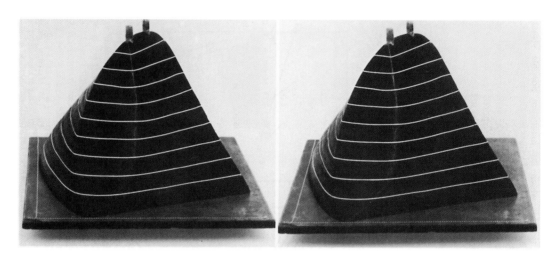

Figure 6.30.
Stereoscopic image pair of the model of MacAdam's (1935) object color solid for daylight illuminant C, 2° observer. Reprinted with permission.

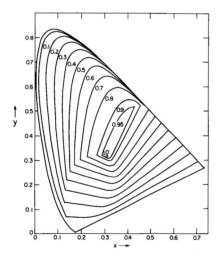

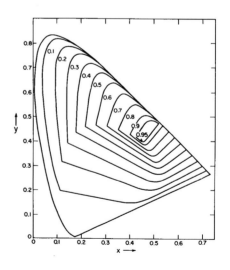

Figure 6.31.
Projections of isoluminant lines of MacAdam's (1935) object color solid for illuminant C and the 2° observer. Reprinted with permission.

Figure 6.32.
Projections of isoluminant lines of MacAdam's (1935) object color solid for illuminant A and the 2° observer. Reprinted with permission.

ABSORPTION SPACE   1944, 1966

R. H. Park and E. I. Stearns, *Spectrophotometric formulation*, 1944
E. Allen, *Basic equations used in computer color matching*, 1966

A version of a color stimulus space that has found extensive application since the middle of the twentieth century is an **absorption space**, a pseudo-color space based on absorption functions derived for opaque materials, not on reflectance or emission functions. Commercial matching of object colors (e.g., of textile fabrics or paints) is an art that has been practiced for more than 2,000 years. Performing it well requires much practical experience. It requires knowledge of and experience with subtractive color mixture (mixture of dyes or pigments) and the specific properties of the colorants involved.

A matching formula is usually a compromise between several requirements: economics, closeness of match, fastness properties in the medium of the final product, and application properties during the coloration process. Metamerism (see glossary) and its related problem of color constancy or inconstancy complicate matching: A formula that one observer finds to match a reference in one kind of light may not be seen to match by another observer in the same light or by both observers in a different light. One or both of reference and matching samples may change appearance when changing the light source in which they are viewed.

Aristotle had already commented on change of colored materials' appearance in different lights. And today, as mentioned in chapter 1, good matches are usually required in three kinds of lights, typically daylight, tungsten light, and a common kind of fluorescent light. This is a formidable problem, and since the 1960s, computers increasingly have been used to determine the optimum colorant formulation. But good matching in three lights does not guarantee constancy of appearance in those lights (the reference material itself may be color inconstant). Color constancy as a match requirement is a relatively new issue.

Matching by mathematical procedure requires knowledge of the relationship between colorant concentration and resulting reflectance. In 1931, German physicists P. Kubelka and F. Munk worked out this relationship for specific conditions. Because of its simplicity, their formula is widely used in computer-assisted recipe formulation (though in most practical conditions, with less than perfect accuracy). The additive function of reflectance as expressed in the Kubelka-Munk formula is the nonlinear inverse of reflectance factor, as illustrated in figure 6.33. Pseudo-tristimulus values are calculated by using the absorption (K/S) values in place of reflectance values.

Several different mathematical methods for calculating a colorant formulation have been developed. The methods used in modern commercial software are usually trade secrets. The earliest technical paper on the subject is from 1944 by American chemical engineers R. H. Park and E. I. Stearns. But applying

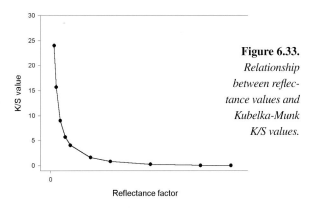

Figure 6.33.
Relationship between reflectance values and Kubelka-Munk K/S values.

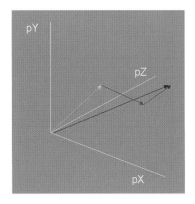

Figure 6.34.
Conceptual sketch of matching of a "brown" pseudo-color stimulus (brown vector) with pseudo-color stimuli of appropriate "yellow," "red" and "blue" colorants (yellow, red, and blue vectors). The vector lengths represent the required colorant concentrations.

their method was not practical until the availability of computers and fast, reliable equipment to measure spectral reflectance.

American chemist E. Allen described an algorithm for solving the matching problem for textile dyes in 1966, and for pigments in paints and plastics in 1974. For the initial solution of the matching problem he applied a matrix algebra method in which the vector of the reference in the pseudo-tristimulus space is matched by vector addition using vectors representing given concentrations of individual dyes (figure 6.34). In a second step, an iteration process is used to improve the formulation until it meets quality requirements or the iteration is stopped after a number of unsuccessful attempts (say, attempting to match a blue color with a yellow and two red dyes).

Qualitatively, the pseudo-tristimulus object color solid is the upside-down version of the solid in the conventional CIE tristimulus space. It originates at zero colorant concentration, that is, at the color of the substrate, usually white. Because the reflectance factors (0 and 1.0) of optimal object colors do not have corresponding K/S values, the optimal object color absorption space can only be approximated. Compression of reflectance produces a solid that is more (upside-down) pear shaped than spindle shaped.

JOZEF B. COHEN   1982

J. B. Cohen and W. E. Kappauf, *Metameric color stimuli, fundamental metamers, and Wyszecki's metameric blacks*, 1982

American psychologist **Jozef B. Cohen** (1921–1995) developed a mathematical procedure for extracting spectral fundamental functions from metameric color stimuli. According to a conjecture by the German-Canadian physicist Günter Wyszecki, metameric stimuli (stimuli of differing spectral composition, but seen as having the same color when viewed by a standard observer) are composed of a spectral fundamental function common to all of them and differing spectral metameric black functions.

A basic component of Cohen's procedure is calculating the invariant matrix R with which the fundamental function can be extracted from any reflectance function. Cohen plotted the cone mantle (also see Ludwig Pilgrim, this chapter) formed by the fundamentals of spectral stimuli. He termed the resulting space "fundamental color space" because it derives from the metric fundamentals of metameric sets. All metameric blacks have tristimulus values of 0, 0, 0.

Figure 6.35 illustrates spectral stimulus vectors of various wavelengths and a smooth connecting line, raised over the corresponding chromatic plane. Figure 6.36 shows the spectral vectors and the lines connecting their endpoints in four different views of the fundamental space. Cohen did not calculate the optimal object color solid for the fundamental color space; Jan J. Koenderink and Andrea J. van Doorn (see figure 8.3) calculated it in 2003.

As a linear transformation of the CIE tristimulus space, Cohen's fundamental color space is not perceptually uniform. But it is the only tristimulus-based space meeting the geometric requirement of orthonormality for Euclidean spaces.

Cohen also projected Munsell chip reflectance functions into the fundamental space (Burns et al. 1990). The chips used for this purpose are those of Munsell Color Company's 1929 *Munsell Book of Color* (pre-Renotations). Figure 6.37 illustrates a selection of Munsell chip reflectance functions in fundamental space. The figure demonstrates that there is no close relationship between perceptual steps as demonstrated in the Munsell system and distances in that space.

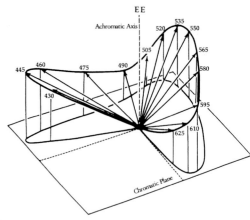

Figure 6.35.

Spectral vectors in Cohen and Kappauf's (1985) invariant fundamental color space and their projection onto the chromatic plane. EE is the equal energy axis. Reprinted with permission.

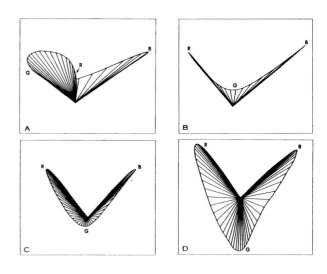

Figure 6.36.

Four different views of the envelope of spectral vectors in fundamental color space (Cohen and Kappauf 1985). Reprinted with permission.

Figure 6.37.

Reflectance functions of Munsell Color Company's 1929 color chips projected into the fundamental color space formed by a red, a violet, and the luminous reflectance axis. The locations of the equal energy and the illuminant C vector are also shown. From Burns et al. (1990). Reprinted with permission.

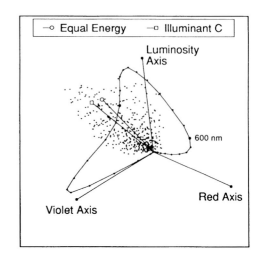

CONE SPACE, CONE EXCITATION, and CONE CONTRAST DIAGRAMS 1979

D. I. A. MacLeod and R. M. Boynton, *A chromaticity diagram showing cone excitation by stimuli of equal luminance*, 1979

Color-matching functions can be linearly transformed in a way that the resulting functions represent cone sensitivity functions: Spectral light absorption functions of the cone types are identified by *L*, *M*, and *S*, for long-, medium-, and short-wave sensitive. The cone functions represent the spectral sensitivity of the three cone types corrected for absorption in eye media. Therefore, the optimal color stimulus solid can also be represented in the *L*, *M*, *S* cone space (figure 6.38). Unlike in the *X*, *Y*, *Z* space, where the *Y* dimension is related to perceptual lightness, none of the dimensions in the *L*, *M*, *S* space is directly related to a perceptual attribute.

To aid data representation in vision research, in 1979, American psychologists Donald I. A. MacLeod and Robert L. Boynton proposed a cone excitation diagram, illustrated in figure 6.39. In the diagram, the curved spectral trace is, as usual, closed by the line of purple colors. The coordinates of the diagram are calculated as follows: $l = L/(L + M)$, $s = S/(L + M)$. *L* and *M* are defined so that their sum equals the CIE luminance function. The diagram is somewhat impractical because the white point W is near the abscissa. In figure 6.39, dashed lines are examples of lines of constant *S* and *L* cone excitations, respectively.

Because the visual color experience depends to a large degree on the surround, the cone contrast space and chromatic diagram was developed. Both are normalized for cone excitations of the surround by dividing the cone excitations of the test field by the related ones of the surround. Figure 6.40 illustrates an example of an *L*, *M* cone contrast diagram with the origin at the *L* and *M* excitations of the surround. The diagram indicates the relative difference between surround and test field. The cone contrast diagram is used widely in color-vision science. All three kinds of representations are stimulus representations only.

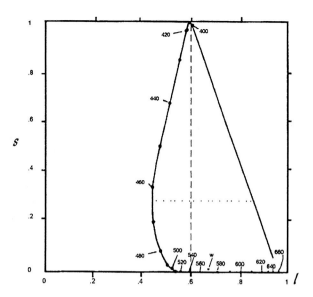

Figure 6.39.
Cone excitation diagram by MacLeod and Boynton (1979), with the trace of spectral stimuli and the connecting line of purple colors. The white point of the diagram (W) is located close to the horizontal axis. Reprinted with permission.

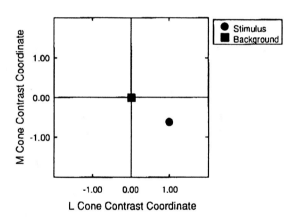

Figure 6.40.
Example of a relative cone contrast diagram normalized for L and M cone activity due to the surround stimulus (center). The cone activity of a central stimulus is shown relative to that of the surround.

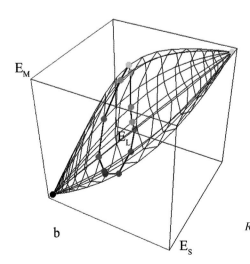

Figure 6.38.
Optimal object color stimulus solid in the L, M, S color space. The colored dots represent the stimuli of a color circle made from colored papers (von Campenhausen and Schramme 2003). Reprinted with permission.

DKL COLOR SPACE   1984

A.M. Derrington, J. Krauskopf, and P. Lennie, *Chromatic mechanism in lateral geniculate nucleus of macaque*, 1984

In the early 1980s, psychologists **Andrew M. Derrington**, **John Krauskopf**, and **Peter Lennie** performed experiments measuring the response to light stimuli of certain neuronal cells in the brains of macaques, which are a genus of monkeys (*Macaca*). The rhesus monkey (*M. mulatta*) has a trichromatic color-vision system considered to be similar to that of humans. The neuronal cells involved are in the lateral geniculate nuclei (LGN), which are two masses of mammalian brain cells that are way stations for signals generated in retinal cells. From the LGN, signals are channeled to the main visual processing center of the brain at the back of the head.

Macaques were exposed to light of given wavelengths, and the electrochemical response to these lights of cells in the LGN was measured. Different cells were found to respond in a different manner to various wavelengths. To be able to represent the results systematically in terms of luminance and chromaticness of the lights, the authors devised a color stimulus space that became known as the DKL space based on the chromaticity coordinate system introduced by MacLeod and Boynton [see Cone Space, Cone Excitation, and Cone Contrast Diagrams, this chapter] "because its axes correspond to the 'cardinal directions' of colour space identified by Krauskopf et al." (Derrington et al. 1984, p. 243).

A luminance axis (luminance expressed as $L + M$) was raised perpendicular to the chromatic plane. The axes of the chromaticity coordinate system are represented by stimuli resulting in constant $L + M$ values (called R and G by the authors), on the one hand, and constant S (called B) values, on the other. "The colour space can be thought of as a sphere with the white point at its centre; the azimuth [angle θ] and elevation [angle φ] of a point can be thought of as its longitude and latitude respectively" (p. 245).

The resulting space is illustrated in figure 6.41. The axes of the chromatic plane are illustrated in the CIE chromaticity diagram in figure 6.42. Derrington and colleagues found that many cells in LGN fall into two classes, one they described as $L - M$ [$R - G$] or the reverse, and the other they describe as $S - (L + M)$ [$B - (R + G)$] or the reverse. Such cells have chromatic opponency of some kind. However, the diagram's axes have not been found to relate to unique hue experiences of the average human.

In the DKL space, the basis of the visual system model has been moved forward from the cones in the retina to cells in LGN. Studies have since shown that this simple separation into two cell types with specific sensitivity in terms of the DKL space no longer exists in the visual center at the back of the brain. The DKL space continues to be used to represent stimulus data in vision science.

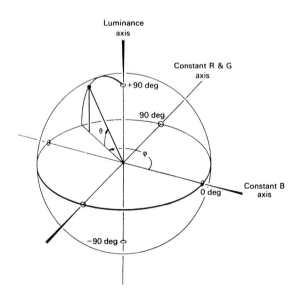

Figure 6.41.

Conceptual sketch of the spherical DKL space (Derrington et al. 1984).

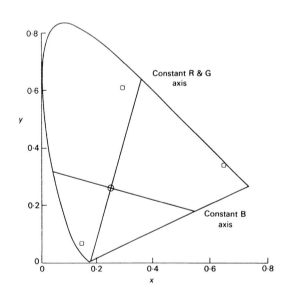

Figure 6.42.

CIE chromaticity diagram illustrating the chromatic axes of the DKL space. The small circles near the periphery indicate the chromaticities of the primary lights used in the experiment (Derrington et al. 1984). Reprinted with permission.

Notes

1. Important early measurements of cone sensitivities were made in Helmholtz's laboratory by König and Dieterici (1886, 1892).

2. The *Rig Veda* is a series of Hindu sacred verses written before 2000 B.C.

References

Allen, E. 1966. Basic equations used in computer color matching, *Journal of the Optical Society of America* 56:1256–1259.

Burns, S. A., **J. B. Cohen**, and **E. N. Kuznetsov**. 1990. The Munsell color system in fundamental color space, *Color Research and Application* 15:25–51.

Campenhausen, C. von, and **J. Schramme**. 2003. Die Entwicklung von ästhetischen Farbsystemen zum physiologischen, in *Zu Bedeutung und Wirkung der Farbenlehre Wilhelm Ostwalds*, Bendin, E., ed., Dresden: Verlag Phänomen Farbe.

Cohen, J. B., and **W. E. Kappauf**. 1982. Metameric color stimuli, fundamental metamers, and Wyszecki's metameric blacks, *American Journal of Psychology* 95:537–564.

Cohen, J. B., and **W. E. Kappauf**. 1985. Color mixture and fundamental metamers: theory, algebra, geometry, application, *American Journal of Psychology* 98:171–259.

Commission Internationale de l'Éclairage. 1986. *Colorimetry*, 2nd ed., No. 15.2, Wien: Commission Internationale de l'Éclairage.

Derrington, A. M., **J. Krauskopf**, and **P. Lennie**. 1984. Chromatic mechanisms in lateral geniculate nucleus of macaque, *Journal of Physiology* 357:241–265.

Descartes, R. 1650. *Compendium musicae*, Utrecht.

Grassmann, H. G. 1853. Zur Theorie der Farbenmischung, *Annalen der Physik* 89:69–84.

Grimaldi, F. M. 1665. *Physicomathesis de lumine, coloribus, et iride, aliisque adnexis libri duo*, Bologna: Benatius.

Helmholtz, H. von. 1856–1866. *Handbuch der physiologischen Optik*, Hamburg: Voss; 2nd ed., 1896, 3rd ed., 1909 (for English trans., see Southhall [1924]).

Ives, H. E. 1915. The transformation of color-mixture equations from one system to another, *Journal of the Franklin Institute* 180:673-701.

Koenderink, J. J., and **A. Kappers**. 1996. *Color space*, Report 19/96, Bielefeld, Germany: Center for Interdisciplinary Research.

Koenderink, J. J., and **van Dorn, A. J.** 2003. Perspectives on colour space, in *Colour perception*, Mausfeld, R., Heyer, D., eds., Oxford: Oxford University Press.

König, A. 1886. Die Grundempfindungen und ihre Intensitäts-Vertheilung im Spectrum (in Gemeinschaft mit Conrad Dieterici), *Sitzungsberichte der Akademie der Wissenschaften in Berlin*, 805–829.

König, A. 1892. Die Grundempfindungen in normalen und anomalen Farbensystemen und ihre Intensitätsvertheilung im Spectrum (in Gemeinschaft mit Conrad Dieterici), *Zeitschrift für die Physiologie und Psychologie der Sinnesorgane* 4:241–347.

Kubelka, P., and **F. Munk**. 1931. Ein Beitrag zur Optik der Farbanstriche, *Zeitschrift für technische Physik* 12:593–601.

Luther, R. 1927. Aus dem Gebiete der Farbreiz-Metrik, *Zeitschrift für technische Physik* 8:540–558.

MacAdam, D. L. 1935. Maximum visual efficiency of colored materials, *Journal of the Optical Society of America* 25:361–367.

MacLeod, D. I. A., and **R. M. Boynton**. 1979. A chromaticity diagram showing cone excitation by stimuli of equal luminance, *Journal of the Optical Society of America* 69:1183–1185.

Maxwell, J. C. 1857. Experiments on colour, as perceived by the eye, with remarks on colour-blindness, *Transactions of the Royal Society of Edinburgh* 21:275–297.

Maxwell, J. C. 1860. On the theory of compound colours, and the relations of the colours of the spectrum, *Proceedings of the Royal Society (London)* 10:57–84.

Mollon, J. D. 2003. Introduction to *Normal and defective colour vision*, Mollon, J. D., J. Pokorny, K. Knoblauch eds., Oxford: Oxford University Press.

Munsell, A. H. 1907. *Atlas of the color-solid*, Malden, MA: Wadsworth-Holland.

Munsell Color Company. 1929. *Munsell book of color*, Baltimore, MD: Munsell Color Company.

Neugebauer, H. E. J. 1935. *Zur Theorie des Mehrfarbenbuchdrucks*, Dissertation Technische Hochschule Dresden, Leipzig: Frommhold & Wendler.

Neugebauer, H. E. J. 1937a. Über den Körper optimaler Pigmente, *Zeitschrift für wissenschaftliche Photographie* 36:43–50.

Neugebauer, H. E. J. 1937b. Die theoretischen Grundlagen des Mehrfarbenbuchdrucks, *Zeitschrift für wissenschaftliche Photographie* 36:51–67.

Newton, I. 1704. *Opticks*, London: Smith and Walford.

Nyberg, N. D. 1928. Zum Aufbau des Farbkörpers im Raume aller Lichtempfindungen, *Zeitschrift für technische Physik* 52:406–411.

Palmer, G. 1777. *Theory of colours and vision*, London: Leacroft.

Park, R. H., and **E. I. Stearns**. 1944. Spectrophotometric formulation, *Journal of the Optical Society of America* 34:112–113.

Pilgrim, L. 1901. *Einige Aufgaben der Wellen- und Farbenlehre des Lichts*, Cannstadt, Germany: Mann.

Rösch. S. 1928. Die Kennzeichnung der Farben, *Physikalische Zeitschrift* 29:83–91.

Schrödinger, E. 1920. Grundlinien einer Theorie der Farbmetrik im Tagessehen, *Annalen der Physik* 63:297–447, 481–520. English trans., Outline of a theory of color measurement for daylight vision, in *Sources of color science*, MacAdam, D. L., ed., Cambridge, MA: MIT Press, 1970.

Southhall, J. P.C. (ed.). 1924. *Helmholtz's treatise on physiological optics*, trans. from the 3rd ed., Washington, DC: Optical Society of America.

Wyszecki, G., and **W. S. Stiles**. 1982. *Color science*, 2nd ed., New York: Wiley.

Young, T. 1802a. The Bakerian Lecture: on the theory of light and colours, *Philosophical Transactions of the Royal Society of London* 92:12–48.

Young, T. 1802b. An account of some cases of the production of colours, not hitherto described, *Philosophical Transactions of the Royal Society of London* 92:387–397.

Young, T. 1807. *A course of lectures on natural philosophy and the mechanical arts*, London: Johnson.

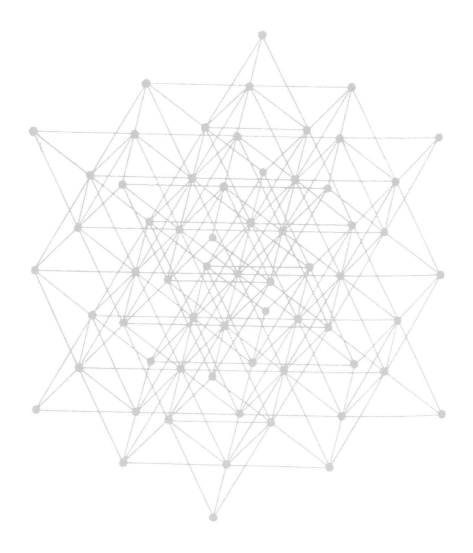

CHAPTER 7

**CONNECTING EMPIRICAL PERCEPTUAL DATA
WITH PSYCHOPHYSICAL SCALES**

A description of attempts to connect empirical perceptual data with psychophysical scaling data in order to predict average perceived differences.

Chapter 6 presents the development of several kinds of object color stimulus spaces and scales adjusted to be in closer agreement with average human difference scaling results. Prediction of human color experiences based on such information has so far been only modestly successful, for reasons given in chapter 6. But such prediction has been the goal of much effort in color science and technology in the twentieth century.

Driving the prediction effort has been the desire to explain color vision and, more important, technological and commercial needs. Of high importance among these needs has been the ability to predict the average perceived difference between two colored materials from knowledge of their spectral properties. Initial theoretical efforts in this matter have been based on so-called line elements, as mentioned in chapter 6 (also see the MacAdam's Empirical Line Element, this chapter). The poor agreement between predictions based on line elements and average perceptual judgments of the size of differences soon resulted in new color vision models and empirical efforts to find high correlations between psychophysical systems and perceptual judgments.

All models presented in this chapter rely on the International Commission on Illumination (CIE) colorimetric system (see CIE X, Y, Z Color Stimulus Space, chapter 6). Modifications to the basic system, such as opponent-color operations, are based on conjectures and empirical fitting. Perceptual data on which the fitting is based have a high level of individual variability and a considerable level of variability between data sets.

The reasons for the variability are not known with certainty. But it is known that the "normal" human color-vision system is complex and is likely to operate, to a degree, based on empirical principles: In certain circumstances, the system assigns color experiences based on analysis of the complete field of vision and greatest likelihood, not on an unchanging relationship between stimulus and response. Consciously perceived color has also been shown to be influenced by high-level "executive" systems of the brain. Recent data also show a surprisingly large variability in color stimuli that different color-normal observers pick as having unique hues. So a standard observer may not be very meaningful.

From Judd to the OSA Uniform Color Scales

Deane Brewster Judd (see entry in this chapter) was a key figure of the twentieth-century effort to empirically relate the colorimetric psychophysical system and perceptual data. In 1932, he assembled several kinds of data related to perceptual color distances and attempted to find a linear transformation of a (pre-CIE system) chromaticity diagram so that geometric distances in the transformed diagram would be equal to judged perceptual distances.

This work became the model for additional efforts by him and other researchers. In 1935, Judd transformed his diagram onto a Maxwell triangle (see figure 7.2) and a year later calculated the form and size (see figure 7.3) of his uniform diagram's equal circles in the newly established CIE chromaticity diagram. The resulting ellipses show the degree by which the CIE chromaticity diagram differs from a perceptually more uniform one. CIE later standardized a slightly modified Judd equation as the CIELUV formula (see CIELAB, CIELUV, this chapter).

American lighting engineer Elliot Q. Adams, who in 1924 had developed a theory of color vision based on an opponent color model, proposed in 1942 that the CIE tristimulus values be used as input data in his theory. To obtain better agreement between perceptual and colorimetric data, the tristimulus values were to use the nonlinear scaling developed for predicting the Munsell value scale from luminous reflectance (tristimulus value) Y.

Based on Adams's proposal, color scientist Dorothy Nickerson wrote a color-difference formula and used it with some other formulas in the first known formal comparison of visually judged differences between colored textile samples and differences calculated from their colorimetric data. Nickerson was also deeply involved in developing the Munsell Renotations (see entry in this chapter), and she calculated the information necessary to build the model of the first attempt at an objective uniform color space, based on the Munsell system (see figure 7.7).

An entirely different approach was taken by David Lewis MacAdam, who, as a physicist, distrusted perceptual distance judgments. He believed that he could use statistical information about the degree of error in an observer's multiple attempts at matching the color of a light field with three primary lights to show the stimuli change needed for just-noticeable differences (JNDs).

MacAdam's empirical line element (as distinct from Erwin Schrödinger's theoretical one; see entry in chapter 6) soon became the basis of color-difference calculation using graphical charts or new, complex color-difference formulas. When tested against sets of perceptual judgments involving sample pairs, these formulas did not perform well, however.

Recently, the cause was determined to be a basic difference between color-matching error and JNDs.

When in the later 1930s reflectance of the samples of Munsell Color Company's 1929 *Munsell Book of Color* was measured and the resulting chromaticity coordinates plotted, it was found that their progress in the diagram was often jagged. This contradicted the expectation that perceptually equally spaced samples would show regularly spaced intervals in a psychophysical diagram.

When during World War II the U.S. National Research Council searched for some efforts not related to the war, Judd and other researchers proposed the development of a perceptually uniform color space and a related color-difference formula. They made the case that the research would benefit several industrial sectors (including, e.g., the producers of cotton). The council delegated the research to the Optical Society of America (OSA), which formed a subcommittee on improved spacing of the Munsell system.

Researchers plotted sample data in the CIE chromaticity diagram and deduced smoothed lines of constant hue and chroma. They produced and visually evaluated new samples representing the smoothed lines. By repeating these steps, researchers obtained averaged smooth lines supported by visual judgments. Locations of colors where samples cannot be realized with pigments were extrapolated along smooth lines to the object color limits.

Based on the final plots, aim colors, known as Munsell Renotations, were determined. Painted samples of about 55% of the total of aim colors, realizable with pigments, were then generated. The resulting atlas, which was produced by the Munsell Color Company in matte and glossy chips, continues to be commercially available. The Japan Color Research Institute also produced atlases of Munsell Renotation colors in different arrangements, one with 5,000 samples (*Chroma Cosmos 5000*, 1978) and one with 707 samples (*Chromaton 707*, 1982).

As an extensive and newly researched collection of samples, the Munsell atlas based on Renotations rapidly became the preeminent color order system. When Nickerson compared the hue, chroma, and value (lightness) scales of the Munsell system, she found variation in the perceptual size of the units. Based on Nickerson's findings, Judd attempted to calculate the implicit space but could not because the total hue angle of the chromatic plane of the space was slightly more than twice that of a circle. Judd created the term "hue superimportance" for this fact.

In the 1930s, an approach different from Munsell's notations was pursued in Germany, where a need was perceived for replacing Wilhelm Ostwald's system (see entry in chapter 10) with a colorimetrically specified one that would be ordered in a general sense according to hue, saturation and lightness. After World War II, this effort was resumed in West Germany under the guidance of Manfred Richter.

Given the long history in Germany of placing the most intense hue samples (full colors) on a single plane, the attributes for the system were selected to be hue, saturation, and relative lightness (relative to the luminous reflectance of the most saturated sample for each hue). Steps along the attributes were selected to be approximately perceptually uniform in a constant hue plane. The system became a German standard (DIN 6164), and atlases with increasing numbers of samples were published in the second half of the twentieth century.

The shortcomings of the Munsell system (not isotropic) came to be understood in the 1940s, and in 1947 the OSA formed a second committee to attempt to develop a system uniform in many directions within the solid, not just along attributes. Thus began the almost 30-year history of development of the OSA Uniform Color Scales (OSA-UCS).

As a result of its experimental work, the committee confirmed the existence of hue superimportance with the consequence that a perceptually uniform color order system in Euclidean form is not possible. Rather than end the effort without fruit, the committee decided to issue an atlas representing the closest a Euclidean system can come to represent a uniform color space. The crystalline interior structure of this system has seven cleavage planes that reveal color series in new ways the committee hoped might be of interest to designers and artists.

Color space and difference formulas

Availability of comparatively inexpensive reflectance measuring instruments and computation devices beginning in the mid-1950s renewed the desire of industrial segments for objective color quality control. New sets of perceptual data were assembled and tested against a variety of color-difference formula proposals. In 1973, more than 15 different formulas were found to have industrial application in the United States. None of the formulas had a predictive accuracy of higher than about 50% for average perceived small color differences. Nevertheless, the CIE was under some pressure to make a recommendation that could reduce the growing confusion regarding color communication.

It was found that there were two major interest groups: lighting engineers interested in a space and formula that is a linear transformation of the CIE chromaticity diagram, and industrial colorists interested in a formula predicting perceived average color differences better than any existing one. As a result, the CIE decided to recommend "for the purpose of uniformity of usage" two different formulas: a space and formula known as CIELUV for lighting engineers, and a space and formula known as CIELAB, a cube-root version of the Adams-Nickerson formula of 1936, for the coloration industries. Both are Euclidean and thereby do not consider hue superimportance. They also do not consider the Helmholtz-Kohlrausch effect (see glossary) and the lightness and chromatic crispening effects. Despite these shortcomings, both formulas achieved the purpose of increasing uniformity of usage as other formulas began to fade away.

The (approximately50%) degree of accuracy of prediction obtained with CIELAB remained unsatisfactory for industrial colorists, and several attempts were made to improve it. The CIELAB formula allows the mathematical splitting of total calculated color differences in terms of basic attributes. In the 1970s, English industrial colorists Keith McLaren and later Roderick McDonald began empirical investigations into the potential to improve accuracy by separately weighting hue, chroma, and lightness difference components calculated from the CIELAB formula.[1]

These measures proved successful and raised prediction accuracy to about 65%. Among the issues they addressed are hue superimportance and chromatic crispening. The resulting Color Measurement Committee (CMC) color-difference formula (Clark et al. 1984) is in wide industrial use today.

Since 1980, several new sets of color-difference judgments have been published and analyzed in terms of prediction by formulas. These experiments have been performed with different methods and under different conditions. Perhaps unsurprisingly, they tend to only modestly agree, and, with rare exceptions, the average predictability has remained at about 65%.

These data sets have been searched for regularities with increasing sophistication, and in 2001, the CIE recommended CIEDE2000, a new color-difference formula, which empirically addresses several of these regularities. However, when tested with historical perceptual data, the formula has not improved prediction accuracy more than by, at best, a minor degree.

Modeling of perceptual color-difference data with psychophysical systems suffers from two issues:

1. Perceptual data vary widely by experiment for reasons that are not understood, but enough is known to show that a set of data is specific to observer panels, surround conditions, and experimental procedures.

2. The psychophysical systems in use (e.g., CIELAB) represent an outdated historical basis.

Further study of the effects of observer panels, surround, and experimental methods on average judgments as well as improved psychophysical models may make it possible in the future to have higher levels of prediction. But they may come at the cost of accurate applicability of the resulting uniform space to increasingly narrower conditions.

D. B. Judd, Chromaticity sensibility to stimulus differences, 1932
A Maxwell triangle yielding uniform chromaticity scales, 1935

American psychologist **Deane Brewster Judd** (1900–1972) worked during most of his professional life at the U.S. National Bureau of Standards, now the National Institute of Standards and Testing, where he was responsible for activities resulting in color standards and color technology recommendations. One of his lifelong interests was the development of a uniform (isotropic) color space. For 25 years, he was chairman of the Optical Society of America Committee on Uniform Color Scales (see OSA Uniform Color Scales, this chapter).

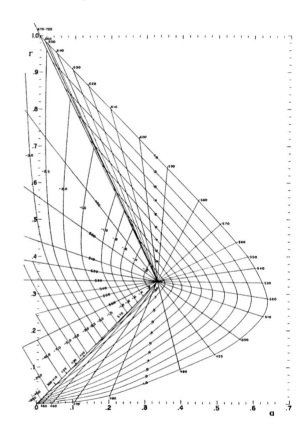

Figure 7.1.

Projective transformation of the CIE chromaticity diagram meant to result in proportionality of geometric distances with visual distances. Radial lines indicate constant dominant wavelengths; curved lines, constant colorimetric purity (Judd 1932). Reprinted with permission

In a 1932 paper, Judd linearly transformed a CIE-like chromaticity diagram based on color-matching functions that had been recommended by the OSA in 1922, before the development of the CIE colorimetric system in 1931. He intended the transformation to represent perceptual differences of his own threshold data and other published data as quantitatively as possible (figure 7.1).

The diagram contains radial lines of constant dominant wavelength (lines of approximately constant hue) and contours of implicit constant colorimetric purity.

In 1935, Judd published a modified version of a uniform chromaticity diagram in a Maxwell triangle (figure 7.2). Transformed into a different reference frame, this diagram became in 1960 the basis for the CIE u, v diagram (in turn modified in 1976 to become the CIELUV color space and difference formula). In 1936, Judd published a graph of the CIE 1931 chromaticity diagram with ellipses that represent in the 1935 diagram circles of equal size, that is, contours of equal distance from their centers (figure 7.3). These ellipses are intended to represent just-noticeable chromatic color differences, enlarged 100 times. They were the first to explicitly illustrate the perceptual nonuniformity of the CIE chromaticity diagram.

Judd discovered what he termed the "hue superimportance" effect: In a Euclidean space, unit difference contours are always elongated in the direction of chromatic intensity regardless of the size of the unit difference. As a result, a perceptually uniform color solid and the associated space cannot be Euclidean in nature (see figure 1.10).

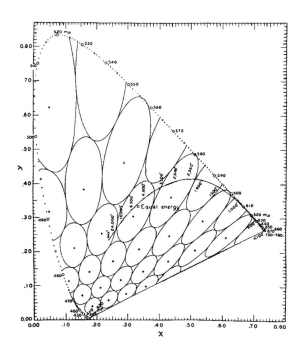

Figure 7.3.

CIE chromaticity diagram with ellipses representing circles of equal size (implied constant perceptual distances from the center points) in the 1935 diagram of figure 7.2. The ellipses are 100 times enlarged from the original ones that represent threshold differences (Judd 1936). Reprinted with permission.

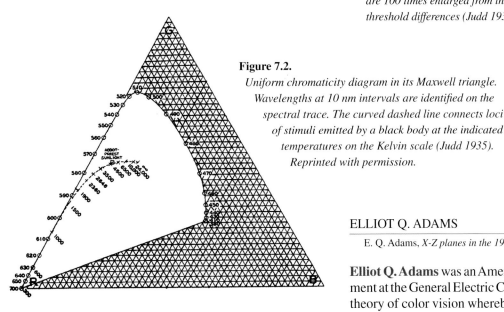

Figure 7.2.

Uniform chromaticity diagram in its Maxwell triangle. Wavelengths at 10 nm intervals are identified on the spectral trace. The curved dashed line connects loci of stimuli emitted by a black body at the indicated temperatures on the Kelvin scale (Judd 1935). Reprinted with permission.

As is now known, the effect means that for a unit perceived hue difference, the change in cone activation needs to be about half that for a unit chroma difference of the same perceptual magnitude. Judd was the first to use ΔE as a symbol for the value of the total calculated difference between two color stimuli. And he was instrumental in developing the National Bureau of Standards color-difference formula of 1939, a formula that did not see much practical application.

ELLIOT Q. ADAMS   1942

E. Q. Adams, *X-Z planes in the 1931 I.C.I. system of colorimetry*, 1942

Elliot Q. Adams was an American engineer in lamp development at the General Electric Company. In 1923, he proposed a theory of color vision whereby inhibitory nerve connections in the retina form opposing chromatic signals. In 1942, he introduced a "chromance" and a "chromatic-value" diagram based on a psychophysical interpretation of his color-vision theory. He took the recently developed CIE color-matching functions (see CIE X, Y, Z Color Stimulus Space, chapter 6) to represent the fundamental color-vision functions.

The chromance diagram is a linear opponent-color diagram plotting the CIE 1931 tristimulus value differences $Z - Y$ and $X - Y$. It updates Robert Luther's chromatic-moment diagram (see entry in chapter 6). In Adams's diagram, achromatic colors "will lie at the origin of coordinates, surrounded at increasing distances by the colors of increasing chroma (saturation)" (Adams 1942, p. 169).

Adams defined the chromatic value diagram as follows: "The ratios X_C/Y_C and Z_C/Y_C depend only on the chromaticity of the sample. A plot of these two ratios may therefore be referred to as a constant-brightness chromaticity diagram" (Adams 1942, p. 170). However, as the Munsell system shows, the relationship between luminous reflectance Y and perceived lightness is not linear. Adams decided to apply the same compression mechanism to all three trichromatic processes:

> While there is a one-to-one relationship between luminous reflectance (albedo) and Munsell value (lightness), the relation is not a mathematically simple one for the conditions under which Munsell colors are compared. Munsell, Sloan and Godlove determined empirically a value (lightness) function on which the present Munsell colors are based. *If this function be applied to all three components of reflectance*, $\rho_X \equiv X_C$, $\rho_Y \equiv Y_C$, $\rho_Z \equiv Z_C$, the resulting V_X, V_Y, V_Z will uniquely define any color. V_Y is the Munsell value (lightness) and the differences $(V_X - V_Y)$, $(V_Z - V_Y)$ will be zero for a physically neutral (achromatic) color. For a chromatic color they may be designated collectively as "chromatic value." (Adams 1942, pp. 170–171, emphasis added)

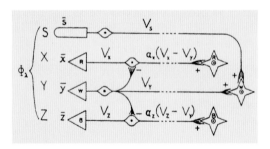

Figure 7.4.
Model of Adams's (1942) color vision theory. The rods and triangles (for cones) represent the four types of primary visual cells in the retina. Interaction between R, W, *and* B *cone types results in the opponent signals and the brightness signal illustrated with starlike symbols. Reprinted with permission.*

The color-vision model source of the diagram is illustrated in figure 7.4. In figure 7.5, the *Munsell Book of Color* (Munsell Color Company 1929) samples at value 5 are plotted in the diagram, with resulting constant hue lines reasonably straight, and constant chroma contours reasonably circular. The normalized, appropriately weighted mathematical interpretation of this system is known as Adams-Nickerson color space and difference formula (see Dorothy Nickerson, this chapter).

Adams conjectured that the mathematical function describing the relationship between tristimulus value Y and the Munsell values should be applied also to the X and Z tristimulus values. That has never been explicitly tested and remains in place in the CIELAB formula but has been found not to be accurate for the Munsell Renotations and the OSA-UCS data (Kuehni 2003).

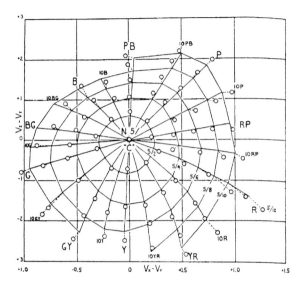

Figure 7.5.
Plot of colors of Munsell value 5 of Munsell Color Company's 1929 Munsell Book of Color *in Adams's chromatic value diagram. The lines represent preliminary smoothing proposals from the OSA Subcommittee on the Spacing of Munsell Colors (Adams 1942). Reprinted with permission.*

DOROTHY NICKERSON   1936, 1943, 1944

D. Nickerson, *The specification of color tolerances*, 1936
D. Nickerson and S. M. Newhall, *A psychological color solid*, 1943
D. Nickerson and K. F. Stultz, *Color tolerance specification*, 1944

Dorothy Nickerson (1900–1985) worked at the Munsell Color Company and the U.S. Department of Agriculture. She was a member of the OSA subcommittee developing the Munsell Renotations and the Committee on Uniform Color Scales (see Munsell Renotations and OSA Uniform Color Scales, in this chapter).

In 1936, in an effort to express light-induced fading of colored textiles in an objective, quantitative manner, Nickerson developed the first color-difference formula. Based on the Munsell system, the formula considered the different unit magnitude of Munsell hue, value, and chroma scales. She weighted hue differences by a chroma-related value that normalizes the increasing distance between neighboring radial lines of constant hue in the Munsell system. The weighted attribute differences were combined by simple addition:

$$I = C/5(2\Delta H) + 6\Delta V + 3\Delta C$$

where I is the difference index value, and ΔC, ΔH, and ΔV are, respectively, the differences in Munsell chroma, hue, and value between reference and test sample. To use the formula, the samples before and after fading were matched with chips of the Munsell system. Then the index was calculated based on the Munsell distances between the matched colors.

Judd later showed that the formula implies a total hue angle more than twice that of a circle and thereby demonstrates what he called "hue superimportance." As mentioned above, the formula is not expressible in a Euclidean system so that geometric distances are proportional to perceptual distances.

In 1943, using data newly available in connection with the development of the Munsell Renotations, Nickerson and Newhall (1943) calculated the dimensions of a psychological uniform color solid. It was to fulfill the following requirements:

> The dimensional scales would be calibrated in perceptually uniform steps; the units of the several scales would be equated; the surface of the solid would represent all colors of maximum saturation; the volume would be representative of all colors which are perceptibly different; the conditions of stimulation or viewing would be prescribed; and finally, the scales would be standardized in terms of a generally recognized psychophysical system. (p. 419)

By normalizing the Munsell attributes as in the above formula and using the extrapolations of Munsell chromatic scales to the limits of the optimal object color solid in the CIE chromaticity diagram (as calculated by MacAdam; see entry in chapter 6), Nickerson and Newhall calculated the contours of the implicit perceptually uniform solid. This was done for two sizes of unit color difference: (1) with the Munsell double chroma step difference as unit, and (2) at the just-noticeable-difference level. The result was expressed as cross sections (figure 7.6) as well as physical models of the solids (figure 7.7).

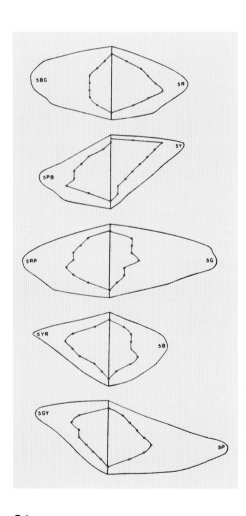

Figure 7.7.
Models of the Nickerson and Newhall uniform color solid: left, at the level of Munsell Book of Color *double chroma step differences; right, at the just-noticeable difference level (Nickerson and Newhall 1943). Reprinted with permission.*

Figure 7.6.
Vertical sections through Nickerson and Newhall's (1943) uniform color solid at the level of Munsell double chroma step differences. The outer contours represent the optimal object color surface; the inner contours the space filled by samples of the 1929 Munsell Book of Color. *Reprinted with permission.*

The two solids are different, showing that the shape of the solid varies as a function of the unit difference size it is based on. These solids are the first attempt at representing object color spaces taken to be perceptually uniform in Euclidean models. From today's point of view, they are not uniform because the hue superimportance effect was not taken into consideration.

In 1944, Nickerson and Stultz, an Army engineer, investigated the usefulness of various proposals of color-difference formulas that use the CIE colorimetric system for object color quality control. Among them was a formula they modified from an earlier version of a color-difference formula based on Adams's color space (see entry in this chapter). It has the form

$$\Delta E = \{(0.23\Delta V_{\mathrm{Y}})^2 + [\Delta(V_{\mathrm{X}} - V_{\mathrm{Y}})]^2 + [0.4\Delta(V_{\mathrm{Z}} - V_{\mathrm{Y}})]^2\}^{1/2},$$

where ΔE is the calculated total color difference and V_{X}, V_{Y}, V_{Z} are, respectively, the Adams chromatic values derived from the related CIE tristimulus values X, Y, and Z (see Elliot Q. Adams, this chapter). The total color difference in this formula is Euclidean distance (square root of the sum of the squares) of the lightness difference (first term) and the two opponent color chromatic differences. For 11 color series of approximately 20 visually graded samples each, the 1936 Nickerson formula provided the best agreement between data calculated from reflectance measurements and average perceived difference data. The formula became influential as the Adams-Nickerson formula. In 1968 in England, its slightly revised form was standardized as that ANLAB 40 formula and became the basis for the CIE 1976 $L^*a^*b^*$ (CIELAB) color space and difference formulas (see CIELAB, CIELUV, this chapter).

MacAdam's Empirical Line Element   1942

L. MacAdam, *Visual sensitivities to color differences in daylight*, 1942
Specification of color differences, 1965

Line elements are mathematical models of the relationship between fundamental color responses (cone responses, sometimes of color-matching functions) and just-noticeable differences (JND) in color. The line-element concept was introduced in 1891 by Hermann von Helmholtz (see entry in chapter 6). In a Euclidean space, a JND can be specified as the square root of the sum of the squares of the three axes' differences. There is the question of the size of increments along the three axes required to result in a JND. Helmholtz assumed fundamental color space to be Euclidean and the JND increments, the so-called Weber fractions, to be identical for all three fundamental processes. His line element had the form

$$ds = [(dR/R)^2 + (dG/G)^2 + (dB/B)^2]^{1/2}$$

where ds is the JND, R, G, and B are fundamental color vision processes, and dR, dG, and dB are the changes in the processes required for a JND. Comparing the predictions of this model with experimental results showed that reality is more complex. Line elements of increasing complexity were developed by Schrödinger in 1920 and Walter S. Stiles in 1946. These efforts began as theoretical models, sometimes modified based on comparison with experimental data. In 1942, the American physicist David Lewis MacAdam

(see entry in chapter 6) published the results of extensive determinations of the color-matching error (CME) of a single observer. He argued that one could achieve a pure, empirical basis for threshold color differences without involving judgmental effects with this method: statistically determining the variability in repeatedly matching the color in one field against a reference color in an adjacent second field by adjusting three dials of a visual colorimeter.

At different times, MacAdam estimated JND to be two or three standard deviations of the CME represented by the results of the experiment. The CME was determined for a single observer at 25 locations in a constant luminance chromaticity diagram. The results were plotted in the CIE chromaticity

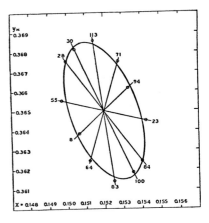

Figure 7.8.
Color-matching errors determined in different directions from the reference stimulus in the center and fitted ellipse, plotted in a portion of the CIE chromaticity diagram (MacAdam 1942). Reprinted with permission.

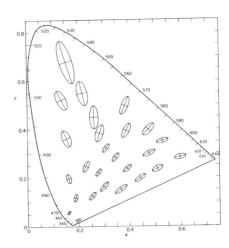

Figure 7.9.
MacAdam ellipses 10 times enlarged, plotted in the CIE 2° observer chromaticity diagram (MacAdam 1942). Reprinted with permission.

diagram (see CIE X, Y, Z Color Stimulus Space, chapter 6), which became known as MacAdam ellipses because its CME contours are elongated ellipses (figures 7.8 and 7.9). Using principles of geometry, MacAdam defined the ellipses as

$$ds = [g_{11}(dU_1)^2 + 2g_{12}dU_1dU_2 + g_{22}(dU_2)^2]^{1/2}$$

where *ds* is the distance between the center point and any point on the ellipse and coefficients g_{ik} are continuous functions of the space coordinates U_1 and U_2. An additional coordinate and three g_{ik} functions are required to describe a three-dimensional line element that includes lightness. MacAdam plotted the three continuous g_{ik} functions representing the ellipses in the chromaticity diagram (figure 7.10).

Chart sets were available on which the reference and the test samples were plotted and the color differences were measured with a ruler. Using scissors and paste, MacAdam assembled a uniform plane in which circles of equal size would represent his ellipses (figure 7.11). But to represent his empirical data with high accuracy requires a highly non-linear and non-Euclidean transformation model of the CIE chromaticity diagram.

In the 1960s, the Dutch psychologist L. F. C. Friele developed two color-difference formulas based on MacAdam's line element and Georg Elias Müller's color vision model[2] (Friele 1961, 1965). They are known as the Friele-MacAdam-Chickering[3] or FMC formulas I and II. The main difference between his two versions is their treatment of lightness differences. These formulas and graphical methods were in use for some 20 years, but they performed poorly in comparison tests of various formulas against experimental average perceptual data of suprathreshold small color differences. As is now known, this was because CME is not an accurate predictor of perceived color difference, including those at the JND level.[4]

In 1965, as a result of stepwise linear regression calculations, MacAdam published the ξ, η geodesic chromaticity diagram, an optimized nonlinear transformation of the CIE chromaticity diagram to produce as nearly as possible circles of equal size from his ellipses (figure 7.12). Lines of constant *x* and *y* chromaticity coordinates are drawn in the diagram, and the purple line is curved.

The MacAdam data remain important for builders of comprehensive models of color vision. Later additional determinations of CME with more observers and different instrumentation have shown considerable individual variation in CME results. MacAdam's ellipses are not universal and do not represent average perceived small color differences between object colors.

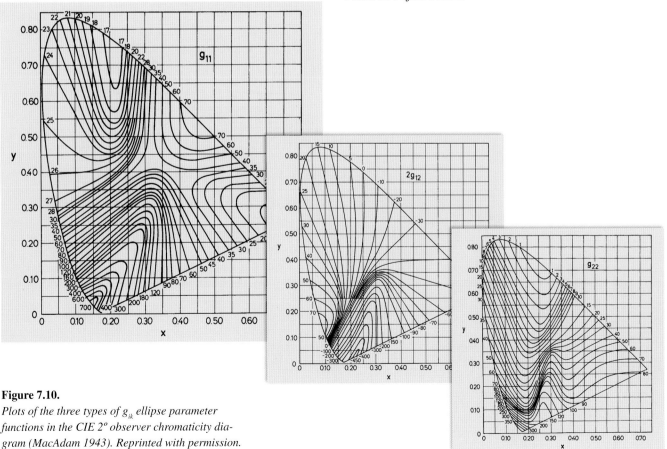

Figure 7.10.

Plots of the three types of g_{ik} ellipse parameter functions in the CIE 2° observer chromaticity diagram (MacAdam 1943). Reprinted with permission.

Figure 7.11.

Scissors-and-paste model of the transformation of the CIE chromaticity diagram required to result in conversion of the MacAdam ellipses into circles of equal size in a plane (MacAdam 1944). Reprinted with permission.

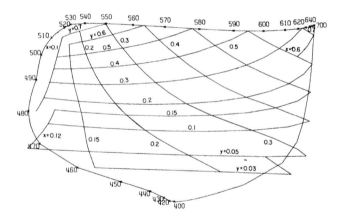

Figure 7.12.

Geodesic chromaticity diagram calculated from nonlinear transformation of the CIE chromaticity diagram to convert the MacAdam ellipses to circles of most nearly identical size (MacAdam 1965). Reprinted with permission.

MUNSELL RENOTATIONS (modern Munsell system)   1943

S. M. Newhall, D. Nickerson, and D. B. Judd, *Final report of the OSA Subcommittee on the Spacing of Munsell Colors*, 1943

The **Munsell Color Company** was formed in 1918, the year of Munsell's death. The U.S. National Bureau of Standards, which had shown interest in the system for color specification, offered research cooperation to improve it. In 1919 and 1926, the bureau made reflectance measurements of Munsell chips from Munsell's 1915 *Atlas of the Munsell Color System*. The bureau also advised on visual evaluation methods and data interpretation.

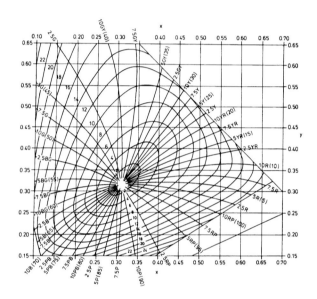

Figure 7.13.

Portion of the CIE chromaticity diagram with radial lines of constant Munsell hue and ovoids of constant chroma at value 6 (U.S. Dept. of Agriculture 1964)

In 1929, a revised and expanded version of the Munsell system was issued as Munsell Color Company's *Munsell Book of Color*, and in 1935, reflectance measurements were made, with results published in 1940 in form of colorimetric data expressed in the 1931 CIE colorimetric system. Plotting the data showed a considerable number of irregularities that could not be substantiated with new visual observations.

Spacing of the series of gray chips that represent the value scale had been reestablished against black, middle gray, as well as white backgrounds for the 1929 edition. The results varied significantly and were averaged into the Munsell value scale, mathematically expressed by Judd with a five-term polynomial formula as a function of the CIE luminous reflectance values Y. For chromatic colors, Munsell value is also based on luminous reflectance; that is, the Helmholtz-Kohlrausch effect has not been taken into account.

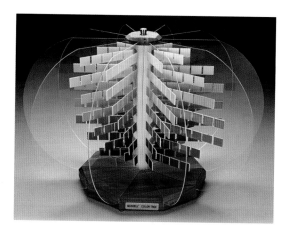

Figure 7.15.

Munsell Color Tree model.
Photo courtesy Gretag Macbeth Corp.

In 1937, the Optical Society of America (OSA) formed a subcommittee charged with improving the spacing of Munsell colors. In 1943, the subcommittee published its final report defining 2,746 chromatic and 9 achromatic color stimuli in the CIE colorimetric system (Newhall et al. 1943). The members arrived at the aim color points first by plotting old and new visual data in the CIE chromaticity diagram. Then, after confirming the results with additional visual evaluations, they smoothed the contours into slightly curved lines of constant hue and ovoids of constant chroma. Both kinds

of lines were extrapolated to the optimal object color limits. Data for one value level are illustrated in figure 7.13.

The final data collection, known as Munsell Renotations, represents the aim values for the Munsell system color chips since that time. A glossy version of the *Munsell Book of Color*, based on the Renotations, contains approximately 55% of the specified colors, the remaining number not physically realizable with pigments. The difference units for the three attributes hue, value, and chroma are the same. As a result of the system's radial organization, as well as of the hue superimportance effect, the system is not isotropic. In addition, the averaged value scale is not accurate for any real surround.

In modern format, the system is available as matte (1,250 chips) or glossy samples (1,550 samples; Gretag-Macbeth n.d.), with an expanded 31- and 37-step gray scale, respectively (figure 7.14). A "nearly neutrals" set of more than 1,100 pastel color samples is also offered. An abbreviated model of the system is available as the Munsell Color Tree (figure 7.15). Under the name *SCOT-Munsell System*, the Renotations have also been issued illustrated in form of 2,034 dyed swatches of polyester fabric.

The Munsell system continues to be one of the most widely used physical implementations of an approximately uniform color order system, despite its mentioned shortcomings.

Figure 7.14.
Modern version of the Munsell Book of Color. *Photo courtesy Gretag-Macbeth Corp.*

MANFRED RICHTER and DIN 6164   1947, 1952

M. Richter, Empfindungsgemäss gleichabständigen Farbsystems, 1947

In the 1930s, members of the German Institute for Standardization (Deutsches Institut für Normung, DIN) recognized the need for "a color-order system more practical than the Ostwald system" in use during the previous 20 years. In 1938, the responsibility for this project was given to **Manfred Richter** (1905–1990), who worked at the organization that preceded today's Bundesanstalt für Materialprüfung (Federal Office of Material Testing).

Discussions with potential users of the system revealed that the majority were interested in an organization according to hue, saturation, and relative brightness with, unlike the Ostwald system (see entry in chapter 10), visually equidistant steps. In addition, the system was to be defined colorimetrically.

Test results from 70 observers evaluating differences between Ostwald color chips were used to arrive at a uniformly scaled hue circle, followed by saturation and lightness scaling (Richter 1950). By 1947, Richter had assembled an atlas with 12 planes of constant hue containing chips painted with pigments dispersed in a glue binder, the so-called Richter system (figure 7.16).

The schematic perceptual diagram of the DIN system is illustrated in figure 7.17. Data for 24 equally distant hues were defined in the CIE chromaticity diagram and designated with the letter *T* and a number from 1 to 24. Lines of constant hue were arbitrarily defined as straight lines between the full hue sample points and the illuminant point in the CIE chromaticity diagram (figure 7.18). As a result, perceived hue of the "constant hue" samples varies slightly for most hues as a function of saturation and lightness.

As it does in the DIN system, the term "saturation" can mean the content of chromatic color in relation to lightness. The result is that, unlike in the Munsell system, where chroma is independent of lightness and the system's geometric form is cylindrical, here the perceptual solid has the form of a conical section (see figure 7.17).

In the DIN system, six levels of saturation were experimentally determined at one level of relative lightness to arrive at the psychophysical definition of constant saturation *S*. The results were extrapolated to other lightness levels. Colors of constant saturation are located on roughly elliptical contours in the CIE chromaticity diagram (see figure 7.18). To preserve the customary color circle with all full colors on the same horizontal plane, DIN 6164 employs a relative lightness scale based on the 1879 formula of the Belgian psychologist Joseph R. L. Delboeuf. The lightness of the optimal object color of a given hue is the reference against which the relative lightness is calculated. Constants used in the formula are based on experimental data. DIN relative lightness *D* has a scale from 0 to 10.

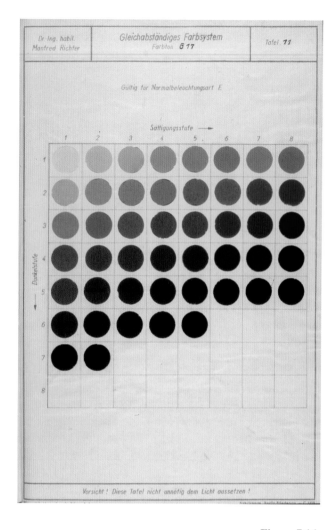

Figure 7.16.
Chart of hue G17 (corresponding to DIN 6164 hue T17) of the Richter (1947) system.

The chip aim colors have been defined for the CIE 2° standard observer and daylight illuminants C and D65. DIN 6164 has also been defined in the CIELUV system (see CIELAB, CIELUV, this chapter), and a corresponding color-difference formula was developed. Figure 7.19 illustrates a cross section (hues T1 and T16) of the DIN 6164 system illustrated in CIELUV color space. The DIN system is intended to be approximately uniform only within its three attributes and not isotropic, as figure 7.19 shows. Like the Munsell system, its chromaticness dimension is open-ended.

The first version of DIN 6164 (Deutsches Institut für Normung 1952), in the form of gelatin filters for use in visual photometers, was issued in 1952 (figure 7.20). In 1962, an edition of 587 matte color chips in 24 constant hue planes and a gray scale was published (figure 7.21). An enlarged version with 1,001 glossy samples appeared in 1978 (figure 7.22). This edition is known for its high level of accuracy of aim color match.

In 1957, Ernst Biesalski published a version of the DIN 6164 system in the form of a special botanical color atlas (figure 7.23). It contains 450 color chips each measuring 20 × 14 mm, in 30 hues, with holes to aid comparison with plant samples. DIN standard 6164 has not found wide use in industry and the arts. Of the versions published, today only the matte version is still available. Officially DIN 6164 remains a German standard. However, the institute now recommends Swedish standard SS 01 91 00 (see Natural Color System, chapter 5).

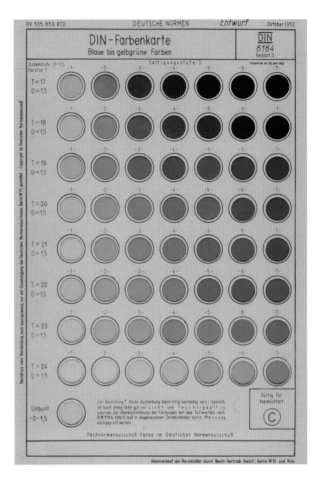

Figure 7.20.
Page with blue to yellow-green color samples (hues T17–T24) of the 1952 gelatin filter version of the DIN 6164 atlas (Deutsches Institut für Normung 1952).

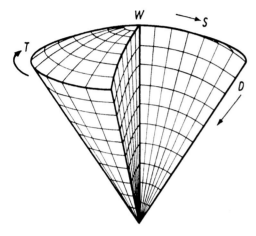

Figure 7.17.
Conceptual perceptual model of the DIN 6164 color solid (Richter 1976).

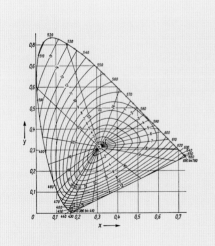

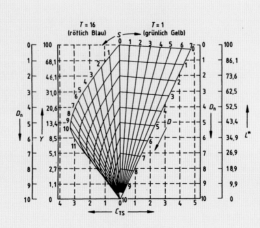

Figure 7.19.
Cross section along DIN 6164 hues T1 (yellow) and T16 (reddish blue) in the CIELUV chroma/L* diagram with additional darkness degree scale, with lines of constant saturation S and constant darkness D (Witt 1983).

Figure 7.18.
Portion of the CIE chromaticity diagram with radial lines of approximately constant hue and ovoid lines of constant saturation of the DIN 6164 system (Richter 1955).

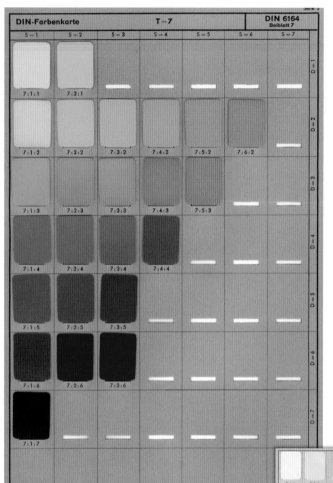

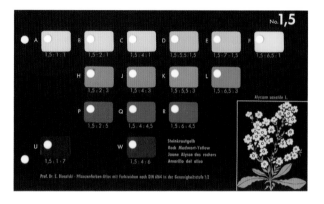

Figure 7.23.
Color samples of hue 1.5 of Biesalski's (1957) botanical color atlas. It represents one of six additional intermediate hues in this atlas.

Figure 7.21.
Page with color samples of hue T7 of the 1962 matte version of the DIN 6164 atlas (Deutsches Institut für Normung 1962).

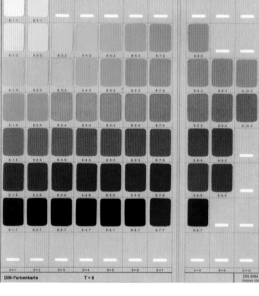

Figure 7.22.
Page with color samples of hue T8 from the glossy 1978 version of the DIN 6164 atlas (Deutsches Institut für Normung 1978).

OSA UNIFORM COLOR SCALES 1977

Optical Society of America, Uniform color scales, 1977

At the suggestion of the U.S. National Research Council, in 1947 the **OSA** undertook to develop an object color space uniform in as many directions as possible (isotropic). Judd (see entry in this chapter) headed the development subcommittee, whose work stretched over 25 years. Color scientist I. A. Balinkin and physicist Ludwik Silberstein advised the subcommittee.

Balinkin proposed a uniform color space based on a perfect space-packing scheme, a concept of crystalline structure that would result in uniformity in 12 directions from a given point. Silberstein demonstrated that a plane in a space can be tested for conformance with Euclidean geometry by evaluating samples in triangles where the perceptual distances between all three legs are the same (and represented by the corresponding geometric distance). According to Silberstein, when in a plane six of these triangles are connected based on the same central color, they should form a regular hexagon. This is the case only if the triangles are equilateral. If not, there is either a gap or an overlap in the hexagon. The former indicates an elliptical (Riemannian) space, the latter a hyperbolic (Lobatchevskian) space.

In the system's key experiment, 76 observers evaluated a series of 43 colored ceramic tiles (figure 7.24), arranged in a grid of triangles, regarding the relative perceptual magnitude of the distances between related sample pairs. The tiles were of nominal luminous reflectance $Y = 30$ (Munsell value 6). None of the samples represents an achromatic color.

Differences between samples are comparatively large, approximately 40% larger than Munsell double chroma steps. For the purposes of the experiment, the samples were displayed in sets of three, labeled A, B, C, about which observers answered these questions: Is the distance between sample A and B smaller or larger than between B and C? The results were statistically evaluated, and mean perceptual distances between samples were calculated.

Additional tests were performed establishing perceptual magnitudes of combined chromatic and lightness differences. The results showed that there were gaps in the Silberstein hexagons and that thereby in the chromatic plane, the underlying color space was Riemannian. The cause is hue superimportance resulting in elongated unit contours in a Euclidean plane.

This finding, corresponding to a similar one in the Munsell system (see Albert Henry Munsell, chapter 5), put the plan of displaying the system in jeopardy. At this point, the committee decided to "force the data into a Euclidean form, despite clear indications that the judgment data required non-Euclidean representation" (MacAdam 1981, p. 167).

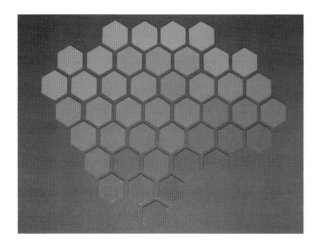

Figure 7.24.

Image of the 43 hexagonal enamel color plates used by the OSA-UCS committee to establish the fundamental perceptual data used in the development of the system.

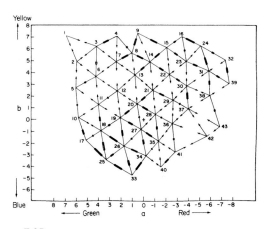

Figure 7.25.

Plot of 102 chromatic differences between 43 samples in the chromatic diagram optimized by the committee to the perceptual data. Numbers identify the samples. Lines with arrows and bars indicate the size of the average visual judgments. A perfectly fitting mathematical model would not result in any gaps or overlaps. Reprinted with permission.

A psychophysical Euclidean formula was then fitted to the data, the best fit formula having moderate accuracy. Perceptual data were plotted in the resulting psychophysical chromatic diagram (figure 7.25) with average perceptual distances shown with arrows or overlap bars. This diagram represents the perceptual results only modestly well.

In addition, the committee incorporated the Helmholtz-Kohlrausch as well as the lightness crispening effects into the system. The final formula was used to calculate aim points for an atlas according to the Balinkin proposal where each color (except on the outer borders) is surrounded in a cubo-octahedron by 12 nearest neighbors (figure 7.26). The

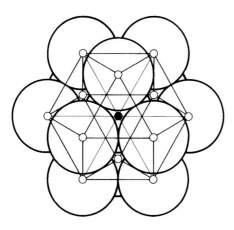

Figure 7.26.
Schematic drawing of 12 samples (represented by balls with open centers) located equidistant from the central sample (ball with black center). The resulting geometric form is a cubo-octahedron (Gerstner 1986). Reprinted with permission.

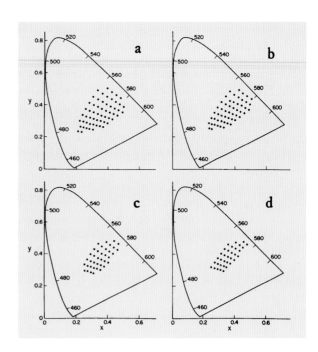

Figure 7.28.
CIE chromaticity diagram with location of aim colors at four levels of OSA-UCS lightness: a, L = 1; b, L = 2; c, L = 3; d, L = 4 (MacAdam 1974). Reprinted with permission.

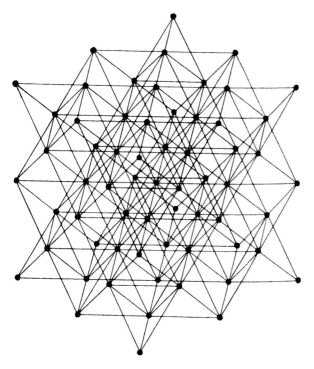

Figure 7.27.
Basic cubo-octahedral unit lattice doubly expanded in all three dimensions (Foss 1978). Reprinted with permission.

crystalline structure of the system is further illustrated in figure 7.27, where the unit lattice has been doubly expanded in all three directions. Aim points at four levels of lightness are shown in the CIE chromaticity diagram in figure 7.28.

Pigment colorations matched to the aim points were prepared and the results displayed in a physical model (figure 7.29).

Figure 7.29.
View of MacAdam's model of the OSA-UCS color solid illustrating several of the seven cleavage planes. Slide courtesy D. L. MacAdam.

This model clearly displays the cleavage planes available in the chosen structure. They result in approximately uniform color scales in many directions that are not explicitly available in any other system (figure 7.30).

The system falls short of being fully isotropic because uniformity is maximally possible only in 12 directions away from a given point rather than in all directions. In addition, as mentioned, hue superimportance has been neglected.

In 1977, the OSA published the atlas of the *OSA Uniform Color Scales*, containing a total of 558 glossy color chips (424 of the regular and 134 of a pastel set). The structure of the system is more complex and less intuitive than that of the Munsell system. This has hampered its use and limited its application.

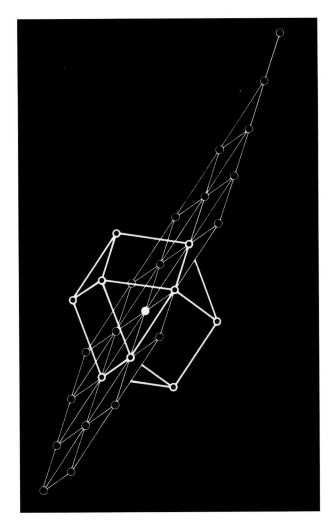

Figure 7.30.
Schematic illustration of aim points in a slanted cleavage plane (Gerstner 1986). Reprinted with permission.

CIELAB, CIELUV 1976

Commission Internationale de l'Éclairage, *Recommendations of uniform color spaces, color-difference equations, psychometric color terms*, 1978

In the 1960s and 1970s, veritable Tower of Babel confusion existed in color technology regarding the use of color space and difference formulas. A 1973 survey in the United States showed more than 15 such formulas in industrial use. Data could not be easily compared because color differences from the various formulas are not related by simple constants. At the same time, new formulas that claimed higher accuracy were frequently being published.

In response, the CIE recommended ("in the interest of uniformity of usage") two sets of formulas that mustered enough support in this international organization. It recommended one set primarily for object colors (**CIELAB**), and another set, a linear transformation of the CIE chromaticity diagram, primarily for lights (**CIELUV**). The recommendations were officially published in 1978.

The CIE 1976 $L*a*b*$ (CIELAB) formula is based on the Adams-Nickerson formula of 1936 (see Dorothy Nickerson, this chapter) and the cube-root version of the Adams-Nickerson formula developed in 1958 by L. G. Glasser and colleagues. The lightness scale is essentially identical to the Munsell value scale:

$$L* = 116(Y/Y_n)^{1/3} - 16,$$

where $L*$ is the metric lightness value, Y is the luminous reflectance value of the sample being assessed, and Y_n is the luminous reflectance of the reference white. This formula does not consider lightness crispening, or the Helmholtz-Kohlrausch effect (see glossary). It is a fair statement that it is not strictly valid for any set of observation conditions. There are two chromatic components fashioned in an opponent color system according to the proposal by Adams (see entry in this chapter):

$$a* = 500[(X/X_n)^{1/3} - (Y/Y_n)^{1/3}]$$
$$b* = 200[(Y/Y_n)^{1/3} - (Z/Z_n)^{1/3}]$$

where X, Y, and Z are the CIE tristimulus values of the reference color and those with subscript n of the illuminant as reflected from the reference white. (Modified equations apply for very low luminous reflectance values.) The three axes are taken to be orthogonal, forming the CIELAB color space. Figure 7.31 illustrates the space with the optimal object color solid and the spectral trace (note the curvature of the line of purple colors due to the nonlinear compression of the tristimulus values).

Total color-difference values ΔE^*_{ab} are calculated as the Euclidean square root of the sum of the squares of the component differences:

$$\Delta E^*_{ab} = [(\Delta L^*)^2 + (\Delta a^*)^2 + (\Delta b^*)^2]^{1/2}$$

The polar coordinate version of the difference formula turned out to be of considerable interest:

$$\Delta E^*_{ab} = [(\Delta L^*)^2 + (\Delta H^*)^2 + (\Delta C^*)^2]^{1/2}$$

where ΔH^* is the hue difference and ΔC^* is the chroma difference, with C^* calculated as

$$C^* = [(a^*)^2 + (b^*)^2]^{1/2}$$

and $\quad \Delta H^* = [(\Delta E^*)^2 - (\Delta L^*)^2 - (\Delta C^*)^2]^{1/2}.$

That is, the hue difference is the Euclidean difference remaining after subtracting lightness and chroma differences from the total difference. It was of interest because the three component differences can be adjusted separately.

At its creation, the CIELAB formula was known to no more accurately predict average perceived small color-difference data than some other formulas; efforts to improve it began immediately. Analyzing small color-difference data showed that unit contours in the a^*, b^* diagram were elongated (due to hue superimportance) and were in most cases aligned in the direction of the diagram origin. It was discovered that correlation between perceptual and calculated data improved when separately weighting the three component differences in the polar form of the equation. Several formulas with increasingly complex weighting of component differences have been offered since (see CIEDE2000, this chapter). When the CIELAB and CIELUV formulas were created, lighting engineers supported a formula with linear transformation of tristimulus values. A linear system simplifies calculation of the result of mixing lights, as demonstrated in the CIE chromaticity diagram. As a result, the committee recommended a second formula, CIELUV, a linear transformation of the chromaticity diagram considered perceptually significantly more uniform. It modifies the linear transformation formula originally proposed by Judd in 1936 (see entry in this chapter).

The lightness formula is identical to that of the CIELAB formula (with the same limitations). The chromatic components are calculated as follows:

$$u^* = 13\, L^*(u' - u'_n),$$
$$v^* = 13\, L^*(v' - v'>_n),$$

where $u' = [4X/(X + 15Y + 3Z)]$ and $v' = [9Y/(X + 15Y + 3Z)]$. Subscript n again refers to the corresponding value for the illuminant. Also in this case, the total color difference is calculated as the square root of the sum of the squares of the component differences. CIELUV space, the corresponding object color solid, and the spectral trace are shown in figure 7.32. As expected in a linear system, the trace of the purple colors created from mixtures of lights of 360 and 750 nm wavelength is a straight line. The optimal color solids of the two spaces are noticeably different. Neither is isotropic.

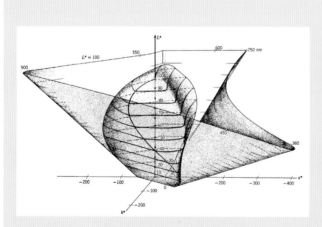

Figure 7.31.

Projective view of the optimal object color solid and the spectral envelope in CIELAB space. CIE 10° observer and illuminant D65 (Judd and Wyszecki 1975). Reprinted with permission.

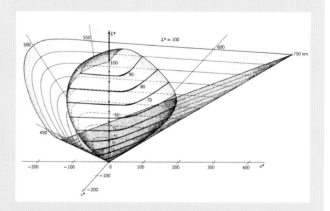

Figure 7.32.

Projective view of the optimal object color solid and the spectral envelope in CIELUV space: CIE 10° observer and illuminant D65 (Judd and Wyszecki 1975). Reprinted with permission.

CIEDE2000   2001

Commission Internationale de l'Éclairage, *Improvement to industrial colour-difference evaluation*, 2001

CIEDE2000 is the CIE's latest recommendation for a CIELAB-space-based color-difference formula. It made previous recommendations for color-difference formulas in 1964, 1976, and 1995. Such formulas are of considerable interest for objective color control in industrial manufacturing of colored goods.

The CIELAB color space and color-difference formulas of 1976 are simple Euclidean constructs based on CIE tristimulus values and the application of cube root compression. In this space, perceptual unit difference contours, are ellipsoids (due to the hue superimportance effect), rather than spheres of equal size. The ellipsoids generally point to the origin of the CIELAB chromatic diagram and vary in size, which increases with distance from the neutral point.

Improvements in the correlation between visual and calculated data were obtained by considering those and other effects (e.g., regularities in the perceptual data as expressed in the psychophysical data), resulting in several formulas based on CIELAB. Adjustment for the hue superimportance effect (unit contour elongation) and the chromatic crispening effect was obtained by separately fitting weights for the calculated hue and chroma difference components. Weights for lightness differences were also fitted.

The latest version of such a formula is CIEDE2000. It has six separate adjustments, optimized against regularities in several sets of experimental data that are considered reliable. A lightness difference weight adjusts for lightness crispening for a surround of $L^* = 50$, and chroma and hue difference weights adjust for chromatic crispening as well as for the basic elongation of the unit contours. Calculated hue differences are adjusted with a hue angle-based function.

In the a^*, b^* diagram, experimental unit contours of blue colors are tilted against the line of constant hue, and an additional function corrects for this effect. The chroma and hue difference weights result in little or no adjustment for ellipses near the origin of the diagram; another function was developed to adjust for this situation. No color space that directly reflects CIEDE2000 exists.

The unit chromatic difference contours for which the formula corrects perfectly, converting to circles of equal size, are shown in figure 7.33. Figure 7.34 illustrates fitted unit difference ellipses in the same diagram for two sets of average visual data. Considerable differences between the two sets of ellipses (and other factors) result in the formula's approximately 65% accuracy in predicting average visual data. The reasons for this moderate level of agreement are not known. Only new, carefully designed replication experiments will show if it is possible to obtain a higher level of accuracy for a specific set of surround and experimental conditions.

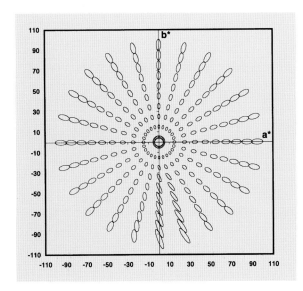

Figure 7.33.
Plot in the CIELAB a^*, b^* chromatic diagram of contours of constant perceptual difference from their center points according to the CIEDE2000 equation (Luo et al. 2001). Reprinted with permission.

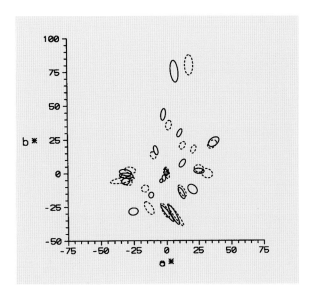

Figure 7.34.
Plot of unit difference ellipses in the a^*, b^* *diagram, fitted to selected average perceptual small color-difference data (Luo and Rigg and RIT-DuPont data sets) (Melgosa et al. 1997). Reprinted with permission.*

Notes

1. For a more detailed recent history, see Kuehni (2003).

2. Müller (1850–1934), a German experimental psychologist and philosopher, included color among his many psychological interests. His color vision theory is described in Müller (1930).

3. In 1967, Ken D. Chickering further optimized the Friele formula of 1965.

4. Recent analysis has indicated a fundamental difference between CME data and JND and suprathreshold color-difference data (Kuehni 2003). It appears that MacAdam's data depend only on the sensitivities of the three fundamental processes and the chromatic crispening effect. Perception of JND or of differences of larger magnitude appears to involve more complex processing at later stages along the visual path, explaining the inferior performance of MacAdam's approach in industrial color-difference calculation.

References

Adams, E. Q. 1923. A theory of color vision, *Psychological Review* 30:56–76.

Adams, E. Q. 1942. X-Z planes in the 1931 I.C.I. system of colorimetry, *Journal of the Optical Society of America* 32:168–173.

Biesalski, E. 1957. *Pflanzenfarbenatlas für Gartenbau, Landwirtschaft und Forstwesen mit Farbzeichen nach DIN 6164*, Göttingen: Musterschmidt-Verlag.

Chickering, K. D. 1967. Optimization of the MacAdam-modified 1965 Friele color-difference formula, *Journal of the Optical Society of America* 57:537–541.

Clark, F. J. J., **R. McDonald**, and **B. Rigg**. 1984. Modification to the JPC79 colour-difference formula, *Journal of the Society of Dyers and Colourists* 100:128–132.

Commission Internationale de l'Éclairage, 1978. *Recommendations of uniform color spaces, color-difference equations, psychometric color terms*, Wien: CIE.

Commission Internationale de l'Éclairage, 2001. *Improvement to industrial colour-difference evaluation*, Wien: CIE.

Deutsches Institut für Normung. 1952. *DIN Farbenkarte, Beiblätter 1–3 zu DIN 6164*, Berlin: Beuth.

Deutsches Institut für Normung. 1962. *DIN Farbenkarte*, Berlin: Beuth.

Deutsches Institut für Normung. 1978. *DIN Farbenkarte*, Berlin: Beuth.

Foss, C. E. 1978. Space lattice used to sample the color space of the Committee on Uniform Color Scales of the Optical Society of America, *Journal of the Optical Society of America* 68:1616–1619.

Friele, L. F. C. 1961. Analysis of the Brown and Brown-MacAdam colour discrimination data, *Die Farbe* 10:193–202.

Friele, L. F. C. 1965. Further analysis of colour discrimination data, in *Proceedings of the International Colour Meeting (Lucerne 1965)*, Richter, M., ed., Göttingen: Musterschmidt.

Gerstner, K. 1986. *The forms of color*, Cambridge, MA: MIT Press.

Glasser, L. G., **A. H. McKinney**, **C. D. Reilly**, and **P. D. Schnelle**. 1958. Cube-root color coordinate system, *Journal of the Optical Society of America* 48:736–740.

Gretag-Macbeth. n.d. *Munsell book of color*, New Windsor, NY: Gretag-Macbeth.

Japan Color Research Institute. 1978. *Chroma cosmos 5000*, Tokyo: Japan Color Research Institute.

Japan Color Research Institute. 1982. *Chromaton 707*, Tokyo: Japan Color Research Institute.

Judd, D. B. 1932. Chromaticity sensibility to stimulus differences, *Journal of the Optical Society of America* 22:72–108.

Judd, D. B. 1935. A Maxwell triangle yielding uniform chromaticity scales, *Journal of the Optical Society of America* 25:24–35.

Judd, D. B. 1936. Estimation of chromaticity differences and nearest color temperature on the standard (ICI) colorimetric coordinate system, *Journal of the Optical Society of America* 26:421–426.

Judd, D. B., and **G. Wyszecki**. 1975. *Color in business, science, and industry*, 3rd ed., New York: Wiley.

Kuehni, R. G. 2003. *Color space and its divisions*, Hoboken, NJ: Wiley.

Luo, M. R., **G. Cui**, and **B. Rigg**. 2001. The development of the CIE 2000 colour-difference formula: CIEDE2000, *Color Research and Application* 26:340–350.

MacAdam, D. L. 1942. Visual sensitivities to color differences in daylight, *Journal of the Optical Society of America* 32:247–274.

MacAdam, D. L. 1943. Specification of small chromaticity differences, *Journal of the Optical Society of America* 33:18–26.

MacAdam, D. L. 1944. On the geometry of color space, *Journal of the Franklin Institute* 238:195–210.

MacAdam, D. L. 1965. Specification of color differences, *Acta Chromatica* 1:147–156.

MacAdam, D. L. 1974. Uniform color scales, *Journal of the Optical Society of America* 64:1691–1702.

MacAdam, D. L. 1981. *Color measurement*, Berlin: Springer.

Melgosa, M., **E. Hita**, **A. J. Poza**, **D. H. Alman**, and **R. S. Berns**. 1997. Supra-threshold color-difference ellipsoids for surface colors, *Color Research and Application* 22:148–155.

Müller, G. E. 1930. Über die Farbenempfindungen, *Zeitschrift für Psychologie*, Ergänzungsband 17 and 18, Leipzig: Barth.

Munsell, A. H. 1915. *Atlas of the Munsell color system*, 2nd ed., Malden, MA: Wadsworth-Howland.

Munsell Color Company. 1929. *Munsell book of color*, Baltimore, MD: Munsell Color Company.

Newhall, S. M., **D. Nickerson**, and **D. B. Judd**. 1943. Final report of the OSA Subcommittee on the spacing of Munsell colors, *Journal of the Optical Society of America* 33:385–418.

Nickerson, D. 1936. The specification of color tolerances, *Textile Research* 6:509–514.

Nickerson, D., and **S. M. Newhall**. 1943. A psychological color solid, *Journal of the Optical Society of America* 33:419–422.

Nickerson D., and **K. F. Stultz**. 1944. Color tolerance specification, *Journal of the Optical Society of America* 34:550–570.

Richter, M. 1947. *Empfindungsgemäss gleichabständiges Farbsystem*. Farbenatlas. 550 Farben in Leimfarben-Aufstrichen nebst zugehörigen trichromatischen Masszahlen des Internationalen Bezugssystems DIN 5033, Berlin.

Richter, M. 1950. Untersuchungen zur Aufstellung eines empfindungsgemäss gleichabständigen Farbsystems, *Zeitschrift für wissenschaftliche Photographie, Photophysik und Photochemie* 45:140–162.

Richter, M. 1955. The official German standard color chart, *Journal of the Optical Society of America* 45:223–226.

Richter, M. 1976. *Einführung in die Farbmetrik*, Berlin: Walter de Gruyter.

Schrödinger, E. 1920. Grundlinien einer Theorie der Farbenmetrik im Tagessehen, *Annalen der Physik* 63:481–520.

Stiles, W. S. 1946. A modified Helmholtz line element in brightness-colour space, *Proceedings of the Physical Society (London)* 58:41–47.

U.S. Dept. of Agriculture. 1964. *Loci of constant hue and constant chroma in the Munsell Renotation System*, Washington, DC: U.S. Dept. of Agriculture.

Witt, K. 1983. Helligkeit und Buntheit im DIN-Farbsystem, *Die Farbe* 31:61–71.

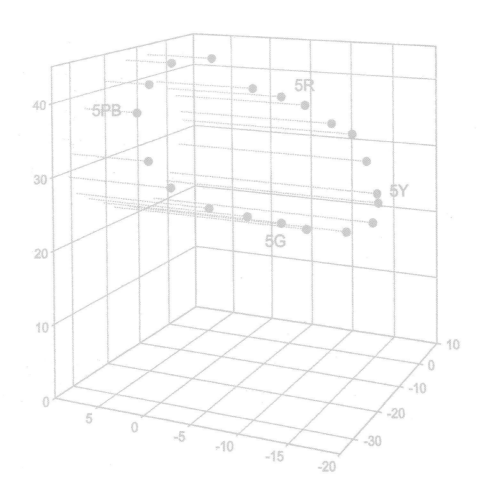

A brief introduction to physical order systems that place color stimuli as a result of mathematical manipulation of purely physical measurements (reflectance data), without any connection to the color-vision apparatus.

To fully exploit the information in a given spectral light signature, one must know it in full spectral detail. For a natural image, this represents a large amount of information, depending on the spectral interval at which the function is measured. Such measurements are of purely physical nature. For the human visual system to measure spectral functions in detail would require many (up to 30 or more depending on the desired spectral resolution) spectrally narrowly tuned sensors for each defined microregion of the image. As Thomas Young pointed out (see entry in chapter 6), this is "almost impossible to conceive," and in reality most of us have only the three sensor types of the trichromatic system.

The problem for the trichromatic system was how to exploit the information in the spectral function with as little loss as possible. As a matter of natural law or mathematical necessity, dimension reduction of a reflectance curve's information from, say, 31 values to three (the tristimulus values) results in metamerism, the fact that differing spectral functions can have the same indexes (locations) in the system of reduced dimensions.

Our visual system may have evolved to pursue one of two possible different strategies toward the goal of exploiting as much of the original information as possible: 1. assuring that the filtering allows the reconstruction of the spectral function with a high degree of accuracy, which makes sense only if brain mechanisms can exploit the full spectral information, or 2. assuring that differences in the spectral functions are represented with reasonable proportionality in distances in the psychological space. The second strategy was pursued, with a result somewhat short of the optimum.

For technical purposes, such as when all spectral functions in a natural image must be accurately and easily captured, the original function should be highly reconstructable. Mathematical methods can optimize the filters' number and functions applied in such a situation.

When the filters represent the three cone or color-matching functions the reflectance function (390–720 nm) is recoverable with an average of 66% accuracy. The error of reconstruction for color stimuli in this case is found to be a function of hue. Recovery of spectral information with an accuracy of 99% or higher requires five or six filters optimized to the set of spectral data involved. The number depends on the nature of the spectral data to be recovered. As mentioned above, metamerism is an outcome of any kind of dimension reduction. However, which spectral functions are metameric depends on the filter functions. Different filter functions result in different sets of metamers. In color technology, this situation is known with respect to the two

CIE standard observers (see glossary): Metamers for one observer are not exact metamers for the other.

Dimension reduction of spectral stimulus functions can be achieved with any continuous filter form. Choosing a filter form depends on the purpose of filtering. Several mathematical techniques achieve filtering that is optimized to recover the original information. In 1964, Jozef B. Cohen (see entry in chapter 6) first used the principal component analysis method (PCA) for the analysis of spectral reflectance functions of Munsell samples. The same method was used in the same year by D. B. Judd, D. L. MacAdam, and G. Wyszecki (1964) to calculate spectral power distributions at various correlated color temperatures from functions fitted to measurement data of the spectral power of sunlight arriving at the surface of the earth.

More recent methods are independent component analysis and nonnegative matrix factorization (NMF). In all methods, the single spectral function that provides the highest degree of reconstruction of the Munsell reflectances set is more or less spectrally flat with all positive values. Except in NMF, the second and third functions have positive and negative values and crudely resemble opponent color functions (figure 8.1).

Functions calculated according to these methods for the same data are of somewhat different shape, resulting in different metamers. These functions are used like cone or color-matching functions for the purpose of calculating the corresponding tristimulus values. These tristimulus values form a space in which any spectral function is represented as a point. More than three functions results in spaces of dimensionality higher than three.

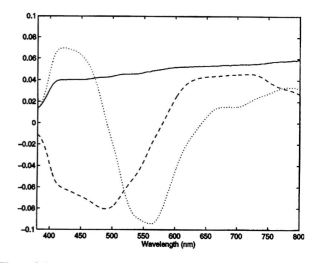

Figure 8.1.

Spectral functions of the first three principal components fitted to a database of 1,269 Munsell chip reflectance functions. Solid line, first principal component (PC); dashed line, second PC; dotted line, third PC (Lenz et al. 1996). Reprinted with permission.

In figure 8.2, the values for 20 Munsell chips varying in hue but of constant value and chroma (every second of the 40 Munsell hue chips of the hue circle) have been plotted. The orthogonal space in which they are located is based on the first three PCA functions calculated from 1,269 Munsell chip reflectances. They are located in the space in ordinal order, but not on a flat plane. Their distances do not agree with the claimed perceptual distances.

The hue circle (on an ordinal level) is one of the automatic outcomes of dimension reduction of spectral color stimuli. Hue circles, therefore, are implicit in the mathematical operation of dimension reduction of spectral stimuli to three dimensions if the applied dimension reduction functions are continuous and curved. This demonstrates the impact of physics on psychophysics and psychology of color.

Spectral dimension reduction functions that apply to color chip collections are influenced to some extent by the choice of the specific pigments (the shape of their reflectance functions) used in their production. In this situation, accurate reflectance recovery filters are a function of pigment choice. So a given dimension reduction method will result in somewhat different optimized filter functions for the atlas-chip set of Munsell than for that of the Natural Color System (see entry in chapter 5), because of pigment difference and variation in the distribution of samples in the corresponding space. Metamers based on mathematical filter functions for the two sets will differ from each other and from visual metamers.

As a result of the dependencies of the filter functions on colorants used for the samples and the structure of the sample set, it is impossible to to optimally recover any and all spectral functions with a small (in numbers) set of filter functions.

The optimal object color solid in three-dimensional PCA space has been calculated. It somewhat resembles Robert Luther's solid (see entry in chapter 6) because of the similarity of his chromatic functions with PCA principal components (eigenvectors) 2 and 3 based on the Munsell chips. Figure 8.3 illustrates the view against the top surface of the solid with white in the center. Even though the space is derived from color stimuli, it is not really appropriate to call it a color space; at best it is a color-stimulus space. It is a completely physical space with no direct relationship to psychology.

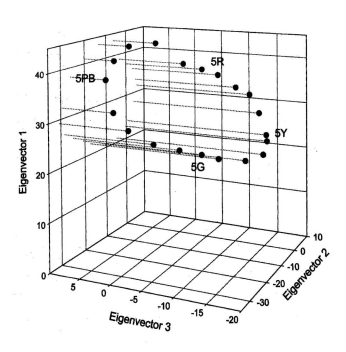

Figure 8.2.

Plot of 20 Munsell chips (every second hue, four identified) at value 6 and chroma 8 in the space formed by the principal components of figure 8.1.

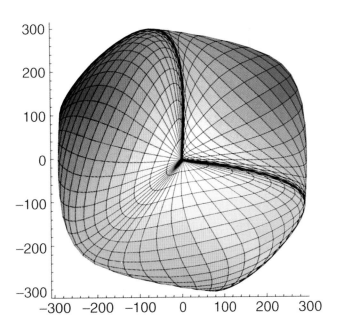

Figure 8.3.

View of the surface of the optimal object color stimulus solid in the singular value decomposition space. The view is toward the top of the space, along the equal energy axis (Koenderink and van Doorn 2003). Reprinted with permission.

References

Cohen, J. 1964. Dependency of the reflectance curves of the Munsell color chips, *Psychonomic Science* 1:369–370.

Judd, D. B., **D. L. MacAdam**, and **G. Wyszecki.** 1964. Spectral distribution of typical daylight as a function of correlated color temperature, *Journal of the Optical Society of America* 54:1031–1040.

Koenderink J. J., and **A. J. van Doorn**. 2003. Perspectives on colour space, in *Colour perception: mind and the physical world*, Mausfeld, R., Heyer, D., eds., Oxford: Oxford University Press.

Lenz, R., **M. Osterberg**, **J. Hiltunen**, **T**. **Jaaskelainen**, and **J. Parkkinen**. 1996. Unsupervised filtering of color spectra, *Journal of the Optical Society of America A* 13:1315–1324.

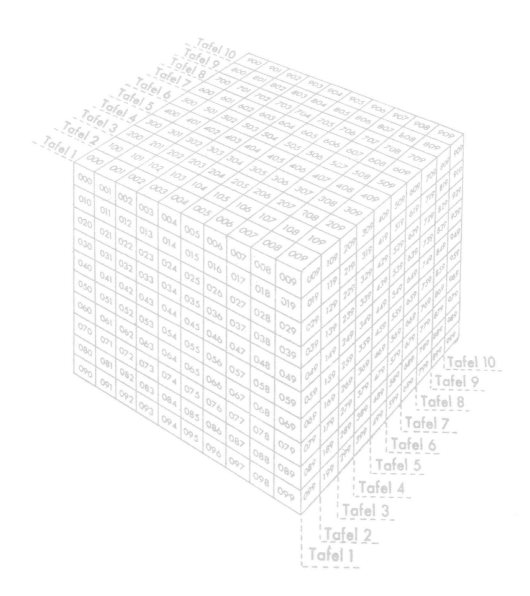

CHAPTER 9

TECHNICAL COLOR SYSTEMS

A description of technical color order systems where the samples of the system are arranged according to systematic mixture of usually three primary colorants (dyes or pigments, typically in form of printing inks) or three lights, systems that most often support technological applications of color stimuli.

Color is usually related to some practical application, rather than treated as a separate subject. The manufacture of colorants and colored products, and the mass media require tight color quality tolerances to produce high-quality products competitively. That, in turn, requires a high technological standard based on extensive research and development. To manage color effectively in these areas, systematic color order is essential, not merely useful.

The halftone printing process

Gutenberg's fifteenth-century invention of letterpress printing with movable letters initiated the development of mass media, finding expression in books, pamphlets, newspapers, and magazines. The inclusion of hand-colored illustrations in manuscripts and books represented a standard of expectation that would take printing technology several more centuries to fully meet at the level of mass media. The inventions of photography, television, and the computer provided new opportunities for mass distribution of images. Today, the computer is a most important tool in the rapid processing and distribution of colored images, supporting or replacing older technologies, including color charts and atlases. In the years immediately following Gutenberg's invention, color printing was sparingly used, and only for initials. Illustrations were carved in wood (later engraved) and colored by hand, if at all. Engraved copper and steel plates and (by the beginning of the nineteenth century) polished stones (lithography) replaced woodcuts.

Since the eighteenth century, printing basically has fallen into two categories: printing of uniform areas (solid printing) and printing of images. In the former, technologically simpler, case, printing inks are mixed like paint to achieve desired colors. The matching process is comparable to that in dyeing or in paint manufacture. But the latter case required the development of technologies to break the image locally into color components that, when printed, reproduced the original image as accurately as possible.

By the beginning of the eighteenth century, it was well known how to reproduce many color stimuli with three primary colorants, as briefly presented in chapter 1. Before 1720, German-born engraver Jakob Christof Le Blon (1667–1741)[1] began to experiment with producing multicolored reproductions by printing the image from three or four plates, each applying one of the three primary chromatic colorants yellow, red, and blue, as well as black (figure 9.1).

Le Blon had the uncanny mental ability to separate a colored image into three or four separate images that he then printed superimposed. To obtain smooth transitions of color, he employed the mezzotint method of engraving (see glossary). In 1719, he obtained a British patent for his invention.

The technology was continued by Le Blon's student and successor Jacques Fabian Gautier d'Agoty (1717–1785), who is best known for colored anatomical images.[2] In 1837, G. Engelmann and his son received a French patent for a lithographic version of Le Blon's process. Their version, which came to be known as chromolithography, often used more than three or four plates and associated printing inks to generate lifelike images in color.

Reproduction of the world as we experience it required a color photography process. In 1861, the Scottish physicist James Clerk Maxwell (see entry in chapter 6) demonstrated that photographic color reproduction is possible if the object is photographed on three plates through red, green, and violet-blue filters, the resulting plates converted to positives, and projected superimposed, using lights that passed through the same three filters.

At about the same time, the French physician and inventor Louis Arthur Ducos du Hauron (1837–1920) began to experiment with color photography. In 1869, he wrote a booklet outlining a program of color photography.[3] The technology he described was prescient, but practically impossible at his time because of the lack of photosensitive chemistry.

In the early 1870s, the German natural scientist H. W. Vogel made progress in this respect. Among the earliest printing methods based on a simple photosensitive chemistry was the collotype process.[4] In 1876, Josef Albert (1825–1886) introduced a commercial version (Albertype) in Germany. A competitive process was introduced a year later in England. Collotype printing was in use until well into the twentieth century but was not rugged enough for large-scale production.

Fundamentally, there are two kinds of photographic color reproduction processes:

1. The additive process, demonstrated by Maxwell, where filtered lights are transmitted through three black-and-white positives made from negatives shot with the same kind of filters; the resulting three images are superimposed on a screen.

2. The subtractive process, in a particular version first described by Ducos du Hauron. In the later versions (e.g., Kodachrome), three layers of the film are differently sensitive to lights. When developed with three appropriate dyes, light passing through the three layers (transparencies) or reflected from the surface of the paper (paper prints) is locally modified by the subtractive process to result in a colored image.[5]

Figure 9.1.
Images of strike-offs of a reproduction by Le Blon. The separate prints from the yellow, red, and blue plates as well as the combined print are shown (ca. 1735, Bibliothèque de l'Arsenal, Paris). Reprinted with permission.

Figure 9.2.
Multiple color rotary printing machine of Maschinenfabrik Augsburg-Nürnberg AG, 1914.

The technology to produce black-and-white separation negatives was established by 1880. Although Le Blon had demonstrated colored reproductions with three inks, in the 1720s he used the mezzotint process to mute the transitions from one color to another. A comparable technology needed to be developed for modern three-color printing based on photography.

That was accomplished in 1885 by the American printing technologist and inventor Frederic Eugene Ives (1856–1937). He invented the crossline, or halftone, screen, consisting of two sheets of glass plate with finely etched (100 lines/inch) lines, with the lines filled with an opaque substance and the two plates arranged with lines crosswise. Images were photographed through such screens, and the first commercial halftone black-and-white images were printed soon after. In the same year, he demonstrated the halftone trichromatic process. Rare commercial applications began to appear in 1893. By the mid-1890s, the process flourished in Chicago, then the print capital of the United States (Regensteiner 1943).

Le Blon struggled to develop suitable inks for his process, and his successors struggled to duplicate the color brightness of some of his prints. Maxwell fundamentally described the optical properties of the pigments required for optimal reproduction. In his efforts, Ives made use of Maxwell's findings (Ives 1902) and described the optimal colorants as "minus red" (peacock blue [a bright greenish blue, cyan]), "minus green" (crimson [a bright bluish red, magenta]), "minus blue" (yellow)."

Soon the problem arose of finding suitable pigments that not only approached the ideal as closely as possible, but also were fast to light and could be formulated as an ink that performed well in the printing process. It was not until the 1930s that pigments similar to those still in use in so-called "process colors" would be invented.

In the early decades of its use, three-color halftone printing was beset by many problems having to do with lack of standardization. Several different photographic methods with different sensitivities for the production of the color separation negatives were in use. Variables included the quality of paper, the inks, the angles at which the separation negatives were shot, the specific printing equipment, the sequence of printing the inks, and more.[6] As Ives and others described it, much "fiddling" was needed to obtain acceptable results. Progress was achieved in individual firms by difficult trial and error.

Lithographic printing (from plates) by the halftone process had the potential of being less expensive than chromolithography, because it used fewer inks, and more robust for high-speed production than was collotype. For these reasons, it gained more and more market share. Today, it is used in perhaps half of all detailed-image printing, with gravure printing (from rollers) responsible for another major share. The color effects of halftone printing are a complex combination of partitive and subtractive mixture.[7]

Until the middle of the twentieth century, color separation in halftone printing relied entirely on photographic processes. Since then, more and more electronic scanning devices or other digital capture of the image are used, today nearly exclusively.

Color control in lithographic halftone printing was a major issue for many decades. A significant positive step was taken in the United States in 1957 with the introduc-

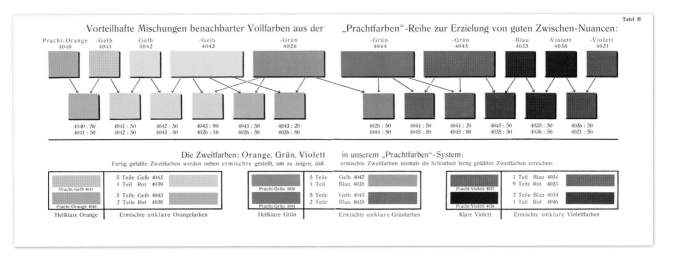

Figure 9.3.

Mixtures of inks of neighboring hues from Farben-Kamerad (Color Buddy; Dörbrand 1937), a booklet for printers. Ten of 16 basic colors are mixed, and the result of the mixture is compared with the pure inks.

tion of the LTF/GATF color chart available in the form of master films (see entry in this chapter).

While in the early twentieth century many different colorants were used in process colors, standardization has today limited their number. Different formulations are required for different mechanical printing processes. Until the 1940s, the chromatic process inks were colloquially referred to as "yellow, red, and blue," regardless of the color of the colorants. Since then, the terms cyan, magenta, yellow have become accepted, today (with black) abbreviated as CMYK.

CMY inks result in an achievable gamut of colors that is somewhat limited in regard to color intensity (chroma). There are also technical problems with the proportionality failure of the cyan and magenta inks in lighter colors.[8] The gamut issue is a matter of the absorption properties of the inks, addressed by pigment chemists, as well as by the use of additional inks (e.g., in Pantone's Hexachrome process that adds orange and green inks) and by other solutions proposed since 1950 (Ball 1950). The proportionality issue has been addressed by use of additional cyan and magenta inks diluted with white (or weaker inks as used in high-quality photo printers, CMYKcm, where the lower case letters refer to additional, weaker formulated inks).

At the beginning of the twentieth century, color printing had become an important industrial activity making use of impressive equipment (figure 9.2). Advertising made magazines with colored illustrations inexpensive for growing numbers of subscribers. At that time, the major volume of color printing continued to be in plain or simply patterned format (solid printing) where the colors are achieved by ink

mixture before the printing process. As many as a dozen basic inks would be used for this purpose (figure 9.3).

The first entry in this chapter and subgroup is devoted to the system of the French botanist Charles Lacouture, published in 1890. The system is based on the then-flourishing technique of chromolithography, using different line widths and over-printing at different angles to achieve multiple perceived colors with three inks. German book printer Hermann Hoffmann, in his "technical color system" published only two years later, stayed mostly away from systematic ink mixtures (due to the unpredictable behavior of most inks of his time). He ordered his samples mainly according to hue, lightness, and grayness but did not proceed beyond a limited systematic approach.

Despite initial difficulties, more and more three- or four-color halftone printing was produced. This was due to the growing desire to reproduce complex colored images and the lower cost of photoengraving rather than the expensive manual work of chromolithography. Color charts began to appear that demonstrated the colors that could be achieved in these processes. One of the first extensive systems of color stimuli achievable with three-color printing is by the French printer Robert Steinheil, who demonstrated 13,300 different color samples obtained with the halftone printing process and how the results depend on paper quality and printing-ink sequence.

Photographer John Cimon Warburg's is an early (1899) English example of a systematic representation in two dimensions and three screen-dot sizes of three-color halftone results. Thirteen years later, Nathan Chavkin continued the circular, two-dimensional format but expanded the number of samples to about 1,500 by using more screen-dot sizes and a black plate. But his reliance on the hue circle precluded the development of a three-dimensional model. In addition, because the technical execution was insufficient and the color samples small, Chavkin's system had only a small practical role. In 1936, the French author A. Painter offered the first complete halftone-printing color order system in the form of a Mayer-like double tetrahedron (see Tobias Mayer, chapter 4).

At the turn of the twentieth century, systematic color order had not yet established itself in printing technology. This fact is demonstrated in the German standard work for color printing of the time, J. Müller and M. Dethleffs's *Praktischer Leitfaden für den Buntbuchdruck* (Practical compendium for color letterpress printing, 1900). In its first three charts, the book contains a total of 72 colors obtained from superimposition of the three (so-called) standard inks, standard yellow, standard red, and standard blue, in three steps as well as with addition of standard gray (figure 9.4).

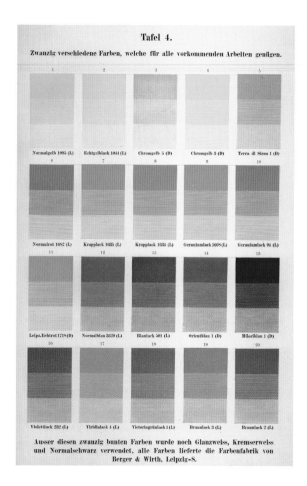

Figure 9.5.

Chart 4 from Müller and Dethleffs (1900) with 20 basic inks from which the colors of the remaining charts are mixed. Each sample (30 × 17 mm) is additionally varied by the halftone method, resulting in 60 samples.

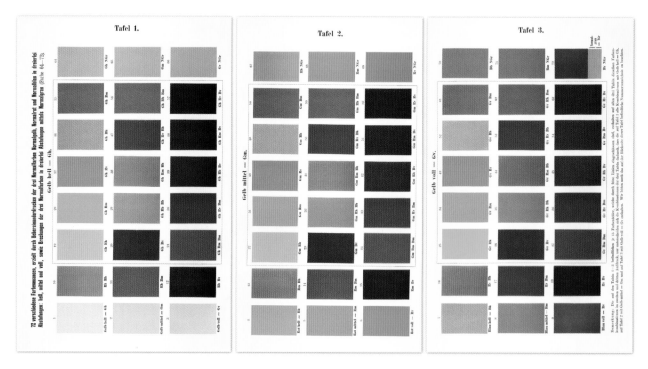

Figure 9.4.

Charts 1–3 from the work of Müller and Dethleffs (1900) with a total of 72 (rather unsystematically presented) mixed inks resulting from the three so-called standard inks of three-color printing.

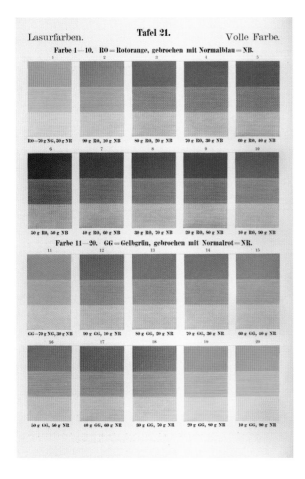

Figure 9.6.
Chart 21 from Müller and Dethleffs (1900) with mixed colors obtained from the inks of chart 4 (figure 9.5).

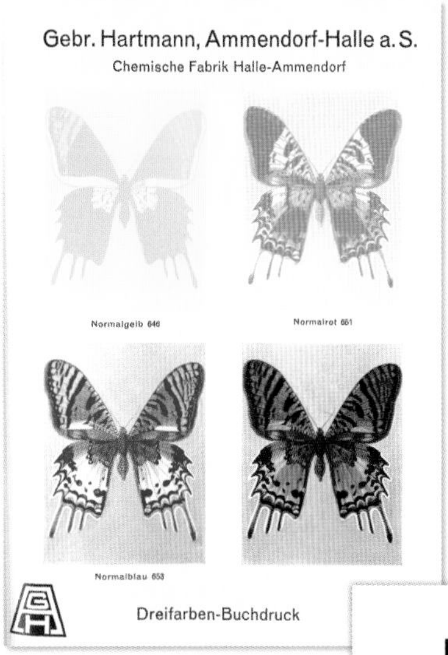

Figure 9.7.
Example of a halftone trichromatic print from the firm of Gebrüder Hartmann from 1914 demonstrates that high-quality reproduction of color photographs was already routinely achieved at that time (Hartmann 1914).

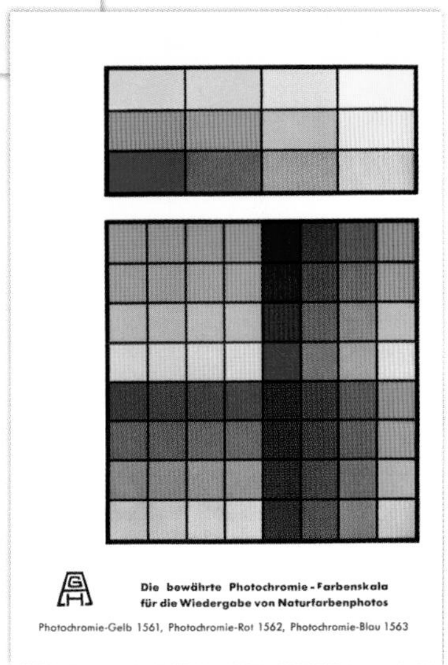

Figure 9.8.
Photochromie Farbenskala *of 1937, widely used at the time, with 64 color samples to be used in connection with nature color photographs produced with the halftone process (Hartmann 1937).*

The major portion of the book consists of 30 additional charts with a total of 600 often highly saturated colors produced from a total of 20 inks (figures 9.5 and 9.6). The quality of work in this book is very high, with every color identified with ink and mixture information. However, one can imagine the problems of print colorists of the time to find the best combination in terms of matching the desired colors and consistently producing prints. High-quality results in three-ink printing were achieved in the second decade of the twentieth century, as figure 9.7 illustrates.

In the 1920s, standardization of printing inks was realized as an important step toward rationalization. By the end of the decade, a range of 13 compatible standard inks were available from several German manufacturers under the designation Nagra inks. Ludwig Gerstacker's 1934 work demonstrates the range of results obtainable with these inks.

However, similar to Lacouture's, Gerstacker's work lacks a fundamental and consistent system, difficult to achieve

with the large number of basic inks. Only certain inks are combined to obtain the most required colors resulting in unsystematic coverage of the achievable gamut. In Germany, a so-called *Photochromie Farbenskala* was used for the reproduction of images of nature, but it had a very limited color range (figure 9.8). In the 1930s trichromatic inks were standarized in the United Kingdom and produced by several manufacturers.

In Germany in 1943, Carl Blecher, using selected Nagra inks to correspond to the typical process inks of the time, developed a process color system significantly more detailed than what was previously available. In some of its systematic aspects, it paralleled Alfred Hickethier's system of the same year. Hickethier, influenced by color photography and perhaps by Hans Neugebauer (see entry in chapter 6), offered the modern form of trichromatic print color space, the cube. In the second edition of Hickethier's atlas (1972), new inks are considerably better attuned, as noticeable in the gray axis (noticeably chromatic in the earlier edition).

Color scales for book and offset lithography printing were standardized in Germany in mid-1950s. Lithographer H. Wezel's system and the FDGB *Farbmesstafeln* demonstrated these scales, printed in atlas form on various paper qualities, as a tool for printer and colorist in 1959. In the United States, the LTF/GATF chart in master film form, first released in 1957, had a major impact on color quality control of halftone printing.

In Europe, the *Euroscale* for book and offset lithography printing was introduced circa 1970, still valid today. The ink mixture tables of the ink manufacturer Siegwerk (figures 9.9 and 9.10), in the arrangement introduced by Hickethier, are examples of realization of this standard. In the Euroscale standard, to make the basic colors cyan, magenta, yellow, and black (CMYK) independent of specific pigments, they are defined as color perceptions only.

As elsewhere, in the United States and England, many printers produced representations of the colors that can be achieved with their in-house systems for advertising and quality control purposes (Field 2004). Methods to systematically present the colors were patented in 1935 by Herbert Eugene Ives (the son of above-mentioned Frederic Eugene Ives), and in 1969 by Cyril J. Wedlake (see entry in chapter 10). American color technologist Carl Foss's Munsell-Foss charts of 1972 represent an innovative arrangement of the samples. Like the LTF/GATF chart of 1957, they were available in the form of color separations.

A 25,000-sample atlas of the four-color process according to the Hickethier model was offered in 1987 by Harald Küppers (see entry in chapter 11). A more recent example with the same number of samples is that by the American designers Michael and Pat Rogondino.

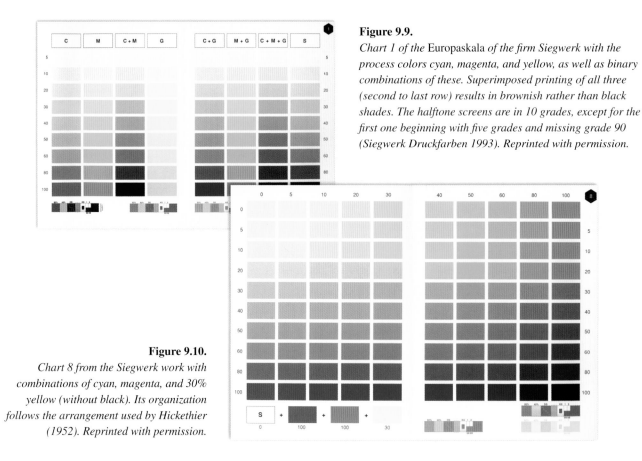

Figure 9.9.

Chart 1 of the Europaskala *of the firm Siegwerk with the process colors cyan, magenta, and yellow, as well as binary combinations of these. Superimposed printing of all three (second to last row) results in brownish rather than black shades. The halftone screens are in 10 grades, except for the first one beginning with five grades and missing grade 90 (Siegwerk Druckfarben 1993). Reprinted with permission.*

Figure 9.10.

Chart 8 from the Siegwerk work with combinations of cyan, magenta, and 30% yellow (without black). Its organization follows the arrangement used by Hickethier (1952). Reprinted with permission.

Today, for solid printing mostly standarized pigment mixtures are used. The halftone technique for printing of complex images typically employs image scanning and four process inks. In both cases, technology has provided standard inks that are largely identical across the world. A well-known example of standardized colors is produced by Pantone Inc. This firm licenses its ink mixing system and offers color reference materials. They include fans and chip books with 1,114 solid Pantone colors, others with more than 3,000 colors based on the four-color halftone process, and guides showing the four-color-process combinations that match (more or less closely) the solid colors.

Today, halftone printing is inseparably linked with the digital computer. The computer is used to manipulate the scanned digital image files and prepare them to be made into plates. Much computer output is in the form of images, viewed on monitors of various technical bases. What monitors have in common is their reliance on three primary color stimuli, red, green, and blue (RGB). These are partitively mixed to create upward of 16 million different kinds of stimuli. It is customary to represent them in a cube space (see Color Display Solids, this chapter). The gamuts of RGB and CMYK do not coincide, as shown in figure 1.20. Software programs are used to calculate perceptually acceptable transformations.

Other color technologies

There are several other color-related technologies. One of these is color television. The first public broadcast of color television took place in the United States in 1954 (figure 9.11), a few years after its invention. Its technology of color stimulus generation became the basis for color computer monitor technology and the RGB and related systems that order the stimuli.

Technological systems based on additive or subtractive color mixture

Colorimetry is the basis of another group of technological color order systems, those of English beer brewer Joseph Williams Lovibond, German botanist Emil Detlefsen, and German illumination engineer Leopold Bloch. These systems are byproducts of the development of colorimetric instruments or measurement techniques, all working with color filters. The resulting order systems were useful to demonstrate the gamuts of the measuring systems and to visually compare them to other systems. Lovibond and Bloch appropriately use the cube form of space, while the form of Detlefsen's system is that of a cone.

Color order systems based on the CIELAB formula

The CIELAB system (see CIELAB, CIELUV, chapter 7) differs from the above-named systems in that it is psychophysical in nature, the result of a nonlinear transformation of the International Commission on Illumination (CIE) colorimetric system. The solids of the systems of Lovibond, Bloch, and Detlefsen are strictly relative to the filters that were used and subtractive in nature. Unlike their systems, psychophysical systems such as CIELAB are capable of showing specific colorant or filter gamuts in terms of the optimal object color space, that is, the maximal object color space of the CIE standard observer. Because the CIELAB transformation is roughly perceptually equidistant, it can be, and is, used to calculate approximate perceptual distances between stimuli (but see CIEDE2000, chapter 7).

A few years after its introduction in 1976, visualization of the numerical CIELAB system by populating portions of it with samples was considered useful. The first atlases based on CIELAB were the Eurocolor system of 1984 designed by Ludwig Gall, the (somewhat modified) RAL Design System, and the Acoat Color Codification system shown in cylindrical coordinates. The Colorcurve system, designed by the color technologist Ralph Stanziola, is shown in Cartesian coordinates and does not have completely systematic sampling.

Figure 9.11.
Advertisement of the American color TV producer RCA, 1965.

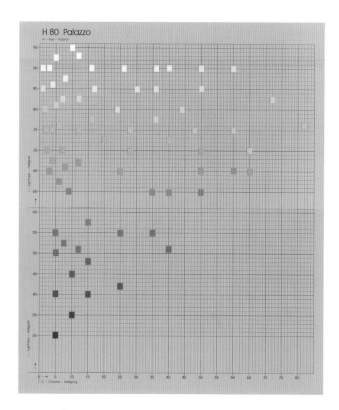

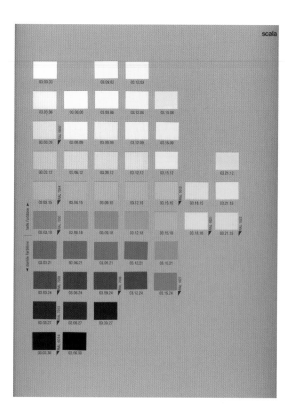

Figure 9.12.
Hue H of 3D-System published by Caparol in 1999 indicating its nature as an unsystematic sampling of CIELAB space. Reprinted with permission.

Figure 9.14.
Image of the Brillux Scala Farbfamilie 03 showing apparent nonuniformity in terms of constant hue. The samples are placed in straight rows and columns implying constant metric lightness and metric chroma (Brillux 2001). Reprinted with permission.

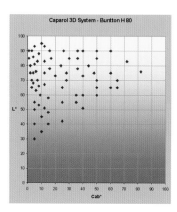

Figure 9.13.
The placement of samples on a constant hue page in 3D-System corresponds exactly to their placement in the CIELAB C*, L* diagram (Caparol 1999).

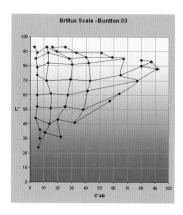

Figure 9.15.
Projection of the Brillux Scala Farbfamilie 03 onto the CIELAB C*, L* diagram, indicating the unsystematic arrangement. (Brillux 2001).

The 1986 *Catalog Cvetov*, a Russian system of G. P. Višniak and I. S. Fajnberg, uses cylindrical coordinates and limits the samples in all cases to eight metric chroma steps.

Two other recent European systems mentioned only briefly here, Caparol 3D-System of the German Caparol Group and Brillux Scala by the German paint manufacturer of the same name (the atlas of the latter receiving a German award for its design), are also marketed as representing CIELAB. In the Caparol 3D atlas, the samples are identified by CIELAB coordinate values that make their unsystematic, marketing-oriented selection obvious (figures 9.12 and 9.13). The Brillux Scala atlas arranges the samples as if they were selected by CIELAB metric hue, metric lightness, and metric chroma. But measurements indicate that they were selected instead unsystematically according to marketing concepts (figures 9.14 and 9.15).

CHARLES LACOUTURE   1890

C. Lacouture, *Répertoire chromatique, solution raisonnée et pratique des problèmes les plus usual dans l'étude et l'emploi des couleurs*, 1890

Charles Lacouture (1832–1908) was a French botanist, Jesuit, and professor of physical and natural sciences at the École Saint-Clément in Metz. While admiring the work of Michel-Eugène Chevreul (to whom the book is dedicated; see entry in chapter 4), he found it wanting in several respects and hoped to fill its gaps in his own color system. His goals were to improve on color nomenclature, establish a theory from a number of fundamental facts, show typical color ranges in charts, and consider possible applications for the system.

Lacouture's work is divided into four books and 29 colored tables. The tables are colored by chromolithography. Full colors are printed as solid surfaces, and whitened colors, as lines of different thickness on white paper. The gray scale and blackened colors are obtained by printing (or over-printing) lines of black ink at varying line width. Secondary and tertiary colors are obtained by over-printing one primary color with another at right angles or a primary color with curved lines.

The charts in the book were printed with lines from 1/12 mm to 5/12 mm in width. They are best viewed from a certain distance to obtain optical mixture of the components. Lacouture wrote that his are the first color charts produced by this method, and credits the chromolithographer M. G. Severeyns with the artistic work.

While using a four-ink printing method, Lacouture's system is meant to be a system of color perceptions. The chromatic circle is derived from three primitive colors red R, yellow J, and blue B, located at the corners of an equilateral triangle. They are defined by wavelengths R = 700 nm, J = 580 nm, and B = 470 nm. The intermediate hues orange (O = 597 nm), green (VE = 526 nm), and violet (V = 400 nm) are placed on a second triangle, with always one intermediate hue interspaced, the whole system in form of a circle of flower petals, affording a general view of the hue arrangement (rose synoptique) (figure 9.16). Complementary colors as represented by the appropriate wavelengths are placed diametrically opposite in this figure. The six chromatic principal colors are toned with white (BL) and black (N).

Lacouture described the three primitives as equidistant, or equally estranged, from each other. He took all other colors to be equidistant from the colors from which they derive; thus, "true green" is equally distant from the blue and yellow from which it derived. He defined tones as the degree of chromatic intensity of a given color. Nuances (hues in our terminology) describe the degree of chromatic parentage of a color. His tones are whitened, grayed, or blackened colors. The gray scale has five grades between black and white.

Lacouture next developed a chromatic notation for all colors based on the letters for the eight principal colors. For steps between principal hues, he added suffixes to the components. Thus, a series of six steps between red and orange is designated as R, R_5O_1, R_4O_2, R_3O_3, R_2O_4, R_1O_5, O. For blackened colors, the amount of blackening is shown by the letter N and a subscript number. This identification system can be expanded to any other level of division.

Lacouture viewed his color tables as representing color standards (*types*). For this purpose, he selected pigments resulting in colors free from any trace of the other two. Regarding tone, he stated that they must be absolutely pure (*franc*), not whitish, nor grayish, nor blackish. But he was aware that any three pigments represent only approximations of perfect primitives.

Lacouture, based on the three primitives plus white and black, aimed to establish the range of colors in a manner comparable to the diapason of sounds. He described his chromatic diapason as fixed, and its composition assured by the possibility of determining the "mean wavelength" of each sample.

Colors obtained by mixing the three primitives only are shown in the *trilobe synoptique* of figure 9.17. It attempts to demonstrate the generation of mixed hues as well as the result of mixing all three primitives. The trilobe is generated from partially overlapping semicircles. Each primitive is shown undiluted in a semicircular band in six grades, and in five levels of dilution with white.

Figure 9.16.
Lacouture's Rose synoptique *hue circle (Lacouture 1890, plate I).*

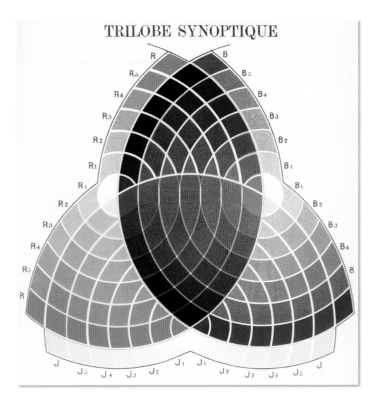

Figure 9.17.
Lacouture's Trilobe synoptique
(Lacouture 1890, plate I bis).

The plate indicates that red was printed first, followed by blue and yellow. Over-printed full colors do not mix well, and the overlapping lines form partially subtractive and partially additive mixtures, showing the limitations of the chosen method.

There are four types of charts with a total of 27 plates: In 12 plates of the first type, the six chromatic principal colors and their 1:1 mixtures are shown in squares of 49 samples, with the tint/shade scales placed in the top row and rightmost column (figure 9.18, red-orange). The 12 plates can be seen as forming a cylindrical system with a common gray scale in the leftmost column (figure 9.19). The remaining fields show intermediate tones.

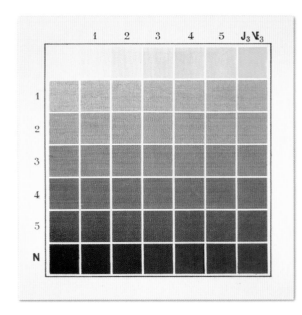

Figure 9.18.
Lacouture's Jaune-Vert, example of type 1
(Lacouture 1890, plate VII).

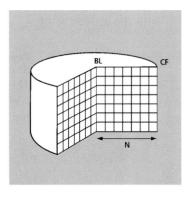

Figure 9.19.
Schematic
representation of
diagram type 1.

The six plates of the second type of table represent 1:1 mixtures of a primitive and an adjoining mixed principal color and their tints toward white. Figure 9.20 shows the blue-violet chart. They can be seen as located on a plane with white in the center, but they do not fit into a hexagonal arrangement of the chromatic principle colors (figure 9.21).

The only three plates of the third type (figure 9.22, green) show the creation of the three intermediates from the three primitives. They also can be seen as located on a plane with white in the center (figure 9.23).

The six plates of the last type show the progressive blackening of mixtures of a primitive and an adjoining

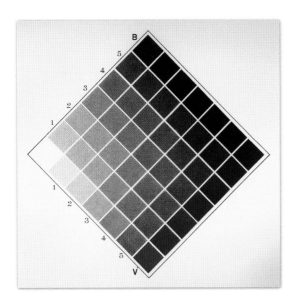

Figure 9.20.
Lacouture's Bleu-Violet (Lacouture 1890, plate XVIII).

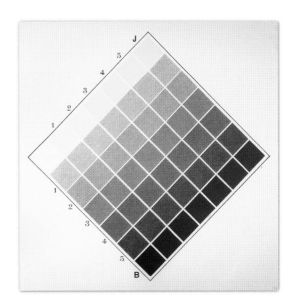

Figure 9.22.
Lacouture's Vert (Lacouture 1890, plate XXI).

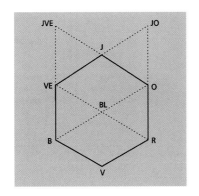

Figure 9.21.
*Schematic
representation
of diagram type 2.*

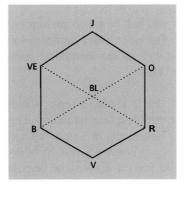

Figure 9.23.
*Schematic
representation
of diagram type 3.*

mixed principal color, in triangular form (figure 9.24, orangé-jaune). This figure illustrates an unintended problem that can occur in such presentations: a very distinct moiré effect (see glossary). It shows up to a smaller or greater extent in all six charts of this type. The charts use the central trilobe principle of figure 9.17, but with the difference that the three principal "colors" are a primitive, an intermediate, and black.

Lacouture did not discover any new facts about color order. His book is a labor of love that shows earnest striving for clarity of relationship and standardization. His figures represent a novel approach to the problem of economical atlas coloration. Unfortunately, the solution he and his engraver developed has considerable limitations.

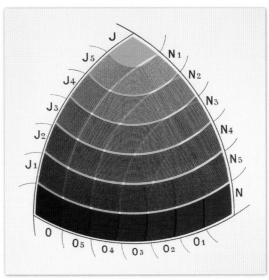

Figure 9.24. *Lacouture's Orangé-Jaune
(Lacouture 1890, plate XXIV).*

189

HERMANN HOFFMANN   1892

H. Hoffmann, *Systematische Farbenlehre*, 1892

Hermann Hoffmann was a book printer in Berlin, with professional interest in color printing. In 1927, he published a book on type foundry work. Hoffmann worked for 10 years on his manuscript of *Systematische Farbenlehre* (Systematic color theory), revising it several times because of rapid changes in scientific knowledge on color. He kept up-to-date by maintaining contacts with Wilhelm von Bezold (see entry in chapter 10) and Artur König, both leading lights of color research in Berlin at the time. His profession also kept him in touch with printing ink manufacturers, where he learned the latest developments in colorant chemistry and technology.

In *Systematische Farbenlehre* Hoffmann described what he called a "technical color system" because it involves colorants rather than lights. He said of his system "the content of colors in our table is to be considered theoretical and one should not believe that the indicated mixture ratio would produce the indicated result with corresponding volumina or weights of colorants" (Hoffmann 1892, p. 24, translated by R.G.K.).

Hoffmann described his system in tabular form only (figure 9.25). It consists of 30 hues, half of which are termed warm and the other half cold. The table lists a lightness value relative to white = 48, as well as the theoretical content of primary yellow, red, and blue.

The circle is a "natural" one in Hering's sense with the colors shown at their presumably highest saturation. The theoretical content varies in increments of 1 on a scale of 0–5. Although Hoffmann implies a continuous circle, he does not show one because the fact that circles have no beginning or end makes them, in his opinion, useless for practical application. The three primary colors are identified with numbers 4, 14, and 24; the intermediate secondaries, with 9, 19, and 29. In addition to a number, every hue has a distinct name based on simple color words.

The lightness values do not represent physical measurements. The lightness sum of two compensating colors always adds up to 48, the value for white, thus indicating a commingling of concepts related to colored lights and objects. Lighter colors are obtained by diluting the full-color ink with binding resin; darker ones, by including black ink. Hoffmann stated: "With small additions of black the color is dulled, with larger ones it is broken" (p. 37).

In the fourth column from left Hoffmann listed the names of colors obtained from a given hue by adding a neutral black. In the color tables (figure 9.26), he showed three levels of dilution with black. In other tables, the primary and intermediate colors are diluted in one step with binding resin and then blackened in two steps. In a gray scale, the white of paper represents grade 1 and black grade 5 (figure 9.27).

No.	Reihenfolge der Farben.	Lichtwerth (Weiss = 48)	Veränderung durch beigemischtes Neutralschwarz.	Theoretischer Inhalt an		
				Roth	Gelb	Blau
1	Roth-Violett	17	Grau-Violett No. 1	5	0	3
2	Violettes Roth . . .	20	do. „ 2	5	0	2
3	Violettfarbenes Roth .	23	do. „ 3	5	0	1
4	**Roth**	26	**Braun-Roth.**	**5**	**0**	**0**
5	Orangefarbenes Roth	27	do. „ 5	5	1	0
6	Orange-Roth . .	28	do. „ 6	5	2	0
7	Roth-Orange . .	29	do. „ 7	5	3	0
8	Röthlich Orange	30	do. „ 8	5	4	0
9	**Orange** . . .	31	**Braun.**	**5**	**5**	**0**
10	Gelblich Orange . . .	32	do. „ 10	4	5	0
11	Gelb-Orange . . .	33	do. „ 11	3	5	0
12	Orange-Gelb	34	do. „ 12	2	5	0
13	Orangefarbenes Gelb	35	do. „ 13	1	5	0
14	**Gelb**	36	**Grau-Gelb.**	**0**	**5**	**0**
15	Grünlich Gelb . . .	34	Oliv-Grün „ 15	0	5	1
16	Grün-Gelb	31	Oliv-Grün No. 16	0	5	2
17	Gelb-Grün	28	do. „ 17	0	5	3
18	Gelblich Grün . . .	25	Grau-Grün „ 18	0	5	4
19	**Grün**	22	**Grau-Grün.**	**0**	**5**	**5**
20	Bläulich Grün . . .	21	do. „ 20	0	4	5
21	Blau-Grün	20	do. „ 21	0	3	5
22	Grün-Blau	19	Grau-Blau „ 22	0	2	5
23	Grünlich Blau . . .	18	do. „ 23	0	1	5
24	**Blau**	17	**Grau-Blau.**	**0**	**0**	**5**
25	Violettfarbenes Blau	16	do. „ 25	1	0	5
26	Violett-Blau . . .	15	do. „ 26	2	0	5
27	Blau-Violett . .	14	Grau-Violett „ 27	3	0	5
28	Bläulich Violett . .	13	do. „ 28	4	0	5
29	**Violett** . . .	12	**Grau-Violett.**	**5**	**0**	**5**
30	Röthlich Violett . . .	14	do. „ 30	5	0	4

Technisches Farben-System. 25

WARME FARBEN — KALTE FARBEN

Anschluss an No. 1.

Hoffmann, Systemat. Farbenlehre.

Figure 9.25.
Tabular representation of Technisches Farbsystem *(Hoffmann 1892).*

The treament of interior colors in the implied color space is less than systematic and based on theoretical mixing ratios.

Hoffmann's system encompasses a total of 143 different color samples: 30 full, 90 broken, 18 tint/shade, and 5 achromatic colors, including paper white. Hoffmann's main purpose was to generate color standards for printing. He was able to induce König to "scientifically determine the wavelengths of the main colors, whereby the scale takes on durable value" (p. 4). Hoffmann proudly listed the names of two German printing ink makers that offered ready-made printing inks with his color number printed on the can and in the price lists.

Hoffmann did not contribute anything new to color order but, with August Ludewig Pfannenschmid (see entry in chapter 4), was among the pioneers that actively worked on standardization of the color samples in the system.

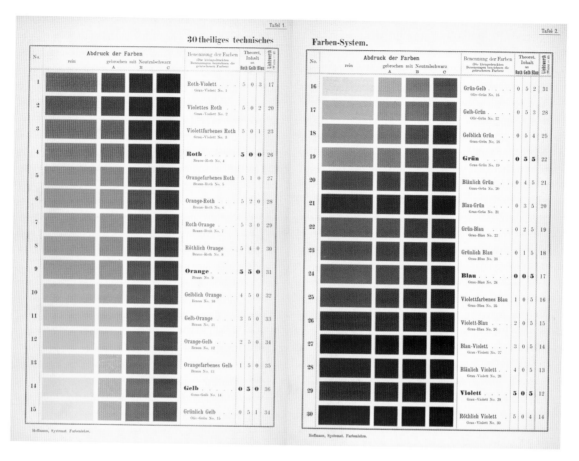

Figure 9.26.
*Charts 1 and 2 from Hoffmann's (1892) book showing full colors
and tonal variations in agreement with the table of figure 9.25.*

Figure 9.27.
*Chart 21 showing the gray
scale (Hoffman 1892).*

ROBERT STEINHEIL   1896

*R. Steinheil, La reproduction des couleurs par la superposition des trois
couleurs simples, 1896*

Robert Steinheil was employed in Paris by the French publisher and printer Berger-Levrault (which today is a publishing and software firm) when he produced *La reproduction
des couleurs par la superposition des trois couleurs simples*
(Color reproduction by superimposition of the three simple
colors, 1896). It is an ambitious work with nearly 15,000
samples obtained from halftone printing with three chromatic primaries yellow, red, blue, and black. It was an attempt
to demonstrate the capabilities of three- and four-color printing with a photo-reproductive technique superior to the lithographic method that was used for Lacouture's plates (see
entry in this chapter).

The samples are shown in circular diagrams on 150 plates
(figures 9.28–9.30). The plates are printed by *similigravure*
(halftone printing process). The resolution, according to the
"American system," is 3,025 points per square centimeter.
Increments in dot size are in 10% steps.

Steinheil listed Chevreul, Ogden Nicholas Rood, and Lacouture (see entries in chapters 4, 10, and this chapter, respectively) as information sources for his work. The printing
inks were selected "to be as much as possible in agreement
with Chevreul's simple colors" (Steinheil 1896, p. 2, trans.
by R.G.K.); that is, red and blue are different from magenta
and cyan. Inks were printed singly (figure 9.28), with partial
overlap as binary combinations (figure 9.29), as well as ternary ones (figure 9.30).

The *gammes lavées* (tint scales) were obtained in 10 grades
by halftone scaling. To mix the three basic colors, yellow
was always printed first, followed by red, and then blue. In
the first printing step, 10 grades of the simple color were
obtained (around the circle). Over-printing with red resulted in 10 grades of red plus 100 mixed grades of yellow
and red (distributed over multiple plates). Adding blue in
the third step resulted in a total of 1,330 color samples on
133 plates.

The samples are identified by three numbers, the first always
indicating the amount of yellow, the second red, and the third
blue, printed at the outer circumference in large type (see
figures). *Gammes rabattues et grisées* (grayed scales) were
obtained by partial over-printing with 10 halftone grades
of black, the latter also as the result of subtractive mixture.

Not satisfied with showing only these effects, Steinheil also
partially over-printed plates with varnish, resulting in intensified colors and some hue changes. When printed on color
fields without black, the resulting scales are called *gammes
brillantes* (bright scales). When printed on the blackened
fields, they are called *gammes glacées* (iced scales).

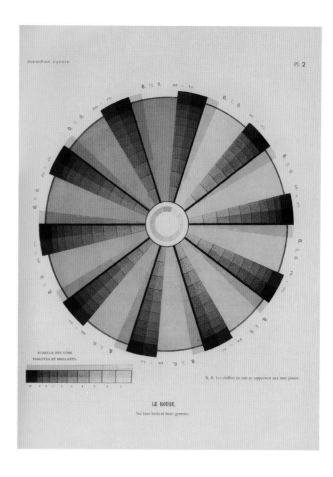

Figure 9.28.
Planche *(plate) 2, colors 0.1.0 to 0.10.0, basic
red ink only in 10 grades, with 10 grades of
blackness added, as well as partial over-printing
with varnish, resulting in the intensified bright
and iced scales (Steinheil 1896).*

The numbers in smaller font, printed along the circumference, identify colors complementary to the color shown
in the segment, using the same identification scheme. The
complementary colors are calculated by subtracting the values of the original color each in turn from 10. The resulting
numbers represent the complementary colors implicit in the
system. Appended tables identify the plates on which colors
with commonly known names can be found. Another table
lists all 1,330 colors by designation and plate location.

The complete set of plates was printed on three different qualities of paper (only one quality per copy of the book), and
Steinheil remarked on the substantial differences obtained as

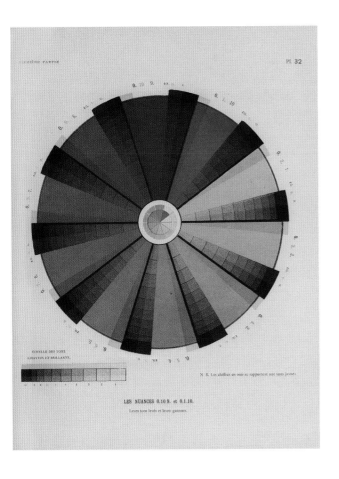

Figure 9.29.
Planche *32, colors 0.1.10, 0.2.1, 0.3.2 . . . 0.10.9,*
combinations of red and blue basic inks only, with 10
grades of blackness and partial over-printing with
varnish (Steinheil 1896).

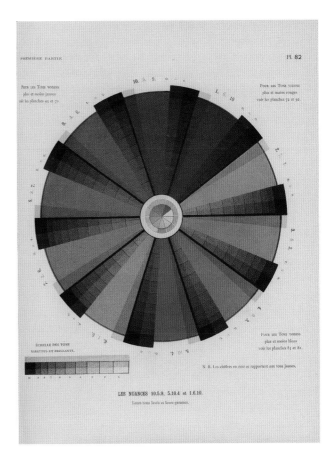

Figure 9.30.
Planche *82, colors 1.6.1, 2.7.2, 3.8.3 . . . 10.5.10,*
combinations of yellow, red, and blue, with black
and varnish over-printing as in figures 9.28 and
9.29 (Steinheil 1896).

a function of paper quality. Based on his practical experience showing that the printing sequence of inks has a considerable effect on the results, he demonstrates results (mostly of two-color combinations) printed in different ink sequences in plates 134–149. He also planned to issue other volumes with the complete set of 1,330 samples printed in different ink sequences (it is not clear if such volumes were actually produced).

Steinheil's work demonstrated the large number of samples of different color obtainable from three- and four-color printing by the (then) recently invented method of halftone printing. Although he was optimistic about the future of this technology, he also clearly showed many of its problems,

dealt with in the industry over the next decades before the process was sufficiently robust for routine industrial usage. As a color order system, it is incomplete, and it contains a degree of duplication of colors achieved with halftone printing with the three chromatic inks as well as with four inks.

JOHN CIMON WARBURG   1899

J. C. Warburg, A new three-colour chart, 1899

Although the principles of three-color printing were developed in the 1880s, widespread commercial application was slow in coming. Toward the end of the nineteenth century, charts began to appear that showed the colors obtainable from systematic application of three-color printing. In 1896, in England, S. Shepherd published a plate with the three process colors in five gradations, all two-color combinations, and some three-color combinations. In the same year, Steinheil published his extensive set of plates (see entry in this chapter).

In 1899, English photographer **John Cimon Warburg** (1867–1931) published a plate with only three gradations per ink, but all possible two- and three-color combinations (figure 9.31), and in systematic form.

> My object in making a new chart was to obtain *all* the possible mixtures which could be made with three colours, by means of one and the same pattern printed in three colors and superposed at different angles. I found that the pattern was necessarily rather complicated, and for this reason I have only used three gradations of each colour. . . . I originally designed this method in order to see the mixture of colours obtainable with three coloured transparencies superposed. (Warburg 1899, p. 13)

Identification of the inks used in different sectors is found in figure 9.32.

Warburg used Martius yellow, erythrosine, and Bleu Carmin N as the three dyes in the transparencies, commenting that they are "nearly complementary to the three primary colours" (pp. 13–14). The chart is printed with "Fleming & Co.'s Photochromic Inks" (p. 16A). Concerning the printing sequence, Warburg commented as follows:

> [W]hen the three colored inks are superposed the last printed is almost bound to show most. In order to minimize this, it is best to print the blue last, as preponderance of this colour in mixtures is less noticeable. If an opaque yellow is used, it must be printed first, but if a transparent yellow lake can be employed it is preferable to print it *over* the red. (p. 14, emphasis original)

Warburg's chart is an early effort to show (limited) systematic results of halftone-printed three-color mixture in a two-dimensional chart.

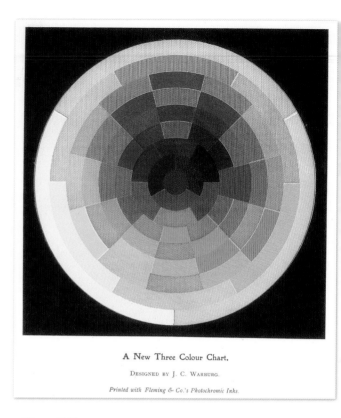

A New Three Colour Chart.

DESIGNED BY J. C. WARBURG.

Printed with Fleming & Co.'s Photochromic Inks.

Figure 9.31.

Warburg's (1899) "New Three Colour Chart."

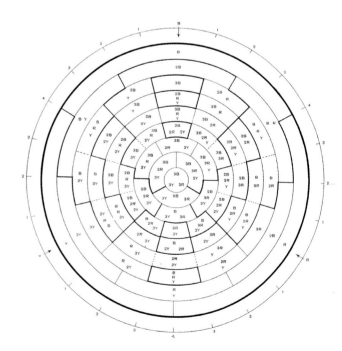

Figure 9.32.

Schematic drawing of the chart with identification of the composition of the color sectors (Warburg 1899).

NATHAN CHAVKIN   1912

Nathan Chavkin, Chavkin's Farbenkreis, 1912

In 1912, German author **Nathan Chavkin** arranged to print his booklet entitled *Chavkin's Farbenkreis* with Munich-based art poster printer Otto Schmidt-Bertsch. The booklet contains a color chart with approximately 1,500 different samples (figure 9.33). The print production of the color chart employed yellow, red, blue, and black process inks, with white paper supplying the whiteness content of the samples.

The *Farbenkreis* (color circle) begins in the center. The innermost circle consists of the three perceptual primary colors, which Chavkin obtained by superimposed printing of chrome yellow, purple-red or carmine red, and ultramarine blue, their combination resulting in blackish brown. In the adjoining ring, the three chromatic inks are printed with partial overlap, so that next to the individual colors the so-called *Doppelfarben* (double colors) dark violet, green, and orange or cinnabar are obtained. Ring 3 contains the three primary colors lightened in each case in five grades to white, numbered from 1 to 5.

Rings 4–8 have six sectors with 25 colors each. Three sectors are mixtures of pairs of the process inks in five grades, each in five grades of lightness. The remaining three sectors are of comparable composition, but over-printed with the third process ink at full color strength. In rings 9–13, all three process inks are printed in various concentration ratios, the numerical values of which are found in the columns of the adjacent ring. The two outer groups of rings, 14–18 and 19–22, contain the chromatic process inks of the inner rings, now shaded with increasing amounts of black, resulting in many grayish tonal colors.

Chavkin's two-dimensional *Farbenkreis* is an early, unique example of a printing ink color system. Black is generated in two ways: Not only is it the result of mixing the three chromatic process inks, but it is also represented by the fourth primary ink.

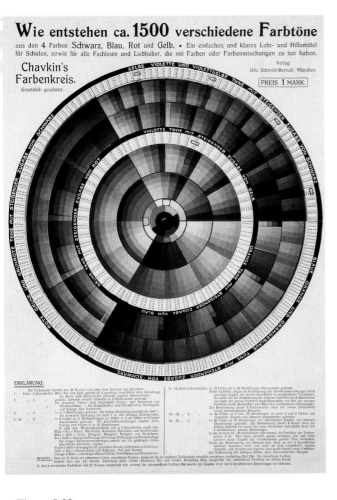

Figure 9.33.

Chavkin's Farbenkreis *(1912).*

(Bayerische Staatsbibliothek München)

A. PAINTER   1936

A Painter, *Tous les coloris. Répertoire des couleurs, tons nuances,
avec teneurs & en couleurs primaires*, 1936

The identity of the author is unknown, and the name may be a pseudonym. In the preface. **Painter** noted that at the time of his publication, French color atlases were regrettably lakking, even though some had been published in France. He was presumably referring to those of Gaspard Grégoire and Chevreul (see entries in chapter 4) and of Steinheil and Lacouture (see entries in this chapter).

The system is colored using the halftone printing process. Painter did not identify the process inks used but described them as "standardized according to the British system" (Painter 1936, p. VIII, trans. by R.G.K.). The process inks blue B, red R, and yellow J are printed in steps of 10% screen density in scales of 100 grades (figure 9.34), as well as in triangular format with systematically changing ratios. The resulting solid is a double tetrahedron consisting of 18 triangles, from near white to near black (figure 9.35).

The basis triangle, designated *triangle chromatique*, is located on plate 9 and contains a total of 121 colors (figure 9.36). Here the primary inks are printed in 11 grades from 0 to 100% as well as in combinations. There is no central gray. Eight triangles are shown from the basis triangle toward white, ending two steps before white, and nine toward black. In both directions, the size of the triangles, and thereby the number of colors, decreases, caused by elimination of one percentage level.

The solid contains a total of 882 colors, including 45 duplications. Some of the duplicated colors differ significantly in appearance, partly due to surround, but also due to variables in the halftone printing process. The work includes a transparent sheet imprinted with the process inks in steps from 10% to 80% to be used as overlay to compare between printed colors on the paper pages. In addition, a separate triangular mask helps to isolate specific colors on the plates (thereby changing their appearance).

Each color is identified by the three halftone screen percentages of the process inks. Plate 9 contains the comments: "The precision of the numbers given is limited by the accuracy of the engraving" and "The tonal scales proceed in a manner not perfectly uniform. That is because the visual impressions are not perfectly proportional to percentages of pigments (Fechner's law)."

In the brief user's guide, Painter recommended the system for "appreciating color differences" by "analyzing both colors compared [in terms of process ink percentages] and visually comparing the two sets" ("General instructions for usage"). Painter saw his system as having application for

"engravers, painters, designers, architects, decorators, dyers and colorists, tradesmen, naturalists, philatelists, writers, linguists, and manufacturers of dyes, pigments and inks" (p. I).

The sequence of printing of the plates was blue, red, and yellow. Most plates appear to be created from overlays of transparent sheets. For unidentified reasons, greens are mostly missing, and saturated oranges and purples/violets are entirely missing. Another problem of the printing sequence used appears on plate 18, where the central triangle is composed of 90% each of blue, red, and yellow. Thus, it has 80% more pigment content than the surrounding three samples. Yet it appears to be the lightest color because yellow partly obscures red and blue.

Painter's system is the first completely systematic halftone pigment mixture system representing three specific process inks. Copies of it are very rare.

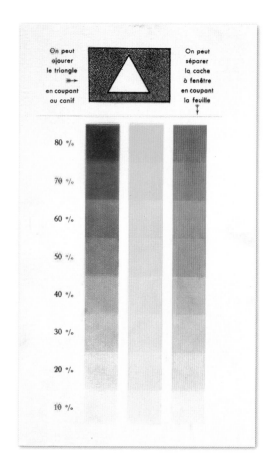

Figure 9.34.

Halftone grades of Painter's process inks blue, yellow, and red (Painter 1936).

(Figures 9.34. to 9.36. Abteilung Historische Drucke, Staatsbibliothek zu Berlin – Preussischer Kulturbesitz / bpk)

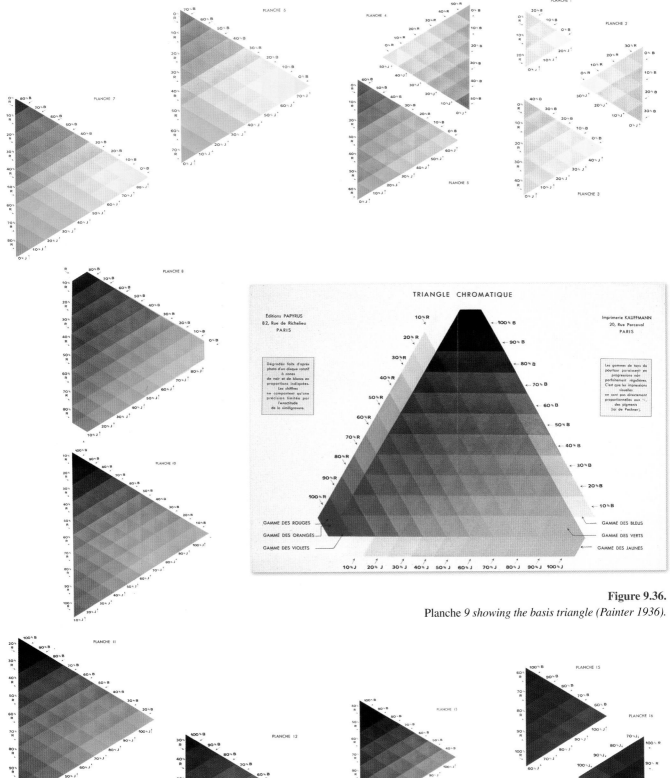

Figure 9.36.
Planche *9 showing the basis triangle (Painter 1936).*

Figure 9.35. *Painter's (1936) complete color order system with 18 planches on 11 charts.*

LUDWIG GERSTACKER   1934

L. Gerstacker, *Das Farbenmischbuch*, 1934

Because of process control problems, general commercial acceptance of color printing with three chromatic inks and black was not achieved until well into the twentieth century. But growing printing volume and cost pressure made the technology of growing interest. Tools were needed for achieving inexpensive, accurate color reproduction in halftone printing. In 1934, German printer **Ludwig Gerstacker** produced such a tool in the form of *Farbenmischbuch* (Book of color mixture), in cooperation with the printing ink manufacturers Förster & Börries from Zwickau, Saxony, and Kast & Ehringer from Stuttgart.

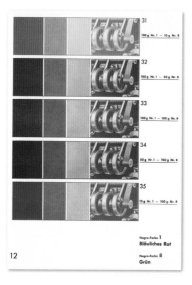

Figure 9.37.

Mixtures in different weight ratios of the Nagra bluish red and green inks (Gerstacker 1934).

A previous book of this kind, Müller and Dethleffs's *Praktischer Leitfaden für den Buntbuchdruck* (Practical compendium for book printing in color, 1900; see figures 9.5–9.7), which introduced Nagra printing inks, no longer applied because of technical advances in the process. Gerstacker used 13 chromatic Nagra inks in his book: bluish red, red, yellowish red, orange, reddish yellow, middle yellow, greenish yellow, green, greenish blue, blue, Milori blue, reddish blue, violet, and neutral black.

Gerstacker mixed pairs of chromatic inks in every possible combination, resulting in 390 chromatic colors. For each of these colors, he showed two tint grades as well as a practical example, with the mixture formula by weight (figure 9.37). Samples measure 2.3 × 2.4 cm and 1.7 × 2.4 cm, respectively. He followed this with a series of tint colors of selected pairwise mixtures, obtained by over-printing with transparent white, resulting in a total of 84 additional colors, each with two lighter versions and a practical example.

Each of the 13 original inks is then printed in the halftone process in five grades reduced to white (figure 9.38, leftmost column). Each member of this tint scale is blackened according to the five grades of the gray scale (figure 9.38, top row), resulting in 36 tonal combinations measuring 1.8 × 1.8 cm. Twenty-eight more charts are similarly generated, but black is replaced by another original ink (figure 9.39). The collection is completed with four unsystematically arranged images of various two- to four-ink combinations (figure 9.40).

Gerstacker's *Farbenmischbuch* is a collection of several more or less systematic system components,[9] not a complete system of colors. In principle, the first part is comparable to Müller and Dethleffs's work, but the remainder far exceeds it. A successor of sorts to Gerstacker's work is Hans Gaensslen's *Das grosse Drei-Farben-Mischbuch* (The great book of three-color mixture, 1959). However, when it was published in 1959, it was already superseded by the work of Hickethier, who finally developed a complete system applicable to halftone printing (see entry in this chapter).

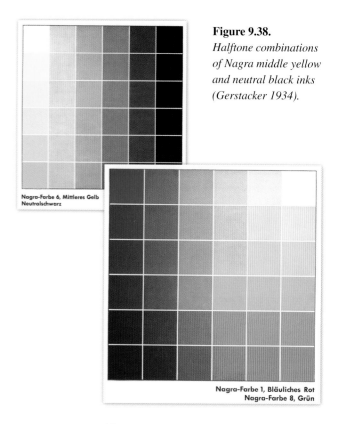

Figure 9.38.

Halftone combinations of Nagra middle yellow and neutral black inks (Gerstacker 1934).

Figure 9.39.

Halftone combinations of Nagra bluish red and green inks (Gerstacker 1934).

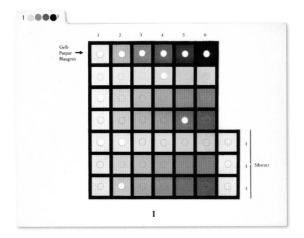

Figure 9.40.

Various halftone combinations of two to four Nagra inks in different ratios: middle yellow, bluish red, greenish blue, and reddish blue (Gerstacker 1934).

CARL BLECHER   1943

C. Blecher, *Die Herstellung mehrfarbiger Reproduktionen nach bunten Durchsichtsbildern durch Druck*, 1943

Carl Blecher was a German professor at the photomechanical institute of the Academy for Graphical Arts and the Book Trade in Leipzig. He was interested in the exact reproduction in magazines or books of images captured on the then-new color transparency film. Connected with this effort, he developed a color order system that has a degree of similarity to one developed at nearly the same time by Hickethier (see entry in this chapter). Like Gerstacker (see entry in this chapter), Blecher used the Nagra printing inks, but only three chromatic inks and black, the inks typically used in the halftone process.

Blecher selected yellow, purple, and blue-green inks (Nagra 6, 1, and 9) as chromatic process inks based on the then well-understood idea that the inks' spectral reflectance functions should correspond as closely as possible to the transmission functions in the color transparency film's three dyes.

The first two charts of Blecher's publication *Die Herstellung mehrfarbiger Reproduktionen nach bunten Durchsichtsbildern durch Druck* (Print production of multicolored reproductions from color transparencies, 1943) show the three chromatic printing primaries in six halftone screen levels scaled with the help of a gray scale (figure 9.41). Next, they are printed in combinations of two and toned down with the addition of the first three levels of gray.

Six additional charts are based on the yellow-purple combinations of figure 9.42. Next, increasing amounts of blue-green are added to the first two chromatic primaries, resulting in halftone mixtures of all three chromatic primaries (figure 9.43). The charts contain a total of 342 color samples measuring 12 × 12 mm, printed on high-gloss art paper. Conceptually, they fit into a cube solid, a fact Blecher did comment on.

Figure 9.41.

Charts 1 and 2 contain the gray scale as well as the chromatic primary inks yellow, purple, and blue-green printed at six halftone levels, subsequently toned down with the first three gray grades (Blecher 1943).

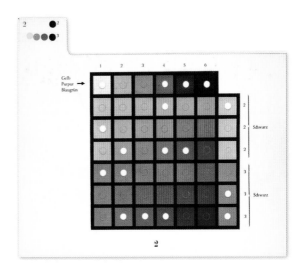

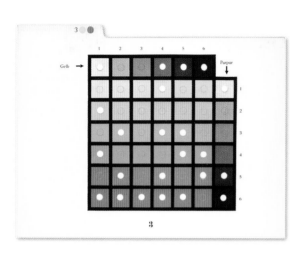

Figure 9.42.

Chart 3 with systematic mixtures of the chromatic primary inks yellow and purple (Blecher 1943).

Figure 9.43.

Chart 9 corresponds to chart 3 (figure 9.42), but with addition of blue-green ink at level 4 to each color (Blecher 1943)

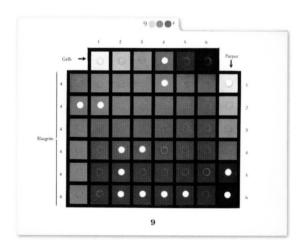

Figure 9.44.

Chart 14 corresponds to chart 3 (figure 9.42) but with addition of gray of level 3 to all colors (Blecher 1943).

Blecher's goal was to produce a practical system rather than a complete and systematic one. For example, he noticed that a full quarter of the toned-down colors were indistinguishable from each other, and thus superfluous duplications. He also noticed that some distances between the remaining samples were large, requiring intermediate grades. He produced an additional 507 color samples (also requiring reduction by one quarter because they lacked perceptual distinction) by toning down the halftone steps of the primary inks as well as the two ink combinations with the first three levels of the gray scale (figure 9.44).

Blecher recommended Wilhelm Ostwald's system (see entry in chapter 10) for identifying the colors on his 20 charts. He designed the charts to be used for matching colors of unknown formulation. For this purpose, samples have perforsted holes of 4 mm diameter (figures 9.41–9.44).

ALFRED HICKETHIER   1943

A. Hickethier, *Der Farbendruck verlangt eine Ordnung der Farben*, 1940
Farbenphotographie, Farbendruck und Farbenordnung, 1943
Farbenordnung Hickethier, 1952
Grosse Farbenordnung Hickethier, 1972

The innovative German color printing technician **Alfred Hickethier** (1901–1967) contributed not only a color order system that is considered useful today, but also color theory materials and publications that were translated into English, French, Spanish, and other languages.

As a printing technician, Hickethier knew well the technical problems when attempting to reproduce color photography with three-color halftone printing. To aid such reproduction, he produced a systematic order system of colors obtainable in the three-color halftone process. In 1940, he described his system of generating 1,000 color samples on basis of halftone printing as follows:

> As already mentioned, the complete order is based on decimal division and the three process colors. This is the reason for designating each color with a three-integer number. The first number describes the yellow content, the second the content of red and the third of blue. For example, the number 804 designates a pure yellowish green consisting of 8 parts yellow, 0 parts red, and 4 parts blue. In this manner exactly 1000 different colors are generated, designated from 000 to 999, with 000 designating pure [paper] white and 999 deepest black. It will be of interest that the pure process colors generate [by dilution with white] 27 colors (3 × 9), for binary combinations the number is 243 (9 × 9 × 3), and those generated from all three process colors add up to 729 (9 × 3 × 9 × 3). (Hikkethier 1940, pp. 110–111, trans. by R.G.K.)

The systematic ordering of the 1,000 colors in a cube was suggested by G. Kujawa (1940) and is found for the first time as a graphical model in Hickethier's paper of 1943 (figure 9.45). Three corners of the cube are occupied by the full-color versions of the primary colors, designated 900 for yellow, 090 for red, and 009 for blue. The full binary mixtures are located on three other corners, designated 909 for green, 990 for orange, and 099 for violet. The gray scale axis runs diagonally through the cube, from corners 000 to 999.

With the generative and order principles established, Hickethier had to solve the technical reproduction problems. The pigment concentrations in the inks needed adjusting to a colorant strength so that neutral grays resulted at equal "concentrations" in the triple combinations. He used an (undefined) 10-grade gray scale and adjusted the screen densities of the three process inks to closely approximate the neutral grays of the gray scale in equal "concentration" at different depths and to obtain a deep black in the highest concentrations (figure 9.46). In the 1943 text, Hickethier no longer used the

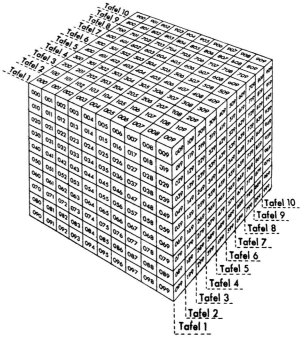

Figure 9.45.
Graphic model of the color cube of 1943. Unlike Küppers and William Benson (see entries in chapters 11 and 10, respectively), Hickethier did not tilt the cube to result in a vertical gray axis (Hickethier 1943).

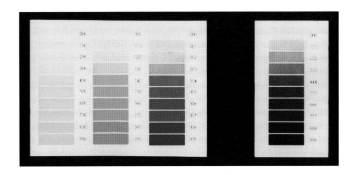

Figure 9.46.
Primary colors yellow, purple, and blue-green in 10 grades that when printed together result in the gray scale (Hickethier 1952).

terms yellow, red, and blue for the primary colors but instead yellow, purple, and blue-green, alternative names for the now customary process color names yellow, magenta, and cyan.

As World War II raged, the first coloring of the system was generated at Druckerei Osterwald in Hannover, where Hikkethier worked. A few hours after the first edition was printed, a bomb struck and destroyed the plant. Only in 1952 Hickethier was able to issue the *Farbenordnung* (Color order system) in the form of a book.

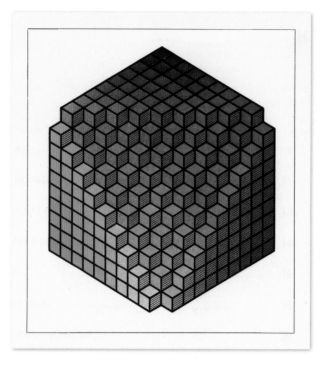

Figure 9.47.
Illustration of a diagonal section of the cube through central gray (Hickethier 1952).

The book contains a general section on color theory as well as a detailed explanation of the system's structure. The system itself is illustrated on 10 pages, each with 100 color samples measuring 1.7 × 1.7 cm. The book contains several images of sections through the cube. An example is shown in figure 9.47.

Impressive for its time, the system was less than perfect in terms of colorant selection and plate densities, a situation most clearly noticeable in the gray axis. Its samples are noticeably tinted.

Hickethier continued intensive work on the system and arranged in 1967, the year of his death, to have a revised version of the color samples printed. It was issued in 1972, supervised by Siegfried Rösch (see entry in chapter 6), whom Hickethier had asked to provide colorimetric data for the system, as *Grosse Farbenordnung Hickethier* (Large color order system Hickethier). Each of 40 removable pages of the much-improved edition contains 25 samples measuring 2 × 2

cm, separated on the page for easier comparison (figure 9.48). For higher depth of color, the samples in both editions are covered with transparent film. Figure 9.49 shows the comparable sections through the cube from the 1952 edition

Building on the ideas of Tobias Mayer and Johann Heinrich Lambert (see entries in chapter 4), Hickethier completed the development of a cubic color order system based on three primary printing inks. It was clear to him that each printer should develop an in-house system according to his principles and based on the process inks in use. But he also saw value in standardization of process inks such as DIN and Euroscale standards, or the Pantone process color-specification system. Hickethier thought his system to be adaptable to different numbers of inks or plate density levels. He also suggested a paint version of the system with the paints mixed according to his scheme, an example being Aemilius Müller's cube of 1951 (see entry in chapter 5).

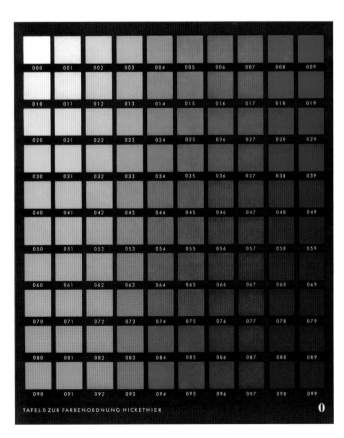

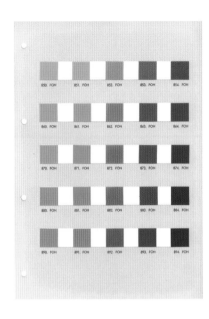

Figure 9.48.
Plate 34 from Hickethier (1972) containing the 25 colors of the lower left quadrant of plate 8 of the 1952 edition, and plate 1 of the 1972 edition containing the 25 colors of the upper left quadrant of plate 0 of the 1952 edition.

Figure 9.49.
Plates 8 and 0 from Hickethier (1952).

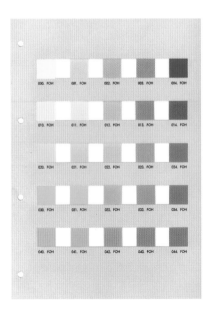

HERBERT E. IVES   1935

H. E. Ives, *Color chart and method of making the same*, 1935/1938
International Printing Ink Corporation, *The Ives trichromatic palette*, 1935

American electro-optical physicist **Herbert Eugene Ives** (1882–1953), son of Frederic Eugene Ives, was a prolific inventor. During his employment at Bell Laboratories, he obtained more than 100 patents and was one of the pioneers of black-and-white and color television and the wire photo process. In 1935, he applied for a patent on a color chart, granted to him in 1938 (U.S. Patent 2,128,676). It describes a hexagonal chart printed with partial overlaps with the three printing primaries. An implementation of the chart was published in 1935 by the International Printing Ink Corporation as *The Ives Trichromatic Palette*.

The color chart is produced using the halftone process in either a continuous or a graded version. Pure full-strength printing inks are located in three corners of the hexagon (figure 9.50, points A, C, and E). Intermediate full-strength colors are found at corners B, D, and F, with gradations along the periphery. Printing is toned down either continuously or in grades toward the center. Figure 9.51 schematically illustrates the print plate for one ink, with the full colors at the periphery (2, 3, 4) toned down toward white at line 1, 0, 5. In the graded version, small rhombi of constant color fill the hexagon (figure 9.52). Ives's color chart is a precursor of the set of charts patented by Wedlake in 1969 (see entry in chapter 10).

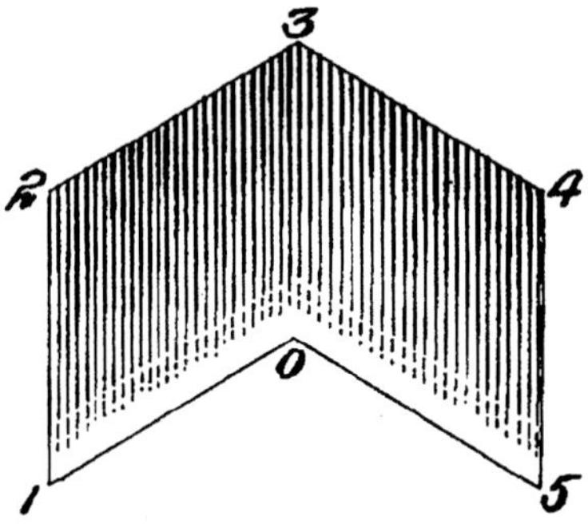

Already in 1935, a continuously toned version of Ives's color chart was published by an important printing ink manufacturer, the International Printing Ink Corporation under the name *The Ives Trichromatic Palette* (figure 9.53). The primary printing inks used are magenta, yellow, and cyan, interestingly named *Achlor* (anti-green), *Zanth* (yellow, from the Greek *xanthon*), and *Syan* (cyan) in patented versions. The chart is of size 9.5 × 12 inches. A separate device consisting of a continuous transparent gray filter with a movable hexagonal mask makes it possible to darken the colors of the chart from white to near black. The system represents all colors obtainable by halftone printing with the three printing primaries involved. Quantitative information about the amount of pigments at any given point can be approximated, as shown in figure 9.50.

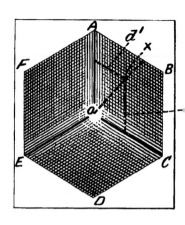

Figure 9.50.
Schematic drawing of Ives's hexagonal color chart (1938), with the pure primary inks located at points A, C, and E, and the results of their mixtures along the periphery. Colors are toned down to white at the center. The components of a given location can be determined by calculation.

Figure 9.51.
Schematic drawing of the area of the hexagon printed with one primary ink, toned down from full strength at the upper edge toward white at the lower edge (Ives 1938).

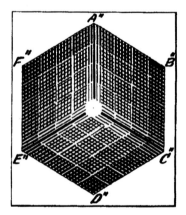

Figure 9.52.
Schematic drawing of the hexagonal chart in the graded version with rhombi having uniform colors (Ives 1938).

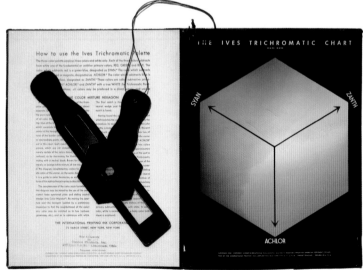

Figure 9.53.
Image of the Ives Trichromatic Palette *system, with the trichromatic chart on the right and the attached darkening device, published by International Printing Ink Corporation, 1935.*

LTF/GATF COLOR CHART   1957

Lithographic Technical Foundation, *LTF color chart*, 1957

In 1957, the **Lithographic Technical Foundation** issued a set of master films representing nine sections through the three-color halftone cube, and seven of these darkened always with five levels of black. The set was based on the research of B. E. Tory, F. M. Preucil, and E. Brody. At the time, the foundation was located in New York and Chicago (it's now called the Graphic Arts Technical Foundation [GATF] and located near Pittsburgh). Each section is a square with six columns and seven rows. Figure 9.54 shows a top surface square with white in the upper right corner, solid process yellow in the upper left, solid cyan in the lower right half-columns, and solid green in the lower left corner. The figure also shows the reduction steps for yellow and cyan.

Cyan is added to the corresponding yellow/magenta square in six grades of screen density (as identified for the columns of figure 9.54), the resulting squares placed into a row. Figure 9.55 shows such a square with additions of black. The amount of cyan added is shown in the upper right corner, the amount of black in the lower right corner. The remainder of the bottom half-row shows the combined result of the additions of cyan and black. In total, there are 44 squares of this kind. The master transparencies were shot at the following angles: yellow 90, magenta 75, cyan 15, and black 45 degrees.

The size of the complete chart is a comparatively handy 29 × 22 inches, useful for in-plant consultation. A strip on the right side of the chart allows notation of all critical production parameter values. The chart was sold in the form of master films so that any company could use standard films to produce in-house charts that demonstrate the results in a comparable fashion for a given set of inks, paper, printing conditions (including room temperature and relative humidity), ink sequence, and other parameters.

The LTF/GATF color chart represents a significant advance in color quality control of halftone printing in the United States. It allowed this printing technique to make significant inroads into territory that had been occupied by letterpress printing. At the same time, the chart was very useful in communication between printers' sales representatives and customers.

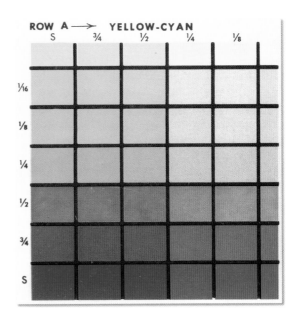

Figure 9.54.
Yellow-cyan process ink mixture square of the LTF chart (Lithographic Technical Foundation 1957). Reprinted with permission.

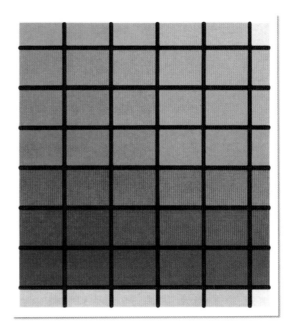

Figure 9.55.
One of the 35 four-ink mixture squares of the LTF chart (Lithographic Technical Foundation 1957). Reprinted with permission.

H. WEZEL   1959

H. Wezel, *Farbmass*, 1959

Like his predecessors Hickethier and Blecher (see entries in this chapter), German lithographer **H. Wezel** was concerned with high color fidelity in the printed reproduction of color transparencies and photographic negatives. To support this goal, he created color charts as a tool for four-color offset printers. He based his charts on DIN 16509, the official German standard color scale for halftone printing, introduced in the mid-1950s. Wezel's *Farbmass* (Color measure; figure 9.56) was to aid rapid and certain identification of specific colors, shade matching, and quality control.

DIN 16509 has three chromatic standard inks, yellow, purple, and cyan, in addition to black, also known as *Tiefe* (depth). In the standard, the printing sequence of the four inks is left to the printer. Wezel settled on the sequence yellow–black–purple–cyan in screen coverage levels of 10%, 30%, 50%, 70%, and full, designated 0 for the paper basis white, 1, 3, 5, 7, and 9. He chose a limited number of levels to assure that individual color patches can be distinguished without difficulty.

Colors are identified according to Hickethier's system (see entry in this chapter), expanded to four digits with the inclusion of black. The digits refer to the screen density values of the standard inks. On basis of the chromatic inks alone, 6 × 6 × 6 = 216 colors result, but adding black in five levels expands this number to 1,296 different colors. The colors are presen-

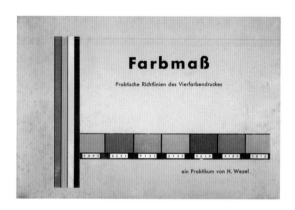

Figure 9.56.
Title page of Wezel's Farbmass *(1959).*

ted on four atlas pages, with each page containing nine fields with 36 samples measuring 14 × 7 mm each (figure 9.57). Wezel repeated this arrangement on different types of paper and with matte and glossy inks, resulting in a wide representation of color appearances achievable in offset printing.

Wezel was among the first to translate the then-new DIN standard for printing (Deutsche Industrie Norm 1955) into the form of a well-executed, practical color atlas. In the atlas, Wezel announced comparable editions for four-color book printing and three-color offset and book printing for which comparable DIN standards had been established. However, they do not appear to have been produced.

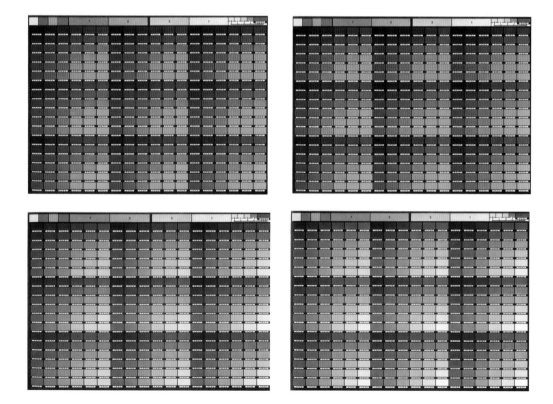

Figure 9.57.
The four atlas pages of the Farbmass *printed on chromo paper (Wezel 1959).*

FDGB FARBMESSTAFELN late 1950s

Freier Deutscher Gewerkschaftsbund, *Farbmesstafeln für Vierfarbenbuchdruck*, n.d.

At about the same time and in a comparable manner to Wezel's system (see entry in this chapter) of (then) West Germany, the Industry Association for Graphical Trade and Paper Processing of Freier Deutscher Gewerkschaftsbund (**FDGB**; Free German Association of Unions) of (then) East Germany issued a series of six color charts for halftone book printing. Like Wezel's system, it was created as a tool for the production of sample prints and for the production control of printed matter. Also like his, the basis for both systems was Blecher's color cube (see entry in this chapter). Unlike Wezel's, the FDGB system did not become an official standard, and the systems differ to some degree.

Chart 1 contains prints of the basic colors red and blue, mixed in 10 grades of increasing halftone density. The mixtures result in a square of 100 colors, with red, blue, violet, and white in the corners. In five additional charts, the original chart is over-printed with five, rather than ten, halftone grades of yellow. The selected halftone grades are 1, 3, 5, 7, and 9. As a result, the implicit cube is reduced to a square box (figure 9.58), and the chromatic basis system consists of six hundred 14 × 14 mm samples.

Each of the six charts is darkened with black at halftone levels 1, 3, 5, and 7, resulting in a total system with 3,000 samples. There is no level 9, as it would consist of 100 black colors. In case of the darkened samples, the color fields have been reduced to half size so that two levels of blackness are shown on one chart (figure 9.59). The charts are cut vertically into strips, fastened on top so that, for matching purposes, they are movable.

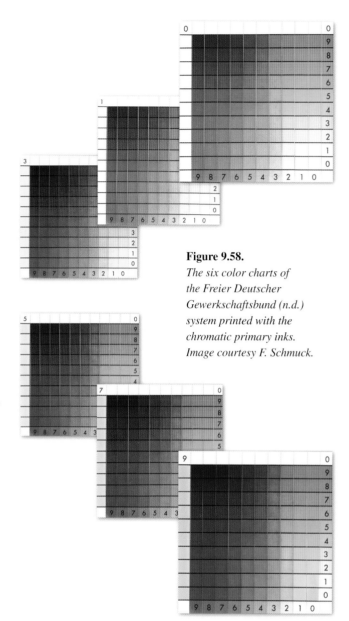

Figure 9.58.
The six color charts of the Freier Deutscher Gewerkschaftsbund (n.d.) system printed with the chromatic primary inks. Image courtesy F. Schmuck.

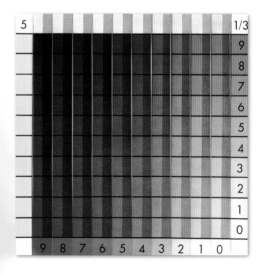 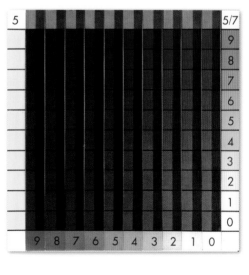

Figure 9.59.
Two charts of the FDGB system (Freier Deutscher Gewerkschaftsbund, n.d.) of yellow grade 5, one with black grades 1 and 3 (left), the other with grades 5 and 7 (right). Image courtesy F. Schmuck.

CARL E. FOSS   1973

C. E. Foss, Foss color order system, 1973

In 1973, **Carl E. Foss** (1906–1986), an American expert on color order and tolerance systems, received a patent for a way to represent a large number of color samples obtained by systematic variation of four-color halftone process inks in a two-dimensional format. He also was instrumental in developing the American version of Ostwald's system (see entry in chapter 10), the *Color harmony manual* (Jacobson et al. 1942; four editions, the last in 1958), and a number of government-sponsored color-reference systems.

An atlas of the cubic system implicit in three-ink mixture consists of many charts. The results (as demonstrated in the atlases by Küppers and the Rogondinos; see entries in chapter 11 and in this chapter, respectively) are difficult to fully comprehended, so it is difficult to find any specific color. Foss attempted to overcome this difficulty by organizing a total of 5,831 different samples on two charts of size 23 × 30 inches in an intuitive arrangement.

The solid produced by mixtures of the three chromatic inks is a conventional cube (figure 9.60). Interior views are shown in the form of smaller interior cubes, as indicated with dashed lines. Foss offered views of rhombic-surface sections of the cubes. In figure 9.60, the corners of cubes identify the three printing primaries yellow Y, magenta M, and cyan C, as well as their mixtures red R, green G, blue B, white W, and Black.

Foss's surface rhombi are shown in figure 9.61. On the left figure, a typical rhombus is formed by the surface area connecting points W, Y, Black, G, and back to W.

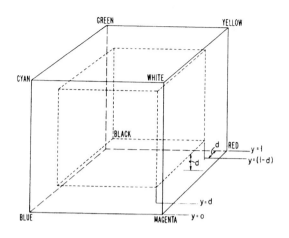

Figure 9.60.

Schematic view of the three-pigment mixture cube, with a reduced interior cube in dashed lines (Foss 1973). Reprinted with permission.

A cube surface is covered by six rhombi of this kind, as shown in two dimensions on the right side.

In the two charts, the six surface rhombi are shown as squares of four diminishing sizes, beginning with eight colors along the side of a square and reducing to six, four and two. One of the resulting charts is shown in figure 9.62. Foss chose the sample grades to result in approximately visually equal steps. The individual colors are identified with a three-digit number, the first representing (in grades from 1 to 8) yellow, the second magenta, and the third cyan.

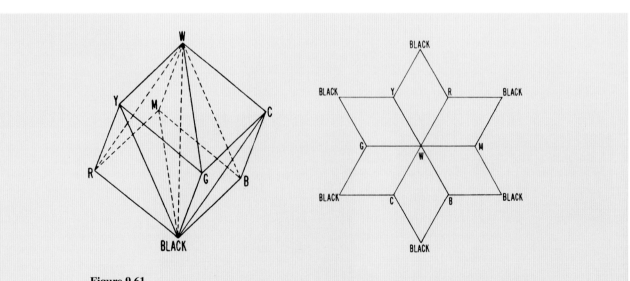

Figure 9.61.

Left: Tilted cube of figure 9.63 with lines bisecting the surface squares. Rhombi such as W, Y, Black, G, W form the basis of the system. Right: Flat representation of the six surface rhombi, with central white (Foss 1973). Reprinted with permission.

Figure 9.62.
Chromatic chart of the Foss system with the six rhombi of figure 9.61 stretched into squares, populated with diminishing numbers of samples as cube size is reduced (Field 2004). Reprinted with permission.

Figure 9.63.
Left: Square with black values 0, 1, 2, and 3; Right: square with black values 4, 5, 6, 7. These squares were used to over-print the color fields of the chromatic chart for part 1 and part 2 of the system (Foss and Field 1973). Reprinted with permission.

Figure 9.64.
Left: rhombus W, C, Black, G, W of part 1; Right: the same square of part 2 (Foss and Field 1973). Reprinted with permission.

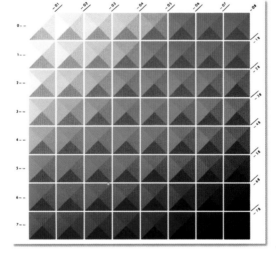 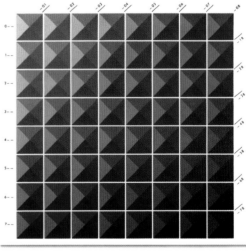

Foss created two figures to demonstrate the effect of over-printing with black (figure 9.63), adding black in eight grades (from 0 to 7 on his eight-grade gray scale). In part 1 of the Foss color chart, individual color squares are over-printed with the left figure of figure 9.63; in part 2, they're over-printed with the right figure. Two resulting squares of the charts are shown in figure 9.64. The sections of a particular unit square are identified by a four-digit number, with black content representing the fourth digit.

A particular advantage of the system was its availability as films. That made it possible to print specific charts representing an in-house printing process, both at the preproofing stage and at the corresponding production machine stage, thereby offering a quality control tool in a partly standardized format (Field 2004). By presenting the samples in a consecutive (implicit) hue circle format, the charts are intuitive, and specific color samples are relatively easy to locate. But the success of the system was somewhat limited because, soon after its appearance, color monitors began to replace color charts.

M. ROGONDINO and P. ROGONDINO   2000

M. and P. Rogondino, Process color manual, 2000

Michael and **Pat Rogondino**'s *Process color manual* of 2000 is one of the most extensive systematic collections of color samples produced by the halftone four-color printing process using standard process printing inks. Its authors operate a design firm in the United States.

The *Process color manual* is computer generated using Adobe Corporation's Illustrator software. It contains a total of 24,480 unique printed color samples of size 18 × 18 mm, 100 to a page. The samples are printed on high-whiteness coated paper stock. Film separation output is at 2,540 dots per inch with a halftone line screen of 175 lines per inch. The screens were printed at the standard angles of 15° for cyan, 45° for black, 75° for magenta, and 90° for yellow. The inks used are standard process inks.

The charts begin with the four process colors cyan, magenta, yellow, and black, each printed at 20 levels from 5 to 100. The remaining charts all have the same basic structure: On a two-page spread, one ink is printed in rows at five screen unit increments, and another one in columns at 10-unit increments. The first six two-page spreads contain all two-color variations of the four process inks (see figure 9.65 for an example).

The next 54 two-page spreads contain three-color combinations with a third process color added to each page. Typically chromatic inks have six levels of increments of the third ink (in units of 10), and black has three levels. The remaining 62 two-page spreads contain four colors in a similar organization, except for the addition of two inks, such as 30% black and 30% yellow (figure 9.66). These additions can vary by combination. The authors do not explain how they chose the printed combinations. Screens to view individual samples against a white, mid-gray, and black surround are included.

In this massive compendium, orientation by color is not easy because samples on many different pages fall into the same color categories. Although every presented sample is constitutionally unique, they are not all perceptually distinguishable. Gradations are often very subtle, which shows the power of modern technology to control the process to a high degree. It is evident that all samples fit within a cubic system formed by systematic printing ink variation. A similar color atlas, also based on the cubic system and with a similar number of samples, was published by Küppers in 1987 (see entry in chapter 11).

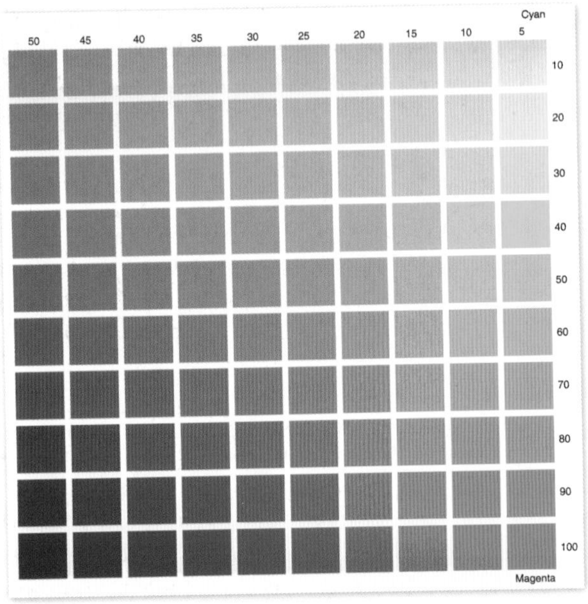

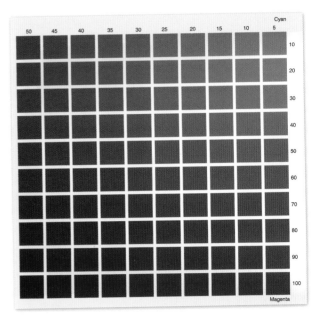

Figure 9.65.
Right half of the cyan/magenta two-color chart (Rogondino and Rogondino 2000). Reprinted with permission.

Figure 9.66.
Right half of the cyan/magenta chart with 30% black and 60% yellow added (Rogondino and Rogondino 2000). Reprinted with permission.

COLOR DISPLAY SOLIDS (RGB, HSB, CMYK)   1970s

Color display units began as color television sets first marketed in 1953 in the United States. Use of color display
units (monitors) in computers began with the introduction
in 1977 of the Apple II computer. The display units were
cathode ray tubes, invented in 1897 by German physicist
Karl Ferdinand Braun (1850–1918), then extended from a
single beam (black-and-white image) to three beams (color
image). Color images are created by additive mixture of
beams of red, green, and blue (RGB) light resulting from
electrical activation of three different phosphors.

The RGB color solid is a cubic representation of the additive
mixtures of these lights (figure 9.67). It can be envisaged as
a tilted cube in the manner of the cubes of Hickethier or Max
Becke (see entries in this and chapter 10, respectively).

At maximum output (relative value 255) of all three phosphors, the experience is "white." At a relative luminance of
128, half of the total scale, the single lights are typically
seen as intense lime green, scarlet, and blue. The spectral
nature of the light output of the phosphor compounds is
broadband and depends on the exact chemical nature of the
compounds. At zero electrical activation, the light output is
also zero and the implied color is black. Intermediate color
stimuli result from intermediate activation levels, a possible
total of 16.6 million. The gamut of color stimuli obtainable
in the system depends on the nature of the phosphor compounds but is smaller than the maximum possible.

Only some two million stimuli, identified by triple numbers,
are distinguishable as separate colors. The appearance of the
stimuli depends on the spectral output of the phosphor compounds, the surround stimuli, and the observer. The scaling
of the stimuli is far from perceptually uniform.

In the RGB cube, the stimuli are arranged in a cubic space
with all three dimensions ranging from 0 to 255. Orientation in an RGB cube is somewhat difficult because of the
nonintuitive effects of additive color mixture (e.g., green
light plus red light equals yellow light). For this reason,
the mixture combination is often displayed in the HSB
(hue, saturation, brightness) format (see Leo M. Hurvich
and Dorothea Jameson, chapter 5). A mid-gray stimulus
has RGB values of 128 each and HSB values of X, 0, 128
(X indicates that at zero saturation the hue number is ir-

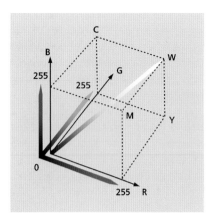

Figure 9.67.
*Schematic representation of the RGB
space in cube form.*

relevant). A saturated lime green stimulus has HSB coordinates of 85, 255, 128, that is, hue 85, relative saturation 255, and relative brightness 128. Once the hue scale
is internalized, a color stimulus with a given identification
can be visualized comparatively easily. A range of stimuli
with identical hue number but differing in saturation or
lightness is rarely of constant perceived hue because of the
nonlinear relationship between stimulus and perception.

It is often desirable to convert stimuli presented on a monitor to object color stimuli in the form of printed images.
For this purpose, conversions are calculated to express the
additive stimuli in form of subtractive printing primaries
CMYK (cyan, magenta, yellow, black) required to obtain
comparable experiences when printed with a particular ink
system. Such transformations are very complex and depend
on the quality of the colorants used and the printing process.
Well-known software programs such as Adobe's Photoshop
display stimuli in form of different solids, in which colors
can be selected. The corresponding RGB, HSB, CIELAB-
like L, a, b and CMYK values are provided simultaneously.
Displays are limited to stimuli achievable with the corresponding primaries of the unit.

JOSEPH WILLIAMS LOVIBOND 1893

J. W. Lovibond, *Measurement of light and colour sensations*, 1893

Joseph Williams Lovibond (1833–1918) was heir to a family that owned a beer brewery in London. At the age of 13, he entered the merchant navy, and in his later teens, he dug for gold in California for three years. Soon after his return to London, he assumed control of the business, adding a brewery in Salisbury.

Lovibond came to believe that the color of beer is a quality indicator. He invented an instrument, the Tintometer, to measure the color of transparent liquids as matched with combinations from three graduated sets of 20 each yellow, red, and blue color filters. The early Tintometer is illustrated in figure 9.68. In 1885, he founded the Tintometer Ltd. Company in Salisbury to manufacture the instrument. The company still produces various types of color-measuring instruments.

In connection with these activities, Lovibond concerned himself with scientific aspects of color, producing three publications: *Measurement of light and colour sensations* (1893), *An introduction to the study of colour phenomena* (1905), and *Light and colour theories and their relation to light and colour standardization* (1921). Near the end of his life, Lovibond conducted investigations into color camouflage for the British War Office.

The first book describes the Tintometer and its uses. The test liquid is placed in a glass cuvette and inserted into the instrument. A series of glass filters can be placed in slots next to it. By visual comparison, looking through the ocular, the color of the test liquid is matched in color appearance against combinations of appropriately selected filters, a task requiring some practical experience. The measured color of the liquid was identified with filter and filter-strength numbers.

A friend proposed a geometrical model to systematically represent the colors that can be measured with the Tintometer: "Dr. Munro, of the Downtown College of Agriculture, who has taken great interest in the progress, and has suggested the mathematical method of representing every possible colour by reference to space of three dimensions" (Lovibond 1893, p. 5). The proposed space is a cube whose three axes represent the filter colors, with the scales in terms of filter strength (figure 9.69). White is placed at the origin O and represents colorless filters (figure 9.70). The neutral axis extends diagonally through the cube to end in black N. By using the subtractive primaries yellow, blue, and red, the subtractive secondaries fall on the diagonal axes in the corresponding plane.

Lovibond's Tintometer was (and in modified form is) widely used for specifying the color of liquids of many kinds. Using a modern instrument, Mark Fairchild (n.d.) produced a modern and growing listing of widely varying yellow, cyan, and magenta Tintometer values of many beers of the world.

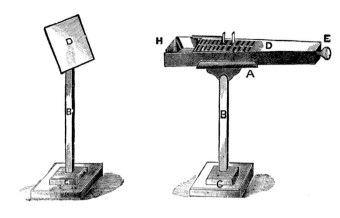

Figure 9.68.
Image of the Tintometer (Lovibond 1893).

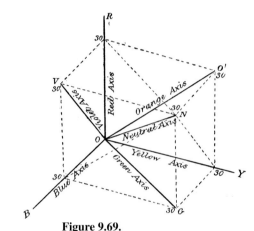

Figure 9.69.
Color cube of the Lovibond (1893) system suggested by Munro.

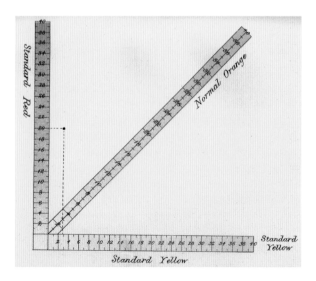

Figure 9.70.
Orange as the result of mixture of white and red shown in grades from the cube origin at white (Lovibond 1893).

EMIL DETLEFSEN   1905

E. Detlefsen, *Blütenfarben*, 1905

In 1905, German botanist **Emil Detlefsen** developed a method for specifying the colors of blossoms. He represented the results in a color order system for which he received a patent in the same year (Detlefsen 1905b). The specification is in three separate gray scale values obtained from viewing the blossoms through three colored filters, and estimating the corresponding lightness against a gray scale. He called the resulting three numbers "measure of color."

The first number relates to the lightness judgment using a red filter made from a railway signaling glass filter; the other two numbers refer to results obtained with green and blue gelatin filters that Detlefsen produced with aniline dyes. Transmittance of the filters was adjusted so that "they exclude each other and add up the white" (Detlefsen 1905a, p. 6, trans. by R.G.K.). The gray scale employed in the measurements consists of 20 grades in logarithmic sequence of luminous reflectance, painted with pigments.

Detlefsen ordered his number triplets according to *Ton* (hue), *Tiefe* (lightness), and *Kraft* (power) or *Sättigungsgrad* (degree of saturation). As a result of the method, colors are identified by the lightness value of the complementary filter (or combination of filters). Thus, the color red is defined in hue by values of the green and blue filters. Hues are arranged in a conventional circle. Figure 9.71 schematically represents the basis plane of the system.

Radial lines on the plane represent colors increasing in saturation from white to the full color (e.g., in the case of a bluish green with values 20 0 0). Detlefsen named the definitions of the colors on this plane *Grundmass* (basic measure). Colors of decreasing lightness are located on additional planes. They are defined by adding appropriate increments to all three numbers; thus, gray at lightness level 4 has values of 4 4 4, and as figure 9.72 illustrates, the fourth lightness level of a color with the basic measure of 0 7 3 has the values 4 11 7. Radial planes in the solid are defined as having constant hue.

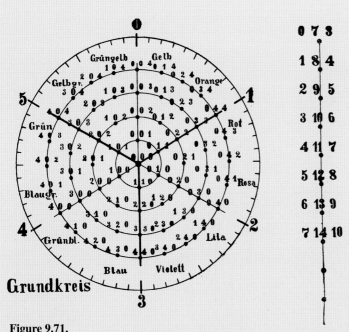

Grundkreis

0 7 3

1 8 4

2 9 5

3 10 6

4 11 7

5 12 8

6 13 9

7 14 10

Figure 9.72.
Schematic representation of a series of colors of constant hue and saturation but differing in lightness (Detlefsen 1905a).

Figure 9.71.
Schematic representation of the hue circle with the basis measures of the colors of the first four saturation grades (Detlefsen 1905a).

Figure 9.73.
Schematic representation of a portion of a constant hue plane with every second specification identified (Detlefsen 1905a).

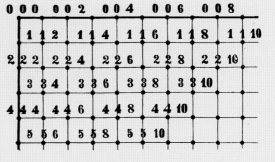

Figure 9.73 is a schematic view of a portion of the constant hue plane of a color of greenish yellow hue. All such constant hue planes arranged around the common neutral axis result in an inverted cone (figure 9.74), although not explicitly identified as such by Detlefsen, with black (values 20 20 20) at the tip and white (0 0 0) in the center of the plane. In Detlefsen's words:

> If one wants to represent the tonal colors the resulting planes are arranged sequentially in space, with colors of equal depth on horizontal planes and those of equal hue in vertical ones that radiate from the vertical line containing the gray colors. All equally strong and equally light colors must fall on horizontal circles. A solid, formed when the representation of figure 9.69 is rotated around the left vertical line of gray colors, conforms to these requirements. One then obtains a solid in which all measures are ordered according to hue, saturation, and lightness. (Detlefsen 1905a, p. 14, trans. by R.G.K.)

Detlefsen defined 1,260 colors for the basis plane. In figure 9.75, the sequences of basic measures *a*, *b*, and *c* serve the purpose of orientation in the solid.

Detlefsen allowed that his filter color-measuring tool is not a precision instrument, but considered it sufficient for the blossom-color specification, especially if the scales are enlarged by using decimal fractions. In his paper, he announced a painted color atlas based on the system, but the atlas was not published.

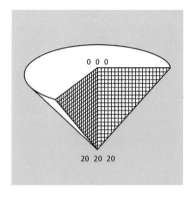

Figure 9.74.

Schematic implicit representation of Detlefsen's color solid (Detlefsen 1905a).

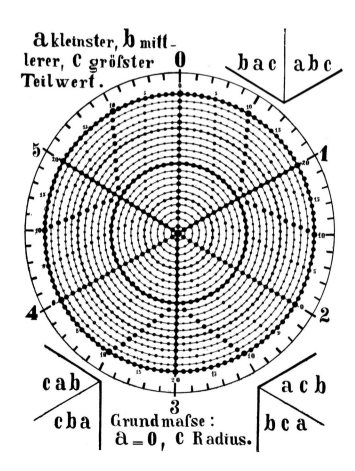

Figure 9.75.

Schematic representation of the complete basis plane with 1,260 basic measures, with white in the center and 20 saturation grades (Detlefsen 1905a).

LEOPOLD BLOCH   1915

L. Bloch, *Die Messung und zahlenmässige Darstellung der Körperfarben*, 1915

Leopold Bloch (1876–?) was a German illumination engineer and author or editor of books on lighting technology (e.g., Bloch 1907). In 1914, he wrote a paper describing the development of a visual colorimeter for lights (built by the optical equipment firm of Schmidt & Hänsch) and the results he obtained with it for daylight and some 50 different artificial light sources.

The measurement values consisted of readings with three colored glass filters: red, green, and blue. Bloch described the filters, produced by the well-known filter manufacturer Schott, as having tightly controlled thickness and color. However, he did not identify the filters, or publish transmittance data, this information presumably having been a trade secret. He expressed the results as three reflectance values in a two-dimensional diagram plotting the red/green and the blue/green ratios.

In 1915, Bloch published a second paper, "Die Messung und zahlenmässige Darstellung der Körperfarben" (Measurement and numerical representation of object colors). Spectral photometers at the time were complex, difficult to operate, and very slow; there was considerable interest in a piece of equipment such as developed by Bloch. He recognized that his colorimeter (figure 9.76) might be modified to measure object colors and saw a large potential for application.

Bloch used the same manually rotated filter disk of the colorimeter in the new instrument. He visually compared the brightness of the light reflected from the object and transmitted through a filter to that reflected from a white magnesium oxide standard. He expressed as three reflectance values in percentages compared to the standard white, which was defined as 100%. He estimated the accuracy of his equipment to be about 2%. He claimed that "color-blind" persons were able to accurately measure color with his equipment.

Bloch understood that graphically representing the three measurement values of object colors required a three-dimensional presentation, so he proposed a color cube (figure 9.77), with black and white in opposite corners and 100% red, green, and blue in the corners adjacent to black. Black objects (theoretically) reflect no light as measured through any of the filters, white objects reflect 100% of all three lights. Middle gray is located on a gray scale from black to white passing through the center. The cube is a simple geometric representation of the data with no quantitative relationship to additive or subtractive color mixture.

Bloch's instrument may be the first version of a tristimulus colorimeter for object colors. Such instruments, with filters matching the CIE standard observer color-matching functions and a photometer for measurement, became popular after the 1956 development of the Hunter photoelectric colorimeter in the United States. Widely employed for measurement of color differences between objects, they continue to be used today.

Figure 9.76.
Farbenmesser *(Bloch 1915).*

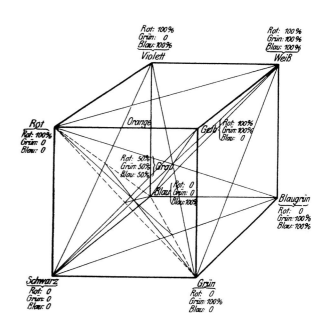

Figure 9.77.
Bloch's (1915) color cube.

EUROCOLOR   1975, 1984

L. Gall, Farben, *Farbnamen und das Eurocolor-System*, 1975
Schwabenmuster Eurocolor-Farbatlas, 1984

German publisher **Hans Kupzyck** and color scientist **Ludwig Gall** developed an ambitious color atlas with samples that were the first to demonstrate the CIELAB color space formula (see CIELAB, CIELUV, chapter 7) (Gall 1975a, 1975b). As the name suggests, the authors hoped that this system would become the official color-designation system of the European Union. By 1979, a box with 85 defined glossy samples was available.

In 1984, Kupzyck and Gall introduced the Eurocolor atlas. It contains a systematic sampling of the CIELAB object color space, to the extent possible with pigments. The geometric system is in cylinder form. The hue circle is separated into 1,000 grades. The chroma scale is open-ended in steps of 10 metric CIELAB chroma units. The lightness scale is in 11 grades from $L^* = 0$ to 100 (black to white).

The atlas contains approximately 1,100 samples identified according to hue H, lightness L and chroma C, that is, identical in principle to the Munsell system (see Munsell Renotations, chapter 7). The hue designation scale runs from 0 to 999 (figure 9.78). Hues 000, 250, 500, and 750 fall on, respectively, the $+a^*$, $+b^*$, $-a^*$, $-b^*$ axes of CIELAB. Originally, hues were illustrated at every 50th grade. Visual evaluation revealed the steps in the first quadrant to be very large, and intermediate grades were introduced in that area. As a result, the system consists of 24 constant hue charts (figure 9.79).

After the death of Kupzyck (also the publisher of the journal *Farbe + Design*) in 1989, his firm was sold and remaining inventories of the atlas were destroyed. A direct successor system of Eurocolor is the RAL Design System (see entry in this chapter).

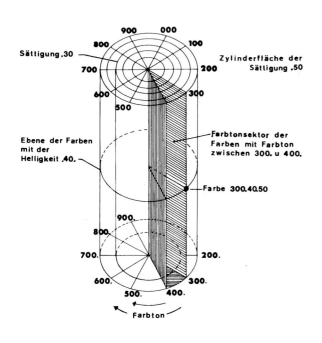

Figure 9.78.
First version of the Eurocolor color solid (Gall 1975a).

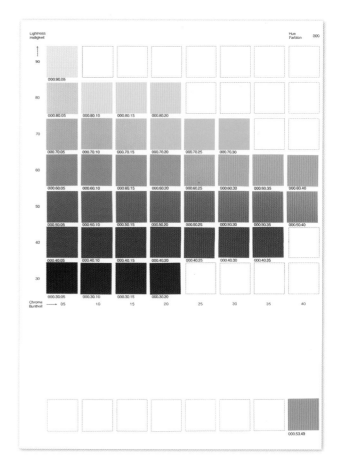

Figure 9.79.
Atlas page of the Eurocolor-Farbatlas, hue 000 (Gall 1984).

RAL DESIGN SYSTEM   1993

RAL Deutsches Institut für Gütesicherung und Kennzeichnung,
RDS Farbenatlas, 1993

The **RAL Design System** is a direct successor of the Eurocolor system (see entry in this chapter). After the latter disappeared from the market, a quality-assurance organization, RAL Deutsches Institut für Gütesicherung und Kennzeichnung,[10] asked Gall, one of the creators of the Eurocolor system, and other experts to design a new system based on the CIELAB formula. The organization wanted an attractive format competitive to the Natural Color System (see entry in chapter 5). In 1993, the RAL Design System (RDS) was unveiled in the form of the *RDS Farbenatlas*, containing glossy lacquer samples.

Comparable to Eurocolor, RDS is a visualization of the cylindrical CIELAB system with the coordinates metric hue H, metric lightness L, and metric chroma C (figures 9.80 and 9.81). Unlike Eurocolor, RDS's hue samples are based on CIELAB hue angle figures (degree), shown in increments of 10 degrees. In addition, hues were added in the yellow-orange region at a five-degree interval level. As a result, the hue circle consists of 39 colors (figure 9.82). Lightness and chroma variation samples were also prepared in increments of 10 units.

For approximately half of the hues, additional intermediate chroma steps were generated to meet practical needs. The atlas consists of 1,688 samples of size 17 × 18 mm. The color samples are also available as fans and sheets of size DIN A4 (210 × 287 mm).

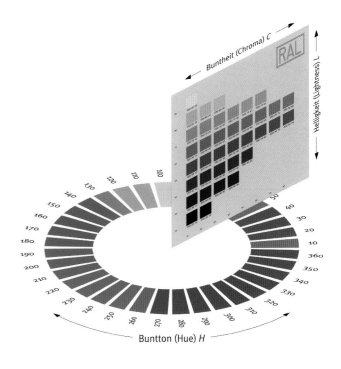

Figure 9.81.
Schematic representation of the RAL Design System, based on the cylindrical CIELAB system Buntton (hue) H, *Helligkeit (lightness)* L, *and* Buntheit *(chroma)* C *(RAL Deutsches Institut für Gütesicherung und Kennzeichnung e.V. 1993). Reprinted with permission.*

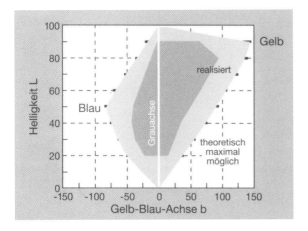

Figure 9.80.
Gamut (dark shaded area) of RDS in the L*, b* *plane of the CIELAB space (RAL Deutsches Institut für Gütesicherung und Kennzeichnung e.V. 1993). Reprinted with permission.*

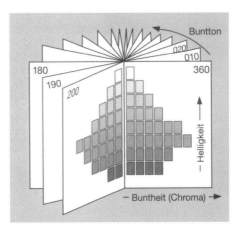

Figure 9.82.
Schematic representation of the RDS hue circle and a constant hue page (RAL Deutsches Institut für Gütesicherung und Kennzeichnung e.V. 1993). Reprinted with permission.

ACOAT COLOR CODIFICATION   1978

Sikkens, *Acoat color codification system – Handbuch für Farbgestaltung*, 1978

The **Acoat Color Codification** system, abbreviated ACC, was developed by the Dutch paint manufacturer Sikkens, now the coatings division of the Dutch chemical company Akzo Nobel. Beginning in the 1970s, the growing trend toward color diversification made it necessary for paint manufacturers to enlarge their offerings. Sikkens's *Kollektion 2001* of 1972 encompassed 120 colors; the *Kollektion 2011* of 1974 already contained 600 colors, identified according to the DIN 6164 system (see Manfred Richter and DIN 6164, chapter 7). Sikkens's was the first commercial color order system based largely on the structure of the CIELAB formula that was introduce by the CIE in 1976 (see CIELAB, CIELUV, chapter 7).

The 1978 version, the *Color Collection 2021*, contained 635 samples. In 1992, *Color Collection 3031* was issued with 1,235 colors, expanded in 2000 to approximately 1,400 colors (now called *Color Collection 3031+*; figure 9.83). Until 1992, the color samples contained in the various editions were selected by the German architectural color designer and educator Friederich Schmuck to fill the needs of exterior and interior architectural design. As a result, a portion of the ACC samples represents systematic sampling of CIELAB, while another portion represents more detailed sampling in various regions to fill specific needs (figure 9.84).

ACC's structure is closely related to CIELAB. The order and spacing of hues, named HC in ACC, match CIELAB exactly (figure 9.85), as does ACC lightness LC. But ACC is different from Eurocolor and RAL Design System (see entries in this chapter) in that it does not simply represent a visualization of CIELAB. ACC has been expanded past L^* = 100 to encompass fluorescent paints.

There is considerable discrepancy with regard to chroma. Here, a compromise was sought between the saturation of the DIN 6164 system and the metric chroma of CIELAB. For every hue represented in ACC, visual evaluations with Sikkens employees as observers were conducted to arrive at the definition of saturation SC.

Lines of constant ACC saturation SC end in a hypothetical black point below the CIELAB space (figure 9.84), not in a common real black point as in DIN 6164. Sikkens engineers have expressed the location of this point with an unpublished formula. As a result of the hypothetical black point, lines of constant SC fall on cone section surfaces (figure 9.86), not on cylindrical surfaces parallel to the lightness axis as in CIELAB. The official format of the system is, however, cylindrical (figures 9.87 and 9.88). An atlas has so far not been produced, but the samples are available as a fan set (figure 9.83) or as loose sheets.

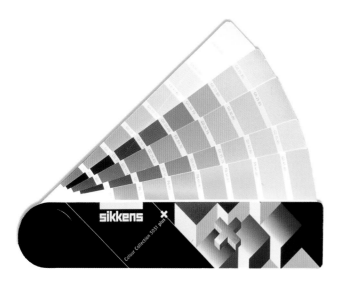

Figure 9.83.
Sikkens Color Collection 3031+ *fan (Sikkens 2000).*
Reprinted with permission.

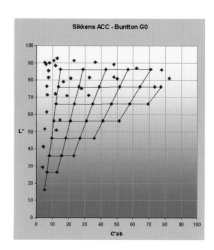

Figure 9.84.
Projection of hue G0 onto the CIELAB C*, L* diagram, showing for this hue sample selections according to a regular grid as well as the special samples required in the trade (Sikkens n.d.). Reprinted with permission.

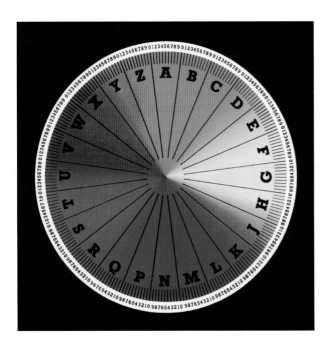

Figure 9.85.
240-grade CIELAB-based ACC hue circle (Sikkens n.d.). Reprinted with permission.

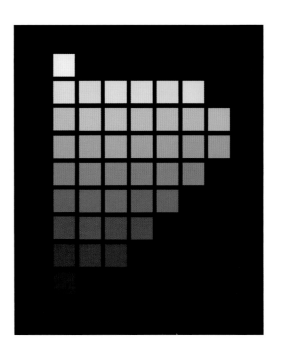

Figure 9.87.
Standard grid colors of hue G0 (see figure 9.84) and the gray scale from Color Collection 3031+ *(Sikkens n.d.). Reprinted with permission.*

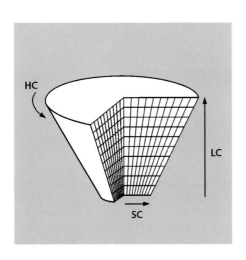

Figure 9.86.
Schematic representation of the ACC color solid in form of a cone section, demonstrating the compromise between saturation and chroma.

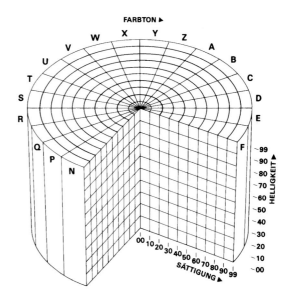

Figure 9.88
Official schematic ACC color solid (Sikkens 1978). Reprinted with permission.

G. P. VIŠNIAK and I. S. FAJNBERG   1986

G. P. Višniak and I. S. Fajnberg, *Atlas cvetov*, 1986

A color atlas representing a modified visualization of the CIELAB system (see CIELAB, CIELUV, chapter 7) was published in the former Soviet Union in 1986, the result of cooperation between two groups of researchers. The first group, headed by **G. P. Višniak**, represented the All-Union Center for the Development of Product Ranges, Light Industry, and Fashion and Clothing Industries.[11] The second group, headed by **I. S. Fajnberg**, represented the All-Union Polygraphical Research Institute (printing in multiple colors). Expected uses of the atlas included color identification, standardization, and selection.

The offset-printed *Atlas cvetov* (Color atlas) contains 46 constant hue planes. The hue circle consists of eight basic hues: yellow, orange, red, purple, violet, blue II (sinij), blue I (goluboy),[12] and green, as well as 38 mixed hues (figure 9.89). The implicit solid of the system is a cylinder (figure 9.90). For each constant hue plane (figure 9.91), there are ten lightness grades in increments of $\Delta L^* = 8$ and eight relative chroma grades. The atlas contains 1,808 samples of size 19×23 mm. Color samples are identified with a six-digit number, in which the first number pair refers to the hue, the second to relative chroma, and the third to lightness.

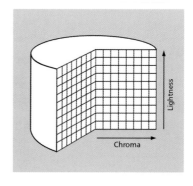

Figure 9.90.
Schematic representation of the implicit color solid in Višniak and Fajnberg's Atlas cvetov.

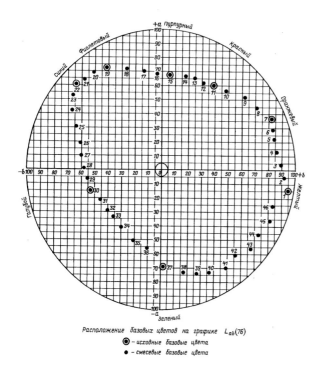

Figure 9.89.
Position of the 46 full colors in the CIELAB a*, b* diagram. The eight basic colors are highlighted (Višniak and Fajnberg 1986; image courtesy Werner Spillmann).

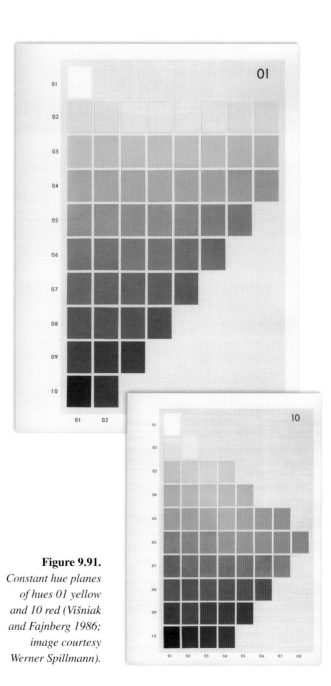

Figure 9.91.
Constant hue planes of hues 01 yellow and 10 red (Višniak and Fajnberg 1986; image courtesy Werner Spillmann).

COLORCURVE   1988

Colorcurve Systems Inc., *Colorcurve*, 1988
R. Stanziola, *The Colorcurve system*, 1992

American color technologist Ralph Stanziola (1931-) conceived the **Colorcurve** system published in the form of two atlases by Colorcurve Systems, Inc. in 1988. It is described as "a system of color communication. It provides a direct means of communication between specifiers or creators of color and the manufacturer of colored products. It is a combination of a visual and a numerical system" (Stanziola 1992, p. 263).

Stanziola specified aim points for the system in CIELAB space (see CIELAB, CIELUV, chapter 7), and represented them as color chips. Data tables list reflectance values of the chips at every 20 nanometers from 400 nm to 700 nm, tristimulus values X, Y, and Z for the CIE 10° standard observer and illuminant D65, and the corresponding L^*, a^*, b^* values of the aim colors.

The system is based on nine starting points defined by their CIE tristimulus values – eight chromatic and one achromatic – for each of 18 lightness levels (L^* 30 to 95 at $L^* = 5$ intervals, with additional levels at $L^* = 87.5, 92.5, 93.75$). The

intermediate steps were calculated by additively blending the stimuli representing the starting points at 20% intervals. For example, the toning from red ($a^* = 60$, $b^* = 0$) toward gray (at 0, 0) might be done in four intermediate steps by changing the tristimulus values at 20% intervals.

Where pigments made it possible, aim colors outside the starting colors were calculated. The resulting aim tristimulus values were converted to CIELAB values, and reflectance functions representing pigment mixtures were calculated that match the aim values. This provides the user of the chips with reflectance data suitable for computer-assisted colorant formulation. Figure 9.92 illustrates the aim points of the system at $L^* = 65$ in the a^*, b^* diagram.

A sample notation system was developed consisting of two components per sample: The first designates the lightness level, and the second, the chromatic position. For example, the designation L65 R2Y7 describes a sample at lightness $L^* = 65$ and located in grid position R2Y7, indicating a dull yellowish orange.

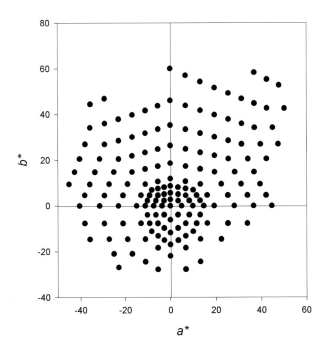

Figure 9.92.
Aim points of the color samples of Colorcurve at L*
= 65 in the a*, b* *diagram.*

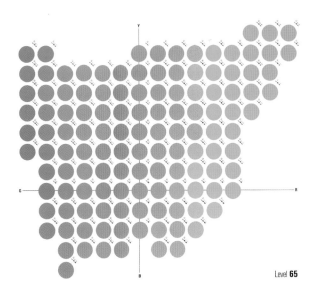

Figure 9.93.
*Color samples at Level 65 (*L* = 65), Master Atlas
(Colorcurve Systems, Inc. 1988). Reprinted with permission.*

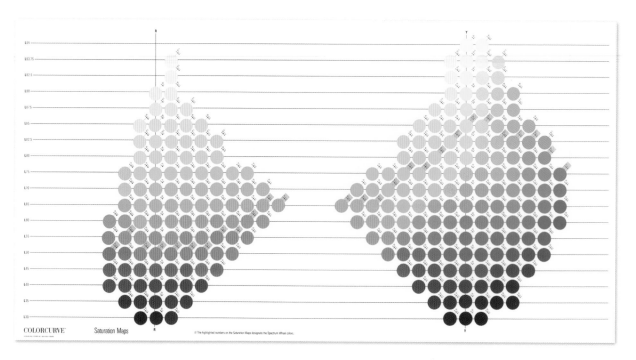

Figure 9.94.

Saturation maps R and Y of the Colorcurve system (Master Atlas) *showing the samples on the surface of the implied solid when facing toward red and toward yellow (Colorcurve Systems Inc. 1988). Reprinted with permission.*

The physical system consists of the *Master Atlas* and the *Gray and Pastel Atlas*. The former contains a total of 1,229 samples (circular, 16 mm diameter, opaque paint) on 18 lightness levels (figure 9.93). There is also a "spectrum wheel" illustrating the most highly saturated samples around the hue circle, as well as four "saturation maps" representing the color samples on the surface of the implicit color solid (figure 9.94). A view of the solid is shown in figure 9.95.

The *Gray and Pastel Atlas* contains 956 unique additional samples. They are intermediate to samples of the *Master Atlas*, at one-half, one-third, or one-quarter steps near gray from L40 to L95 (figure 9.96). A gray scale illustrates the 18 lightness levels. White and black isolation masks allow viewing single chips or arrays of nine chips in isolation. Although the system is based on the CIELAB system, it is not arranged in equal increments of CIELAB as are Eurocolor and the RAL Design System (see entries in this chapter). The Colorcurve system has found considerable use around the world but is no longer being marketed.

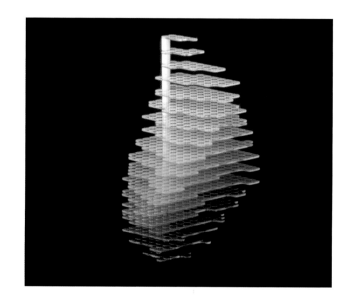

Figure 9.95.

Illustration of the solid formed by the samples of the Colorcurve system (image courtesy Colorcurve Corp.). Reprinted with permission.

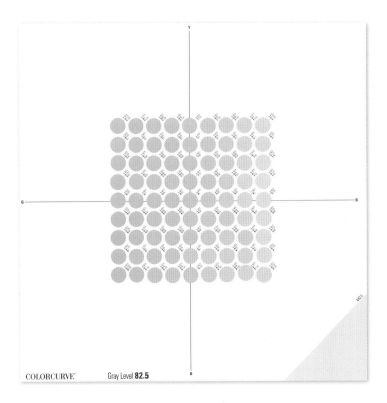

Figure 9.96.
Samples at gray level 80, Gray and Pastel Atlas *(Colorcurve
Systems Inc. 1988). Reprinted with permission.*

Notes

1. Le Blon is related on his mother's side to the engraver and publisher Matthaeus Merian (1593–1650). Le Blon is the author of *Il Coloritto* (ca. 1721), a book about painting in which he affirmed his belief that any color can be obtained from yellow, red, and blue, the three together forming black. In the 1710s, he lived in Amsterdam, and it is likely that, as a painter of miniatures, he knew the book attributed to Boutet in which an unidentified writer made the same assertions in a Dutch 1707 edition (see Anonymous, chapter 3). On Le Blon's life and work, see Lilien (1985).

2. d'Agoty apprenticed for six weeks with Le Blon and later claimed that three- and four-color printing was his invention, an assertion denied, however, by Antoine Gauthier de Montdorge, believed to have been the author of *L'art d'imprimer les tableaux, traité d'après les écrits, les operations & les instructions verbales de J. C. Le Blon* (1756), a detailed description of Le Blon's technique.

3. Ducos du Hauron expanded on the subject in *Les couleurs en photographie* (1870).

4. The process of photosensitizing gelatin with potassium bichromate was invented in 1855 in France by A. L. Poitevin. In the collotype process, glass plates coated with sensitized gelatin are covered with a photographic negative and exposed to light. The plate is then soaked in water. The gelatin absorbs some water and swells in inverse ratio to the light exposure. After drying the plate, printing ink can be applied to the hard, but not to the soft water-swelled, areas of gelatin, followed by printing with the plate. The process is very sensitive to conditions and requires an inordinate amount of manual work. Only 200–400 impressions can be taken from a plate.

5. The history of photography is very complex. For a survey, see Rosenblum (1997).

6. As a result of image structures and minor misalignments of the plates, moiré effects may appear. They troubled some of Charles Lacouture's charts (see entry in this chapter). The problem is exacerbated when printing with four plates. The usual printing angles are 45, 75, 90, and 105 degrees. A moiré pattern is often obtained involving the ink printed at 90 degrees and its two angular neighbors. However, because yellow is usually printed at 90 degrees, the moiré pattern is rarely discernible.

7. A single ink in small dot size printed on white paper produces locally partitive mixture with paper white. When two or more inks are printed more or less overlapping, subtractive mixture takes place, depending on the degree of overlap and the transparency of the pigments in the inks. E. Demichel in 1924 and Hans Neugebauer in 1935 (see entry in chapter 6) proposed mathematical equations to calculate the combined effect in terms of the colorimetric system. However, the results are only approximate due to the complexity of the total process.

8. Proportionality failure refers to the situation that for certain inks the relationship between coverage of the paper by halftone dots and the corresponding appearance of depth of color is nonlinear to an unacceptable degree, requiring adjustments in dot coverage.

9. There is a second edition of Gerstacker's book, published in 1954 by the same publisher, having little or nothing in common with the first edition.

10. The original RAL organization was founded in 1925 and was concerned with establishing rules for terms of delivery. In 1927, it issued the RAL Colors, an unsystematic collection of color samples, still available and now numbering 194. This collection coexists with, but is different from, the RAL Design System.

11. The term "All-Union" refers to institutes with responsibility in certain matters for the complete Soviet Union.

12. The Russian language has two basic color terms for blue, *sinij* and *goluboy*, the former having a focal color of (comparatively) slightly reddish, darker blue, the latter one of a slightly greenish, lighter blue.

References

Anonymous (Antoine Gauthier de Montdorge). 1756. *L'art d'imprimer les tableaux, traité d'après les écrits, les operations & les instructions verbales de J. C. Le Blon*, Paris: Mercier, Nyon, Lambert.

Ball, J. A. 1950. Process of multicolour reproduction, U.S. Patent 2,507,494.

Blecher, C. 1943. *Die Herstellung mehrfarbiger Reproduktionen nach bunten Durchsichtsbildern durch Druck*, Leipzig: Instituts für Farbenfotografie an der Staatlichen Akademie für Graphische Künste und Buchgewerbe zu Leipzig.

Bloch, L. 1907. *Grundzüge der Beleuchtungstechnik*, Berlin: Springer.

Bloch, L. 1914. Die Farbe der künstlichen Lichtquellen, *Die Naturwissenschaften* Bd. 2, 85–91.

Bloch, L. 1915. Die Messung and zahlenmäßige Darstellung der Körperfarben, *Die Naturwissenschaften* 3.26:333–339.

Brillux. 2001. *Brillux Scala* (Farbenatlas, Farbmuster, Farbenfächer, Cd-Rom), Münster, Germany: Brillux.

Caparol. 1999. *Caparol 3D-System* (Farbenfächer, Farbmuster, Index, Cd-Rom), Ober-Ramstadt, Germany: Caparol.

Chavkin. 1912. *Chavkin's Farbenkreis. Wie entstehen ca. 1500 verschiedene Farbtöne aus den 4 Farben: Schwarz, Blau, Rot und Gelb. Ein einfaches und klares Lehr- und Hilfsmittel für Schulen, sowie für alle Fachleute und Liebhaber, die mit Farben oder Farbenmischung zu tun haben*, München: Schmidt-Bertsch.

Colorcurve Systems Inc. 1988. *Master Atlas* and *Gray and Pastel Atlas*, Minneapolis, MN: Colorcurve Systems Inc.

Detlefsen, E. 1905a. Blütenfarben. Ein Beitrag zur Farbenlehre, addendum to *Programm der Grossen Stadtschule zu Wismar*, Wismar, Germany: Eberhardt.

Detlefsen, E. 1905b. *Verfahren zur Feststellung der Farbwerte*, German Patent no. 162838.

Deutsche Industrie Normen. 1955. German Standard DIN 16509, *Farbskala für den Offsetdruck*, Berlin: Beuth.

Dörbrand, G. 1937. *Wirtschaftliches Farbenmischen* (beigebunden der) *Farben-Kamerad. Farbenwahl und Farbenmischen*, Celle/Hannover: Chr. Hostmann-Steinberg'sche Farbenfabriken.

Ducos du Hauron, A. L. 1869. *Les couleurs en photographie – – solution du problème*, Paris: Marion.

Ducos L. and **A. L. Ducos du Hauron**. 1870. *Les couleurs an photographie et en particulier l'héliochromie au charbon*, Paris: Masson.

Fairchild, M. n.d. Lovibond beer color scale. Available at www.cis.rit.edu/fairchild/lovibond.html.

Field, G. G. 2004. *Color and its reproduction*, 3rd ed., Sewickley, PA: GATF Press.

Foss, C. E. 1973. Color order system, U.S. Patent 3,751,829.

Foss, C. E., and **G. G. Field**. 1973. *The Foss color order system*, Research Report 96, Pittsburgh, PA: Graphic Arts Technical Foundation.

Freier Deutscher Gewerkschaftsbund. n.d. (late 1950s). *Farmesstafeln für Vierfarbenbuchdruck*, Leipzig: FDGB, Industriegewerkschaft Grafisches Gewerbe und Papierverarbeitung.

Gaensslen, H. 1959. *Das grosse Drei-Farben-Mischbuch*, Ravensburg: Maier.

Gall, L. 1975a. Farben, Farbnamen und das Eurocolor-System, *Farbe + Design* 1:49–50.

Gall, L. 1975b. Das Eurocolor. Farbmetrische Definition, *Farbe + Design* 2:23–24.

Gall, L. 1984. Die Realisierung des CIELAB-Systems im Eurocolor-Atlas, *Farbe + Design* 29/30:4–9.

Gerstacker, L. 1934. *Das Farbenmischbuch*, Zwickau/Sachsen: Förster & Börries.

Hartmann, Gebrüder. 1914. *Druckfarben—ihre Erzeugung und Verwendung*, Ammendorf bei Halle-Saale: Gebrüder Hartmann.

Hartmann, Gebrüder. 1937. *Photochromie-Farbenskala*, Ammendorf bei Halle-Saale: Gebrüder Hartmann.

Hickethier, A. 1940. Der Farbendruck verlangt eine Ordnung der Farben, *Klimschs Jahrbuch des graphischen Gewerbes* 33:110–113.

Hickethier, A. 1943. Farbenphotographie, Farbendruck und Farbenordnung, *Deutsches Buchgewerbe* 1:140–145.

Hickethier, A. 1952. *Farbenordnung Hickethier*, Hannover: Osterwald.

Hickethier, A. 1972. *Die grosse Farbenordnung Hickethier*, Ravensburg, Germany: Mayer.

Hoffmann, H. 1892. *Systematische Farbenlehre*, Zwickau, Germany: Förster & Borries.

Ives, F. E. 1902. The half-tone trichromatic process, *Journal of the Franklin Institute*, Jan., 43–49.

Ives, H. E. 1938. Color chart and method of making the same. U.S. Patent 2,128,676.

Jacobson, E., **W. C. Granville**, and **C. E. Foss**. 1942. *Color harmony manual*, Chicago: Container Corporation of America.

Kujawa, G. 1940. Über den praktischen Farbkörper, *Archiv für Buchgewerbe und Gebrauchsgraphik* 77:140–145.

Lacouture, C. 1890. *Répertoire chromatique, solution raisonnée et pratique des problèmes les plus usual dans l'étude et l'emploi des couleurs*, Paris: Gauthier-Villars.

Le Blon, J. C. ca. 1721. *Il Coloritto. Or the Harmony of Colouring in Painting reduced to Mechanical Practice and Infallible Rules*, London; facsimile ed., Birren, F., ed., New York: Van Nostrand Reinhold, 1980.

Lilien, O. M. 1985. *Jacob Christoph Le Blon*, Stuttgart: Hiersemann.

Lithographic Technical Foundation. 1957. *LTF color chart*. New York: Lithographic Technical Foundation.

Lovibond, J. W. 1893. *Measurement of light and colour sensations*, London: Gill.

Müller, J., and **M. Dethleffs**. 1900. *Praktischer Leitfaden für den Buntbuchdruck. Ein wirkliches Hilfsbuch für den Farbendruck und Farbenmischung im Buchdruck*, Berlin: Müller.

Painter, A. 1936. *Tous les coloris. Répertoire des couleurs, tons nuances, avec teneurs & en couleurs primaires*, Paris: Librairie Polytechnique Béranger.

RAL Deutsches Institut für Gütesicherung und Kennzeichnung. 1993. *RDS Farbenatlas*, Sankt Augustin, Germany: RAL Deutsches Institut für Gütesicherung und Kennzeichnung.

Regensteiner, T. 1943. *My first seventy-five years*, Chicago: Regensteiner Corp.

Rogondino, M., and **P. Rogondino**. 2000. *Process color manual*, San Francisco: Chronicle Books.

Rosenblum, N. 1997. *A world history of photography*, New York: Abbeville.

Schwabenmuster. 1984. *Eurocolor system (atlas)*, Gaildorf, Germany: Schwabenmuster.

Siegwerk Druckfarben. 1993. *Farbmischtafel für Bogenoffset*, Siegburg, Germany: Siegwerk Druckfarben.

Sikkens. 1972. *Kollektion 2001*, Hannover: Sikkens.

Sikkens. 1974. *Kollektion 2011*, Hannover: Sikkens.

Sikkens. 1978. *Acoat Color Codification System – Handbuch für Farbgestaltung*, Hannover: Sikkens.

Sikkens. 1992. *Kollektion 3031*, Hannover: Sikkens.

Sikkens. 2000. *Kollektion 3031+*, Hannover: Sikkens.

Sikkens. n.d. *Mit Farbe leben*, Hannover: Sikkens.

Stanziola, R. 1992. The Colorcurve system, *Color Research and Application* 17:263–272.

Steinheil, R. 1896. *La reproduction des couleurs par la superposition des trois couleurs simples*, Paris and Nancy: Berger-Levrault.

Višnjak, G. P. and **Fajnberg, I. S.** 1986. *Atlas cvetov*, Moscow: Ministry for Light Industry and USSR Committee for Book Publication and Trade.

Warburg, J. C. 1899. A new three-colour chart, in *The process year book for 1899*, London: Penrose.

Wezel, H. 1959. *Farbmass. Praktische Richtlinien für den Vierfarbendruck*, Frankfurt: Polygraph.

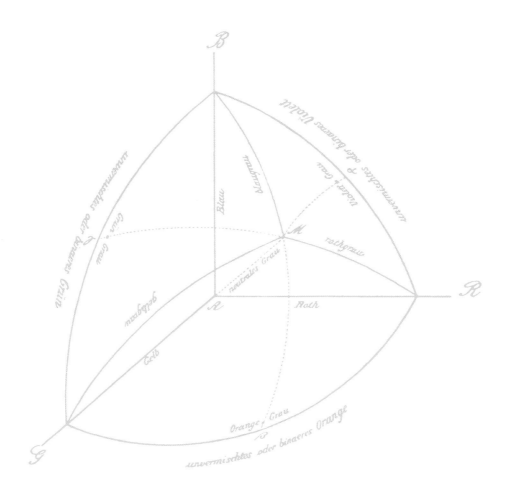

Descriptions of four kinds of color order systems that do not easily fit into the relatively narrow definitions of the other chapters: (1) largely conceptual systems from the early years of psychophysics, (2) disk mixture systems, (3) cubic systems not based on three-color printing, and (4) colorant mixture systems. These systems range from the nineteenth century to the present.

This chapter contains, on the one hand, the ideas of largely forgotten authors such as French theoretician Jean d'Udine and, on the other hand, the work of such well-known authors as German chemist Wilhelm Ostwald. The latter produced a large legacy of writings in books and journals as well as several atlases and pieces of equipment. His work resulted in extended discussions and several related color order systems.

Early psychophysical systems

The entries in this chapter begin with a system by the astronomer and mathematician Christian Doppler that in a way stands in a class by itself. His solid is located in an orthogonal space with the axes representing primaries yellow, red, and blue, with black at the origin. It is the result of a curious mixture of additive and subtractive concepts. It has the form of a sphere octant, later also offered in a more quantitative fashion for object colors by Erwin Schrödinger (see entry in chapter 6).

The following systems are qualitative expressions of psychophysics from the early days of experimental psychology. To a smaller or larger degree, they are influenced by the work of Ewald Hering and of Hermann von Helmholtz (see entries in chapters 5 and 6, respectively). The lack of quantitative data resulted in a variety of geometrical forms of the color solids, from cone (German physicist Wilhelm von Bezold, psychologist Wilhelm Wundt, and neurologist Georg Theodor Ziehen), to double cone (Wundt and psychologist August Kirschmann), to sphere (Wundt and German physiologist Ernst Wilhelm von Brücke). Several authors presented more than one model.

Disk-mixture systems

In the second group are systems based on disk mixture. This technology, popularized by James Clerk Maxwell (see entry in chapter 6), made it possible to achieve partitive mixture, a form of additive mixture with technically much simpler means than those required for quantitative light mixture.

In disk mixture, different segments of a disk have different colors, for example, three segments with primary colorants yellow, red, and blue, or five where segments of white and black have also been added (figure 10.1).[1] The disk is then

Figure 10.1.
Image of a nineteenth-century disk mixture apparatus. The disk, a composite of multiple disk segments painted in given colors, is attached to the apparatus and rapidly spun, resulting in a uniform color appearance (Guignet 1889).

spun rapidly, resulting in optical mixture of the color stimuli of the segments in the retina. The results of disk mixture depend entirely on the choice of the colorants with which the disks have been painted. Without standardization of the primaries, the results are not directly comparable.

Maxwell learned about disk mixture from one of his teachers, James David Forbes (see entry in chapter 11), who pursued various approaches to color order. Forbes's effort to fill Tobias Mayer's triangle (see entry in chapter 4) with results of disk mixture influenced Maxwell's own important efforts in this respect.

Disk-mixture technology became very popular. American physicist Ogden Nicholas Rood made use of the technology in the nineteenth century. French chemist Daniel-Auguste Rosenstiehl devised a system directly related to Maxwell's. D'Udine's system related to the musical scale, relies on disk mixture as well.

American zoologist Robert Ridgway also used disk mixture to develop his atlas. To arrive at an atlas with color chips, he had industrial colorists match his disk-mixture colors with inks. He attempted to make the resulting scales perceptually uniform by applying the Weber-Fechner law (see glossary) to increments, as Ostwald did a few years later. Measurements of samples show that Ridgway's system ended up as a compromise between ideas and practical reality.

Early in the twentieth century, Ostwald offered a completely systematic approach that combined disk mixture, a theoretical treatment of reflectance functions, psychophysics in form of the Weber-Fechner law, and a hue circle that placed all full colors onto the same horizontal plane. His system was copied with some minor modifications almost immediately by his commercial competitor, German painter Otto Prase (*Atlas II*, 1922).

Ostwald's system directly influenced the two German systems, DIN 6164 (see Manfred Richter and DIN 6164, chapter 7) and TGL 21 579 (see entry in this chapter). The latter was an attempt to combine the advantages of Ostwald's and other systems, which might be why it remained largely conceptual in nature. French visual artist André Lemonnier, who placed disk mixture into the hands of the user, invoked Ostwald, and the similarity of their constant-hue triangles is obvious.

Several additional European systems were developed that were either related to Ostwald's or in opposition to it. An example is Austrian textile chemist Max Becke's consciously different approach. Others borrowed ideas from him without naming him explicitly. The Coloroid system of Antal Nemcsics (see entry in chapter 5) was developed with the help of disk mixture and with full knowledge of Ostwald's system. E. B. Rabkin's double cone (see entry in chapter 11) is suggestive of Ostwald's, even though its author claimed to have developed the system on a purely perceptual basis.

Cubic systems

The third group of systems is based on the form of the cube. English architect William Benson introduced the cubic system derived from the results of additive color mixture and illustrated with watercolors. Later, French scientist J. Carpentier proposed a similar, but purely conceptual, system. Both systems can be viewed as prototypical of the modern monitor RGB space (see Color Display Solids, chapter 9).

On the other hand, the cube of Joseph Williams Lovibond's color measuring system (see entry in chapter 9) represents the result of subtractive mixture based on filter measurements of liquids' transmittance. Other cubes are based on colorant mixture. Prase planned a cube based on paint (no copy could be located), and Aemilius Müller populated his cube with dyed paper samples (see entry in chapter 5). Becke's cube was to be filled with wool dyeings (no copy could be located).

Colorant mixture systems

The final group of systems in this chapter is based on subtractive color (colorant) mixture. The commonality of these systems is that they are derived from a series of pure basis colorants that form the hue circle. Tint/shade and tonal variations were obtained by mixing the basis colorants with appropriate amounts of white and black pigments.

The most consistent approach is outlined by Joseph Marcel Vogel and Carl Sali Plaut, who published an atlas of 3,600 samples. Among creators of such systems are Prase, E. Fellowes, Càndido and Julio Villalobos, Gustave and Gladys Plochere, Louis Cheskin, and Andreas Kornerup and Johan Henrik Wanscher. Systems include the Advanced Ink Mixing System (AIMS), EuColor, Nu-Hue, and Colorizer.

Three basic approaches have been pursued in these systems:

1. The full color is shaded toward black, and the resulting grades were whitened. This approach is most common.
2. In the opposite approach, the full color is tinted toward white, and the resulting grades are darkened toward black. This method was used by Cheskin, Kornerup and Wanscher, and others.
3. Full colors are mixed with other full colors to achieve tonal colors. Used by Gericke & Schöne (see chapter 11), this approach results in a kind of planetary system where any color can be used as the center of local order.

Prase initiated the first approach. Before World War II, the Plochere family imported Prase's system for sale in the United States. Because of the ban on German imports during the war, the Plocheres produced a slightly modified version that maintained Prase's whitening method. Prase's system was no longer produced in divided postwar Germany, so Stephan Eusemann applied for a patent for EuColor using Prase's method.

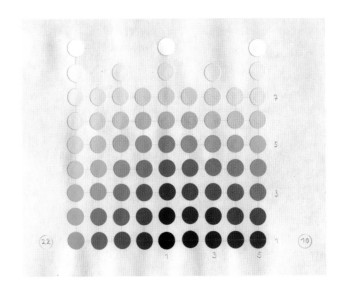

Figure 10.2.
Two opposing constant-hue pages of Zeugner's (1990) prototype of Farbenkarte 90. Image courtesy G. Zeugner.

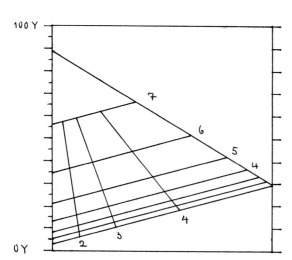

Figure 10.3.

Schematic representation of the constant-hue planes of Farbenkarte 90 *with the typical color scales in direction of white obtained by the whitening method (Zeugner 1990). Image courtesy G. Zeugner.*

CHRISTIAN DOPPLER   1847

C. Doppler, *Versuch einer systematischen Classification der Farben*, 1847

Austrian astronomer, physicist, and mathematician **Christian Doppler** (1803–1853) is best known as the discoverer of the Doppler effect. It applies to any kind of energy taken to have waveform, such as light or sound (Doppler demonstrated the effect for sound). If the wave source and the receiver approach each other, the frequency of the waves will increase (wavelength will decrease). As a result, to a human observer, sounds will have a higher pitch, and light will appear bluer. The effect reverses when wave source and observer are moving apart. An example is the sound of a train moving at constant speed, approaching a stationary observer, passing by her, and disappearing. The effect is well known in astronomy as red shift (or blue shift) of light from celestial bodies, an indicator of the speed at which they recede from (or approach) the earth.

Doppler's interest in color order appears to have derived from his astronomical work. One of his main interests was double stars. In Doppler's time, they could be distinguished only by the usually different color of the light from the two stars. Doppler assumed the different color to be due to the effect named after him. In order to be able to describe it quantitatively, Doppler required a quantitative description of color order.

Doppler believed colors to reside in objects. "However the intensity of a color perception is mainly dependent on the distance and intensity of the light source, the reflective properties of the object, and the current state of sensibility of our visual organs" (Doppler 1847, para. 3, trans. by R.G.K.). Doppler also was aware that a given object can appear as having different colors in different lights.

But Doppler did not have a clear understanding of the difference between additive (lights) and subtractive (colorants) color mixture. "Of all color perceptions that we know only red, blue and yellow can be viewed as actual primary colors because they can never be mixed from others in their purity, while they can generate all other colors" (para. 4). This seemingly refers to lights. But he continues: "Black and the various levels of gray are levels of intensity of white, and the latter can be viewed as composed from the above three primary colors" (a mixture of concepts related to objects and lights).

Doppler's knowledge of mathematics resulted in a mathematics-based conceptual proposal of a color identification system. His three primary colors form orthogonal axes. The origin of the system is the black point A in figure 10.4 and 10.5. The proposed scaling of the three primaries is such that when mixed in equal amounts, they result in white. The neutral axis of the system therefore rises from the system axes (line A–M in figure 10.6), separated by equal angles.

Eusemann's EuColor system saw its demise when Stocolor acquired it. This firm already had an in-house system, loosely derived from Ostwald's. More recently, there have been efforts in Germany to bring Prase's system back to life. In 1990, Gerhard Zeugner produced a small edition of a prototype, the so-called *Farbenatlas 90* (figures 10.2 and 10.3) but could not find a commercial publisher.

The system of Fellowes, independently developed, also uses the whitening method, as does Hilaire Hiler's conceptual system. The Nu-Hue system is closely similar to a whitening system. In the early Colorizer system, full colors were mixed with white and light gray but without achieving systematic color series. The Villalobos system takes a different approach: It uses light and darker versions of inks representing full colors, which made it possible to produce a large number of samples.

Some systems have been designed with a very specific application in mind, such as that by Frantz Braun. Braun was concerned with mixture results on textiles from specific dyestuffs. Many years earlier, Ostwald had prepared a version of his complete color solid with samples dyed on different kinds of textile fabrics (see figure 10.45; Schwarz 2003). An abbreviated version of the Munsell system, the SCOTDIC system (Standard Color of Textile Dictionary, Kensaikan International, Japan), also has been prepared on textile fabric (see Munsell Renotations, chapter 7).

But the primaries are not taken as vectors because their mixture is not by vector addition; rather, one part of blue plus one part of yellow results in one part of green. As a result, constant "intensity" lines are circle segments, and the resulting geometric solid is a sphere octant (figure 10.6). Geodesic lines, identified with yellow-gray, blue-gray, and red-gray, run from the achromatic point M to the primaries. Doppler's system does not distinguish between object colors and lights.

Doppler provided algebraic formulas for calculating the magnitude of differences in his space. He distinguishes between three classes of colors: The first class consists of the three primary colors. The second class consists of the three secondary colors and their variation toward the two primary colors involved (mixtures of two primaries only and falling on the plane between them). The third class consists of seven species of tertiary colors where all three primaries are activated (those colors that fall in the interior space of the sphere octant).

The conceptual nature of his proposal was clear to Doppler. In his concluding remarks, he pointed to an instrument recently invented by François Arago (1786–1853) that he hoped would make it possible to create practical applications of his proposal.[2] Doppler's proposal can be considered a proto-psychophysical color space.

Despite the shortcomings of Doppler's system, sphere octants proved an important idea in color science. Schrödinger (see entry in chapter 6) described a stimulus space, presumed to be in agreement with perception, based on color fundamentals measured by König and Dieterici (König 1892). He did not consider the logarithmic compression of the Weber-Fechner law to be applicable. Instead, he used a square root compression (a compression he also called unsatisfactory, however). The resulting space, named ξ (xi) space, has the form of a sphere octant (see figure 6.18).

Doppler's paper on color order was published in the *Proceedings of the Royal Bohemian Society of Sciences* and did not find much interest elsewhere. It was soon overshadowed by the work of Maxwell and Helmholtz (see entries in chapter 6).

Figure 10.4.
Linear intensity scale for "blue" light (Doppler 1847).

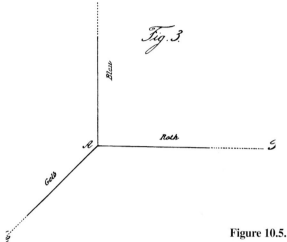

Figure 10.5.
Orthogonal space formed by the axes of "yellow," "red," and "blue" light (Doppler 1847).

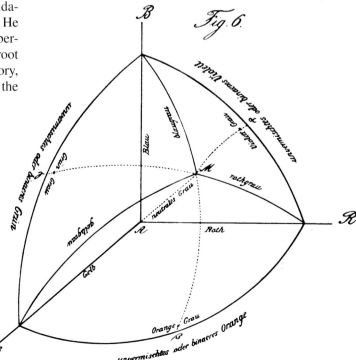

Figure 10.6.
Sphere octant space formed by mixtures of the lights. The neutral axis proceeds from A to M. The results of binary mixtures are indicated by dashed lines (Doppler 1847).

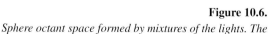

ERNST WILHELM VON BRÜCKE   1866

E. W. Brücke, *Die Physiologie der Farben*
für die Zwecke der Kunstgewerbe, 1866

German physiologist and son of a painter, **Ernst Wilhelm von Brücke** (1819–1892) was a close friend of Helmholtz. He became head of the physiological laboratory at the University of Vienna, where he supervised Sigmund Freud from 1886 to 1892. He co-discovered the Bezold-Brücke effect, according to which most color stimuli change apparent hue as a function of brightness. He also had considerable interest in art.

In 1866, Brücke published a book with the ambitious title *Die Physiologie der Farben für die Zwecke der Kunstgewerbe* (Physiology of colors for art and design). As a result of the practical orientation of this book, he limited himself largely to object colors. In this book, he also presented a conceptual color order system in the form of a sphere (figure 10.7). Brücke arranged the hue circle so that complementary colors are diametrically opposed. Complementary colors do not allow arrangement of hues according to perceptual differences. As a result, the hue circle (figure 10.8) has a large range of greenish colors.

Concerning saturation, Brücke commented:

> We . . . can think of object colors always as a combination of a saturated color and a larger or smaller amount of gray, considering that the gray can have various degrees of lightness, and it therefore can also be white. . . . I call the saturated color full color. I designate all colors as belonging to one and the same full color, those occupying a section from any point on the meridian toward the axis, but only to that point. . . . I also consider the sphere to be composed of superimposed ellipsoids, generated from rotation of ellipses around their common axis, as shown in accompanying figure [10.9]. The extreme ellipse is a circle and the solid generated by its rotation around the axis is the sphere itself. (Brücke 1866, p. 62, trans. by R.G.K.)

When viewed in a vertical section (figure 10.9), the ellipses are seen as concentric rings. At different levels of lightness, the distance ratio of a line from the central axis to ellipse surface always equals that of a line from the axis to the sphere surface. Brücke's saturation is therefore a function of lightness. He designated the colors falling on the constant saturation lines of figure 10.9 as tints and shades.

Regarding lightness he wrote: "The lightness of a color is determined by the strength of the light sensation it generates in us. In case of local colors, that is colors we attribute to a solid, we do not consider changes in illumination but compare the colored objects, respectively pigments, under one kind of light" (p. 25) However, his equatorial color circle contains full colors, regardless of their lightness.

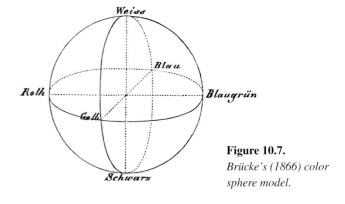

Figure 10.7.
Brücke's (1866) color sphere model.

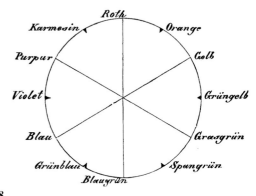

Figure 10.8.
The hue circle with approximately complementary colors opposing each other (Brücke 1866).

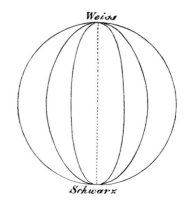

Figure 10.9.
The color sphere that, according to Brücke, has on its surface (as well as on the surface of the internal ellipsoids) colors of constant saturation (Brücke 1866).

According to Brücke his color sphere contains all colors "that can be generated with pigments or those for which it is likely that it will become possible as a result of enlargement of our chemical capabilities, therefore pigment colors in their widest sense." But he understood that for an actual placement of colors obtained from pigments, he would require objective data "because as yet we do not have the means to make the required measurements" (p. 64). Following a long tradition, Brücke considered all of his full colors as having the same saturation. In reality, his ellipsoids show saturation relative to hue as well as to lightness.

The different angular increments between the 10 named colors are due to W. Dobrowolsky's results of incremental wavelength sensitivity. Mixture of opposing lights (in proper intensities) results in the central white. Saturation decreases from the periphery to the center. When adding brightness to the chart, a three-dimensional solid results. In addition to the normal white found at the center of the chart, Wundt also posed a dazzling white, the absolute white, located above the plane.

According to Wundt:

The complete system of light sensations can be represented in form of a double cone where the central plane contains colors of highest saturation. In place of a double cone one can also use a double pyramid or, as the simplest form, a sphere. Here colors of highest saturation and the lower grades obtainable from mixture are located on the equatorial plane. One of the poles corresponds to the most intensive white, the other to the darkest black, neither further changeable with increase or decrease of light intensity. On the line connecting the two poles all possible light intensity variations are located, from absolute white to absolute black. If in place of the circular color chart a triangular chart based on the law of color mixture [e.g., Helmholtz's representation of the spectral trace in a Maxwellian triangle; see figure 6.10] is used a complete system of representation of color sensations in a double tetrahedron constructed from the mixture table is possible. (Wundt 1874, vol. 1, pp. 464–465, trans. by R.G.K.)

Wundt here envisaged the kind of color stimulus space (not including the dazzling region) later calculated by Sigfried Rösch (see entry in chapter 6). He did not settle on a particular color solid. When discussing pyramid and sphere, he mentioned Johann Heinrich Lambert and Phillip Otto Runge (see entries in chapter 4). His sphere of figure 10.14 appears to be related to colorant mixture because the center of the equatorial plane is designated as gray.

In the second edition of *Vorlesungen über die Menschen- und Thierseele* (Lectures on human and animal soul, 1892), Wundt offered a Helmholtz-style cone (figure 10.15) without considering superwhite, and in his last discussion of the subject in *Grundriss der Psychologie* (Outline of psychology, 1896), he proposed a double cone with the four Hering fundamental chromatic colors (figure 10.16).

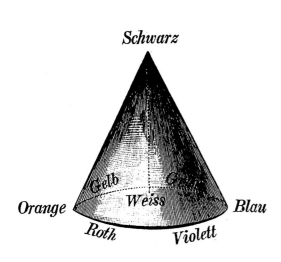

Figure 10.15.
Schematic representation of Wundt's (1892) color cone according to Helmholtz.

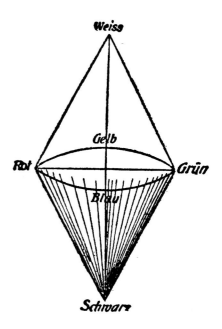

Figure 10.16.
Schematic representation of Wundt's double cone based on Hering's psychological hue circle (Wundt 1922).

GEORG THEODOR ZIEHEN   1891

G. T. Ziehen, *Leitfaden der physiologischen Psychologie*, 1891

German neurologist, psychiatrist, and philosopher **Georg Theodor Ziehen** (1862–1950) co-discovered the Ziehen-Oppenheim syndrome, progressive muscle torsion spasms caused by a brain lesion. He was head of physicians at the psychiatric clinic in Jena, Germany, and the author of several neurological and psychiatric textbooks. Later, he taught philosophy.

Ziehen's *Leitfaden der physiologischen Psychologie* (Textbook of physiological psychology) saw 12 editions, the last in 1924. In it, he described conceptual color order of lights in the general manner of Helmholtz (see entry in chapter 6). The seven-hue circle (figure 10.17) is remarkable for its extended region of blue that, with violet, occupies nearly half of the circle.

As the spectral colors decrease in intensity, they become darker and change: Blue turns into *Graublau* (grayish blue), green into olive-brown, yellow into brown, and red into brown-red, while white turns into gray (figure 10.18). Ziehen called this "melanotropic" (in direction of black) desaturation.

Ziehen demonstrated no clear separation of the concepts of light and object colors. White light can never turn into gray by itself. If it is reflected from a white surface, its grayness, if any, depends on the surround. This commingling of light and surface colors is even more evident in figure 10.19. Here, he used colors of objects to describe the loss of saturation of the spectral lights when mixed with white light: *Strohgelb* (straw yellow), *Fleischfarben* (meat colored), and *Rosa* (rose colored, pink). This kind of desaturation is called "leukotropic" (in the direction of white). Ziehen continued the tradition of intermixing results that apply to lights and to object colors.

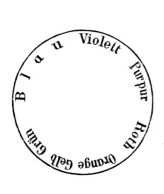

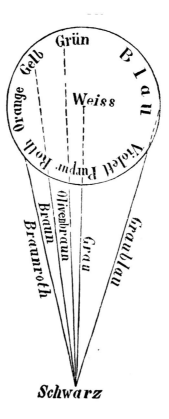

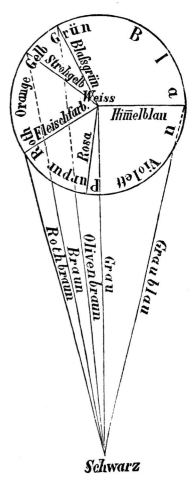

Figure 10.17.

Circular diagram of spectral colors with expanded blue region (Ziehen 1891).

Figure 10.18.

Implicit cone with melanotropic desaturation of spectral colors toward black (Ziehen 1891).

Figure 10.19.

Cone of figure 10.18 with leukotropic desaturation on the basis plane (Ziehen 1891).

OGDEN NICHOLAS ROOD   1879

O. N. Rood, Modern chromatics with applications to art and industry, 1879

In 1879, the American physicist **Ogden Nicholas Rood** (1831–1902) wrote a book on color science that became very influential. It saw several editions, was translated into French and German, and was studied by postimpressionist painters, such as Seurat and Signac. Rood himself was a life-long naturalistic painter and watercolorist. Three years before his death, he contacted Albert Henry Munsell (see entry in chapter 5) and exchanged ideas about color order.

Based on extensive experiments with disk mixture (see glossary, and the description in the introductory text in this chapter), Rood developed a color-contrast diagram with many colors defined by pigment names (figure 10.20) and wavelength of spectral light with the same hue. Regarding color order, he commented:

> With the colours of the spectrum, and a purple formed by mixing the red and violet of the spectrum, we can match any colour whatsoever, provided we are allowed to increase or diminish the luminosity of our spectral hues, and to add the necessary amount of white light. . . . In making the colour-chart we can place complementary colours opposite each other and white in the middle . . . every possible hue and tint belonging to the adopted grade of illumination will be found somewhere within the circle [figure 10.21]. . . . In the construction of this colour-chart we imagined the brilliant colours of the spectrum to be situated on the circumference of the circle, and as we advanced toward the centre a larger and larger portion of white light was to be mixed with them. Suppose now we diminish somewhat the luminosity of our spectral colours; this will change every tint in the chart correspondingly, and also the central white; all will be darkened proportionally; we shall obtain a new colour-chart . . . as we continue the process, we go on accumulating new colour-charts, each being darker than the last. . . . If we arrange this whole series of charts . . . we shall obtain a cylinder which will contain within itself an immense series of colours. The axis of the cylinder at the top will be white, and as we descend it will pass through a great series of darkening greys, finally to end in black. If we make a vertical section of the cylinder, its appearance will then be of the nature roughly indicated in figure [10.22 left]. Now we know it to be a fact that, as coloured surfaces are more and more feebly illuminated, so does the number of tints which we can distinguish on them constantly decrease . . . and we may as well reduce our cylinder to a cone, as indicated in figure [10.22 right]. (Rood 1879, pp. 213–216)

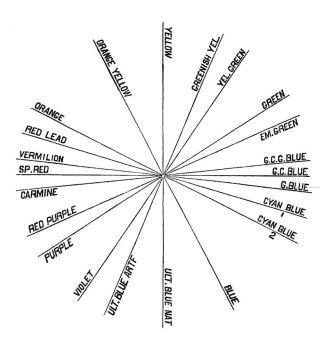

Figure 10.20.
Rood's (1879) contrast diagram.

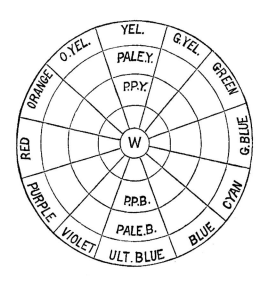

Figure 10.21.
Color chart with spectral colors on the periphery and mixtures with increasing amounts of "white" light toward the center occupied by white (Rood 1879).

Kirschmann, *Color-saturation and its quantitative relations*, 1895

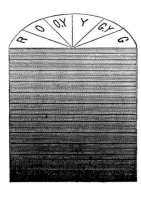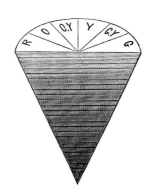

Figure 10.22.

Left: Section of a pile of color charts with the most luminous one on top. Right: Section of a color cone (Rood 1879).

Rood continued that if the first chart consists of spectral colors and their mixtures with white, more charts are possible with spectral lights at higher intensity. The "tints" of these also diminish as intensity increases, ending in a second cone with the brightest imaginable white on top of the double cone.

Rood saw several problems with his double cone. Pigment (object) colors would occupy only a small portion of the lower half of the cone. He also realized that he did not have a rational order for distributing the hues along the circumference: "[W]e do not know if the yellow is to be placed 90° from the red or at some other distance; the same is true with regard to the angular distribution of all the other colours" (p. 218).

In 1902, in their last conversation, Munsell demonstrated his newly developed Value 6 color chart to Rood, who commented favorably on its even gradation of color. But Rood expressed concern about the varying amount of white light he had found in all pigment colorations that "disturbs all their relations." Thus, the physicist Rood, not considering the possibility of psychological order, could not see a fully satisfactory answer to object color order.

August Kirschmann (1860–1932), born in Germany, studied in Leipzig under the celebrated experimental psychologist Wundt (see entry in this chapter) and was one of his earliest Ph.D. candidates. After his studies, Kirschmann moved to Toronto, where in 1893, he became director of the psychological laboratory at the university, a position he held until 1909. During these years, he experimented extensively with color. He designed a disk-mixing apparatus that became standard equipment in many psychological laboratories.

Kirschmann used results of disk mixing experiments as a basis of his conceptual form of a tilted double-cone color solid (figure 10.23). The seven-color spectral circle (including purple) is tilted to show the intrinsic brightness of these colors. Colors of constant brightness are located on horizontal circular planes. As described by Wundt (see entry in this chapter), the colors range from 0 to infinite intensity, not from black to white.

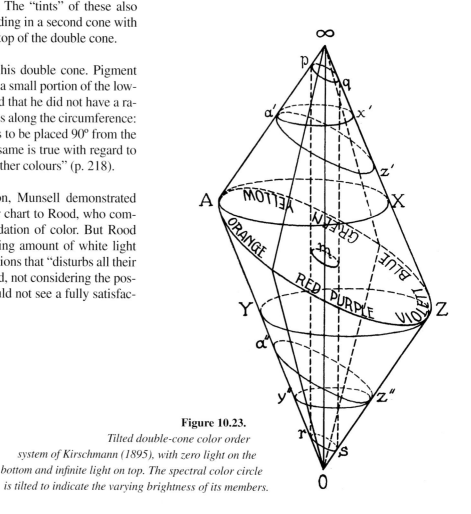

Figure 10.23.

Tilted double-cone color order system of Kirschmann (1895), with zero light on the bottom and infinite light on top. The spectral color circle is tilted to indicate the varying brightness of its members.

JEAN D'UDINE   1903

J. d'Udine, *L'orchestration des couleurs, analyse, classification et synthèse mathématique*, 1903

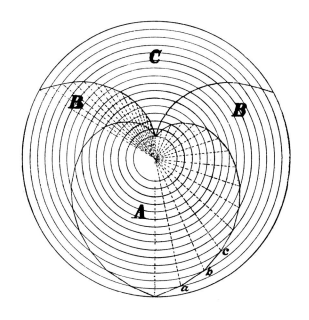

Figure 10.24.

Spinning disk arrangement to result in 15 concentric circles of equal luminous reflectance with (according to the Weber-Fechner law) equally different steps of saturation of the chromatic color displayed on shape A. B is the black and C the white disk (Kirschmann 1895).

Kirschmann commented: "It is a mistake to set at the ends of the axis of a color-sphere or a double cone 'white' or 'black' for these expressions do not designate correctly the extremes of the achromatic series of light sensations" (Kirschmann 1895, p. 391). Like Ziehen's cones (see entry in this chapter), Kirschmann's double cone mixes concepts related to lights with those related to object colors.

Kirschmann was interested in perceptual spacing of disk mixtures' saturation; he invented a disk form resulting in a Fechnerian logarithmic saturation scale of the chromatic disk color in use (figure 10.24). The heart-shaped section A provides the chromatic color; sections B and C are designed to produce the appropriate level of gray to be added to the full color of A so that all saturation steps have identical lightness (when seen as object colors). The exact shapes of B and C depend on the photometrically measured luminosity of the color of section A.

Jean d'Udine was the *nom de plume* of the French music theoretician/critic, philosopher, and educational children's books author Albert Cozanet (1900–1999?). In 1903, d'Udine wrote an essay on the correlation of sounds and colors in the arts. To deduce rules of color harmony, he developed a color order system based on disk mixture. The title of his book translates as "The orchestration of colors: analysis, classification and mathematical synthesis."

D'Udine appears to have been unaware of work before and during his time in this field. His text contains no references and no mentions of previous authors, with the exception of Isaac Newton (see entry in chapter 6), whose name appears two or three times. In this sense, the work with all its strengths and weaknesses seems self-contained.

D'Udine defined color as "the quality of bodies, independent of their form, that provide for us visual sensations under the influence of light" (d'Udine 1903, p. 11, trans. by R.G.K.). Colors have two attributes and four powers.

The first attribute is *hauteur* or height and makes possible distinction between light and dark colors, somewhat similar to relative lightness. Typical related words are "light" and "pale," or "middle," "dark," or "somber." But *hauteur* is not conventional lightness because, as d'Udine stated, "a yellow of height 23 is much lighter than a green of height 23 and that in turn appears lighter than a violet of height 23" (p. 12).

The second attribute, hue (*teinte*), is "the chromatic element that permits distinction between the various colors, even those of the same height" (pp. 12–13).

The powers are defined as follows:

1. Chromatic substance power, which applies to material substrates, e.g., a given dye when applied to silk is brighter and more intense than when dyed on wool
2. Chromatic power of colorants
3. Chromatic power of *hauteur*
4. Chromatic lightness power, which depends on the ability to reflect light

There are only two relations among colors: optical mixture and juxtaposition. Optical mixture produces various kinds of scales, among them height scales and hue scales. The basic height scale is the gray scale, with a maximum of 24 different grades distinguishable between white and black. There are also whitish scales and blackish scales. The height of a given chromatic color is determined by mixing it with a gray of a given height in equal parts; the result, by disk mixture, does not differ from the gray of that height alone.

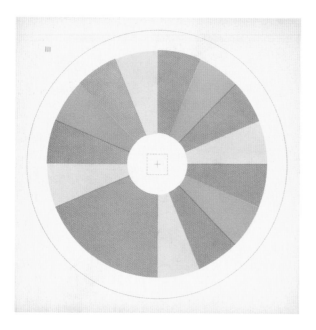

Figure 10.25.

Disk 3 with sectors of the three primaries at equal hauteur resulting on rotation in the appearance of a ternary color (d'Udine 1903).

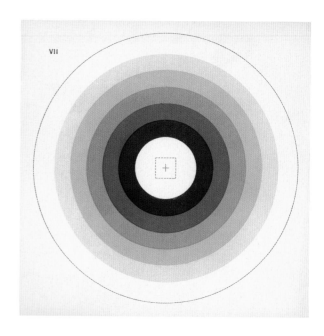

Figure 10.26.

Disk 7 with five grades of hauteur *of primary blue, used in the determination by disk mixture of* hauteur *of other stimuli (d'Udine 1903).*

All colors can be mixed (by disk mixture) from three primaries, with white and black. D'Udine's hue scales consist of colors having the same height as determined in the way just described. He defined the three chromatic primaries used in his disk mixtures as rose (pink, *rose de carthame*, a synthetic lake), yellow in the form of chrome yellow no. 1, and blue in the form of Prussian blue.

Binary mixtures are composed of primaries in various ratios; ternary mixtures are those involving all three primaries. Figure 10.25 is an example of a disk used to generate a ternary mixture of the three primaries at equal *hauteur*. Figure 10.26 is a disk with five grades of *hauteur* of primary blue. They are used, according to d'Udine's method, in the determination by disk mixture of the *hauteur* of a given stimulus.

D'Udine's hue triangle is shown in figure 10.27. Mixing binary mixtures results in ternary mixtures as shown in figure 10.28. They can be achieved with multiple combinations. The scales forming the triangles were to be built to result in perceptually uniform steps. D'Udine did not discuss how to achieve this.

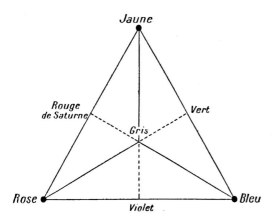

Figure 10.27.

D'Udine's (1903) chromatic triangle, based on the primaries pink, yellow, and blue, with the binary mixtures. Primaries and secondaries are considered balanced so that in mixture they form gray.

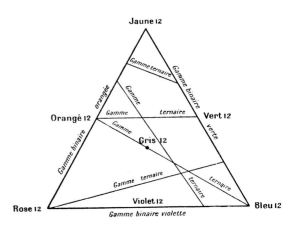

Figure 10.28.

The chromatic triangle at height 12 showing binary and several possible ternary scales (d'Udine 1903).

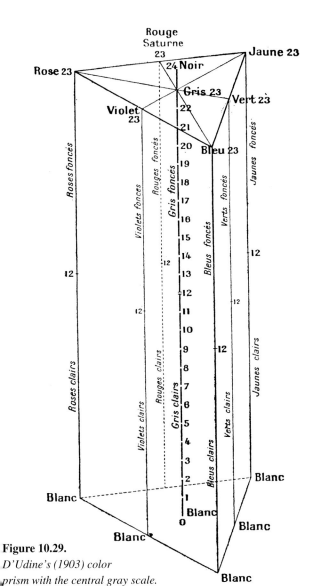

Figure 10.29.
*D'Udine's (1903) color
prism with the central gray scale.*

D.-A. Rosenstiehl, *Détermination de la distance angulaire des couleurs*, 1881
De la construction chromatique dans l'espace, 1910
*Traité de la couleur au point de vue de physique, physiologique,
et esthétique*, 1913

Daniel-Auguste Rosenstiehl (1839–1916) taught dyestuff chemistry at the École de Chimie in Mulhouse, France, and worked as a chemist for a textile dyer/printer and for a dye manufacturer. He was well acquainted with Michel Eugène Chevreul (see entry in chapter 4). Beginning in the early 1880s, he presented a number of papers before the French Academy of Sciences on color, the nature of Thomas Young's (see entry in chapter 6) three fundamental colors, and color order.

In *Traité de la couleur au point de vue de physique, physiologique, et esthétique* (Treatise on color, from the physical, physiological, and aesthetic point of view, 1913), Rosenstiehl interpreted Young's three fundamental colors as a yellow-green, a reddish blue, and an orange. He arrived at this by studying, using disk mixture, complementary color pairs, and color triads that produced white at equal intensity. The colors were plotted according to the color disk segment ratios in an equilateral triangle (figure 10.30).

Only with the mentioned primaries did colors have geometrical distances from each other along the triangle outline that seemed to reasonably agree with perceptual distances. When plotting complementary and triad colors in Maxwell's (see entry in chapter 6) triangle based on yellow, red, and blue primaries, Rosenstiehl found some perceptually similar colors to be widely separated and some distinctly different colors to be close.

Such triangles can be built at each of the 23 levels of height, the result forming a prism, its central axis being the gray scale, as illustrated in figure 10.29. The basis triangle is completely white while that the one at level 23 contains the darkest possible hues, none of which can be black. Black, therefore, occupies a lone position on top of the gray scale. The prism form does not show that d'Udine had a central concept of saturation or chroma.

D'Udine was an unconventional thinker. His descriptions of the interactions of colors and their intrinsic powers are equally unconventional. He was well aware of contrast effects and attempted to describe these quantitatively. He used such findings to propose rules of color harmony. But d'Udine did not know many facts established at the time. His system appears to have had no noticeable impact on the arts and crafts or on the scientific treatment of color.

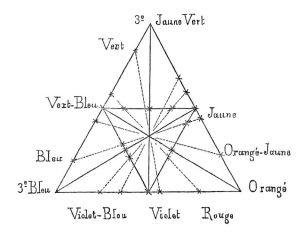

Figure 10.30.
Equilateral (presumed perceptually equidistant) triangle based on disk mixture of compensatory colors. The primaries are 3ᵉ jaune vert (third yellow-green), orange, and the third blue. Compensatory colors are connected by dashed lines (Rosenstiehl 1881).

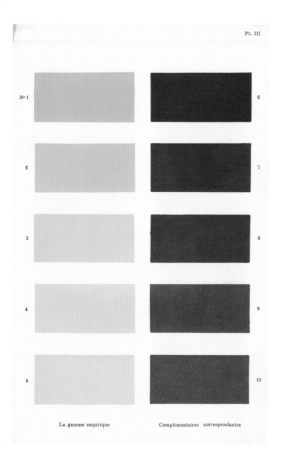

Figure 10.31.

Chart 3 from Traité de la couleur *(2nd ed., 1934)*
demonstrating from disk mixture the nonlinear
relationship between dominant wavelength of a color
stimulus and the corresponding perceived color.

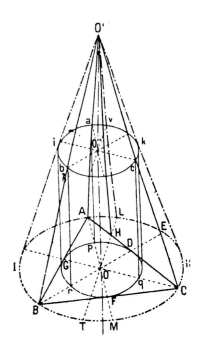

Figure 10.32.
Color stimulus space based on Rosenstiehl's
(1910) equilateral triangle.

Rosenstiehl defined the wavelengths in terms of the Fraunhofer lines in the spectrum: orange, three-quarters from C toward D; yellow-green, three-quarters from D toward E; blue, one-third from F toward G. These definitions correspond approximately to wavelengths of 605 nm, 545 nm, and 468 nm. He believed these colors to be perceptually equidistant. He expected a direct relationship between stimuli and perception and attempted to find the perceptually uniform arrangement based on disk mixture.

Rosenstiehl was aware of the nonlinear relationship between dominant wavelength of stimuli and hue perception. Chart 3 of *Traité de la couleur* (1913) represents a series of oranges of constant dominant wavelength but different saturation (figure 10.31, left column). It also shows the change in hue of the blues (right column) required to neutralize by disk mixture the orange stimuli, changing from a dark greenish blue to a light reddish one.

In 1910, Rosenstiehl proposed a color solid based on his disk-mixture research (figure 10.32). The three primaries A,

B, and C are located on the corners of an equilateral triangle. Their complementary hues are created from equal mixtures of the opposite primaries, for example, intermediate D for primary B. Their luminous intensity is shown by the circle passing through the primaries, in the case of color D by point E. In the center of the triangle is point O, indicating black.

The luminance of the three primaries combined is shown by the point **O**, three times the distance from a primary to the center of the triangle. Optimal tint colors fall on lines such as **BO**. Colors of equal intensity are located on cylindrical layers shown by the cylinder. Rosenstiehl understood that real color stimuli produced by disk mixture would fall well within the surface of this cylinder.

Rosenstiehl's book was influential in France, where it was printed in a second edition in 1934.

ROBERT RIDGWAY   1912

R. Ridgway, *Color standards and color nomenclature*, 1912

Robert Ridgway (1850–1929) was an American zoologist and ornithologist with several books on American birds to his credit. In the late nineteenth century, naturalists saw a pressing need for a set of color standards and a related naming scheme. In 1886, Ridgway published *A Nomenclature of Colors for Naturalists and Compendium of Useful Knowledge for Ornithologists* containing 186 color samples and their generic composition from pigments and lakes.

Ridgway recognized his book's shortcomings due to the "altogether inadequate number of colors represented, and their unscientific arrangement" (Ridgway 1912, p. ii). (Figures 10.33 and 10.34 are images from the 1886 book.) So in the first decade of the twentieth century, he began to develop an improved standard to represent the best scientific knowledge. In 1912, he self-published *Color Standards and Color Nomenclature*, with 1,115 color samples on 53 plates and a corresponding list of names.

Based on his judgment, Ridgway selected painted samples representing six fundamental hues, red, orange, yellow, green, blue, and violet. Comparing their dominant wavelengths to those determined by other authors, he found a considerable variation. He mentioned "[D]ifferences in the conception of different persons as to just where the reddest red, greenest green, etc., are located" (p. 7 note).

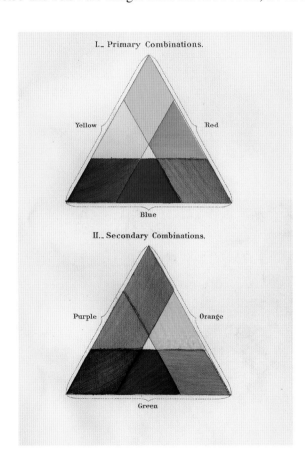

Figure 10.33.

Plate 1 from A Nomenclature of Colors *(Ridgway 1886) illustrating subtractive mixture from the three primary colors yellow, red, and blue. The pigments used are "light cadmium" for yellow, "scarlet vermilion and madder carmine" for red, and "ultramarine" for blue. In the secondary mixtures, "orange cadmium" was used for orange and "aniline violet" for purple because the corresponding mixtures of the primary pigments were found to be too dull. The tertiary mixtures have been obtained from mixtures of the secondaries.*

Figure 10.34.

Plate 9 from Nomenclature of Colors *with 23 named blue colors (Ridgway 1886).*

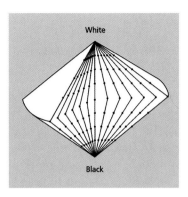

Figure 10.35.

Loosely conceptual model of the double cone implicit in Ridgway's 1886 color system, with the tint-shade and tone colors of constant hues.

Ridgway then matched these colors using disk mixture. His mixture primaries were disks dyed with the most saturated dyes and lakes he could get. Following Chevreul (see entry in chapter 4), he conceived a hue circle of 72 grades, always 12 steps between primaries. Half of these were created by disk mixture "so as to represent an apparently even transition from one to the other . . . as well as the eye alone could judge" (p. 7). He recorded and plotted the relative amounts of always two of the six fundamentals in each of the full colors, and he smoothed the resulting curves. The smoothed values became the aim standards of the full colors.

In his earlier book, Ridgway already had shown a logarithmically scaled gray scale. As a result of discussions with color experts of the U.S. Bureau of Standards and others, he chose the Weber-Fechner law as the scaling tool for tint-shade and tonal scales. Figure 10.35 is a loosely conceptual sketch of the perceptual organization of these scales in a double cone.[3]

For each hue a tint/tone scale of the full color was prepared, as well as up to five tonal scales. The full color, ' (single prime) and " (double prime) tonal scales are complete with 36 hues each. The '" and "" tonal scales exist only for every second hue, beginning with hue 1 (red). Tonal scale """ was executed only for six irregularly spaced hues. Only hues 1 and 49 (blue) have all five tonal scales.

Between white and the full color, and black and the full color are always three grades whose stimuli have been graded logarithmically. The same applies to the tonal scales between the full and tint-shade colors and the central gray. This gray had been matched from different

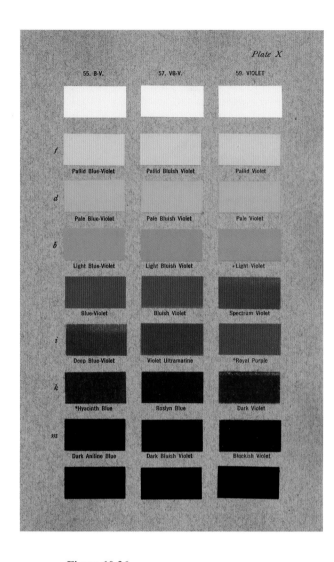

Figure 10.36.

Plate 10 of Ridgway's (1912) atlas with tint-shade scales from white via full color to black of three blue-violet hues.

mixtures of three of the six primaries. The implied, but never explicitly mentioned, three-dimensional structure of the color solid is a double cone (figure 10.35).

The arrangement of colors in Ridgway's book is in the form of linear scales, however. Three tint-shade scales of blue-violet and red colors are illustrated in figures 10.36 and 10.37. The 1 × 1/2 inch samples are of painted paper, matched by a professional colorist to the disk-mixed references.

From many contemporary lists, Ridgway selected color names that he felt most appropriate for his samples. That

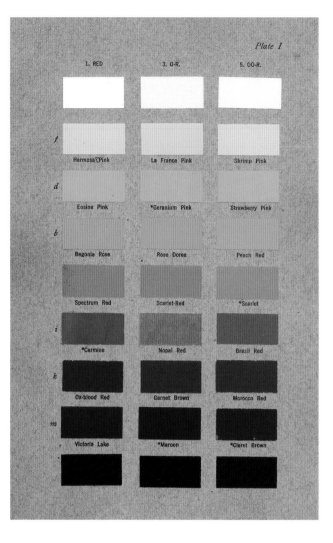

Figure 10.37.

Plate 1 of Ridgway's (1912) atlas with tint-shade scales from white via full color to black of three red and orange-red hues.

his often-unusual names could not be considered a standard is shown in an excerpt of the alphabetical listing: Eugenia Red, Eupatorium Purple, Fawn Color, Ferruginous, Flame Scarlet, Flax-flower Blue, Flesh Color.

Ridgway invested a large personal effort into his proposed color standard. As mentioned by Adelbert Ames (see entry in chapter 5), Ridgway's system was considered the leading one in the United States in the 1920s. Absence of an objective definition of his color samples detracts from the value of the work as a true standard. Ridgway's interests did not extend to a full-fledged color system.

WILHELM OSTWALD   1917

W. Ostwald, *Der Farbenatlas—2500 Farben auf über 100 Tafeln*, 1917
Mathetische Farbenlehre, 1918
Physikalische Farbenlehre, 1919

In 1906, the German chemist and Nobel Prize winner **Wilhelm Ostwald** (1853–1932) decided to end his academic career to pursue topics that intensely interested him. High among them was color order, a previous interest that became a major preoccupation after he retired to his estate in Grossbothen near Leipzig.

In 1905, Ostwald lectured on color at leading eastern American universities, where he met with Munsell (see entry in chapter 5) and became familiar with his work. While initially supportive, Ostwald proceeded on a much different path. As one of the leading scientists of his time, he approached the problem of color and color order in a commensurate manner. In 1918, he began to publish an extensive *Farbenlehre*, planned to encompass five volumes. He wrote three, with the third published posthumously; Hans Podestá (see entry in chapter 5) wrote the fourth, and the final volume was never published.

Fundamental to Ostwald's approach is his definition of the *Farbenhalb* (semichrome). The reflectance function of a perfect white object has a constant value of 1 across the spectrum. For an object with a given spectral hue, the (ideal) reflectance function has a value of 1 across a certain portion of the spectrum and a value of 0 for the remaining portion. The absorbed portion of the spectrum is equal to the compensatory color (the two add up to 1 across the spectrum, thus semichrome).

If the color is veiled in Hering's sense (see entry in chapter 5), it also contains a spectral portion of black and white (segments *h*, *r* and *s*, *i* in the idealized figure 10.38). Using the tool of disk mixture, Ostwald determined 100 hues in a hue circle by judging perceptually equal distances and compensating with the color diametrically opposite (figure 10.39). Although the results showed considerable variation, he took them as confirming the identity of an organization based on perceptual uniformity and compensation.

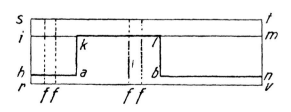

Figure 10.38.

Conceptual reflectance function (h, a, k, l, b, n) of a grayed blue-green color, with black segment on bottom, the chromatic segment in the center, and the white segment on top (Ostwald 1923a).

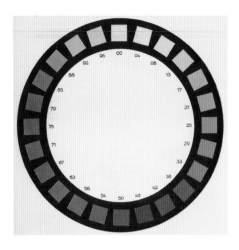

Figure 10.39.
Ostwald's (1924a) 24-grade hue circle.

Figure 10.41.
Commercial model of *Grosser Farbenkörper* with 2,500 samples (Ostwald 1917).

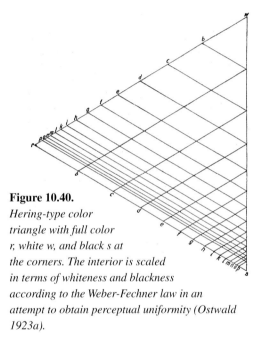

Figure 10.40.
Hering-type color triangle with full color r, white w, and black s at the corners. The interior is scaled in terms of whiteness and blackness according to the Weber-Fechner law in an attempt to obtain perceptual uniformity (Ostwald 1923a).

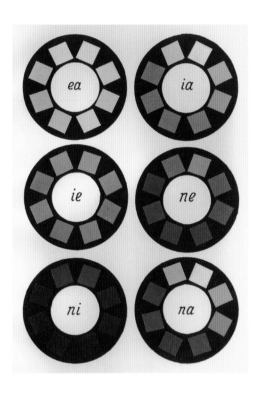

Figure 10.42.
System presentation as Wertgleiche Kreise *(varying hue numbers but constant letter combinations) (Ostwald 1923b).*

In his atlas, Ostwald reduced the number of hues to 24. Realizing that no colorants had purity comparable to spectral colors, he selected the best available as *Vollfarben* (full colors, undiluted, representing semichromes) and used filters to measure their lightness. Using the Weber-Fechner logarithmic relationship between stimulus and response, he scaled the Hering triangle accordingly (figure 10.40). The result is a presumed perceptually uniform color triangle in which colors are identified with double letter codes (in addition to hue).

Like Hering's implicit color solid, Ostwald's is a double cone (figure 10.41). Color samples are arranged according to four different scales: hue, equal whiteness, equal blackness, and equal chromaticness. The *Farbkörper* (color solid) is the first system with extensive systematic coloring in which all colors of a hue are shown on one page, and most colors are members of three separate attribute scales.

Ostwald published his color solid in various forms, among them *Der Farbnormenatlas, Der grosse Farbkörper, Der kleine Farbkörper, Farbenleitern, Farbenfächer,* and *Wertgleiche Kreise* (figure 10.42) (color standards atlas, large and small color solid, color ladders, color fan, and constant value circles consisting of hue circles with varying constant letter combinations). There were two different-

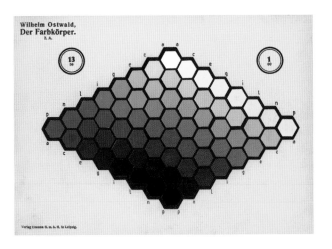

Figure 10.43.
Chart from Ostwald's (1919a) Farbkörper *with the cross section through the color solid along hues 13 (blue) and 1 (yellow).*

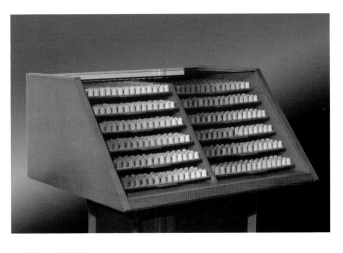

Figure 10.44.
Die Farborgel *(color organ), containing standardized pigment cakes guaranteeing constancy of color stimuli (Ostwald 1919b).*

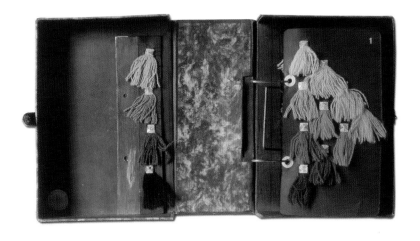

Figure 10.45.
Left: Image of Kleiner Wollfarbenatla *(Ostwald 1924b), Right: Ostwald's* Grosser Wollfarbenatlas *from the advertising pamphlet of Simon & Dorias in Chemnitz.*

sized atlases with dyed wool samples. *Farbkörper* has 12 plates with a total of 672 unique chromatic and eight gray color chips (figure 10.43).

Ostwald also offered the so-called *Farborgel* (color organ, figure 10.44), which contains all colors of his system in form of small pigment cakes that required dispersion in water before applying. Further, he developed a system of color harmony and published *Harmothek*, a separate collection of grayish samples. He also issued his system as dyeings on wool (figure 10.45).

Although Ostwald had his detractors, he significantly influenced color education and the systematic use of color, specifically in the textile industry, and in homeopathy, botany, and other areas of natural science. Several specialty color charts based on Ostwald's system were produced for various practical purposes. In the United States, Ostwald's system found strong support from Container Corporation of America. This company issued its *Color Harmony Manual*, based on Ostwald's system, in three editions from 1946 to 1948 (available until 1972). And he had several followers, in particular Aemilius Müller (see entry in chapter 5), who refined and expanded his work in the form of both atlases and applications of color harmony.

But by the early 1930s, the German Standards Institute saw the need for a more practical system expressed in terms of the new International Commission on Illumination (CIE) colorimetric system (see Manfred Richter and DIN 6164, chapter 7).

DDR-STANDARD TGL 21 579   1965

DDR-Standard, Farbkarte Grundsystem TGL 21 579, 1965

In the mid-twentieth century, the development of the West German DIN 6164 system led by Manfred Richter (see entry in chapter 7) was paralleled by that of the East German **TGL 21 579** system led by Manfred Adam (1901–1987).

Adam, the last of Ostwald's assistants, headed an institute for color and colorimetry research in Ostwald's converted estate Grossbothen near Leipzig. He aimed to integrate the perceived advantages of the systems of Ostwald, Prase, and Munsell and to ground the result in the CIE colorimetric system. The system was designed so that different parameters would allow it to be adapted for different purposes. Adam's system was declared a standard in the German Democratic Republic (Deutsche Demokratische Republik, DDR) in 1965.[4]

The system defines attributes hue (*L*) in terms of dominant or complementary wavelength, purity (*R*) in terms of spectral color content, and lightness (*H*) in terms of luminance (figure 10.46). Distances between the 24 hue lines are considered perceptually equidistant. The object color system is a tilted double cone with a curved central plane (figure 10.47) with optimal object colors on the surface and a 24-color hue circle (figure 10.48). Its internal structure is shown in figure 10.49.

Unlike Ostwald's and Robert Luther's definitions of optimal object colors (see entries in this and chapter 6, respectively), Adam's definition places the *Vollfarben* (full colors) all at equal distance *R* = 10 from neutral gray in the center, with spectral colors defined as *R* = 14. Colors of a given hue can be defined by lightness *H* and saturation *R* or by *Farbklarheit* (clarity) *K* and *Farbtiefe* (depth) *T*. Of the 55 possible samples per hue (see figure 10.49), 36 were to be displayed in an atlas. Together with a nine-grade gray scale (not including black and white), this would have resulted in 873 color samples.

The atlas was only partially completed. Its absence and the DDR's limited influence resulted in limited acceptance of the system.

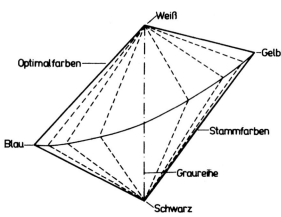

Figure 10.47.
Object color solid of TGL 21 579 (Arnold 1988).

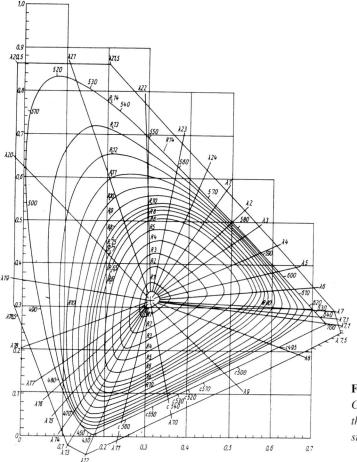

Figure 10.46.
CIE chromaticity diagram with radial lines defining the 24 hues and circular lines defining the 14 purity steps of the system (DDR-Standard 1965).

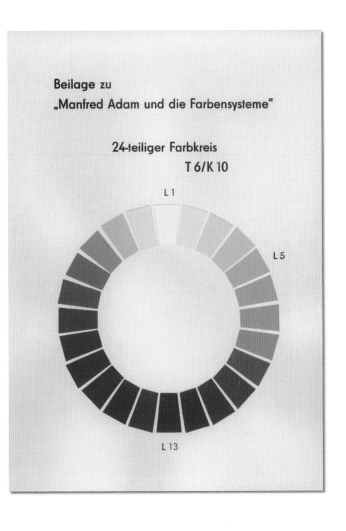

Beilage zu
„Manfred Adam und die Farbensysteme"

24-teiliger Farbkreis
T 6/K 10

Figure 10.48.
*Circle with 24 hues as color samples (Adam 1989)
and conceptually (Zeugner 1969).*

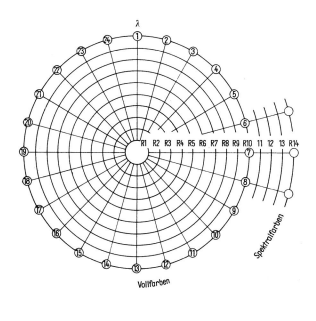

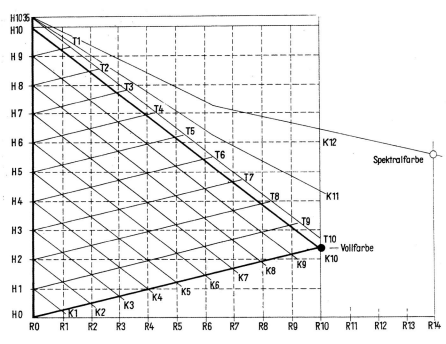

Figure 10.49.
*Conceptual structure of the constant-
hue diagram for hue 14 (ultramarine).
A given color can be defined in terms
of purity R and lightness H or clarity
K and depth D. The spectral color has
purity 14 (Zeugner 1969).*

249

ANDRÉ LEMONNIER   1968, 1976

J. Fillacier and A. Lemmonier, *Intégrateur universel de couleur*,
French patent 1,554,843, 1968
A. Lemonnier, *Couleur, échelles et schémas*, 1976

In 1988, French visual artist **André Lemonnier** and his colleague Jacques Fillacier, patented a disk-mixture system named Polyton. With it, more than 180,000 colors could be reproduced and codified nearly instantaneously. The system consists of a circle of 24 full-color hue disks of 80 mm diameter, each tinted toward white in 12 grades, resulting in 288 disks. In addition, 40 masks from white to black with increasingly larger black segments can be mounted with a color disk on a handheld motorized spinning device for rapid rotation resulting in disk mixture (figure 10.50). For the patent, the mixtures are arranged schematically in the form of a cylinder (figure 10.51). A notation system helps to specify each generated color stimulus.

In 1976, Lemonnier exhibited a series of artistic color scales derived from a color order system in double-cone form with more than 1,500 samples. He based his effort on those of Ostwald, Chevreul, and Munsell (see entries in this and chapters 4 and 5, respectively) and on the attributes hue, lightness, and saturation. The internal division was to represent perceptual uniformity. The motivation was ambitious: a system that would be useful to manufacturers of colored products, architects and designers, natural scientists, teachers, and theoreticians.

The central axis is formed by a 19-grade gray scale (figure 10.52). Twenty-four constant-hue planes contain 66 samples each. The full colors were arranged according to their

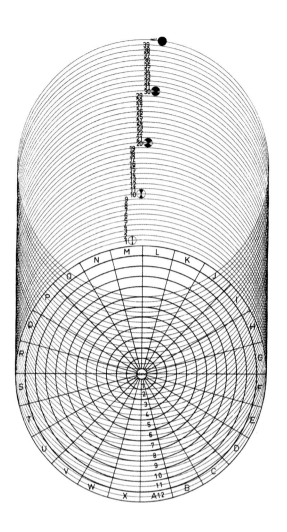

Figure 10.51.

Schematic representation of the Polyton hue circle with the full colors on the periphery and tint declines toward white in the center. Shade and tonal colors are achieved by disk mixture with black as shown on the remaining circles (Fillacier and Lemonier, 1968). Reprinted with permission.

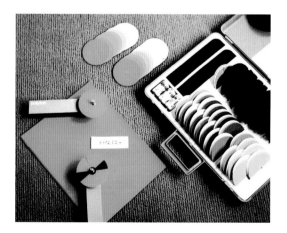

Figure 10.50.

The Polyton disk-mixture system (undated advertising image). Reprinted with permission.

lightness, as shown by the black dots of figure 10.53. As a result, that figure also shows that the line connecting full colors is irregular. A colorimetric definition of the system samples is absent.

Lemonnier developed a series of "harmonizers" that allow systematic creation of harmonic combinations of two and more colors using various schemes of color harmony. He used the solid to create several kinds of systematic sequences (based on his solid) of color samples and exhibited them internationally. Sequences involved the surface of the solid or its interior, or both. An example of such a sequence is illustrated in figure 10.54.

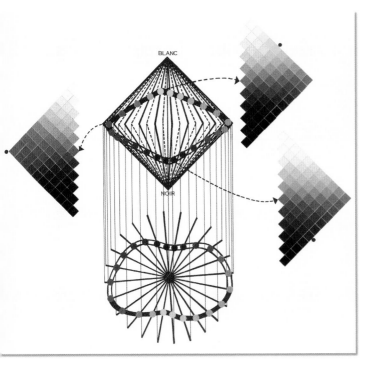

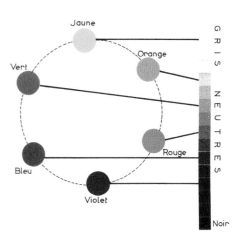

Figure 10.52.
The three primary colors and their secondary mixtures related to the 19-grade gray scale (Lemonnier 1976). Reprinted with permission.

Figure 10.53.
Schematic model of the double cone solid, three constant-hue planes (with the location of the full color shown by black dot) and the curved and bent line connecting the full colors in the system (Lemonnier 1976). Reprinted with permission.

Figure 10.54.
Example of a systematic sequence of colors (échelle) *in Lemonnier's (1976) color solid. Reprinted with permission.*

WILLIAM BENSON   1868

W. Benson, *Principles of the science of colour, concisely stated to aid and
promote their useful application in the decorative arts*, 1868
*Manual of the Science of Colour, On the true theory of colour-sensations
and the natural system*, 1871

The English architect **William Benson** developed a color
system for practical application in the decorative arts. He
kept well informed on the scientific findings in the color
field, as described, for example, in Lambert's book on
photometry and in Helmholtz's and Maxwell's works (see
entries in chapters 4 and 6, respectively). With experience
in pigment mixture as well as his own experiments with a
prism and mixtures with Lambert's mirror, Benson fully understood the difference between light and colorant mixture.

In 1868, Benson published *Principles of the Science of Colour*, which describes a cubic color system. Based on this
system, he derived rules of color harmony for color-design
use. Later editions appeared in 1872, 1876, and 1886. In
1871, he followed that book with *Manual of the Science
of Colour*, a further description of his color order system,
as well as discussions of psychological effects of color and
color blindness. In a later edition of his first book, he advertised *Colour-Lore*, his planned crowning effort in terms of
size and detail, but it was never published.

Familiar with the color order systems of Mayer, Runge, and
Chevreul (see entries in chapter 4), Benson strove to avoid
their perceived errors. He attempted to cover the totality of
color sensations in an appropriate geometric model named the
Natural System of Colours, whose principle, he acknowledged,
John Herschel (1830) had earlier proposed.

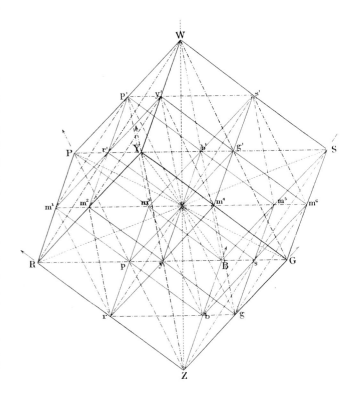

Figure 10.56.
*Graphical representation
of the color cube (Benson 1868).*

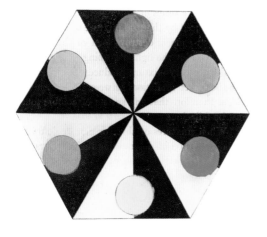

Figure 10.55.

*Benson's (1868) primary and secondary colors
as determined with the help of edge spectra
(symbolically represented by the background).*

Benson's system is a conceptually additive one. He considered
spectral colors to best approximate pure color sensations:

> [T]he best Red, Green, and Blue, of the solar spectrum
> give the nearest possible approach to the three primary
> colours; and therefore that their complementary colours
> Seagreen, Pink, and Yellow, give the nearest possible ap
> proach to those binary compounds of the primaries which
> are usually termed secondary colours. The brightness of
> each of the last (in its greatest intensity) must be equal
> to the sum of the brightness of the full intensities of their
> single components. (Benson 1868, p. 13)

In their binary mixtures, the primary colors red, green, and
blue (rather than Young's violet; see entry in chapter 6
form the secondaries, taken to complement the primaries
as determined with the help of edge spectra (figure 10.55
similar to Johann Wolfgang von Goethe; see entry in chap
ter 3). Their brightness is the sum of the brightness of the
constituting primary lights, as required by additivity. Thus
the traditional purple of colorant-based systems is replaced

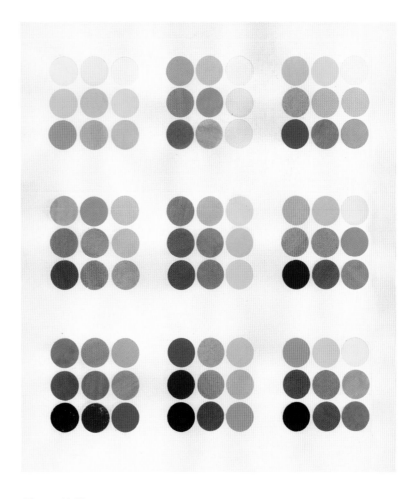

Figure 10.57.

Illustration of cross sections through the cube from corner to corner, respectively, between complementary colors (Benson 1868).

Figure 10.58.

Illustration of three surface planes bordering white (top row), the three that border black (bottom row), and three sections through central gray (middle row) (Benson 1868).

by pink: "The necessary double brightness of the secondary colours, when of full strength, on which so much of their beauty depends, and which is neglected in schemes founded on the mixture of pigments, is here particularly noticed; and the importance of bearing this in mind has induced me to use the term Pink, instead of Purple" (p. i). All possible hues can be generated from the primaries, and all tints and shades, with varying light intensity.

The form of Benson's *Natural System* is a tilted cube, with black at the bottom corner and white at the top (figure 10.56). The three primaries are located on the cube corners adjacent to black; the three secondaries are on the corners adjacent to white directly opposite their complementary primaries. All other possible color perceptions are located on the cube's surface and in its interior.

Benson considered the form of the cube and his selection of primary and secondary colors not only to represent the natural relations of colors, but also to offer new and improved principles of color design:

There are thirteen principal diameters or axes in the cube, and they may be divided into three classes. The three which join the middle points of the opposite sides may be called primary axes, because on them there is a change in one of the primaries only. The six which join the middle points of the opposite edges may be called secondary axes, because on them there is an equal change in two of the primaries, either in the same or in contrary directions. The four which join the opposite corners in like manner may be called tertiary axes, because there is an equal change of all the three primaries, either all in the same direction, or two in the same and one in the contrary direction. Those axes along which one colour diminishes while another increases in intensity, may be distinguished in cross axes. In like manner there are thirteen medial planes, being those which bisect the cube at right angles to the thirteen axes. The advantage of distinguishing these principal lines and planes in the cube of colours will soon be seen in the facility with which the principal gradations, contrasts, and harmonious arrangements of colour may, by means of them, be classified and remembered. (p. 19)

J. Carpentier, Définition, classification et notation des couleurs, 1885

In various editions of the *Principles*, Benson illustrated the cube with 27 color samples, either hand-colored or in the form of tipped-in printed paper roundels of 15 mm diameter (figures 10.57–10.59). They show various views of the cube's surface demonstrating new possibilities of color combinations. Of interior colors, they show only the central gray. Such illustrations make clear how difficult it was for system users to orient within the cube.

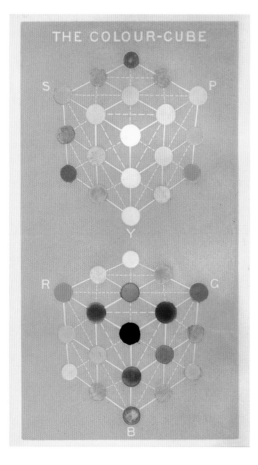

Figure 10.59.

Two surface views of the color cube (Benson 1871).

In 1885, French engineer **Jules Carpentier** published a short paper under the title "Definition, classification and notation of colors." Here he stated:

The color of a surface, as an optical property, will be completely defined if one indicates its reflectance power for all the simple light rays. Such a definition would be much too complex in its formulation, but experience shows . . . a color is defined by its reflective powers for three given simple light rays, conveniently selected in the spectral range. In other words, two surfaces to which our eye attributes the same color possess the same reflecting powers for the same three test rays, and two surfaces that differ in these powers at least for one ray appear to us in distinct colors. (p. 808, trans. by R.G.K.)

Carpentier was well aware that objects with given reflectance properties can look very different, depending on its illumination. To measure the reflective powers, Carpentier conceived a measurement apparatus that may not have existed as more than an experimental model. To express the results of measurement, he proposed the geometrical model of a cube, with axes representing the three reflectance values (figure 10.60).

Maximal reflection at the three wavelengths indicates white, with scale values of 9, 9, 9. All other colors fall within the cube, with black at its origin. With black having the coordinates 0, 0, 0, the cube encompasses 1,000 different colors. Carpentier selected the letters X, Y, and Z as symbols for the three coordinates, thereby anticipating the CIE. Each color is defined by its own three numbers: "The name of a color is nothing but the number that symbolizes it" (p. 810). Should the division into 1,000 colors prove insufficient, Carpentier recommended expansion by adding a decimal.

Carpentier was a forerunner of Leopold Bloch (see entry in chapter 9) and, like him, was not aware of the issue of metamerism.

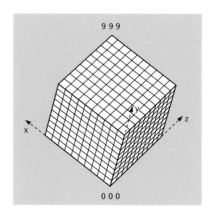

Figure 10.60.

Schematic representation of Carpentier's color cube.

MAX BECKE   1923

M. Becke, Verfahren der Farbengebung nach dem natürlichen Dreifarbensystem, Austrian Patent no. 92344, 1923
Einführung in die natürliche Farbenlehre, 1923/1924

In the early 1920s, Austrian **Max Becke**, then director of the Research Institute of the Textile Industry in Vienna, offered an unusually ambitious proposal for color order. For Becke, the developing synthesis of Helmholtz's and Hering's views on color order was not fundamental enough.

Using "the law of unity, the idea of functionality and the conviction that nature always reaches its goals with the simplest means and by the shortest possible route," Becke developed a theory of the "geometry of the natural three-color order . . . Three ordinates are sufficient to determine a point in space—three fundamental color concepts are also sufficient for nature as ordinates in color space" (Becke 1923/1924, p. 5, trans. by R.G.K.).

However, Becke attempted to reach further. According to him, "the universe is infinite . . . a sphere with its center anywhere and its borders nowhere." He believed that the universe has a soul and a will and that humans represent the crown of development of living organisms and move ever closer to the final goal of the universal will. Our purpose is to be "the executor of the natural will. . . . Any problem, including the color-order problem, can only be solved by reducing them to the causes and actions of the world will . . . identical with hypothetical fundamental matter, the world ether" (p. 8).

More of Becke's philosophy: The spherical units of world ether are ordered so that they have 12 nearest neighbors (see figure 10.61 for a two-dimensional representation). Our vi-

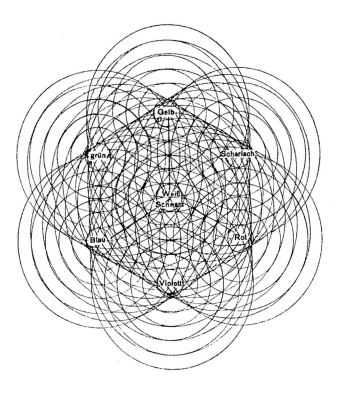

Figure 10.62.
Interacting kernels of world will generate primary and secondary colors (Becke 1923/1924).

sual organ is closely tuned to its activities. The packing of units of world ether results in the organization of perceptual color space as a tilted cube, as illustrated in figure 10.62.

Becke's cube contains all colors. Its axes are red, green, and blue, departing from white on top and located exactly one-third of the distance down the central gray axis that ends in black. The 1:1 mixtures of the primaries, scarlet, green, and violet, are located two-thirds along that line. Maximum intensity of each primary is designated as 120. The dimensions of the sphere change linearly while the weights of the reduced spheres formed by reduced dimensions change according to the third power. Becke concluded from this that the weights of colorants used to implement the natural color order need to change by power 3 to result in a perceptually uniform version of natural color space.

Becke next identified three commercial textile dyes he believed to come closest to the three fundamental colors: Chinolin Yellow (Color Index Acid Yellow 3), Sulphorhodamine B extra (C.I. Acid Red 52), and Patent Blue Supra Fine (C.I. Acid Blue 7). He determined the amounts of these three dyes required to produce a good black on wool yarn and, "in November 1920 at 11 AM in the Michelbeneurn street [presumably in Vienna]: 8.649% Yellow, 2.448 % Red and 2.448% Blue were required" [!].

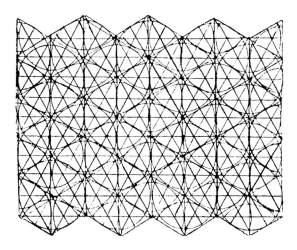

Figure 10.61.
Two-dimensional packing of the smallest entities of world will (spheres) (Becke 1923/1924).

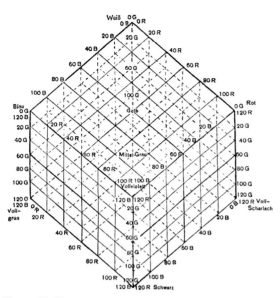

Figure 10.63.

A 120-unit version of the natural three-color solid with white on top and black at the bottom (Becke 1923/1924).

Using 120 parts for each primary produces 120 layers of always-identical yellow, red, and blue content. They can be appropriately reduced in number, for example, to six, to obtain a small uniform color cube (figure 10.63). After calculating such color layers and then "blindly dyeing them," Becke could "point to the nearly triumphal confirmation of the law of conservation of energy" (p. 27).

Colors (in the 120-unit system) are identified by numbers in triangular order, for example, $_{60}{}^{60}{}_{60}$ for a middle gray, where the upper number refers to the yellow primary, the lower left figure to the red, and the right figure to the blue primary. The resulting color sphere has 120 levels of identical darkness (or lightness), and layers of identical color purity: spheres of varying size concentrically located around the middle gray. Becke did not further explain any of these terms.

Becke also produced abridged tables of the 120-unit system in which many colors are named. A full system at the 120-unit level represents more than 1.7 million colors. He also discussed a seven-unit version with 343 samples. The key claim in his patent relates to determining the "optical equivalent" of the primary colorants (they must result in Hering-type unique hue colors). He achieved this by mixing them to form a deep black. Another claim relates to the cubic relationship between colorant concentration and perceptually uniform spacing of the "natural three-color system."

Becke quoted physicist Max Planck's dictum "First understand, then apply" as the maxim for his own work. He believed that by having discovered the relationship of color order to the deeper world order, he had solved all issues of color order once and for all. Although wrapped in an elaborate metaphysical coat, his color sphere is a successor of Benson's (see entry in this chapter)

OTTO PRASE   1912, 1922, 1945

P. Baumann, *Baumanns neue Farbtonkarte, System Prase*, 1912
Baumanns Farbtonkarte, Atlas II, 1922
Baumanns Farbkörperdurchschnitte zum Farbenatlas II, 1922b
P. Baumann, *Der Schlüssel zur Farbenharmonie*, 1924
O. Prase, *Experimentalstudien zur Farbenlehre, 1–4 Teil*, 1941–1945

German painter **Otto Prase** (1874–1956) extensively examined color order from a theoretical as well as practical point of view. The solutions he offered changed over time. His system resembles that of Ostwald (see entry in this chapter), even though Prase sought to demonstrate his independence from Ostwald.

In 1945, Prase proposed a universal color chart, UNIFAKA, to be used around the world. He was most successful with his first proposal, the *Farbtonkarte*, which was widely used in architectural and interior painting and available in Germany into the 1950s.

Prase's success was based on his collaboration with Saxon color-charts manufacturer Paul Baumann (1869–1961). To improve on an earlier award-winning color chart for interior decorators and painters, in 1912 Baumann issued a chart based on Prase's proposals. Its structure results from paint mixture, in agreement with the thinking and working methods of architectural painters. Intermediate steps extended a 24-hue visually uniformly spaced hue circle to 48 hues, consisting of the purest pigments commercially available at the time (figure 10.64). The resulting 48 full colors were scaled according to lightness and shaded in varying numbers of steps (depending on the lightness of the full color) toward black.

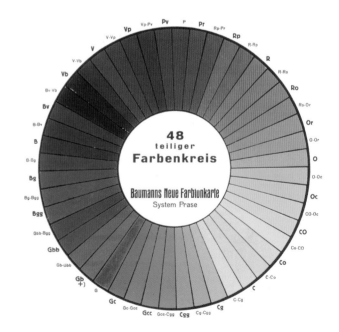

Figure 10.64.

The 48-hue circle of the Prase system (Baumann 1912).

Figure 10.65.
Schematic construction of the constant-hue plane of hue Gb in the Prase system (Baumann 1912). Intensivster Ton, full color; Hellklare Reihe, tint scale; Dunkelklare Reihe, shade scale; Graureihe, gray scale.

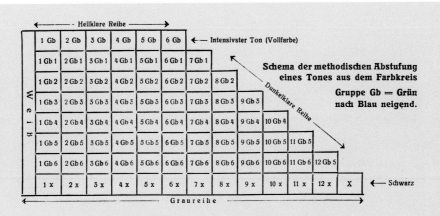

Figure 10.66.
Representation of the colors of the scheme of figure 10.65 (Baumann 1912).

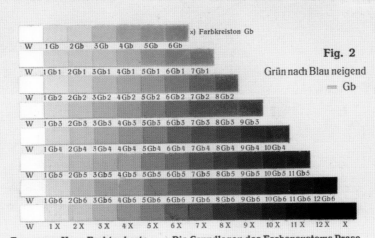

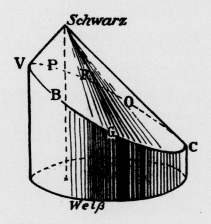

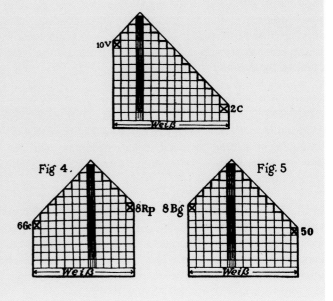

Figure 10.67.
Schematic view (left) and cross sections (right) of the color solid implicit in the Prase system (Baumann 1912).

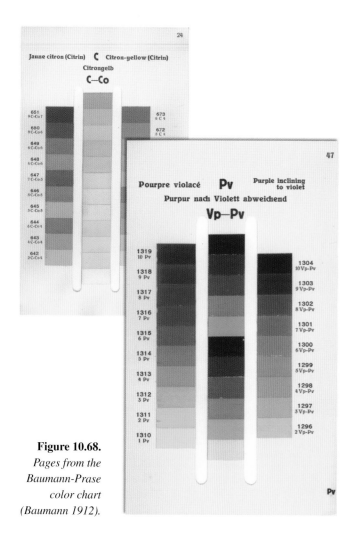

Figure 10.68.
Pages from the Baumann-Prase color chart (Baumann 1912).

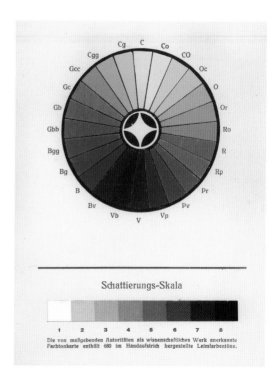

Figure 10.69.
Hue circle and gray scale of the Baumann Atlas II (Baumann 1922a).

The full color, the shaded colors, and black were subsequently tinted toward white, according to the scheme shown for a bluish green hue in figures 10.65 and 10.66, resulting in 63 chromatic colors for this hue. The corresponding color solid is a tilted cone placed on a cylinder, with black located in the point at the top of the cone and white on the basis plane of the cylinder (figure 10.67).

The standard edition of the color chart contains 1,359 painted color sections of size 18 × 10 mm (in addition to a few gold and bronze color samples). Figure 10.68 illustrates the arrangement of three series of violet-purple hue, with cutouts for direct comparison, and the corresponding pigment formulations on the left.

In the 1920s, competition developed between the Saxon systems of Ostwald and Baumann-Prase. Ostwald wanted Prase to use the Ostwald system for designating his samples, but Prase instead developed a new system. For each of the original 24 hues, Prase developed 28 tints, shades, and tones using disk mixture (figure 10.69).

The Baumann-Prase system differs from Ostwald's by using a trigonal rather than a logarithmic scale to relate stimulus values to assumed perceptually equal distances. The systems are alike in that constant-hue planes are shown in triangular form (figure 10.70). Prase also described a version of the system in which he considered the lightness of the full colors, resulting in a double cone.

The Baumann-Prase system, first published in 1922 as *Baumanns Farbtonkarte, Atlas II*, contained 680 hole-punched color samples of size 20 × 10 mm. Following Ostwald, it issued cross sections through the solid (figure 10.71) and color fans (figure 10.72), the latter under the designation "key to color harmony," based on Ostwald's rules of harmony. *Atlas II* and the implicit system remained in the shadow of Ostwald's system.

As Prase described it, between 1917 and 1925 he developed and hand-colored a 1,000-color cube (figure 10.73). He also was responsible for the coloration of the cubic system of Becke (see entry in this chapter).

The significant impact of Baumann and Prase was the 1912 system's trade-based methodology of whitening full and shade colors, a method used by Prase's successors until the present.

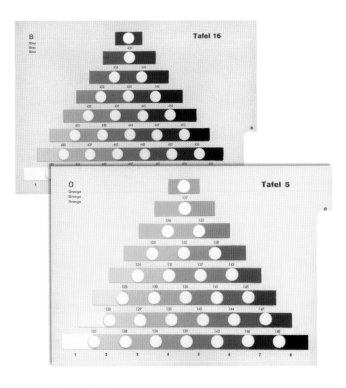

Figure 10.70.
Constant-hue pages for hues B and O from Atlas II *(Baumann 1922a).*

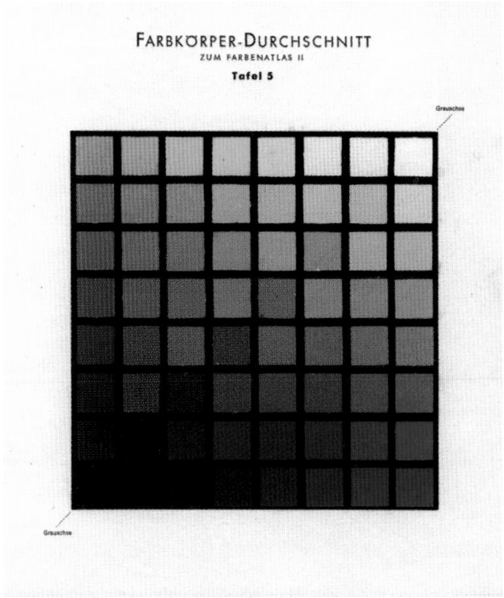

Figure 10.71.
Page 5 C-Co from the cross-section version of Atlas II *(Baumann 1922a). The gray scale is located along the left bottom to upper right diagonal.*

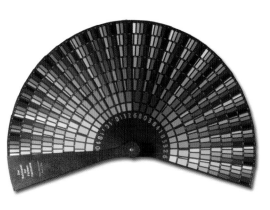

Figure 10.72.
The 680 colors of Atlas II *in fan form (Baumann 1924).*

Figure 10.73.
A hand-colored page from the 1,000-color cube, with decimal color classification,1945 (Prase 1). Image courtesy E. Bendin.

E. FELLOWES   1922

*E. Fellowes, Colour: charted and catalogued
– a key reference for pigmental colour, 1922*

Records about Englishman **E. Fellowes** are sparse, and although his color system has been issued in several editions, it is now rare. Fellowes saw color science and color measurement as being concerned almost exclusively with light colors to the exclusion of the field of pigment coloration. So his primary goal was to develop a color order for object colors, specifically pigment colors, in which individual samples are measured and specified accurately; he saw himself as a pioneer in this field.

All samples of Fellowes's atlas have a name and Lovibond Tintometer (see Joseph Williams Lovibond, chapter 9) designations. In addition, a quarter of the samples were measured spectrophotometrically at Eastman Kodak in Rochester, New York, with the results tabulated.

Fellowes clearly distinguished between light and object colors, and developed for the latter a psychological order, independent of any specific primary colorants: "[A]s far as the painter is concerned, theories about primary colours are of little practical importance, because three bright pigments occurring at such large intervals as red, blue and yellow, cannot be made to produce very brilliant intermediate colours; it is sufficient for him to recognize that all the bright colours are of equal importance" (Fellowes 1922, p. 5).

However, Fellowes provided little or no information on how he arrived at his specific choices. The samples are obtained by hand application of paint, probably watercolor (the samples submitted to Kodak for measurement are identified as "executed by the author in watercolour").

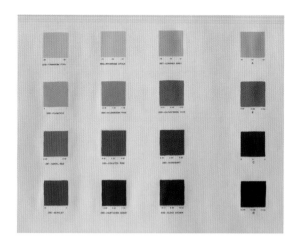

Figure 10.75.
Red constant-hue page with opened gray scale (Fellowes 1930).

Fellowes redefined the primary color attributes as follows:

In relation to pigmental colour the term "colour" has no exact meaning, for it has never yet been standardized or regarded as a measurable factor, while the terms "purity" and "luminosity" have been used in so indiscriminate a fashion by painters and others that they are valueless as technical terms; therefore in case of pigmental colour it would be better to express the three colour-constants in terms of "colour," "brightness" and "tone." The present system of charts, it is suggested, should enable 'colour' to be regarded as an exactly measurable factor. "Brightness" should be regarded as expressing the extent to which any hue is free from the influence of neutrality; it should not imply that the hue is necessarily intense. "Tone" expresses the relative strength of a hue and is equally applicable both to a bright colour and a neutral hue; thus the tone of a white surface will be lowered if any colour whatever is applied. (p. 6)

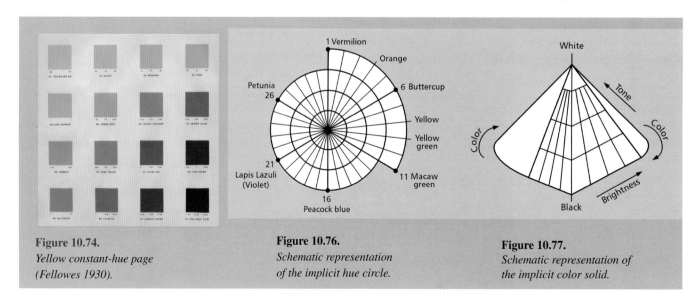

Figure 10.74.
*Yellow constant-hue page
(Fellowes 1930).*

Figure 10.76.
*Schematic representation
of the implicit hue circle.*

Figure 10.77.
*Schematic representation of
the implicit color solid.*

However, in the atlas description, Fellowes primarily used the terms "strength" and "dullness" (typical dyer's terms).

The atlas consists of 30 pages with samples representing 30 hues, beginning with vermilion and proceeding via orange to yellow, macaw green, peacock blue, lapis lazuli, and petunia, back to red.

> As regards the plan of the individual charts, the bright colours are represented in each case in the left-hand column in four degrees of strength. The duller hues of the same bright colours are given in the other columns, the dullness increasing by regular gradations from the left to the right of the chart; they are also shown in four degrees of strength. The weak hues are at the top of the chart, increasing by regular degrees in strength down the columns. (p. 7; figure 10.74)

Twenty of the 30 hues have only three levels of dullness, resulting in a total of 400 color samples. In every atlas page, a four-grade gray scale can be opened up on the right side (figure 10.75). As a result, Fellowes's system consists of 404 measured color standards of size 34 × 34 mm. Hues are generally arranged in complementary order as determined by dominant wavelengths (tabulated), whereby there are only 11 complementary pairs. He saw no need to represent his system graphically in the corresponding hue circle (figure 10.76) or color solid (figure 10.77).

Fellowes saw his atlas's primary use in commercial transactions and in art and crafts. His terminology and arrangement was strongly influenced by industries that produced and used colorants. But because of the CIE's growing efforts in defining color stimuli, and other systems on a perceptually better founded basis, Fellowes's system had a brief life span.

In 1927, Swiss-Americans **Joseph Marcel Vogel** and **Carl Sali Plaut** received a Swiss patent for their color order system, titled simply "Farbentafel" (Color chart). The system is based on the principle of subtractive mixture of nine basis colorants, designated yellow, orange, red, purple, violet, blue, cyan blue, turquoise, and green. Between each colorant are four mixtures, resulting in a 36-grade hue circle, with gray at its center (figure 10.78). That represents the authors' belief that opposing colorants neutralize each other to gray.

The same principle is also the basis for the generation of tonal colors. These they presented in two charts, each divided into 10 × 10 = 100 fields, in mirror symmetry (figure 10.79). Opposing full colors are located opposite in the lower right corner of the upper chart and the upper right corner of the lower chart. The two are mixed in nine ratios each, resulting in neutral gray at positions 10/10. Each of the resulting 20 colors in the rightmost columns is lightened toward left with increasing additions of white. The system contains 18 of these double charts, resulting in a total of 3,600 colors.

In 1928, Vogel published an atlas based on the system under the title *The universal color meter*. Unfortunately, no copy could be located.

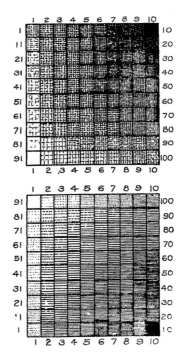

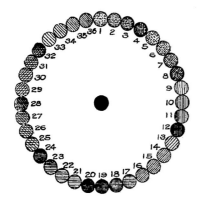

Figure 10.78.
Systematic representation of the 36-grade hue circle with gray in the center (Vogel and Plaut 1927).

Figure 10.79.
Systematic representation of two mirror-symmetrical constant-hue planes with colors opposing each other in the hue circle (Vogel and Plaut 1927).

HILAIRE HILER   1937, 1942

H. Hiler, *Hiler color system*, 1937
Color harmony and pigments, 1942

Hilaire Hiler (1898–1966) was an American painter who "had been for years" working on a color system for painters. He believed that it is necessary to paint in "color" rather than in "pigment." At the same time, he was unhappy about what he perceived as color science's preoccupation with colored lights. He believed that in painting, color is a psychological, not a physical, problem.

Hiler's color solid is conceptual only and cylindrical in nature. Its color circle consists of 30 hues, in 10 groups of three: yellows, oranges, orange-reds, reds, violets, blues, green blues, sea greens, greens, and leaf greens (figure 10.80). Each of the 30 hues has a specific name; a few are pigment trade names (e.g., Monastral Blue). The primaries are identified as 1 lemon yellow, 11 magenta, and 18 cyan. The selected pigments at their highest saturation are called "hues."

Opposite hues on the circle are complementary in that, when mixed additively in disk mixture, they produce gray. A movable disk in the color circle's center allows the determination of certain color harmony types. Colors obtained by adding white to the "hues" are called tints, those obtained by adding black are called tones, and those obtained by adding gray are called shades (figure 10.81). A 10-grade value (gray) scale between white and dark (black) is based on the logarithmic scale. Four shade grades are between the central gray scale and the outermost layer of the cylinder. Hiler chose an identification system based on hue, value, and shade degree (figure 10.82). The system contains 1,812 identified colors.

Except for the hue circle, a hue plane, and the gray scale, Hiler did not color his system. However, he offered the *Hiler Color Piano* (figure 10.83), a painter's box with 20 pigment tubes ("with only loose relationship to the system") on top, approximately producing "hues," and 15 tubes representing shades or tones, and five representing grays I to V on the bottom.

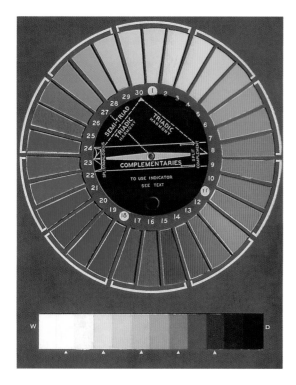

Figure 10.80.

Hiler color chart illustrating 30 hues and a 12-grade gray scale. A movable disk in the center identifies complementaries and certain harmonies (Hiler 1937).

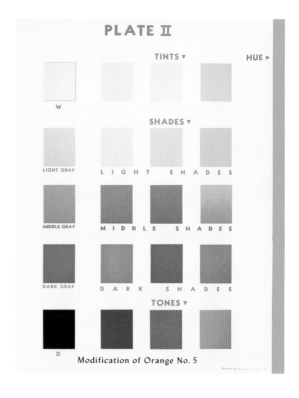

Figure 10.81.

Illustration of a constant-hue plane (orange no. 5) (Hiler 1942).

Plate IV

Hiler Color Solid Sliced to Show Construction

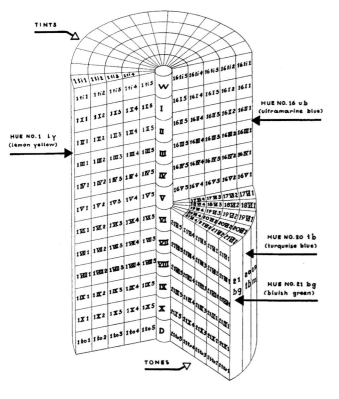

Figure 10.82.

Internal view of the Hiler color solid with identification scheme (Hiler 1942).

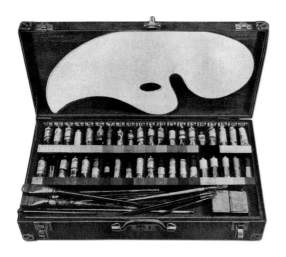

Figure 10.83.

Hiler's (1942) color piano.

Martin-Senour Company,
Nu-Hue custom color system, 1946

In 1946, the Martin-Senour Company of Chicago (now owned by Sherwin-Williams) introduced a paint color order system designed by the American color expert Carl E. Foss (see entry in chapter 9) for selecting interior and exterior paint colors. The company introduced this system together with store paint-mixing automats based on it.

The **Nu-Hue Custom Color System** was available in two forms: as disks mounted in a systematic array between transparent foils, and as 1,000 3 × 5 inch cards in a plastic case. It represents a colorant mixture system based on eight base paints: six chromas, white, and "near-black."

Fifty-four differently hued samples are on the periphery of the disk with the darkest colors. There are nine different levels of admixture of white. Nine rings show different levels of admixture of the near-black paint. Each succeeding chart has one ring less, ending up in white (figure 10.84).[5] The system is dependent on the particular paints used as the base colorants. Each sample's amounts of the base paints by weight and volume were known, making it easy to reproduce the paint in paint stores.

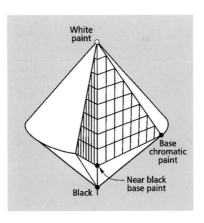

Figure 10.84.

Schematic representation of the solid represented by the Nu-Hue paint mixture system.

CÀNDIDO and JULIO VILLALOBOS   1947

C. Villalobos-Dominguez and J. Villalobos, *Atlas de los colores*, 1947

In 1947 in Buenos Aires, Argentine art instructor and politician **Càndido Villalobos-Dominguez** (1881–1954) and his son, architect Julio Villalobos (1905–?), published what remains one of the most extensive color atlases, with 7,279 printed glossy samples.

The system is arranged in a modified double cone. Its inventors developed special pigment formulations for each of the 38 hues of the system. These were additionally modified to obtain high saturation at lightness levels 7 and 15 (see figure 10.85). Samples in direction of white and black were obtained by halftone effect. Hue steps are approximately perceptually uniform, and diametrically opposed hues are approximately colorimetric complements.

The fundamental ("simple") hues are scarlet, green, and ultramarine (thus representing lights at the beginning, middle, and end of the spectrum). They are arranged in an equilateral triangle, as used by Young (see entry in chapter 5). The "principal double colours," yellow, turquoise, and magenta (comparable to the printing primaries yellow, cyan, and magenta) are intermediate to the fundamental hues, and when combined with them, form the "Chromatic Hexagon."

Between adjacent simple and double hues are "transitive hues": ruby, orange, lime, emerald, cobalt, and violet. The terms "blue" and "red" are not used because they are considered too broad in meaning. Interleaved between these 12 hues are so-called "intercalary hues" (figure 10.86; only 24 hues are shown in this figure). Each of the 38 hues is represented on one chart of the atlas in 191 gradations of lightness value and degree of chromaticity (see figure 10.87 for an example of a constant-hue chart).

The lightness scale of the central gray has approximate cube-root compression of the luminous reflectance of the samples, and all samples in a lightness plane have approximately identical luminous reflectance. Each hue has 12 uniform degrees of saturation, so saturation steps vary in magnitude depending on hue and on lightness (the latter evident from figure 10.87). There are fewer saturation steps near white and black.

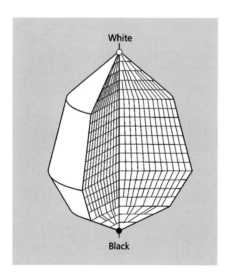

Figure 10.85.

Schematic color solid showing the organization of the samples of an isotint plane in a Munsell-like value–chroma plane (after Judd 1952).

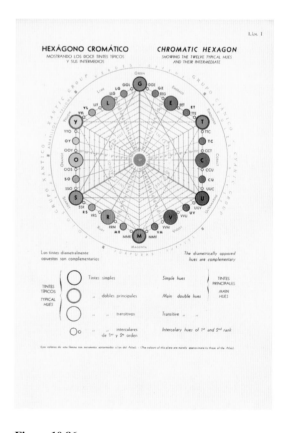

Figure 10.86.

Villalobos chromatic hexagon illustrating the hue circle of the system (Villalobos-Dominguez and Villalobos 1947).

Samples are printed in 10 × 10 mm size and have an off-center hole of 4 mm diameter to visually compare them to other colored materials. Individual color samples are identified by name letters to show hue, a number from 0 to 20 to show lightness, and a number from 1 to 12 with a degree sign attached, to show relative saturation. For example, YL-6-8° represents a sample with a hue between yellow and lime at lightness 6 and degree 8.

The Villalobos color atlas represents a highly ambitious attempt to create the "complete" color order system to satisfy most or all practical needs. Among its goals were complementarity, perceptual equidistance, constant-hue planes, constant perceptual lightness planes based on an experimentally determined gray scale, constant relative saturation surfaces, practical completeness in terms of available samples, and comparatively low cost by use of offset printing. Recent reflectance measurements of many samples show that some of those goals have only been reached approximately, and lack of colorimetric specification of the samples limited its value (Brown et al. 2006).

GUSTAVE and GLADYS PLOCHERE   1948

G. Plochere and G. Plochere, Plochere color system, 1948

The Plochere Color System was developed by artists **Gladys** (1897–1990) and **Gustave Plochere** (1889–1983), Gustave emigrated from Alsace and married Gladys in the United States. They settled in Los Angeles, where the system continues to be published by their son Gustave ("Bud"). Gustave Plochere imported copies of the Baumann-Prase color atlas (see Otto Prase, this chapter) until importation from Germany became illegal. In 1940, the Plocheres published the *Plochere Color Guide*, which was well accepted among decorators and interior painters. Its success encouraged the Plocheres to develop an improved tool in the form of the *Plochere Color System*.

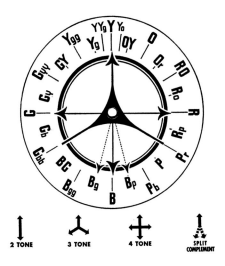

Figure 10.88.
Plochere and Plochere hue circle (1965 edition) with harmony selector derived from Ostwald's system. Reprinted with permission.

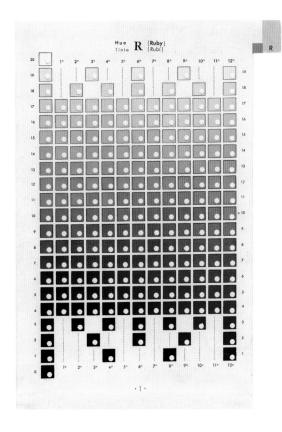

Figure 10.87.
Constant-hue plane of hue R of the Villalobos atlas (Villalobos-Dominguez and Villalobos 1947).

Hue Chroma Levels

	R1a	R2a	R3a	R4a	R5a	R6a
	R1b	R2b	R3b	R4b	R5b	R6b
	R1c	R2c	R3c	R4c	R5c	R6c
Value Levels	R1d	R2d	R3d	R4d	R5d	R6d
	R1e	R2e	R3e	R4e	R5e	R6e
	R1f	R2f	R3f	R4f	R5f	R6f
	R1g	R2g	R3g	R4g	R5g	R6g
	R1h	R2h	R3h	R4h	R5h	R6h

R (red) Hue Scale

Figure 10.89.
Schematic construction of the red constant-hue plane with identification system (Plochere and Plochere system, 1965 edition). Reprinted with permission.

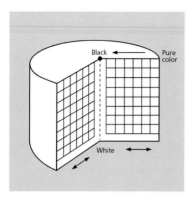

Figure 10.90.

Schematic model of the color solid of the Plochere system. Reprinted with permission.

The Plocheres cite Ostwald (see entry in this chapter) as an important inspiration; however, they adopted only his hue circle. On both sides of pure yellow, they added an intermediate hue grade resulting in a circle of 26 hues (figure 10.88).

The systems also differ in other respects. But the Plochere system has considerable similarity to Prase's system. Like Prase's and later Eusemann's (see EuColor, this chapter) systems, it is a pure pigment-mixture system with tint scales.

All colors of the system are mixed from nine highest-purity pigments, the so-called "Plochere Spectra Hue Colors," as well as white and black pigments. After mixing the 26 base paints of the hue circle, each is shaded with black in five grades. Each of the six grades of the shade scales is then tinted in seven grades toward white (figure 10.89). As a result, each constant-hue plane consists of 48 colors, and there

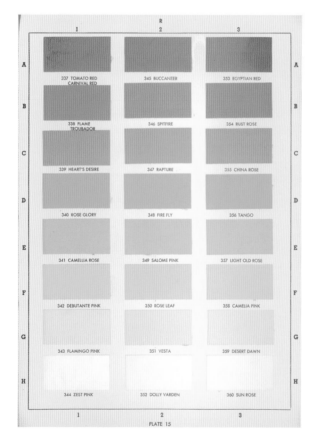

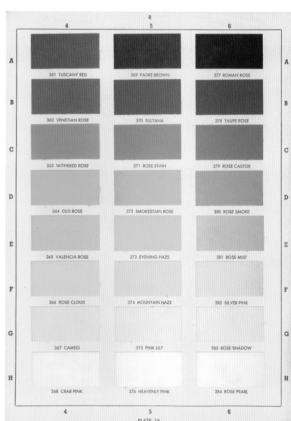

Figure 10.91.

Samples of the constant-hue plane R on two pages of the 1948 Plochere and Plochere atlas. Reprinted with permission.

is no central gray axis. The system consists of 1,248 color chips, arranged by the Plocheres into a cylinder, the inverse of Eusemann's later form, with the top plane formed by the base paints and their shade grades, and the bottom plane entirely white (figure 10.90).

The Plocheres are somewhat vague concerning the attributes in their system. First, they quote Ostwald's view of only three attributes: hue, blackness, and whiteness. Later, because of their broader usage, they cite tint (any clear hue with addition of white), tone (any hue with addition of gray), and shade (any hue with addition of black), but not specifically related to their system. Later editions of the system mention value and chroma levels (see Albert Henry Munsell, chapter 5), even though they do not directly apply to the Plochere system.

Each hue is defined by a capital letter, mixtures in direction of black with numbers from 1 to 6, and their mixtures toward white with small letters from *a* to *h*. In addition, all colors are sequentially identified with numbers from 1 to 1,248 as well as with a specific name. A separate table lists the mixture ratios of all colors and provides, following Ostwald's rules of color harmony, information concerning harmonious combinations.

The individual color chips were produced by hand silk-screening. The 1948 edition was available as an atlas with chips 1 × 2 inches in size as well as loose samples 3 × 5 inches. In the atlas, all chips of a given hue are found on two opposing pages (see figure 10.91). Beginning in 1952, a *Gray Supplement* with 208 "subdued hues" was added, resulting in a total of 1,456 colors. As of 1980, a second version of the system was made available with half-sized chips, the samples of a given hue fitting on a single page (figure 10.92).

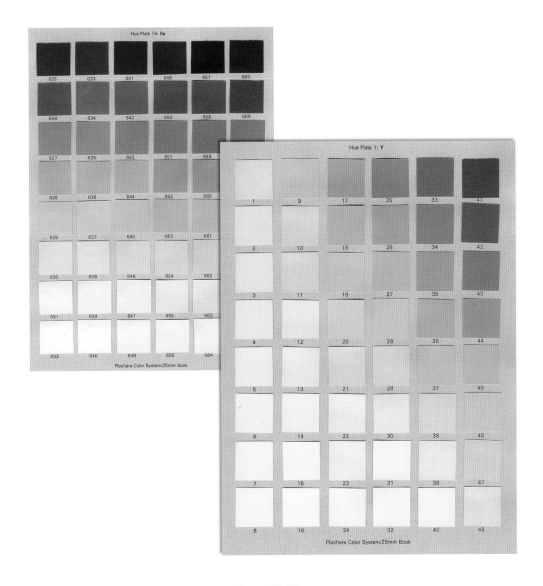

Figure 10.92.

Yellow and blue constant-hue pages from the current edition
(Plochere and Plochere). Reprinted with permission.

LOUIS CHESKIN   1949

L. Cheskin, Cheskin color system, 1949
Cheskin color charts for color planning, 1954

American **Louis Cheskin** (1907–1973) was one of the gurus of psychology-based marketing in the mid-twentieth century, having created several advertising icons, such as the Marlboro Man. He wrote several books on the use of color and founded what was originally called Color Research Institute of America, today Cheskin Research.

In 1949, he published the *Cheskin Color System*, a binder with 48 color plates, displaying a total of 4,800 color samples and including notation on a color system for planning, color identification, color matching, and printing with color. He claimed that gradations beyond 4,800 are indistinguishable. In 1953, he published *Color Wheel for Color Planning*, and the next year, a reduced version of his 1949 *Color System* was printed as *Cheskin Color Charts*.

On each chart of Cheskin's Color System, 100 related colors were placed so that complementary colors face each other. Hues were mixed on basis of specific bulk units of primary colors. Tints were produced by adding white pigment or the halftone process. Shades were produced by adding black pigments or halftone printing with black ink. And tones were produced by adding white and black pigments or corresponding halftone treatment. The samples in the atlas are colored by halftone printing and are of size 19 × 32 mm.

The hue circle is based on three primary pigments, with always 15 intermediate grades between them. Color samples opposing each other in the hue circle are taken to be complementary (figure 10.93). In the smaller version, the *Cheskin Color Charts*, the number of hues is reduced to 12, and there are only 25 samples of each hue, of size 18 × 32 mm (figure 10.94). Color samples are identified by a hue number, a letter denoting whiteness, and a second number denoting blackness. For example, a medium tone of yellow is designated as 1-d-5.

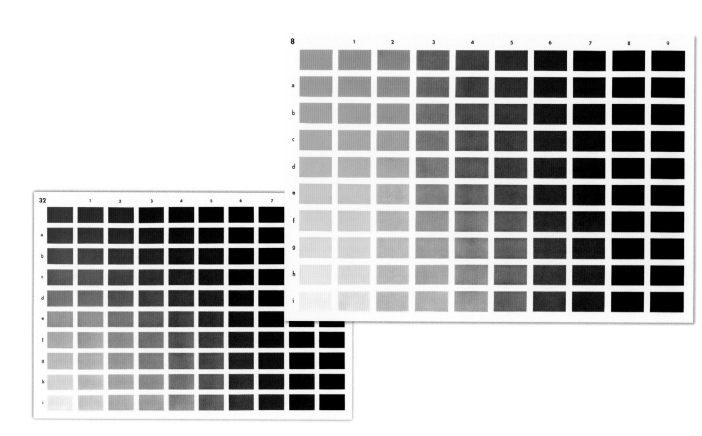

Figure 10.93.

Opposing hue pages 8 (orange) and 32 (blue) of the Cheskin Color System *(1949).*

Colorizer Associates, *Colorizer system*, 1949
New Colorizer system, 1956

Cheskin was not concerned with placing his samples in a three-dimensional space. The arrangement of the color charts is implicitly cylindrical with full colors located at the upper outer edge (figure 10.95). A similar arrangement was used later by Kornerup and Wanscher (see entry in this chapter).

The original **Colorizer system** was developed in the 1930s by two (anonymous) American entrepreneurs in Salt Lake City, Utah. In 1949, the system was introduced nationwide as a rapid paint-formulation system in a story in *House and Garden Magazine*. The original system consisted of 1,322 paint colors derived from 31 concentrated colorants as well as white and light gray base paints into which the concentrates were mixed. The samples were 32 × 32 mm, with 24 near-white samples of larger size.

The well-known American color expert Faber Birren (1900–1988) was later hired to redesign the system to meet competitive pressures. Introduced in 1956, the new system consisted of 1,368 color samples in matte paint and 360 samples in glossy paint, for a total of 1,728 samples. The new system's gamut expanded upon the original system's limited one (the appearance of dark color samples with identical pigment formulation is quite different in matte and glossy finish). The expanded gamut was achieved despite a reduction from 31 to only 17 concentrated colorants that a dealer needed to keep in stock. The new system made use of white, mid-tone, and deep-tone base paints.

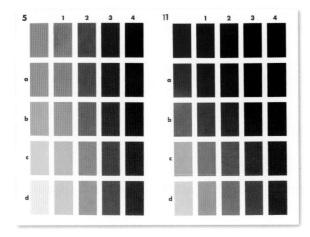

Figure 10.94.

Double page from Cheskins Color Charts *with "complementary" hues (1954).*

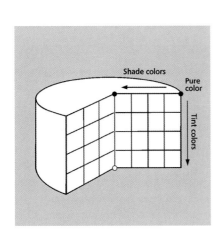

Figure 10.95.

Schematic view of the implicit cylindrical arrangement of the Cheskin Color System, *for clarity illustrated for the reduced* Color Charts.

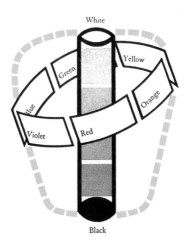

Figure 10.96.

Schematic sketch of the child's top form of the Colorizer implied color space (Deutsch 1956).

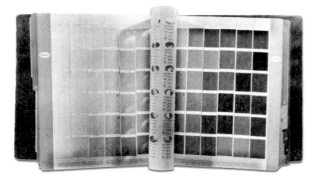

Figure 10.97.

View of a double page of the
Colorizer album (Birren 1965).

To avoid similar-appearing samples that result from mixing according to systematic procedures, samples were matched in an effort to cover as much as possible of the object color space with sensibly spaced samples. Thus, the new system was not a colorant mixture space so much as it was based on psychological order tempered with marketing requirements. The system borrowed from Munsell's, but added light colors that customers preferred.

The shape of the space filled by the samples is described as that of a child's top, but tilted and with a larger area above the plane of the highest chroma colors (figure 10.96). In the "album" of the system, the shape of the top was distorted "to achieve a straight-line layout" of columns and rows with the glossy samples filling out the rectangular form of the charts (figure 10.97).[6] The new system also contained a circular color-harmony selector, allowing the choice of harmonious combinations.

FRANTZ BRAUN   1956, 1957

F. Braun, *Manuel d'initiation à l'étude de la science des couleurs*, 1956
Moderne Gesichtspunkte der Farbmessung und ihrer
industriellen Anwendung, 1957

In an introductory booklet on color science, Belgian lecturer **Frantz Braun** also discussed color reproduction. At the time he wrote the text, industrial color matching in the textile industry was done exclusively by visual means. Color matchers empirically learned the relationship between dye concentration and mixture, and its resulting appearance. They also maintained collections of previous matches as starting guides for new, similar formulations. Dyestuff companies aided these efforts by producing various kinds of sample charts with individual and combination dyeings.

Braun recommended a modified approach in his booklet. Two dyes representing a particular hue in a binary combination (e.g., 87.5% yellow dye and 12.5% red dye in figure 10.98) are dyed at the concentration at which they reach maximum chroma (in figure 10.98, at 3.2% of dye on weight of material). The resulting dyeing (bottom right in figure 10.98) represents the most intense coloration of this hue obtainable with the selected dyes and fabric.

Without changing the total dye concentration, the original dyeing is dulled down by adding a blue dye in eight grades, ultimately resulting in a black (bottom left in figure 10.98). The total dye concentration is reduced stepwise by a factor of 2 in seven grades, so that (left scale in figure 10.98) a gray scale is obtained from white (to sample). Along the hypot-

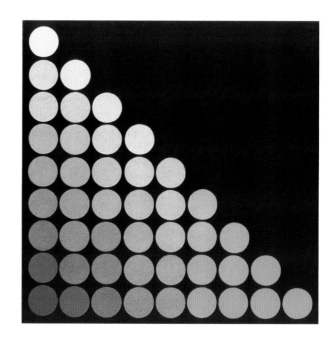

Figure 10.98.

Example of a trichromatic arrangement of dyeings with
the nine-grade gray scale on the left and the full color
on the bottom right. The diagonal forms a tint scale; the
horizontal scales are tonal (Braun 1956).

nuse are dyeings of a tint scale from the full color to white. The interior of the triangle is filled with tonal dyeings. Intermediate colors can be approximated by interpolating chromaticity coordinates (once all the dyeings are measured) and the corresponding dye concentrations. The locations in two views of the CIE *x*, *y*, *Y* color stimulus space of the dyeings of figure 10.98 (as given by Braun) show a hue change as well as irregular spacing in the interior (figures 10.99 and 10.100), perhaps as a result of the dyes' nonlinear behavior or dyeing errors.

In 1957, Braun described efforts toward a different color atlas, under development in Belgium. The concept was based on Deane Brewster Judd's work (see entry in chapter 7) that resulted in the National Bureau of Standards (NBS) color difference formula, considered perceptually approximately uniform. The formula was used to define 24 "perceptually equidistant" hues around the hue circle. All colors of a given hue were arranged in a triangle, differing in form depending on hue (figure 10.101), resulting in an irregular tilted double cone.

Samples in this diagram differed from neighboring ones by 10 NBS units. Samples were defined by a lightness value, a saturation grade S_n, and values that defined the hue of a sample in the hue circle. Braun considered the system to be "chromatic-harmonic," with colors opposing in hue being compensative, and at the same time perceptually uniform. A copy of such an atlas has not been discovered.

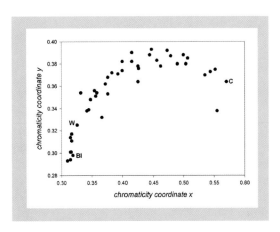

Figure 10.99.
Plot of the x, y chromaticity coordinates of the dyeings of figure 10.98 in a portion of the CIE chromaticity diagram. The locations of the white (W), black (Bl), and full color dyeings (C) are shown.

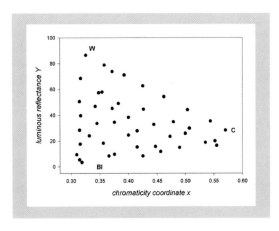

Figure 10.100.
Plot of the locations of the dyeings of figure 10.98 in the x, Y *diagram.*

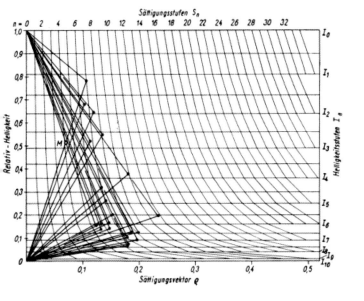

Figure 10.101.
Plot of the triangles of the 24 pages of the planned chromatic-harmonic color atlas in the diagram with saturation vector and related saturation grade on the abscissa and relative lightness (relative in terms of scale) of the planned Braun system on the ordinate (Braun 1957).

ANDREAS KORNERUP and JOHAN H. WANSCHER   1961

A. Kornerup and J. H. Wanscher, *Farver i Farver*, 1961
Reinhold color atlas, 1962
Methuen handbook of colour, 1963
Taschenlexikon der Farbe, 1963

In 1961, two Danish printing engineers, **Andreas Kornerup** and **Johan Henrik Wanscher**, published a small atlas containing a complete color system. Its text was translated into several languages, and it was published in several editions internationally. The American and English editions are named after their publishers: *Reinhold Color Atlas* (1962) and *Methuen Handbook of Colour* (1963), respectively. A brief text introduction precedes the atlas section and an extensive addendum with color names and designations related to the atlas. The system is halftone-printed on high-gloss art book stock. However, the atlas is based on perceptual distances, not on systematic variation of printing inks.

The hue circle consists of 30 hues beginning with greenish yellow designated from 1 to 30 (figure 10.102). In the system, the angles between approximate average unique hues are Y-R 90°, R-B 140°, B-G 65°, G-Y 65°, showing approximate perceptual uniformity. The hues are separated into 19 categories. The planes of constant-hue are 6 × 8 samples with darkness on the abscissa, designated from A to F, and saturation on the ordinate, designated from 1 (gray scale) to 8 (figure 10.103). The saturation steps have been achieved with halftone effects, and the darkness steps, via over-printing with translucent gray. The steps across darkness and saturation scales within a hue plane appear reasonably even, as judged by the authors. No colorimetric documentation of the system has been published. The solid of the system is cylindrical (figure 10.104). Including the gray scale on every hue page, it has 1,440 samples (1266 unique) of size 15 × 30 mm. Very popular in the 1960s and 70s, the atlas is no longer produced.

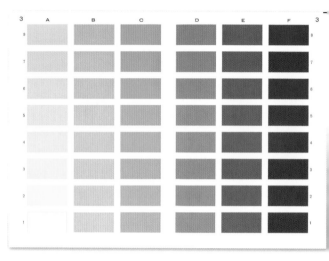

Figure 10.103.
Constant-hue page of the Kornerup and Wanscher (1962) atlas. Reprinted with permission.

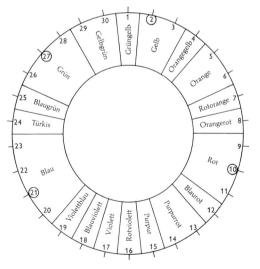

Figure 10.102.
Thirty-grade hue circle (Kornerup and Wanscher 1962). Reprinted with permission.

Figure 10.104.
Constant-hue pages can be arranged in form of a cylinder model (Kornerup and Wanscher 1962). Reprinted with permission.

ADVANCED INK MIXING SYSTEM (AIMS)   1968

A. Kornerup, *AIMS advanced ink mixing system*, 1968

AIMS is a printing-ink mixing system to achieve a considerable range of colors. It was devised by the Danish printing engineer Andreas Kornerup (also see Kornerup and Wanscher entry in this chapter) and published by the Danish Paint and Ink Research Laboratory in Copenhagen and by seven Danish ink manufacturers.

The purpose of AIMS was "to create an ink system in which the colours are arranged in accordance with modern colorimetric principles so that a color appears only in one place in the system, and additionally, in which the visual difference between adjacent shades is constant" (Kornerup 1968, preface).

Twenty-four binary mixtures were generated from eight selected primary inks – yellow, orange, red, bright red, magenta, purple, violet blue, and blue – for a total of 32 hue grades. Although the pigments are not identified, the standardized inks were available from manufacturers. The primary inks and their mixtures vary considerably in saturation.

The resulting 32 inks were mixed in different ratios with colorless transparent ink or one of three grays of decreasing lightness. This resulted in approximate saturation scales of the original ink toward white or the three grays. The mixtures were printed as continuous strips (figure 10.105), and individual colors are viewed behind gray masks (figures 10.106 and 10.107).

The steps were calculated to approximately agree with the CIE 1964 $U^*V^*W^*$ color difference formula (no longer in use). There are four to eight steps between grades from pure color to gray, depending on the saturation of the pure color (figure 10.108). The 32 pure colors and 552 "variants" are illustrated in 128 color series.

Identification of the colors is by a three-number system. The first number identifies the hue, the second, the amount of neutral dilution (01 white, 03 light gray, 05 medium gray, or 07 dark gray), and the third, the saturation grade ranging

Figure 10.105.
Continuous printed strips with primary ink mixture C40. The top row is printed from ink mixture 40 with transparent white, the second row with light, the third with middle, and the fourth with dark gray (Kornerup 1968).

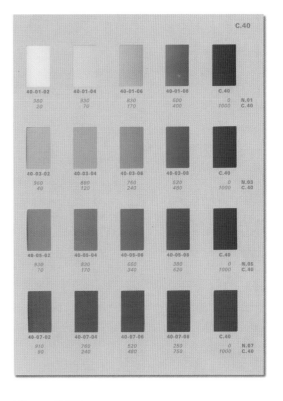

Figure 10.106.
The strips of figure 10.105 with overlaid mask identifying 20 individual colors (Kornerup 1968).

from 02 to 14 (in steps of 2, the pure ink being grade 16). For each color, the ink-weight mixing ratios are provided. For example, color 20-05-08 consisted of 560 grams of medium gray N05 and 440 grams of red mix C20. Intermediate formulas can be calculated by linear interpolation.

Lightness of the pure hue inks varies considerably: That of pure yellow C64 is approximately equal to that of N03, while that of reddish blue C38 is about equal to N07. Therefore, the arrangements of AIMS and the Ostwald system (see entry in this chapter) are somewhat similar, except that in AIMS, steps form the CIE $U^*V^*W^*$ scale, not a logarithmic scale.

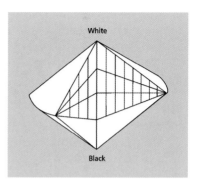

Figure 10.108.
Schematic color solid with four (left) and eight (right) grades from the full ink color to the gray scale.

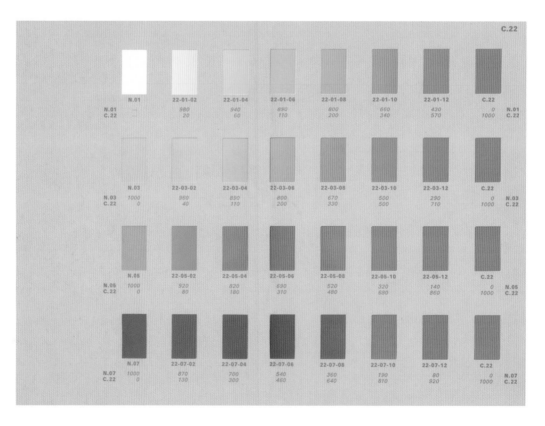

Figure 10.107.
Hue C22 with mask in eight grades toward the gray axis, with a total of 32 samples (Kornerup 1968).

CYRIL J. WEDLAKE   1969

C. J. Wedlake, *Visual arts matching charts,* 1969

In 1969, 12 years after introduction of the LTF/GATF half-tone-printing chart and four years before the Foss charts (see entries in chapter 9), American printing technician **Cyril J. Wedlake** received a U.S. patent for his color order system based on halftone printing. As he wrote, previous color-matching charts failed to provide a convenient systematic presentation that would help to quickly find or match a desired color.

Wedlake's system is based on standardized printing process inks yellow, magenta, cyan, and black. An implemented system could not be found, but it was to consist of eight charts with 468 color samples each (figure 10.109) for a total of 3,744 samples. The patent does not specify how additive color mixture through optical fusing of individual color dots is exactly related to subtractive effects through over-printing.

The samples are arranged on hexagonal charts (figure 10.110). The first chart is printed on white paper, with white located in the central triangle. The full (presumably 100% of the screen) primary colors are located, as shown in figure 10.110, at the three dented corners of the hexagon. They are weakened toward white in regular screen density steps, identified by letters. The intermediate full colors red, green, and blue are located at the other three corners, and like all colors along the periphery, are obtained by over-printing the two adjacent primaries. As the saturation declines in 12 steps toward the center, the system changes slowly from subtractive to partitive mixture (additive optical fusion of dots and white background).

Subsequent charts are printed in identical manner onto a hexagon preprinted with black of increasing density (from 10% to 70%). The form of the implicit space is a hexagonal prism. The system represents a technical simplification in that any color sample is printed from two chromatic inks only, with or without black under-printing. The color identification consists of the two primary ink names, their saturation steps in the scheme, and the density of the black undertone. An example is cyan F, yellow I, 30%.

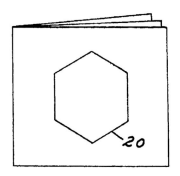

Figure 10.109.

Schematic arrangement of the color atlas of Wedlake's (1969) system.

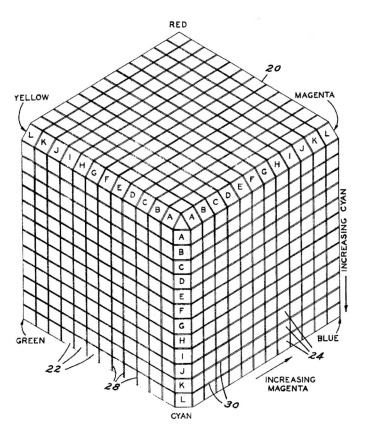

Figure 10.110.

Schematic representation of the organization of individual charts of the system (Wedlake 1969).

IMPERIAL CHEMICAL INDUSTRIES COLOUR ATLAS   1971

Imperial Chemical Industries Ltd., *ICI colour atlas*, 1971

Color atlases in the mid-twentieth century typically contained between several hundred and 1,500 samples. Such numbers were not considered sufficient to support the work of industrial colorists later in the century. In 1969, the English firm **Imperial Chemical Industries Ltd.** (ICI) patented a new system design with an atlas containing 1,379 samples, but with the possibility of viewing 27,580 different colors. In 1971, the Butterworth Group published the atlas.

The system contained 19 neutral filters of different optical density. These could be placed on the 19 × 19 mm samples to reduce the amount of light reflected from them, and thus make them appear darker. The system is based on three primary colorants (red, yellow, blue), mixed in pairs in various ratios to result in intermediate colors of various hues and depth of color. The principle of presentation is illustrated in figure 10.111, where primary yellow runs horizontally in 16 grades and primary red runs vertically in 26 grades. The upper left corner is occupied by white. Colors thus run from low-saturated and light in the second quadrant, to saturated and dark in the fourth (figure 10.112). Colors are identified by the neutral filter employed and the coordinates on the atlas chart, such as N11 Y8 R13.

In the world of the industrial colorist, the atlas had limited value because it offered only a reference system in form of the color-identification code, but neither a reflectance function specification nor colorimetric values.

Figure 10.111.

Schematic representation of the organizing principle behind the ICI Colour Atlas (Imperial Chemical Industries Ltd. 1971).

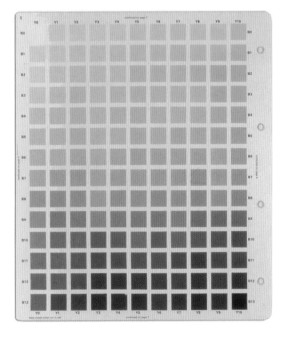
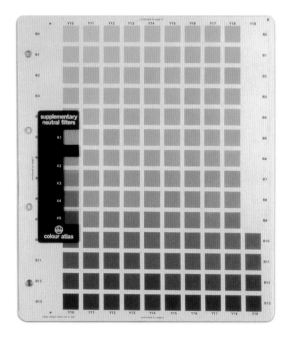

Figure 10.112.
Two pages of the ICI color atlas with neutral filter (Imperial Chemical Industries Ltd. 1971).

EUCOLOR SYSTEM   ca. 1979

S. Eusemann, *Farbklänge im Innenausbau*, 1975
Vorrichtung zur Bestimmung harmonischer Farbkombinationen,
German Patent DE 3033797C2, 1980
Ispo EuColor-System 2001, Systembeschreibung und Farbtafeln, ca. 1995

German designer **Stephan Eusemann** (1924–2005) taught at the Hochschule für Textilgestaltung und Flächendesign in Nürnberg. One of his main interests was the effect of colors and colored textiles in interiors. In 1975, to make the design process transparent and effective, he developed a purely conceptual color order system. Being merely conceptual, the original system was never executed in form of color chips.

The system was based on an 18-grade hue circle derived from primaries yellow, red, and blue (figure 10.113). Opposing full colors are taken to be complementary, that is, to desaturate each other perfectly. The form of the system is a tilted double cone, centered on a nine-grade gray scale (figure 10.114), with attributes hue, lightness, and saturation (figure 10.115).

In the late 1970s, collaborating with the architectural paint manufacturer Ispo, Eusemann considerably modified the system, and changed from a conceptual model based on visual attributes to a paint-mixing system. In 1980, Eusemann was granted a German patent for the system and the design of a set of masks for finding harmonic color combinations in the system.

The revised system is based on the original hue circle. The full color (now called *Grundton*) is shaded in several steps toward black, and the full color and shades are whitened in the number of steps of the gray scale (figure 10.116).

Aim color values of the chips are defined colorimetrically. Eusemann believed his system to be perceptually uniform, and quoted Manfred Richter (see entry in chapter 7) in support of this belief. However, in Richter's DIN system, perceptual uniformity exists within single scales only, but not between them. The complete EuColor system forms a tilted cone (figure 10.117) where full colors are at the same vertical distance from the basis plane as the gray scale grade of the same luminous reflectance. Individual chips are identified in a proprietary scheme.

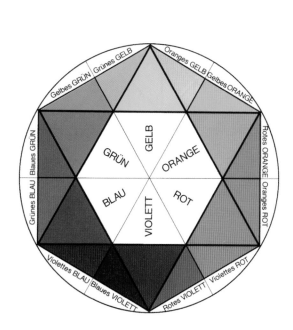

Figure 10.113.
Eusemann's (1975) 18-grade hue circle.

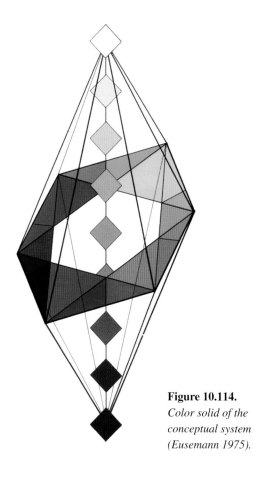

Figure 10.114.
*Color solid of the
conceptual system
(Eusemann 1975).*

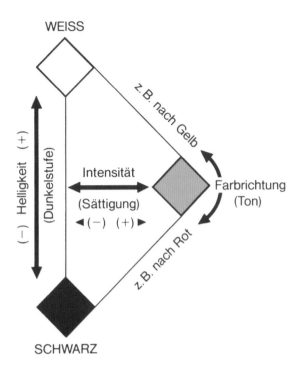

Figure 10.115.

Conceptual view of a constant-hue plane, with colors ordered according to saturation and lightness (Eusemann 1975).

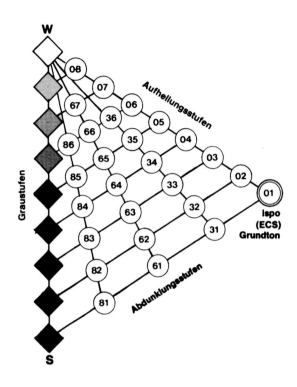

Figure 10.116.

Conceptual drawing of a constant-hue plane of the Ispo EuColor system. The full color (Grundton) is in this version shaded in three steps toward black, and the three shade grades are whitened in seven steps, as shown in figures 10.118 and 10.119 (Ispo ca. 1982a).

To simplify the presentation of the system and the harmony masks, the atlas pages later were converted from triangular to rectangular where the shade scale falls on the abscissa and the tint scale on the ordinate. Samples in opaque paint are of size 25 × 25 mm. The resulting atlas can be unfolded in a manner that makes the structure of the color solid evident (figure 10.118).

Since 1980, Ispo has issued the system in several versions with varying numbers of tint and shade steps to meet different market demands and other requirements. The final version, *Ispo EuColor-System 2001*, contains 410 samples (figure 10.119). As a result of a change in ownership of Ispo in 2002, the EuColor system is no longer produced.

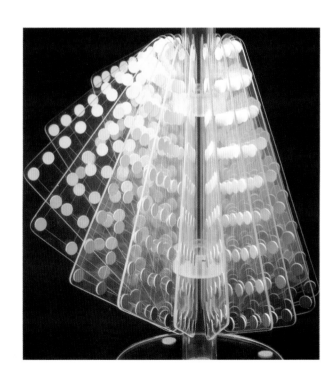

Figure 10.117.

Three-dimensional version of the EuColor system (Ispo ca. 1985).

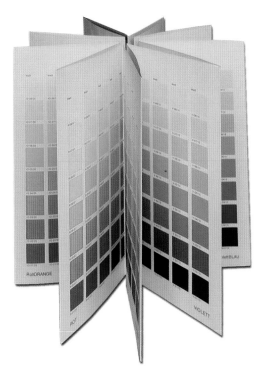

Figure 10.118.
Unfolded Ispo EuColor atlas in the rectangular format
demonstrating the structure of the implicit color solid
(Ispo ca. 1982b).

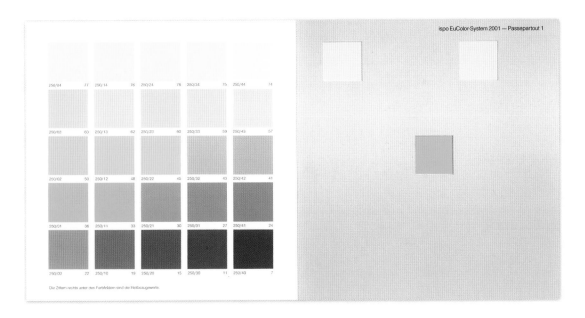

Figure 10.119.
Two neighboring constant-hue planes of the 2001 version of
the Ispo EuColor atlas, the right page covered with mask #1
disclosing a harmonic triple combination (Ispo ca. 1995).

ICI PAINTS MASTER PALETTE   1993

ICI Paints, *Master palette*, 1993

In 1993, the Paints division of the international chemical manufacturer **ICI** Group, sister division of the textile colorants group that created the 1971 *ICI Colour Atlas* (see entry in this chapter), introduced a paint color-specifying system with 6,134 samples, representing one of the largest extant color order systems. It is used in its paint stores around the world for selecting and specifying colors for architectural and other uses.

The *Master Palette* atlas is based on the attributes hue, lightness, and chroma. Hue is specified in a circle with 47 hues (figure 10.120), identified with one of eight double-letter designations. The four primary hues are RR, YY, GG, and BB for, respectively, red, yellow, green, and blue. Intermediate hues are identified as YR, GY, BG, and RB. Each of these major hue families has from 5 to 10 members, with the most members in the YR and YY families.

Lightness is specified by the lightness reflectance value (LRV; luminous reflectance Y; see CIE X, Y, Z Color Stimulus Space, chapter 6). The lightness scale (excluding white and black) has 14 grades with a lightness value compressed from the luminous reflectance scale by power 0.4 (between square

and cube root). The chroma scale is not explicitly defined, is open-ended, and is perhaps based on CIELAB C^* values.

The arrangement of the atlas is cylindrical (figure 10.121). On each of the 47 constant-hue plane pages, there are 118 samples placed in an identical pattern (figure 10.122). However, the arrangement does not follow that of the Munsell system (see Munsell Renotations, chapter 7). In figure 10.123, the locations of the samples for the two hues 40YY and 10BB are shown in the identical chroma–LRV diagram. They indicate the irregular, hue-dependent arrangement of the samples. Figure 10.124 is an image of the atlas page of hue 10BG.

The samples are screen-printed in matte paint on white board. Samples are of size 1 × 1 inch. In addition to the main atlas's 5,546 samples are 252 gray tones representing the gray scale itself as well as tinted grays. There are also 336 "Brights" color samples of 19 × 25 mm: 28 pure hues whitened in 12 steps. They are not continuous with the hues illustrated in the main atlas. The *Master Palette* is a technically well executed, extensive commercial atlas of color stimuli for giving worldwide customers a wide range of paint choices. Colorimetric specifications of the system have not been made public.

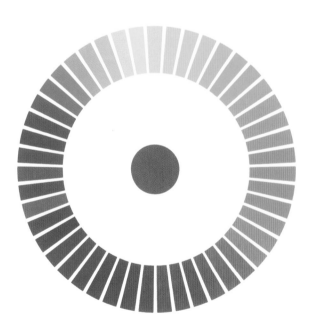

Figure 10.120.

Hue circle with 47 hues of the Master Palette *(ICI Paints ca. 2002). Reprinted with permission.*

Figure 10.121.
Cylindrical arrangement of the Master Palette *atlas
(ICI Paints ca. 2002). Reprinted with permission.*

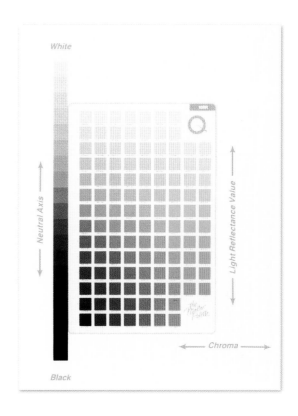

Figure 10.122.
Representation of the Master Palette *atlas sample
arrangement for hue 10RR (ICI Paints ca. 2002).
Reprinted with permission.*

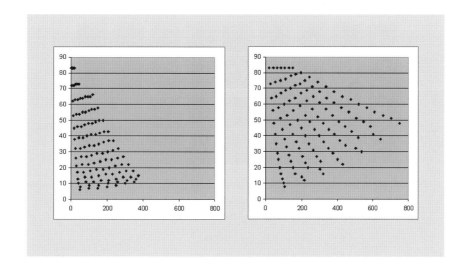

Figure 10.123.
*Left: Plot of the arrangement of the samples of hue
40YY in the chroma–lightness reflectance value
diagram. Right: Plot for hue 10BB.*

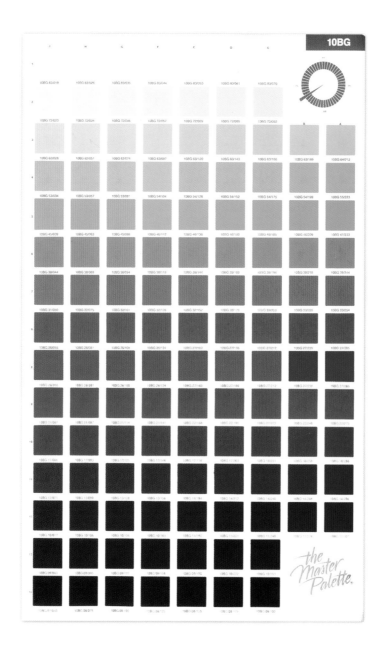

Figure 10.124.
Image of the page of hue 10BG of the Master Palette *atlas (ICI
Paints ca. 2002). Reprinted with permission.*

Notes

1. Color mixture from a rapidly spinning disk with segments of different colors was first described by the Greek geographer and astronomer Claudius Ptolemy (ca. 90 - ca. 168) in his five-part work on optics (Lejeune 1956).

2. The instrument referred to is probably Arago's polarization photometer suitable only for broadband, but not spectral intensity measurements of light.

3. Based on recent unpublished reflectance measurements, only the three most highly saturated scales agree with the methodology described by Ridgway in the text. Ridgway's book also contains two plates (22 and 24) with duplicate colors from the tonal scales set, identified as "made at an early stage in the preparation of the work and discarded."

4. TGL is the abbreviation for *Technische Güte- und Lieferbedingungen* (technical terms for quality and delivery).

5. Despite best efforts, it was not possible to locate a copy of the Nu-Hue Custom Color System.

6. Despite best efforts, it was not possible to locate a copy of the original or the new Colorizer system.

References

Adam, H. 1989. Manfred Adam und die Farb-systeme, Westewitz, Germany: H. Adam.

Arnold, W. 1988. *Farbgestaltung*, 2nd ed., Berlin: VEB Verlag für Bauwesen.**Baumann, P.** 1912. *Baumanns neue Farbtonkarte, System Prase*, Aue, Germany: Baumann.

Baumann, P. 1922a. *Baumanns Farbtonkarte, Atlas II*, Aue, Germany: Baumann.

Baumann, P. 1922b. *Baumanns Farbkörperdurchschnitte zum Farbenatlas II*, Aue, Germany: Baumann.

Baumann, P. 1924. *Der Schlüssel zur Farbenharmonie*, Aue, Germany: Baumann.

Becke, M. 1923. *Verfahren der Farbengebung nach dem natürlichen Dreifarbensystem*, Austrian Patent no. 92344.

Becke, M. 1923/1924. Einführung in die natürliche Farbenlehre, reprint from *Wollen- und Leinen-Industrie*, Reichenberg (Bohemia): Stiepel.

Benson, W. 1868. *Principles of the science of colour, concisely stated to aid and promote their useful application in the decorative arts*, London: Chapman & Hall.

Benson, W. 1871. *Manual of the science of colour. On the true theory of colour-sensations and the natural system*, London: Chapman & Hall.

Bezold, W. von. 1873. Ueber das Gesetz der Farbenmischung und die physiologischen Grundfarben, *Poggendorf's Annalen* 150:71–93, 221–247.

Bezold, W. von. 1874. *Die Farbenlehre im Hinblick auf Kunst und Kunstgewerbe*, Braunschweig, Germany: Westermann; American ed., *The theory of color in its relation to art and art-industry*, Boston: Prang, 1876.

Birren, F. 1965. *Color systems & color standards*, Fort Wayne, IN: General Color Cards.

Boring, E. G. 1929. *A history of experimental psychology*, New York: Century.

Braun, F. 1956. *Manuel d'initiation à l'étude de la science des couleurs*, Bruxelles: Edition Librairie Publicité Internationales.

Braun, F. 1957. Moderne Gesichtspunkte der Farbmessung und ihrer industriellen Anwendung, *Die Farbe* 6:151–170.

Brown, M., R. G. Kuehni, and **D. Hinks**. 2006. The Villalobos atlas: an analysis, *Color Research and Application* 31:109–116.

Brücke, E. W. 1866. *Die Physiologie der Farben für die Zwecke der Kunstgewerbe*, Leipzig: Hirzel.

Carpentier, J. 1885. Définition, classification et notation des couleurs, *Comptes Rendus des Séances de l'Academie des Sciences* 100:808–810.

Cheskin, L. 1949. *Cheskin color system*, New York: Macmillan.

Cheskin, L. 1954. *Cheskin color charts for color planning*, New York: Macmillan.

Colorizer Associates. 1949. *Colorizer system*, Salt Lake City, UT: Colorizer Associates.

DDR-Standard. 1965. *Farbenkarte Grundsystem*, TGL 21 579, Blatt 1, Leipzig: Buchhaus.

Deutsch, C. 1956. *The Colorizer story*, Salt Lake City, UT: Colorizer Associates.

Doppler, C. 1847. Versuch einer systematischen Classification der Farben, *Kràlovskà ceskà Spolecnost Nauk* (Abhandlungen der königlichen böhmischen Gesellschaft der Wissenschaften) 5:401–412.

D'Udine, J. 1903. *L'orchestration des couleurs, analyse, classification et synthèse mathématique,* Paris: Joanin.

Eusemann, S. 1975. *Farbklänge im Innenausbau,* Würzburg, Germany: Stürtz.

Eusemann, S. 1980. Vorrichtung zur Bestimmung harmonischer Farbkombinationen German patent DE 3033797C2.

Fellowes, E. 1922. *Colour: charted and catalogued, a key reference for pigmental colour,* London: Geographia Ltd.

Fillacier, J., and **A. Lemmonier.** 1968. *Intégrateur universel de couleur,* French Patent 1,554,843.

Guignet, C.-E. 1889. *Les couleurs,* Paris: Librairie Hachette.

Herschel, J. 1830. Light, *Encyclopaedia metropolitana,* 2nd div., *Mixed sciences,* vol. 2, pp. 341–586 (article dated Dec. 12, 1827).

Hiler, H. 1937. *Hiler color system,* Chicago: Favor, Rule

Hiler, H. 1942. *Color harmony and pigments,* Chicago: Favor, Rule.

ICI Paints. ca. 2002. *Master palette,* London: ICI Paints. Imperial Chemical Industries Ltd. 1969. British patent no. 1,160,673.

Imperial Chemical Industries Ltd. 1971. *ICI colour atlas,* London: Butterworth Group.

Ispo. ca. 1982a. *Das ispo EuColorsystem* (advertising brochure), *Der Farbenfächer,* Kriftel, Germany: Ispo.

Ispo. ca. 1982b. Das *ispo EuColorsystem nach Prof. Eusemann* (color atlas), Kriftel, Germany: Ispo.

Ispo. ca. 1985. *ispo EuColorsystem 840 nach Prof. Eusemann* (advertising brochure), *Das System und seine praktische Anwendung,* Kriftel, Germany: Ispo.

Ispo. ca. 1995. *Ispo EuColor-System 2001, Systembeschreibung und Farbtafeln,* Kriftel, Germany: Ispo.

Judd, D. B. 1952. *Color in business, science, and industry,* New York: Wiley.

Kirschmann, A. 1895. Color-saturation and its quantitative relations, *American Journal of Psychology* 7:386–404.

König, A. 1892. Die Grundempfindungen in normalen und anomalen Farbensystemen und ihre Intensitätsvertheilung im Spectrum (in Gemeinschaft mit Conrad Dieterici), *Zeitschrift für die Physiologie und Psychologie der Sinnesorgane* 4:241–347.

Kornerup, A. 1968. *AIMS Advanced ink mixing system,* Copenhagen: Danish Paint and Ink Research Laboratory.

Kornerup, A. and **J. H. Wanscher,** 1963a. *Methuen Handbook of Color,* London: Methuen.

Kornerup, A. and **J. H. Wanscher,** 1963b. *Taschenlexikon der Farbe,* Göttingen: Muster-Schmidt.

Lejeune, A. 1956. *L'optique de Claude Ptolémée,* Louvain, Belgium: Publications universitaires de Louvain.

Lemonnier, A. 1976. *Couleur, échelles et schemas,* Paris: Centre Georges Pompidou.

Ostwald, W. 1917. *Der Farbenatlas—2500 Farben auf über 100 Taf. + 23 S. Gebrauchsanweisung und wissenschaftliche Beschreibung,* Leipzig: Unesma.

Ostwald, W. 1918. *Die Farbenlehre, I. Buch: Mathetische Farbenlehre,* Leipzig: Unesma.

Ostwald, W. 1919a. *Der Farbkörper und seine Anwendung zur Herstellung farbiger Harmonien,* Leipzig: Unesma.

Ostwald, W. 1919b. *Die Farborgel,* Leipzig: Unesma.

Ostwald, W. 1923a. *Die Farbenlehre, II. Buch: Physikalische Farbenlehre,* 2nd ed., Leipzig: Unesma; 1st ed., 1919.

Ostwald, W. 1923b. *Farbkunde,* Leipzig: Hirzel.

Ostwald, W. 1924a. *Die Farbenfibel,* 10th ed. Leipzig: Unesma.

Ostwald, W. 1924b. *Wilhelm Ostwalds Wollatlas,* Großbothen: Laboratorium W. Ostwald.

Plochere, G., and **G. Plochere.** 1948. *Plochere color system in book form, a guide to color and harmony.* Los Angeles: Plochere.

Plochere, G., and **G. Plochere.** 1965. *Plochere color system,* 3rd ed., Glendale, CA: Mascon.

Prase, O. 1945. *Experimentalstudien zur Farbenlehre, Teil 4 Der tausendteilige Farbenwürfel,* Lössnitz, Germany: Prase.

Ridgway, R. 1886. *A nomenclature of colors for naturalists, and compendium of useful knowledge for ornithologists,* oston: Little, Brown.

Ridgway, R. 1912. *Color standards and color nomenclature,* Washington, DC.

Rood, O. N. 1879. *Modern chromatics with application to art and industry,* New York: Appleton.

Rosenstiehl, A. 1881. Determination de la distance angulaire des couleurs, *Compte Rendus des Séances de l'Academie des Sciences* 92:207–210.

Rosenstiehl, A. 1910. Conséquence de la théorie de Young. De la construction chromatique dans l'espace, *Compte Rendus des Séances de l'Academie des Sciences* 150:350–352.

Rosenstiehl, A. 1913. *Traité de la couleur au point de vue de physique, physiologique, et esthéthique,* Paris: Dunod et Pinat; 2nd ed. 1934.

Schwarz, A. 2003. Zur Anwendung der Ostwald'schen Farbenlehre in der Textilindustrie, *Phänomen Farbe* 9:22–29.

Villalobos-Dominguez, C., and **J. Villalobos.** 1947. *Atlas de los colores,* Buenos Aires: El Ateneo.

Vogel, J. M., and **C. S. Plaut.** 1927. *Farbentafel,* Swiss patent no. 120551.

Wedlake, C. J. 1969. *Visual arts matching charts,* U.S. patent no. 3,474,546 of Oct. 28.

Wundt, W. 1874. *Grundzüge der physiologischen Psychologie,* Leipzig: Engelmann.

Wundt, W. 1892. *Vorlesungen über die Menschen- und Thierseele,* 2nd ed., Hamburg: Voss.

Wundt, W. 1896. *Grundriss der Psychologie,* Leipzig: Engelmann.

Wundt, W. 1922. *Grundriss der Psychologie,* 15th ed., Leipzig: Körner.

Zeugner, G. 1969. *Farbenlehre für Maler,* 2nd. ed., Berlin: VEB Verlag für Bauwesen.

Zeugner, G. 1990. *Farbenkarte 90,* Leipzig: Selbstauflage (Prototyp).

Ziehen, T. 1891. *Leitfaden der physiologischen Psychologie,* Jena: Gustav Fischer.

CHAPTER 11

MISCELLANEOUS SYSTEMS II
Incomplete and Unconventional Systems

Entries of two further groups of miscellaneous systems: various systems of an incomplete nature, and systems based on more or less unconventional ideas that do not fit into preceding chapters.

This chapter is a kaleidoscope of ideas and manifestations. Of all the chapters in this book, this one most representatively demonstrates the variety and heterogeneity of the historical and cultural development in color order systems.

Some of the systems described in the chapter represent mixed forms of systems introduced in other chapters. Others reflect ideas that may seem odd. Systems include rudimentary beginnings toward a system, as well as systems systematically thought through. They also include controversial proposals. Historically, the systems in this chapter range from the eighteenth century to the recent past. Because of their wide variety, they are placed in chronological order within their respective general groups.

Incomplete systems

This group consists of nine systems seemingly in a preliminary stage, or purposefully incomplete. French physician Edmé-Gilles Guyot demonstrated a toylike tool to produce many colors systematically, but did not offer a complete order system. Scottish physicist James David Forbes had several ideas for color order, none of which he completed. He is best known for having taught James Clerk Maxwell (see entry in chapter 6), exposing him to disk mixture and the problems of color order. In the same year in which Forbes's most detailed paper was published, Frenchman J.-C.-M. Sol published a book on another incomplete color order system.

The systems of French horticulturalists René Oberthür and Henry Dauthenay and of their compatriots botanist Paul Klincksieck and chemist Theodore Valette mainly addressed natural scientists' needs, and they saw no need for complete systems. Both of these systems, as well as French entomologist Eugène Séguy's, were influenced by the work of Michel-Eugène Chevreul (see entry in chapter 4). Chevreul's system approach became paradigmatic and spawned a number of derived systems, mostly in France.

English author Robert Francis Wilson also developed an incomplete system for horticulturalists. The Swiss textile firm Setarti had an entirely new idea: It built a two-dimensional color order system by weaving different-colored yarns in various ratios, thereby generating a systematic large range of color stimuli when viewing the fabric from a certain distance. But again, the idea was incomplete as an ordering system.

Miscellanea

The second group of this chapter consists of systems of the most miscellaneous kind. German physicist Hermann Scheffler, a contemporary of Ewald Hering and Hermann von Helmholtz (see entries in chapters 5 and 6, respectively), believed he had a better ordering system involving nonlinear scaling of the spectrum by frequencies rather than wavelengths.

Americans Denman Waldo Ross and Arthur Pope were artists and teachers, the former influencing the latter. Pope expanded Ross's concept to a self-consistent order, its construction deriving from combining Pope's colorant mixture experience as a painter with perceptual attributes, a system that survives today for didactic purposes in art and design circles.

German philosopher Ernst Barthel's color screw is a color order derived in its entirety from musical theory, to a degree in agreement with Castel's. An earlier, less complete version was offered in 1915 by Carl Hensel (figure 11.1).

Russian ophthalmologist E. B. Rabkin devised a system with nine basic colors and an organization based on hue, lightness, and relative saturation, all defined in terms of the International Commission on Illumination (CIE) colorimetric system. In the efforts of German philosopher Eckart Heimendahl and psychologist Heinrich Frieling, as in those earlier of Johann Wolfgang von Goethe (see entry in chapter 3), real or imagined psychological effects are a fundamental source of color order. Such approaches were novel and interesting but cannot completely avoid the label of ideologies.

German designers Lothar Gericke and Klaus Schöne developed what they named a planetary system in which any sample can assume the role of a star with other colors arranged around it in planetary fashion. French artist Michel Albert-Vanel also conceived a planetary system, with only this designation in common with the system of Gericke and Schöne.

German printing technologist Harald Küppers's two atlases of halftone-printing results use the cube model. Küppers also developed a conceptual color order system in a cube distorted to a rhombohedron, seemingly derived from Hans Neugebauer's rhombohedron of idealized printing ink mixture in the halftone process (see entry in chapter 6). French color harmonist Christian Richardière placed his halftone-printing system into the form of a cone rather than a cube.

Dutch color theorist Frans Jan Gerritsen and German software developer Gerriet Hellwig used psychophysical findings in their systems but in a purely conceptual and qualitative sense.

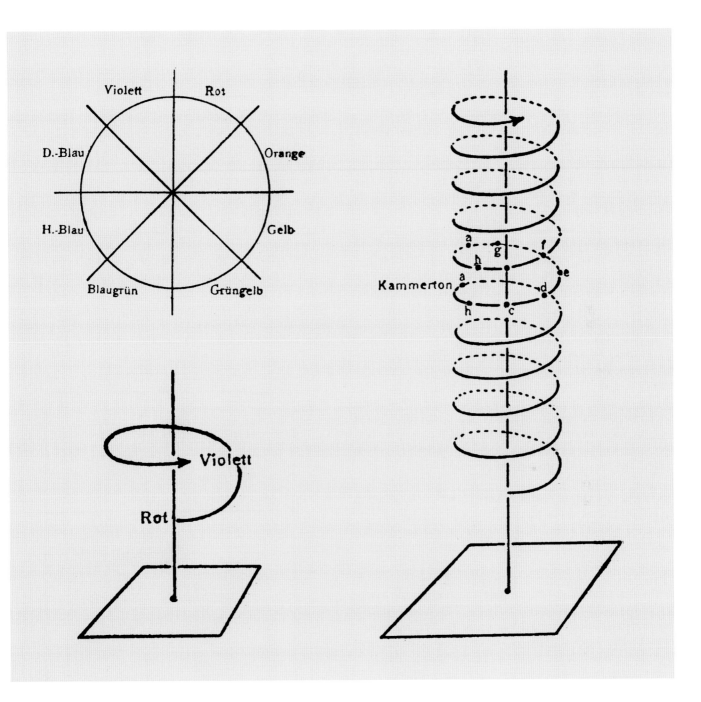

Figure 11.1.
*Hensel's (ca. 1915) representation of the development of the color
spiral from the color circle (left), first opened to form a spiral
(center), and finally expanded in several steps, respectively octaves,
upward in direction of light and downward in direction of dark.*

EDMÉ-GILLES GUYOT 1770

[E.-G. Guyot], *Nouvelles récréations physiques et mathematiques*, 1770

Edmé-Gilles Guyot (1706–1786) was a French geographer, physician, and director of the Paris postal office. He also was the presumed author of the four-volume work *Nouvelles Récréations physiques et mathematiques* (Novel physical and mathematical diversions) published in 1769–1770, a popular work of scientific entertainment that saw several editions and foreign translations.[1] Volume 3 is devoted to optical diversions and contains 52 optical projects ranging from magical mirrors to a simple recreation of Castel's (see entry in chapter 3) ocular musical instrument (see figure 11.2, bottom half).

The final chapter in volume 3 is titled "On generation of colors" and contains projects on mixing colors from three primaries, changing those mixtures with light and shadow to create all natural nuances, producing a rainbow in a room, painting on glass, imitating marble, and the mentioned ocular music.

Project 47 consists of "producing by simple mixture of three colors, blue, yellow, and red, all those visible in the image of a prism and all colors that that can be intermediate to them" (Guyot 1770, p. 212, translated by R.G.K.). Three sheets of highly translucent paper (*serpente d'Hollande*, Dutch snakeskin paper) are lightly and uniformly painted on both sides, the first with a very thin watercolor of the best Prussian blue, the second with gamboge with a bit of saffron added, and the third with carmine.

The sheets are cut into strips of varying width and glued to cardboard frames so that the first horizontal strip has a single layer, the second, two layers, and the fifth, five layers of one of the three primary colors (figure 11.3, top and center right).

The next step calls for building a cardboard box that is open on two sides and painted black inside, and inserting the frame. Inserting a single frame and viewing it against sunlight allows viewing four tint colors and five layers of the "full" color, say, blue. By superimposing the yellow and the blue frame, five layers of green appear, and so on. In this manner, the three intermediary hues and related tint colors are generated.

In the following project, Guyot showed how "with light and shade all natural nuances of the colors producing the prism image can be produced" (p. 217). Seven frames are prepared and the seven (Newtonian) colors of the prism are applied to translucent paper in the described manner. The intermediate colors green, orange, violet, and indigo "are composed from the colors taken to be primitive, blue, yellow, and red." A new piece of translucent paper is painted lightly on both sides with bistre (described as produced from chimney ash). Strips are mounted in a fashion described above except that the first strip is of plain paper. By inserting the resulting frame

at a right angle to a colored frame, the colors are seen in five unshaded depths and 20 versions of various depths and shadings.

The total system encompasses 175 colors in seven hues. The seven hues are shown at five depths and each of these is grayed in four grades. Guyot commented, "There are other colors that can also be produced by means of all three primary colors; their number is large, but mostly they are false colors, nevertheless used in painting for imitation of colors with which they agree. These colors cannot be seen in the image produced by the prism" (p. 219 note).

Despite references to prismatic colors, Guyot's color selection is based on subtractive mixture of three primaries and four levels of gray. The projects' goal was recreation, not standardization. But it is an interesting method for displaying a multitude of colors before Johann Heinrich Lambert's color pyramid (see entry in chapter 4) had been published.

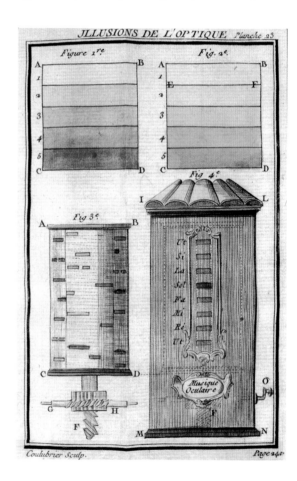

Figure 11.2.

Plate 23 from Guyot's (1770) Récréations *showing the inner cylinder (left) and the outer appearance of the recreation of Castel's ocular music instrument.*

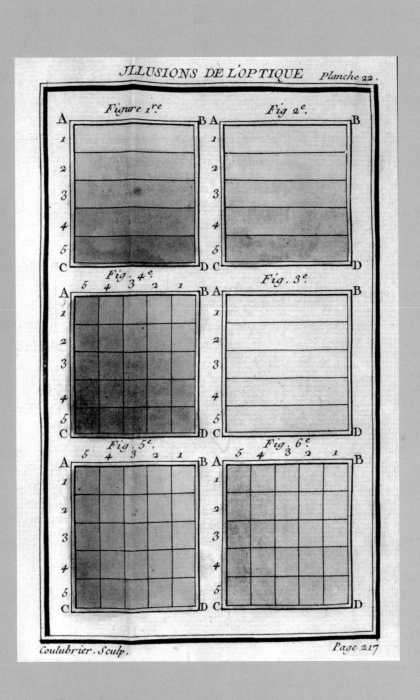

Figure 11.3.
Plate 22 from the Récréations *with the filter arrangement
for blue, red, and yellow in figures 1, 2, and 3. The results of
superposition of pairs of filters are shown in figures 4–6, with
one filter always rotated 90 degrees (Guyot 1770).*

JAMES DAVID FORBES   1849

J. D. Forbes, Hints towards a classification of colours, 1849

James David Forbes (1809–1868) was a Scottish physicist primarily known for his work in glacier science. He was principal of St. Andrews University in Edinburgh. He experimented with disk color mixture. Maxwell (see entry in chapter 6) was one of his students who assisted him. In his paper on color classification, Forbes offered a valid explanation of the difference between mixture of lights and mixture of colorants, anticipating the more complete explanations of Helmholtz (see entry in chapter 6).

Using experimental data provided by Lambert (see entry in chapter 4) and Isaac Newton's gravimetric rule of light mixture (see entry in chapter 6), Forbes estimated the angle segments of primary yellow, red, and blue painted disks required to result in the experience of neutral gray. He showed that it must be possible to match any color in Tobias Mayer's (see entry in chapter 4) basis triangle with appropriate angle segments of three primaries. Forbes realized that the pigments he used for his disks vary in "purity of quality" (saturation) and in "lucidity" (lightness) and bemoaned the absence of scales for either property.

Forbes drew his own conceptual version of Mayer's basis triangle, as derived from Newton's rule and normalized disk sectors of the three primary pigments, in which balanced neutral gray falls on the gravimetric center *W* (figure 11.4). Balanced binary mixtures result in the secondary colors orange, purple, and green. Using Mayer's terminology for individual colors, Forbes showed that mixtures of the primaries with appropriate amounts of gray must result in the same colors obtained from appropriate mixture of the primaries alone (e.g., citrine [Ci], olive [Ol], or russet [Ru]):

> Thus we arrive at this conclusion, that *all combinations of three primary colours* (as far as difference in *quality* is concerned) *may be represented by transitions from the primary and secondary colours into gray*. . . . Hence a classification of colours may be made, which, although redundant in some parts, has the advantage of pointing out clearly the composition of each in this point. . . . This diagram, like Mayer's triangle, includes colours varying in *quality*, but of standard *intensity* [relative lightness] and of the highest attainable *purity*. (Forbes 1849, p. 175, emphasis original)

Forbes envisaged a hue circle of 24 colors, each with a varying number of steps toward medium gray. Additional layers in the direction of white and black result in a three-dimensional color-classification system.

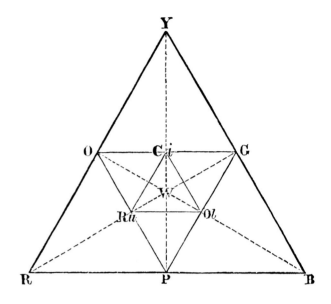

Figure 11.4.
Forbes's interpretation of Mayer's triangle in terms of disk mixture. The primaries R, Y, and B in binary mixtures form O, G, and P. They in turn form rubine (Ru), citrine (Ci), and olive (Ol). With appropriate sector width, the central neutral gray W is obtained either with mixtures of the primaries, secondaries, or tertiaries (Forbes 1849).

A different, two-dimensional version discussed by Forbes has the full colors on one concentric ring, and tint/shade reductions along radial lines toward central white and toward black on the periphery. Such a system is based on the work of Moses Harris (see entry in chapter 3) and had been described in 1839 by Chevreul (see entry in chapter 4). Forbes believed that painted or printed paper samples were not sufficiently stable for a color order system, so he traveled to Rome for variously colored mosaics called *tesserae*. However, he apparently could not find pieces with enough properly spaced colors to build a corresponding color order system.

Forbes's paper is essentially theoretical in nature. While he understood the difference between light mixture, and colorant mixture, he did not appear to realize that colorant mixture is highly nonadditive. Nor did he understand the nature of the nonlinearity of the relationship between color disk-mixture data and perceptually uniformly spaced colors. Using disk-mixture technology learned from Forbes, his pupil Maxwell made major advances in color theory.

J.-C.-M. SOL   1849

J.-C.-M. Sol, *La palette théorique ou classification des couleurs*, 1849

J.-C.-M. Sol lived in the Bretagne, France, and wrote a volume of prose and poetry, in addition to his short book on color order. He noted that nobody before him had treated the subject in an appropriately logical manner. Sol knew of the treatise on color by Goethe but had been unable to consult it.

Sol's book consists of an "Analytical text" and a larger section called "A few insights." He approached the issues from a qualitative logical, rather than a physical, point of view. He placed colors into three categories, the first consisting of white only, being unique. The second category consists of the primary colors yellow, red, and blue and their binary and tertiary mixtures including black, the "isomeric" (equally weighted) mixture result of the primary colors. The third category, named hybrid colors, consists of binary and ternary mixtures of the primaries with white, the isomeric color being middle gray. A conceptual sketch of the full hue plane is shown in figure 11.5.

Sol believed that, in addition to the three primaries and their isomeric mixtures orange, violet, and green, only six additional intermediate hues were necessary in the hue circle, more than that being excessive. He named three grades between full color and white and only one between full color and black (see figure 11.6 for an interpretative example).

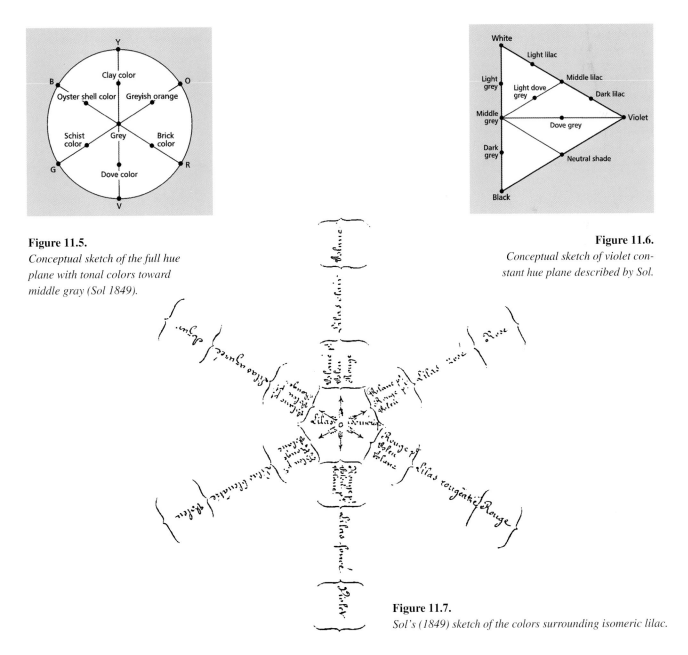

Figure 11.5.

Conceptual sketch of the full hue plane with tonal colors toward middle gray (Sol 1849).

Figure 11.6.

Conceptual sketch of violet constant hue plane described by Sol.

Figure 11.7.

Sol's (1849) sketch of the colors surrounding isomeric lilac.

Mixtures of an isomeric tint color with various surrounding colors are shown in figure 11.7 and in the interpretative sketch of figure 11.8. The isomeric mixtures between primaries or secondaries and white are named straw color, apricot, pink, lilac, azure, and pistachio. Those in the direction of black are named sepia, blackish orange (carmelite), brown, blackish violet (*teinte-neutre*), indigo, and olive (figure 11.9). Sorting by lightness, Sol placed white on top followed by yellow, then orange and green on the same level, red and blue on the next level, then violet, followed by black. He argued against Newton that it is impossible for the mixture of the three primaries to be white; even in disk mixture, they produce gray.

In the insights section, Sol discussed the essential properties of colors and harmonic and disharmonic combinations and, like Goethe before him, attributed moral character to colors. He claimed an additional categorization of colors: They are either attractive or sad. Finally, colors are sorted into Immanuel Kant's categories of modality, quality, and relativity.

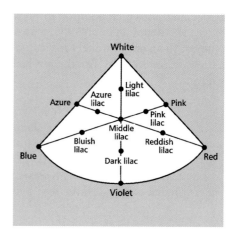

Figure 11.8.

Interpretative sketch of figure 11.7.

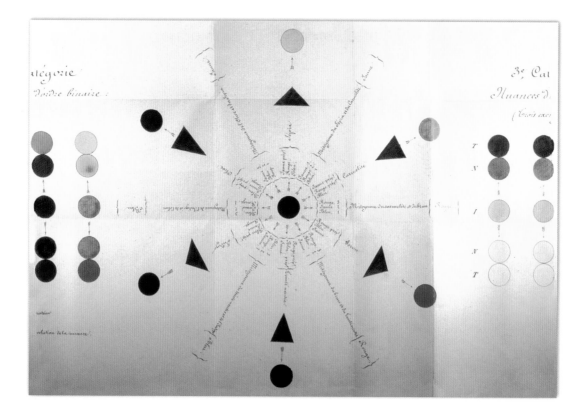

Figure 11.9.

Generation of the blackish isomeric mixtures around the hue circle. Always in two named colors yellow, red, or blue, respectively, are preponderant. For the intermediate colors, the integer ratios of primaries are given (Sol 1849).

RENÉ OBERTHÜR and HENRI DAUTHENAY   1905

R. Oberthür and H. Dauthenay, *Répertoire de couleurs pour aider à la determination des couleurs des fleurs, des feuillages et des fruits*, 1905

At the end of the nineteenth century, members of the French Horticultural Congress agreed on the need for a standard set of color samples for describing the colors of flowers, leafs, and fruit of plants. The French Society of Chrysanthemists and one of its members, the book printer **René Oberthür**, agreed to undertake the work, with **Henri Dauthenay** and other collaborators. With printing-ink manufacturer Lorilleux et Companie, Oberthür produced a first version of 1,700 samples. It was found unsatisfactory by the oversight committee.

In 1905, the final version, *Répertoire de couleurs pour aider à la determination des couleurs des fleurs, des feuillages et des fruits* (Color repertory to help in the determination of the colors of flowers, foliage, and fruit) was published in two volumes. It contains a total of 1,397 individual samples, and the authors claimed that it was inspired by Chevreul's work (see entry in chapter 4). However, there is no direct relationship between the two systems.

Volume 1 contains plates 1–182; volume 2, plates 183–365. On most plates, there are four rectangular color samples of size 32 × 45 mm in a column (figures 11.10 and 11.11). Twenty-one plates have only one color sample. They illustrate agreed-upon colors of a particular category name. Supplied information includes the source of the category name, its French synonyms, German, English, Spanish, and Italian translations, and the sources of the four tonal variations. The variations are derived from colors similar to those found in nature, not from color order theory.

Colors are sorted into 12 categories: white and shaded whites (15 cases), yellows (38 cases), oranges (22 cases), reds (42 cases), pinks (37 cases), purples (17 cases), violets (29 cases), blues (34 cases), greens (62 cases), browns and ochres (34 cases), chestnut browns (13 cases), blacks and grays (22 cases).

A separate table provides examples of the sources of the chosen color names: chemical industry and dyeing (e.g., Hoffmann's violet), typographical inks (e.g., magenta), painting and watercolor (e.g., ultramarine blue), silks and fashion (e.g., spring blue), wool and cotton dyeing (e.g., mauve), and the vegetable kingdom (e.g., geranium red).

For the direct comparison of the colors of the atlas and plant colors, the authors recommend a mask with two lozenge-shaped cutouts. The color of the paper of the mask is not specified, thus presumably white. No information concerning the pigments used in printing or other standardization references have been provided.

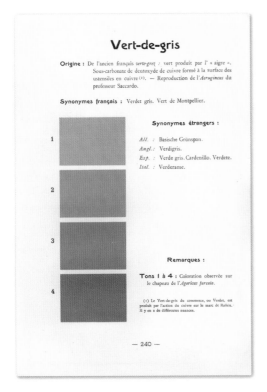

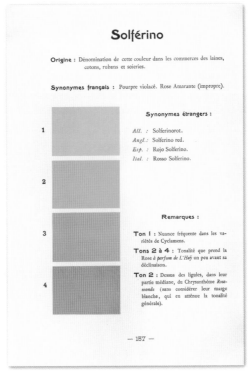

Figure 11.10.
Table 240, Vert-de-Gris (Oberthür and Dauthenay 1905).

Figure 11.11.
Table 157, Solférino (Oberthür and Dauthenay 1905).

PAUL KLINCKSIECK and THEODORE VALETTE   1908

P. Klincksieck and T. Valette, *Code des couleurs, à l'usage des naturalists,
artistes, commerçants et industriels*, 1908

Only three years after publication of Oberthür and Dauthenay's *Répertoire de couleurs* (see entry in this chapter),
French botanist and publisher **Paul Klincksieck** and chemist **Theodore Valette** issued *Code des couleurs, à l'usage
des naturalists, artistes, commerçants et industriels* (Color
code, for use by naturalists, artists, commercial and industrial people). Valette succeeded Chevreul (see entry in chapter 4) at the Manufacture Nationale des Gobelins.

Published as part of Klincksieck's Pocket Library for Naturalists, the also-pocket-size book contains 48 pages and 720
color samples of size 20 × 25 mm, executed in painted paper
cut to size. Klincksieck believed the pocket format to be essential for ease of use. The text also includes a table of color
names of the 10 principal colors, in 18 European languages
"to give the code international character."

The authors described the purpose and contents of the code
on 33 pages of text. A mycology expert, Klincksieck wrote
that mushrooms led to the beginnings of the *Code des Couleurs*, but that the book's application is universal. On the

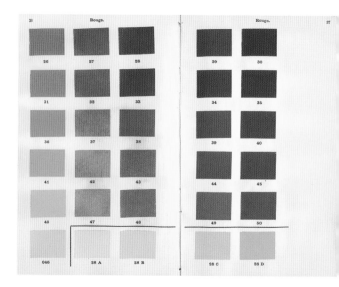

Figure 11.13.

*Color samples of the principal red hue
(Klincksieck and Valette 1908).*

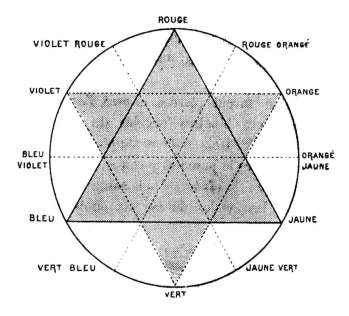

Figure 11.12.

*Conceptual sketch of Klincksieck and Valette's hue
circle limited to 12 colors. In the* Code des couleurs
*there is always an additional hue between those
shown (Klincksieck and Valette 1908).*

title page, Klincksieck and Valette wrote that they selected colors according to a simplified version of Chevreul's
method. Although Chevreul's method was used, the authors
considered his sample designation hopelessly complicated.
They wrote that they took great care to select the fastest
colorants available even if it meant that, for example, pink
colors could not be displayed in as saturated a form as desired. Valette wrote that he thought it likely that colors exist
only in our eyes and that they are a function of the eye's
organization. Valette's basic hue circle is shown in figure
11.12. For the *Code des Couleurs*, he selected every third
hue from Chevreul's 72-hue circle for a total of 24, evenly
spaced. The hue circle begins with Chevreul's red, followed
by the 3° red, the orange red, and so forth. However, similarity to Chevreul's system ends with the hue circle.

Colors associated with a particular hue are found on a double page (figures 11.13 and 11.14). The full color is located
in the upper left corner. In the first column, each succeeding
color is half the strength of the one above it, forming a tint
scale. Columns 2, 3, 4, and 5 contain tonal colors with ad-

EUGÈNE SÉGUY 1936

E. Séguy,
Code universel des couleurs, 1936

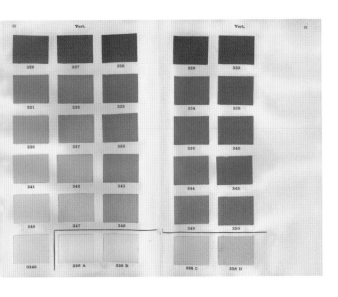

Figure 11.14.

*Color samples of the principal green hue
(Klincksieck and Valette 1908).*

French entomologist **Eugène Séguy** (1890–1985) was active in the Muséum National d'Histoire Naturelle in Paris where, in his later life, he was chair of the entomology department. His 1936 *Code universel des couleurs* (Universal color code) was another attempt to solve the problem of precisely designating colors found in nature. Séguy mentioned that 25,000 color names exist with uncertain meaning, making them of little use for color specification.

In the accompanying text, Séguy discussed colors and their mixture, and offered a color-mixture multiplication table for additive (light) mixture (figure 11.15). His atlas, consisting of 720 color samples on 48 plates, is executed in halftone printing with four pigments (sample size 21 mm × 26 mm). He stated that a "scientific" color code should have several thousand samples, but practical limitation kept the atlas at 720.

Each plate contains 15 samples. Seven series of plates illustrate major hue categories (red, orange, yellow, green, blue, violet-blue, and violet); two additional ones show brown and red-brown colors. There are 11 red hues, 4 orange, 5 yellow, 6 green, 5 blue, 4 violet-blue, and 5 violet, as well as 1 brown and 1 red-brown.

ditions of 1%, 2.5%, 7.5%, and an unlisted percentage of black (percentage based on full color content), amounts that appeared "visually satisfactory" to the authors. As a result, each full color has 25 tint and tone versions.

In addition, there are four lighter colors located in row 6 (separated with a black line). Their composition is described as having been chosen "somewhat arbitrarily." They are derived from the first color of column 3 by lightening it "considerably and successively." They carry the number of the original color with letters A–D added. All colors are identified by a simple number, from 1 to 720.

The authors realized that their system was incomplete, but cost considerations precluded additional samples. The collection lacks pure grays, because those are absent in nature.

Klincksieck and Valette's *Code des Couleurs* represents a limited number of color samples, arranged in a less than fully systematic manner. Its value as a standard was limited due to lack of quantitative description of the samples.

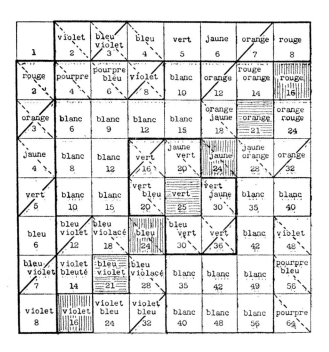

1	violet 2	bleu violet 3	bleu 4	vert 5	jaune 6	orange 7	rouge 8
rouge 2	pourpre 4	pourpre bleu 6	violet 8	blanc 10	orange 12	rouge orange 14	rouge 16
orange 3	blanc 6	blanc 9	blanc 12	blanc 15	orange jaune 18	orange 21	orange rouge 24
jaune 4	blanc 8	blanc 12	vert 16	jaune vert 20	jaune 24	jaune orange 28	orange 32
vert 5	blanc 10	blanc 15	vert bleu 20	vert 25	vert jaune 30	blanc 35	blanc 40
bleu 6	bleu violet 12	bleu violacé 18	bleu 24	bleu vert 30	vert 36	blanc 42	violet 48
bleu violet 7	violet bleuté 14	bleu violet 21	bleu violacé 28	blanc 35	blanc 42	blanc 49	pourpre bleu 56
violet 8	violet 16	violet bleu 24	violet bleu 32	blanc 40	blanc 48	blanc 56	pourpre 64

Figure 11.15.

Additive color mixture table (multiplication table) for seven lights mixed in various binary ratios. Red 2 mixed with violet 2 results in purple 4; orange 3 mixed with blue-violet 3 results in white 9, because they are complementary, and so on (Séguy 1936).

The pages are organized as follows: The full color of a given hue is located in the upper left corner. It is progressively lightened toward white in the left column, in steps that are not explicitly defined. In the middle and right columns, the resulting colors are toned down in a horizontal direction by the use of black, red, blue, yellow or its own complementary color (figures 11.16 and 11.17). Séguy did not explain how he did this, either. The text contains a list of the atlas's complementary colors. The color samples are successively numbered, and most of the samples are named in six or seven languages.

Séguy was well aware of some of the technical problems in attempting to achieve constant hue series by halftone printing with yellow, red, blue and black pigments. On the other hand, he confidently stated: "All colorists using the Code are assured of having in their hands, on the whole, nearly identical samples to work with" (p. 37). The system is clearly incomplete, and he has made no attempt to demonstrate its structure in a color solid.

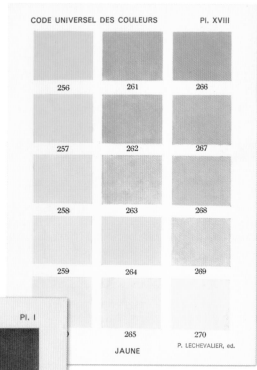

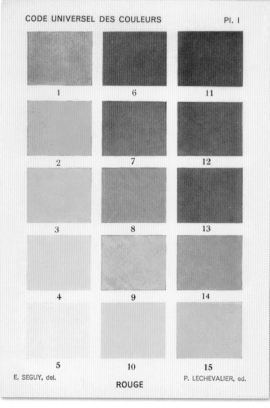

Figure 11.17.
Plate 18 Jaune is treated in comparable fashion as shown in figure 11.16 (Séguy 1936).

Figure 11.16.
Plate 1 of Séguy's Code universel des couleurs, *with the full color of process red (named* lislas*) in position 1. Samples 2-5 are progressively whitened. The resulting samples are desaturated in rows from left to right (Séguy 1936).*

ROBERT FRANCIS WILSON   1938

R. F. Wilson, *Horticultural colour chart*, 1938, 1942
The practical Wilson colour system, 1957

Robert Francis Wilson was art director of the British Colour Council (an organization that issues seasonal color charts for apparel and home furnishings). He developed a color order system primarily for horticulturalists. Two volumes of the *Horticultural Colour Chart* were published under the auspices of the Colour Council, the first in 1938, the second in 1942. The impact of World War II appears to have prevented publication of additional volumes. In 1957, Wilson published a slightly modified, much abbreviated version of his system.

Using colorants producing highly saturated colors, Wilson created a 24-grade hue circle, extended by mixture to 64. Similar to Robert Ridgway's scheme (see entry in chapter 10), the concept for each full color involves tint and shade scales with three grades each and three levels of grayed "tonal colours," resulting in 31 veiled colors per hue (figure 11.18). A gray scale is absent.

Each of the colors is identified in a complex system. The implicit color space is a double cone (see figure 11.19). The system provides logical space for 2,048 colors, but only 200 of these were realized in the first two volumes. However, each was additionally desaturated in three grades by halftone printing (figure 11.20). The result is a total of 800 samples of size 44 × 32 mm. The colors are sorted by horticulturally influenced names. For each color, the system lists equivalents in four other color order systems, synonyms in other languages, and specific horticultural examples. Because pages are not ordered by hue, one must search through a summary table to find all pages with samples of a particular hue.

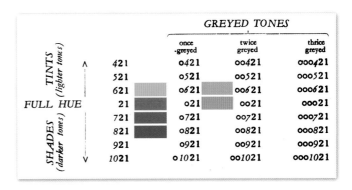

Figure 11.18.

Constant hue plane of Wilson color 21 (carmine rose) with five of the possible 31 veiled colors and their identification scheme (Wilson 1938).

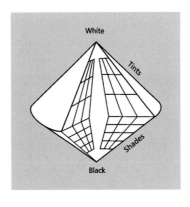

Figure 11.19.

Sketch of constant hue planes in the double cone space implicit in the Wilson system.

Figure 11.20.

Two pages from the Horticultural Colour Chart with colors carmine rose and spinach green (Wilson 1938).

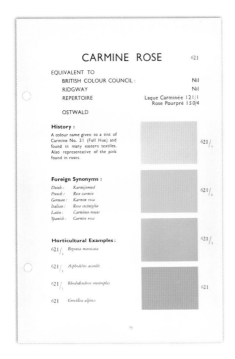

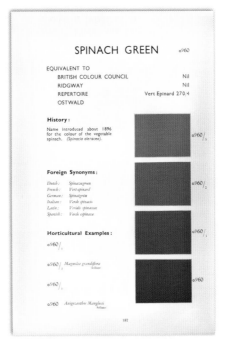

The Practical Wilson Colour System of 1957 is a small booklet with samples neatly executed in opaque paint. The original 24-grade hue circle is reduced to 12 grades and described by Wilson as perceptually uniform. Here, perceptually uniform 10-grade gray and brown scales have been added (figure 11.21). Wilson believed that for certain hues, a brown scale permitted a much more nuanced veiling than a gray scale, as illustrated for orange in figure 11.22. Further, Wilson illustrated the possibility of creating veiled colors by mixing colorants of complementary hue (figure 11.23). Seemingly uninterested in three-dimensional representation, Wilson was vague about the system behind his approach in both versions.

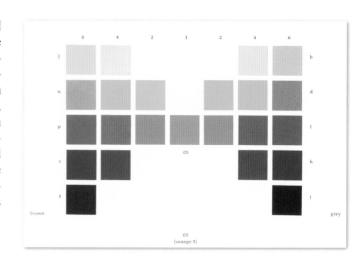

Figure 11.22.
The full color orange mixed with colors from the gray (right) and the brown (left) scale (Wilson 1957).

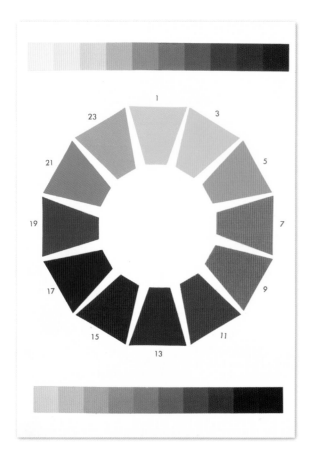

Figure 11.21.
Hue circle and gray and brown scale from The Practical Wilson Colour System *(Wilson 1957).*

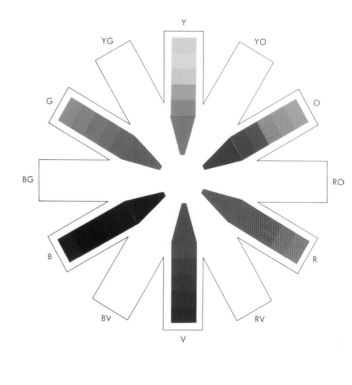

Figure 11.23.
Mixture at different ratios of paints of full colors opposing each other in the hue circle (Wilson 1957).

SETARTI 1947

Setarti AG, *Farbentafel*, Swiss Patent 243891, 1947

In 1947, the Swiss textile company **Setarti AG** obtained a patent for a color system in the form of a woven fabric. According to the patent, the system consists of a large square cloth, woven from differently colored warp and filling threads. They produce a large number of different-colored fields in which each field color is the result of optical mixture of the warp and filling thread colors. The color sequence in warp and filling threads is always identical.

An example of such a chart, produced by the Swiss textile firm Siłor, is shown in figures 11.24 and 11.25. Figure 11.24 shows an enlarged section, each square measuring 6 × 6 mm in the original. The complete piece is shown in figure 11.25. The sequence of colors proceeds from left to right and simultaneously from the bottom to the top. Groups of colors are varied according to lightness (see figure 11.24). With this method, 193 × 193 colors are mixed, resulting in a total of 37,249 different color patches of 6 × 6 mm size in a two-dimensional arrangement. The color table itself is 116 × 116 cm; the complete piece with borders and heading is 124 × 135 cm.

The patent describes the manufacture of such charts with different kinds of textile fibers in some detail with recommended color and colorant selections. Such color charts were used to select colors when ordering goods and to aid communication between manufacturer and retailer.

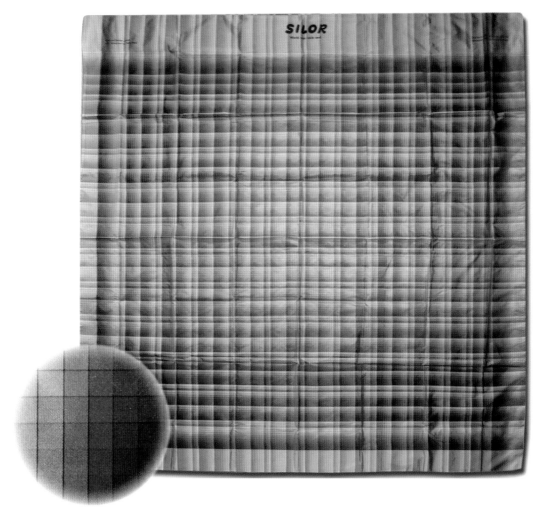

Figure 11.24.

Detail of color chart showing the fabric structure and the resulting mixed patches of 6 × 6 mm. Image courtesy Friedrich Schmuck.

Figure 11.25.

The complete Setarti textile color chart. Image courtesy Friedrich Schmuck.

JIRI PACLT   1958

J. Paclt, *Farbenbestimmung in der Biologie*, 1958

Jiri Paclt was a Slovakian entomologist and author of several books on insects. As a result, he became interested in color order, and in 1958 published in Germany *Farbenbestimmung in der Biologie* (Color determination in biology). The topic interested several earlier members of the entomology community, for example, Ignaz Schiffermüller, Harris, and Ridgway (see entries in chapter 3 and 10). Paclt was well informed about earlier and contemporary color-sample collections and order systems such as those of Albert Henry Munsell and Wilhelm Ostwald (see entries in chapter 5 and 10, respectively) and about the field of color science. Like other entomologists, he was interested in fixing names to color stimuli.

Paclt believed that the decimal system offered considerable advantages for color classification so his hue circle has 10 members (figure 11.26). They are illustrated at two levels of intensity superimposed on a psychophysical diagram in an equilateral triangle version. That connects the samples in terms of hue to a psychophysical system.

Paclt used the general colorimetric term of chromaticity to express hue and saturation in combined fashion. As a result, three-dimensional color space is reduced to two dimensions. The Slovakian color scientist E. Pavlovsky provided Paclt with the original tint/shade scales for each of the ten hues (figure 11.27). On basis of this "decimal color system," Paclt developed what he hoped to be an "international color terminology" for natural scientists.

For each of the hues in the chromatic scale and a gray, samples are identified by color name with a preceding number from 1 to 9 (figure 11.28), in Latin, German, and English, and with space for other languages. The system contains a total of 101 different samples of size 1.5×2 cm, printed in the halftone process in moderate quality. Neither Paclt's sample collection nor his terminology has been found to be sufficiently detailed for exact color definition in the natural sciences.

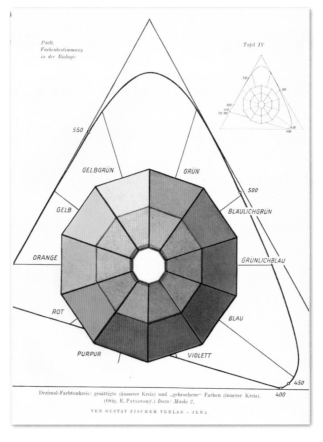

Figure 11.26.
Paclt's (1958) 10-hue color circle at two levels of chromaticity.

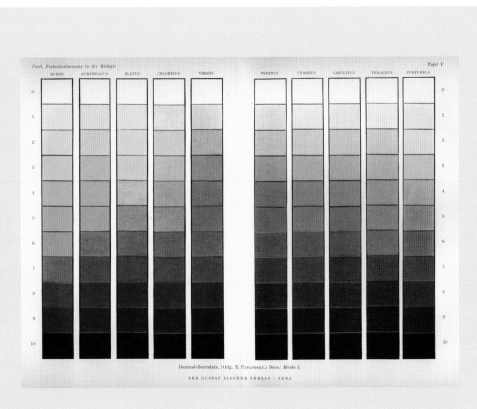

Figure 11.27.
Decimal color scale (Paclt 1958).

lateinisch	deutsch	englisch	usw.
0 ... ALBUS	... WEISS	... WHITE	... usw.
1 ... 1-GRISEUS	... 1-GRAU	... 1-GREY	... usw.
2 ... 2-GRISEUS	... 2-GRAU	... 2-GREY	... usw.
3 ... 3-GRISEUS	... 3-GRAU	... 3-GREY	... usw.
4 ... 4-GRISEUS	... 4-GRAU	... 4-GREY	... usw.
5 ... 5-GRISEUS	... 5-GRAU	... 5-GREY	... usw.
6 ... 6-GRISEUS	... 6-GRAU	... 6-GREY	... usw.
7 ... 7-GRISEUS	... 7-GRAU	... 7-GREY	... usw.
8 ... 8-GRISEUS	... 8-GRAU	... 8-GREY	... usw.
9 ... 9-GRISEUS	... 9-GRAU	... 9-GREY	... usw.
10 ... NIGER	... SCHWARZ	... BLACK	... usw.

lateinisch	deutsch	englisch	usw.
0 ... ALBUS	... WEISS	... WHITE	... usw.
1 ... 1-RUBER	... 1-ROT	... 1-RED	... usw.
2 ... 2-RUBER	... 2-ROT	... 2-RED	... usw.
3 ... 3-RUBER	... 3-ROT	... 3-RED	... usw.
4 ... 4-RUBER	... 4-ROT	... 4-RED	... usw.
5 ... 5-RUBER	... 5-ROT	... 5-RED	... usw.
6 ... 6-RUBER	... 6-ROT	... 6-RED	... usw.
7 ... 7-RUBER	... 7-ROT	... 7-RED	... usw.
8 ... 8-RUBER	... 8-ROT	... 8-RED	... usw.
9 ... 9-RUBER	... 9-ROT	... 9-RED	... usw.
10 ... NIGER	... SCHWARZ	... BLACK	... usw.

Figure 11.28.
*Paclt's (1958) proposal for a systematic color terminology based on
his decimal scale illustrated with lists for gray and red.*

HERMANN SCHEFFLER 1883

H. Scheffler, *Die Theorie des Lichtes*, 1883

German engineer, mathematician, and physicist **Hermann Scheffler** (1820–1903) is the author of more than two dozen books, including the two-volume work *Die physiologische Optik* (Physiological optics), published in 1864-1865 and a four-part text *Die Naturgesetze und ihr Zusammenhang mit den abstrakten Wissenschaften* (Laws of nature and their relationship to the abstract sciences, 1876–1883). Scheffler added three supplements to this latter work, the third named *Die Theorie des Lichtes* (Theory of light, 1883). It should be recalled that Helmholtz (see entry in chapter 6) published the first edition of his three-volume *Handbuch der physiologischen Optik* between 1856 and 1866, beginning well before the first Scheffler volume.

Scheffler believed the Young-Helmholtz trichromatic theory[2] to be insufficient and incapable of explaining the optical processes. He also thought that Hering restructured his (Scheffler's) ideas, expressed in "Physiological Optics," into the assimilation-dissimilation (opponent-color) theory.

One of the topics of "Theory of light" is "Symbolic representation of colors in form of a geometric image." Here Scheffler described a color order solid related to the frequency of spectral lights. From August Beer's frequency measurements of lights made in the 1850s, he selected those he took to represent the three primary hues (red, yellow, and blue) and their intermediates (orange, green, and violet).

Scheffler realized that the relationship between light frequencies of colors and associated hue differences was not linear. At the same time, he saw value in a circular representation (figure 11.29) in which the vector AR for red is opposed by the vector AR' that was to represent the complementary color to red (and comparably for the other two primary colors). He was familiar with Hermann Günter Grassmann's vector addition law of lights (see entry in chapter 6). He knew that additions of lights are additive, not "light destructive," but additions of pigments are light destructive. From this, he concluded that a three-dimensional representation is required to represent the complete color order correctly.

By placing the circular contour as a logarithmic spiral onto a cone, Scheffler was able to even the angular distances between the three primary colors when projected on a plane. Thus, he brought them in line with perceptual distances (figure 11.30). However, he had to apply an adjustment factor to obtain equal spacing along the circle, rationalized with the idea that Beer's measurements likely were not accurate.

In the resulting cone (figure 11.31), colors are represented by vectors from origin A to point b (for blue) on the spiral. (In this manner, the vector lengths are inversely related to the energy content of the light rays.) Vector addition, say, between Ar and Ag, results in desaturation toward the achromatic line AW and increased vector length. The cone form can be selected in a manner that the angles between the vectors of the three primary lights are 90° each; that is, they are orthogonal. But he also postulated a geometric analog of brightness that does not agree with perceptual brightness data measured at the time by Artur König, Helmholtz's assistant and collaborator (see chapter 6) and others.

The origin of the cone at A represents the location of a hypothetical black that has no luminosity or absorption: According to Scheffler, the black of a body has no luminosity but, when illuminated by a "white" or a colored light, would diffuse the light without absorbing it. To be able to represent the results of pigment addition, Scheffler proposed a line from A to W_1 on which blacks of increasing intensity of absorption are located, representing bodies "capable of absorbing white light of the corresponding intensity, that is, absorbing the related amount of the white light illuminating it and diffusing the remainder."

The academic color community of Scheffler's time did not accept his frequency-based color solid or his attempt at reconciling the results of additive and subtractive color mixture, and today is a curiosity.

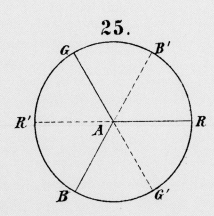

Figure 11.29.
Circular model of spectral colors in which the three primary colors red R, yellow G, and blue B are opposed by their complements green R, violet G, and orange B (Scheffler 1883).

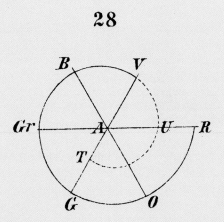

Figure 11.30.
Spiral segment along which the distances are in a degree of linear agreement with the frequencies of the lights representing the major hues and placing complementary colors opposite. U is ultraviolet (Scheffler 1883).

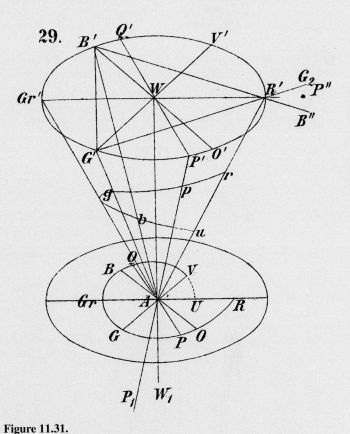

Figure 11.31.
Cone model of the geometric space where the color spiral segment of figure 5.30 hugs the surface. AW is the achromatic axis from the hypothetical nonabsorbing black A to white. Absorbing blacks are located on the axis segment AW₁ (Scheffler 1883).

DENMAN WALDO ROSS   1907

D. W. Ross,
A theory of pure design, 1907

American **Denman Waldo Ross** (1853–1935) was a historian of economics, a painter, and an avid oriental-art collector. He was professor of design at Harvard College in Cambridge, Massachusetts, and wrote several books on painting and design. He wished to infuse these topics with a degree of science, considering *A Theory of Pure Design* "a contribution to science rather than to art."

In line with this goal, Ross attempted to clarify the relationships among colors, which he called the "tone relations." For him, the term "tone" encompassed all three attributes hue (for which he used the term color), value, and intensity or its reciprocal "neutrality." His organization of tones is summarized in the "General Classification of Tones" chart (figure 11.32). The implied but not explicitly used hue circle has 12 grades based on the painter's primaries yellow, red, and blue. The lightness scale has seven qualitatively designated grades plus white and black, and the full colors are shown at the appropriate lightness level.

For each full color, Ross schematically sketched three "color-neutralization" levels (1/4, 1/2, 3/4) ending in neutral gray. He showed the corresponding reciprocal intensity level on the scale at the top of the diagram. There are uniformly four intensity steps for all hues. Ross's recommendation for pigments for painting in oil are Blue Black, Madder Lake (Deep), Rose Madder, Indian Red, Venetian Red, Vermillion, Burnt Sienna, Cadmium Orange, Yellow Ochre, Pale Cadmium, Aureolin, Cremnitz White, Emeraude Green, Cobalt Blue, and French Ultramarine Blue.

Wt															Wt
HLt	R	RO	O	OY	**Y**	YG	G	GB	B	BV	V	VR			HL
Lt	R	RO	O	**OY**	Y	**YG**	G	GB	B	BV	V	VR			Lt
LLt	R	RO	**O**	OY	Y	YG	**G**	GB	B	BV	V	VR			LLt
M	R	**RO**	O	OY	Y	YG	G	**GB**	B	BV	V	VR			M
HD	**R**	RO	O	OY	Y	YG	G	GB	**B**	BV	V	VR			HD
D	R	RO	O	OY	Y	YG	G	GB	B	**BV**	v	**VR**			D
LD	R	RO	O	OY	Y	YG	G	GB	B	BV	**V**	VR			LD
Blk															Blk

(DIAGRAM OF VALUES, COLORS, AND COLOR-INTENSITIES)

Figure 11.33.

Hue–lightness diagram, with the full color of a given hue printed in larger type (Ross 1907).

Unlike his friend Munsell (see entry in chapter 5), Ross placed the totality of object colors only on a plane, and without use of a hue circle, because he wanted to represent color relationships in painting, which are two dimensional. Another schematic diagram (figure 11.33) demonstrates the relationship between hue and value, with the full colors in larger type. His pupil and colleague Pope (see entry in this chapter) later expressed the system in three dimensions.

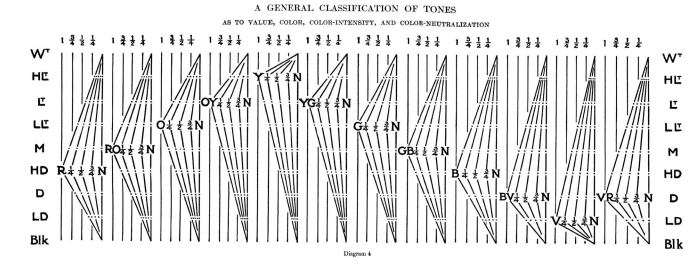

Figure 11.32.

"General Classification of Tones" diagram (Ross 1907).

ARTHUR POPE   1922

A. Pope, *Tone relations in painting*, 1922
An introduction to the language of drawing and painting, 1929

American **Arthur Pope** (1880–1974) was a student and later a colleague of Ross (see entry in this chapter) at Harvard College. They both believed that many people's knowledge on color was "at the level of analphabetism" and attempted to rectify this situation. While Ross decided that color order should remain two-dimensional, Pope extended Ross's approach to a three-dimensional, didactic color order model, which he called the "tone color solid" (figure 11.34).

Pope designed his color solid to aid painters, teachers, and students. Pope placed much import on the relationships in the interior of the color solid, structured with the three attributes hue, value, and intensity. Hue is classically based on the primaries yellow, red, and blue, and their binary mixtures orange, green and violet. Pope recommended the pigments rose madder or alizarine crimson, vermillion, cadmium orange, aureolin or pale cadmium yellow, viridian, and cobalt blue to represent the six basic hues.

The value scale has nine grades including white and black. Like his teacher Ross, Pope used the term "tone" for combinations of the three attributes. Figure 11.35 is his only illustration of tonal colors. Although Munsell (see entry in chapter 5) defined the saturation dimension as chroma in perceptually uniform steps, Pope, as several other authors before him, chose instead to define intensity in relative terms with, for all hues, the same number of grades between full color and neutral gray.

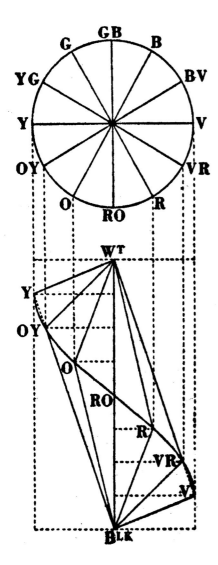

Figure 11.34.
Projected cross section and side view of Pope's (1922) color solid.

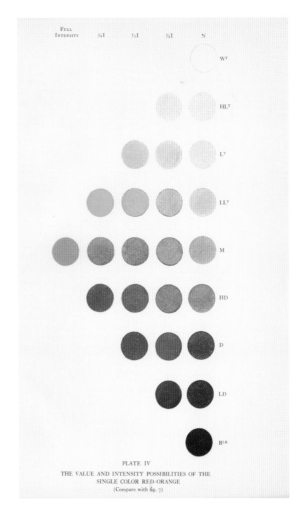

Figure 11.35.
Illustration of 15 tonal colors of the full color RO and the central gray scale (Pope 1929).

In addition to defining intensity (figure 11.36, left), Pope also defined "degree of neutralization" (right), another relative concept that, like Ross, he considered important for painters. Pope believed the regular form of his solid to be superior to the irregular form of Munsell's solid, allowing artists to orient themselves more easily. Figure 11.37 illustrates the symmetric outlines of the system's sampling points on the nine levels of the value scale (e.g., level LL has 30 sample points, the intersection points from that level to the bottom of the diagram).

The system is conceptual, and Pope never published an atlas. Its impact was modest and largely limited to his students, who not only appreciated it, but also critically commented on it (Carpenter and Fisher 1974). It continues to have supporters.

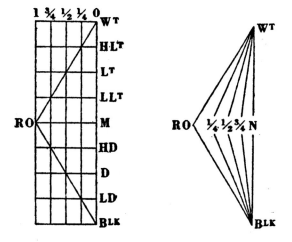

Figure 11.36.
Left: Schematic cross section of the color solid at color RO with intensity levels (Pope 1922). Right: Schematic cross section at color RO with the definition of degree of neutralization.

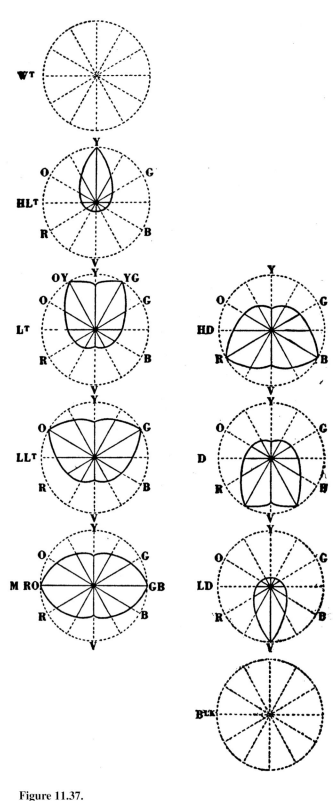

Figure 11.37.
Schematic views of horizontal cross sections of the Pope color solid at nine value levels. The outlines illustrate the regions covered by samples in the system (Pope 1922)

ERNST BARTHEL   1933

E. Barthel, *Die Farbenschraube, Vorschlag einer internationalen Farbennormung*, 1933

Alsatian **Ernst Barthel** (1890–1953) was a philosopher teaching at the University of Cologne until he lost his position in 1940 as a result of his pacifist views. He also opposed the developing mathematical-scientific world view and admired Goethe (see entry in chapter 3) (Barthel published three books related to Goethe). He believed colors to exist objectively in the outside world. In 1917, he authored the paper *In Sachen des absoluten Farbsystems* (In the matter of the absolute color system) containing basic ideas on which he expanded in 1933 in "Die Farbenschraube, Vorschlag einer internationalen Farbennormung" (Color screw, proposal for international color standardization).

Barthel was fully aware of the work of Robert Luther, Sigfried Rösch, and Erwin Schrödinger (see entries in chapter 6) but pursued his own idiosyncratic and ambitious system. He accepted the three-dimensional form of color space but, following Newton (see entry in chapter 6), used the musical idea of the octave to express, in his view, the true nature of color:

> With my color screw the three-dimensionality of color order is now completely expressed. But it is expressed in a more exact manner than in any other existing three-dimensional color order, due to the fact that the three dimensions are traversed on an octave-like spiral (spiral staircase). The spiral represents a skeleton in exact and conceptual analogy to a sequence of tones and makes possible a very exact logarithmic mathematisation of color order. My proposal of a color screw does not negate the three-dimensionality of sequence but raises it onto a plane of accuracy on which optics and acoustics meet. . . . One can consider the color cylinder around which the color spiral is wound to be filled continuously with colors and obtain thereby a color solid without exact intervals. . . . The spiral line does not deny the color solid but completes it with an important mathematical nerve, if I may be allowed this expression (Barthel 1933, pp. 215–216, trans. by R.G.K.; see figures 11.38[3] and 11.39)

The basis of Barthel's concept is a 12-grade hue circle that corresponds to the 12 chromatic tones of the piano octave and is arranged according to complementarity. The sequence is as follows: yellow (j), orange (o), brick red (z), rubine (r), purple (p), violet (v), deep violet (t), indigo blue (i), sky blue or cyan blue (b), sea green (m), grass green (g), spring green = yellow green (f). Barthel proposed prismatic yellow (j) as the basis for standardization. Each of the 12 hues can appear at different levels of density whereby the circle takes on spiral form. He defined density as follows:

> If one dissolves in a certain volume of water 1 g of an aniline dye and in another identical volume 2 g of the same dye a pair of colors results not different according

to purity or wavelength or different addition of white or black, but only in regard to empirical density. Therefore, there are identical colors of different density as there are identical tones at different octave levels. Every color of the color circle can be intensified or diluted. (p. 52)

Barthel defined five color attributes: color circle grade or color quality (hue), color density or height, saturation (defined as ranging from gray of equal height to the maximum expression of a color), lightness (as photometrically defined), and lastly, tone color (in analogy to music, distinguishing among light, transparent, and opaque color).

A physical model of Barthel's color screw was never commercially produced.

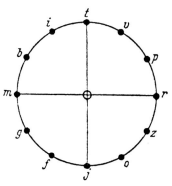

Figure 11.38.
Barthel's (1933) hue circle. See note 3.

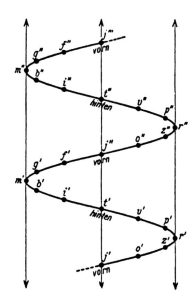

Figure 11.39.
Schematic representation of the color screw (Barthel 1933). Vorn means at the front; hinten means at the back.

E. B. RABKIN   1956

E. B. Rabkin, *Atlas cvetov*, 1956

E. B. Rabkin was a professor at the Institute of Ophthalmology in Moscow. He produced an atlas, published in 1956, to eliminate insufficiencies he perceived in previous Russian systems that lacked definition in terms of the CIE system (see CIE X, Y, Z Color Stimulus Space, chapter 6). Having established this connection, Rabkin saw his system as having the function of a simple colorimeter, which is of practical use in physiology, physiological optics, light and color technology, medicine, and biology. Aspects of the atlas were developed at the Institute of Sensory Neurology in Moscow.

The system behind *Atlas cvetov* (Color atlas) is based on the attributes hue, lightness, and saturation. The hue circle has nine basic colors, red, orange, yellow, green, blue I, blue II,[4] blue-violet, violet, and purple, with always four additional intermediate hues. That results in a total of 45 hues (figure 11.40), ordered according to visual equidistance. The implicit color solid is a tilted cone with the lightness axis running horizontally (figure 11.41, white on the left, black on the right).

For the atlas, the circle was reduced to 12 hues: red, red-orange, orange, orange-yellow, yellow, yellow-green, green, green-blue, blue, blue-violet, violet, and purple (figure 11.42). According to Rabkin, they suffice for most practical purposes. The individual atlas pages contain constant hue triangles with 45 color samples, as well as a 10-grade gray scale (figure 11.43). The full gray scale has 19 grades of which he prepared samples for only every second grade. Saturation increases in nine steps from the gray scale to the full color.

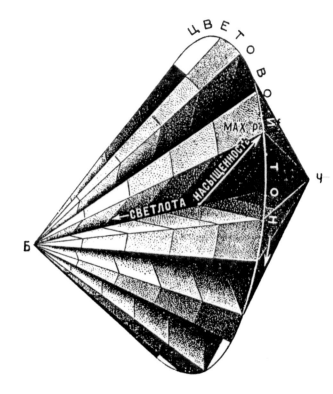

Figure 11.41.
Schematic representation of Rabkin's (1956) tilted color solid with horizontal lightness axis.

The lightness scale in the system is relative, changing according to hue so that the central grade 10 always coincides with the lightness of the full color. As a result, gray scales for constant hue triangle pages vary, even though the designations are identical (figure 11.44). According to Rabkin, the variable gray scale results in an increase in the number of light colors, important for practical applications. Another result is the asymmetric conceptual form of the color solid (see figure 11.41).

The atlas contains 660 printed color samples in 12 hues (figure 11.45). The samples have a diameter of 12 mm. For each sample, Rabkin listed colorimetric values x, y, Y of the 1931 CIE standard observer and illuminant C.

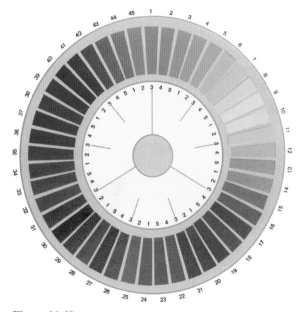

Figure 11.40.
Schematic hue circle with 45 grades (Rabkin 1956).
Image courtesy Friedrich Schmuck.

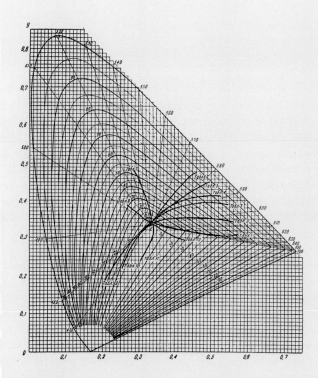

Figure 11.42.
Location of the 12 hues of the atlas in the CIE chromaticity diagram, shown as curved solid lines originating at the illuminant point (Rabkin 1956).

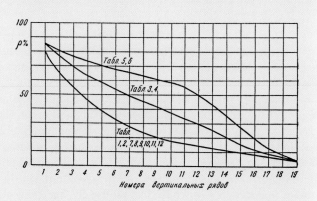

Figure 11.44.
System gray scale number versus luminous reflectance Y for the three different lightness scales. The top function applies to hues 5 and 6 (yellow and yellow-green), the middle one to hues 3 and 4 (orange and orange-yellow), and the bottom function to the 8 remaining hues (Rabkin 1956).

Figure 11.43.
Structure of the constant hue triangles with designation of saturation (before slash) and lightness (after slash) (Rabkin 1956).

Figure 11.45.
Constant hue page 5 for yellow from the atlas (Rabkin 1956). Image courtesy Friedrich Schmuck.

ECKART HEIMENDAHL   1961

E. Heimendahl, *Licht und Farbe*, 1961

In 1961, the doctoral thesis of German social philosopher **Eckart Heimendahl** was published in revised and enlarged form as a book. It was influenced by his teacher, influential art historian Wolfgang Schöne, who had published *Über das Licht in der Malerei* (On the role of light in painting, 1954). Schöne believed that the known color order systems did not adequately represent human perception and ideas about color phenomena.

Heimendahl's work is written in dense philosophical jargon, at times using uncommon meanings for common terms. He takes color perceptions to be "vivid data of states of energy," a position influenced by Goethe (see entry in chapter 3), whose color order also influenced Heimendahl's. But unlike Goethe, Heimendahl gave green a primary role in his system. He also studied in detail the work of Hering, Helmholtz, and Ostwald (see entries in chapters 5, 6, and 10) and their followers. Some of his conclusions bear reconsideration.

Heimendahl developed his system in logical progression from axioms that the reader must accept if they are to appreciate the work. He began with a psychological order in the sense of the actions of colors on the human mind as a result of tensions and polar contrasts. In this order, the three basic color experiences are purple, green, and gray, named the *Grundtriade* (basic triad). The triad is located on the vertical bisector of an equilateral triangle (figure 11.46) having purple, white, and black at the corners. The vertical axis is called chromatic; the axis represented by the basis line is called lightness.

Even though Heimendahl used simple geometric forms, he viewed his system as dynamic/rhythmical. He believed that a color perception can not be properly represented as a geometric point in a space. As a consequence, he turned most lines into curves, as shown in figure 11.47. Here we also find several other colors in the system. The left side of the figure contains positive, active colors, intensifying from yellow to purple; the right contains negative, passive colors declining from purple to blue. Green, located near the center, has a large power region, so it belongs to the basis triad, while purple represents culmination. The choice of purple and green as primary chromatic colors indicates a degree of relationship with Frieling's system (see entry in this chapter).

The color order of figure 11.47 is reflectively duplicated, resulting in a rhomboid (figure 11.48). The lower half represents color reconciliation and is dominated by brown. According to Heimendahl, the rhomboid can be used for measuring degree of lightness or "chromaticness," determining complementarity (as illustrated in figure 11.49), and identifying the power regions and mixture ratios of colors.

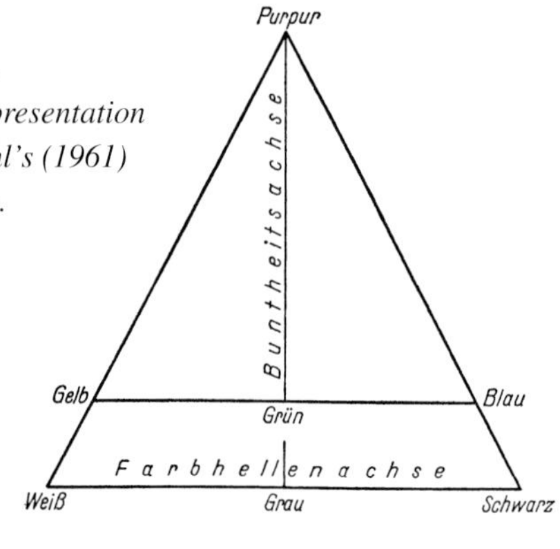

Figure 11.46.

Schematic representation of Heimendahl's (1961) color triangle.

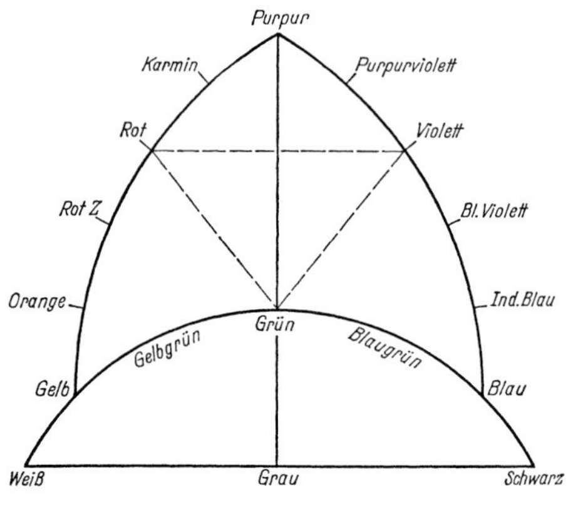

Figure 11.48.

Illustration of the color rhomboid derived from figure 11.47 (Heimendahl 1961).

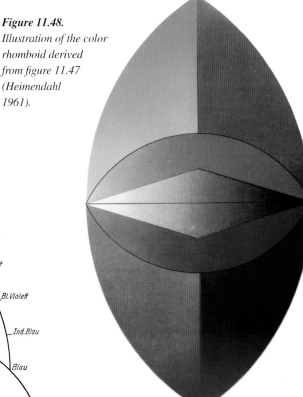

Figure 11.47.

Schematic representation of the curved-line version of the triangle of figure 11.46 (Heimendahl 1961).

Because colors are phenomenal images of lights, Heimendahl also offered a *Licht-Farben-Kreislauf* (light-color-circulation system) with complementary colors playing a key role. Figure 11.50 illustrates the separation of daylight into the complementary pair green and purple, both neither warm nor cool. The light, positive, and active colors are located on the upper semicircle; the dark, negative, and passive colors, on the lower one. The one-sided character of this representation is balanced by rotation and the introduction of a second system (figure 11.51).

According to Heimendahl, "[A]ll positive-negative tensions of colors in the oscillation system are now harmonized all the way to complementary reconciliation. The structural form of the oscillatory field is a symbol of the continuity of the color world and of the living harmony in the organically constructed color order representing the life forces of light" (Heimendahl 1961, p. 155, trans. by R.G.K.). To complete the picture, the oscillatory system is tripled with room for lighter and darker colors (figure 11.52), and the color reconciliation system centered on brown is schematically shown in greater detail (figure 11.53).

Heimendahl also attempted an "elementary-ontological color order as functional ordering of lights" and a "psychophysical light-color-function order," both considered outside the scope of this text.

Figure 11.51.
Illustration of the rotated and duplicated light-color oscillation system of figure 11.50 (Heimendahl 1961).

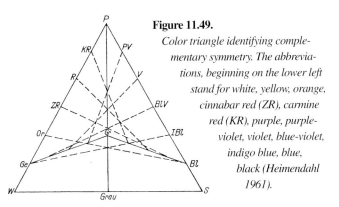

Figure 11.49.
Color triangle identifying complementary symmetry. The abbreviations, beginning on the lower left stand for white, yellow, orange, cinnabar red (ZR), carmine red (KR), purple, purple-violet, violet, blue-violet, indigo blue, blue, black (Heimendahl 1961).

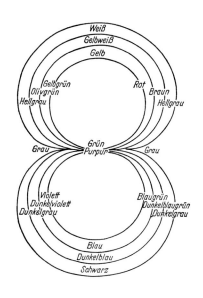

Figure 11.52.
Schematic representation of the three-step color power field (Heimendahl 1961).

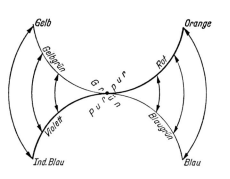

Figure 11.50.
Schematic representation of the basis position of the light-color oscillation system (Heimendahl 1961).

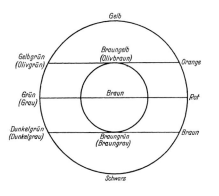

Figure 11.53.
Schematic representation of the color reconciliation circle, with brown in its center (Heimendahl 1961).

HEINRICH FRIELING   1968

H. Frieling, *Die Sprache der Farben*, 1939
Gesetz der Farbe, 1968

German **Heinrich Frieling** (1910–1996) was a color psychologist and founder and director of the Institute for Applied Color Psychology in Marquardtstein, Germany. Goethe and Ostwald primarily influenced his thinking on color. In *Die Sprache der Farbe* (Language of colors, 1939), Frieling derived a qualitative color system from darkness and light in the form of a pentagon. Its primary purpose was to represent dynamic color relationships. In the 1968 book *Gesetz der Farbe* (Law of color), the pentagon is called *Color Activ Fünfeck* (color active pentagon) and expanded to three dimensions in the forms of a cylinder and a cone. Frieling did not color the pentagon or its expanded cylindrical version, but he recommended specific pigments for this purpose.

Frieling considered the fundamental color perceptions derived from darkness and light to be white, green, gray, purple, and black. Their relationship and the approximate positions of yellow and blue were illustrated graphically (figure 11.54). Circles representing the five primary colors partially overlap in a lightness scale. In figure 11.54, yellow, green, blue, and purple are located on a square. He considered yellow, red, and blue to be dynamic colors while black,

white, purple, and green to be comparatively static. Bringing dynamic and static colors into their correct relationship requires a reconstruction of the square as a pentagon.

Yellow and blue of the square are redefined as slightly greener, red as slightly yellower, and purple is replaced by violet located halfway between red and blue (figure 11.55). The result is the balanced pentagon (figure 11.56) with middle gray in the center and the gray axis shown as a dashed line, indicating yellow-green as the lightest and violet as the darkest color. This line passes at an angle through the plane of the pentagon. According to Frieling, the five primary colors are derived from the spectrum, producing uniform sensations. He argued against Hering (see entry in chapter 5), claiming that if green is a primary sensation, violet also must be one. The pentagonal arrangement of the primary colors is seen to represent their harmonic relationships.

In his book of 1968, Frieling presented the color active pentagon (figure 11.57), a further development of figure 11.56 and basis for a three-dimensional color solid. Its cylindrical/conical form makes possible the placement of

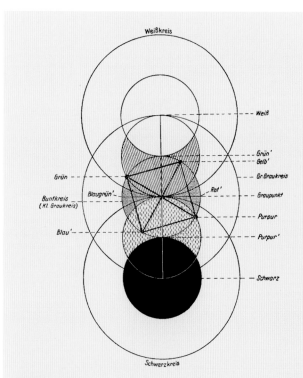

Figure 11.54.

Derivation of fundamental color perceptions white, green, gray, purple, and black from darkness and light (Frieling 1939).

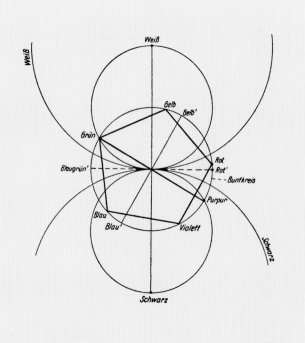

Figure 11.55.

Derivation of the basic color pentagram (Frieling 1939).

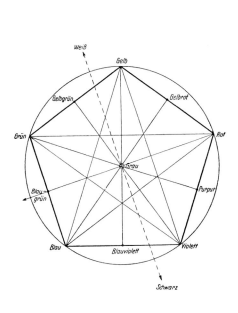

Figure 11.56.
The balanced pentagon with middle gray in the center (Frieling 1939).

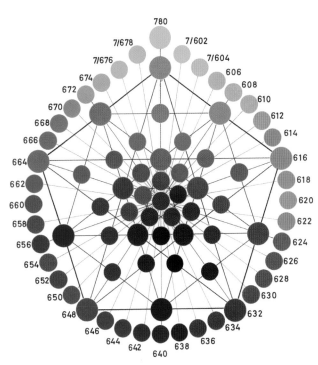

Figure 11.57.
Color illustration of the Color Activ Fünfeck *(Frieling 1968).*

many tonal colors missing in the pentagon (figure 11.58). Here he stressed the exact compensatory nature of the pentagon. The cylinder has 10 lightness grades with grade 0 represented by a dark gray of 6% reflectance and grade 10 with 80% reflectance. Below grade 0 is black and above 10 is white. The six saturation levels were achieved by appropriate mixture of compensatory colors.

The cone of the "fundamental pentagon" is located within the cylinder with the pentagon on level 6 (except for pure yellow located on level 7). In the cone, the fundamental colors of the pentagon are mixed according to logarithmic laws with white and black.

Frieling also introduced a color-naming scheme in which, for example, full red (red neither containing black nor white) has the designation 6/616. That means that it is located on the sixth level of the cone, on the surface of the cylinder (maximum purity 6, hue 16). A designation of 10/416 indicates a red of hue 16 and purity 4, but whitened to be located on gray level 10.

Frieling, well aware of the work of Helmholtz, Hering, and others, disregarded their findings in favor of a more humanities-oriented interpretation of color order, but with a degree of specification with reflectance values.

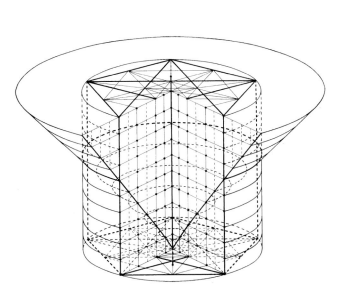

Figure 11.58.
The cylinder surrounding the hexagonal prism converted to an inverted cone (Frieling 1968).

LOTHAR GERICKE and KLAUS SCHÖNE   1969

L. Gericke, R. Schumitz, O. Richter, and K. Schöne,
Farbenkatalog für die Gestaltung, 1969
L. Gericke & K. Schöne, *Das Phänomen Farbe*, 1970

In the late 1960s, Germans **Lothar Gericke** and **Klaus Schöne**, working in the (then) German Democratic Republic, developed a color atlas to help designers. They intended their so-called *Planetensystem* (planetary system) of sampling color space to let designers make rapid, targeted, and considered decisions concerning color combinations. They believed that conventional color order systems based on three attributes were not adequate for this task. In their system, a given color can be varied in multiple, not only three, directions radiating from a given central color, which the inventors believed essential for successful color design.

In the planetary system, each color, so to speak, can take the place of a sun around which other colors are arranged in "planetary" fashion (figure 11.59). The planetary colors are mixed from the standard colorants of the system in a visually harmonious manner. The basis of the atlas's organization is Phillip Otto Runge's color sphere (see entry in chapter 4), interpreted as having a 24-grade hue circle (figure 11.60).

Another influence was Ostwald (see entry in chapter 10), from whom they borrowed the 21-grade gray scale.

Although systematic sampling of object colors according to Gericke and Schöne is possible, the *Farbenkatalog für die Gestaltung* (Color atlas for design, 1969) contains only limited sampling. Each atlas page contains 20 painted samples of size 40 × 40 mm (figure 11.61), representing half the samples of a planetary system (e.g., the top half of the samples in figure 11.62 left; the samples of the bottom half are located on the next atlas page).

The Gericke and Schöne atlas consists of 27 planetary systems with a total of 540 colors, including the samples in a supplementary atlas of 1978. Each sample is defined in terms of the CIE colorimetric system (see CIE X, Y, Z Color Stimulus Space, chapter 6), and the area occupied by the samples on an atlas page is shown in the corresponding chromaticity diagram (figure 11.63).

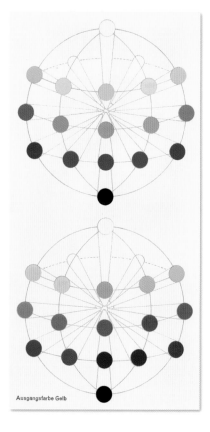

Figure 11.59.
Two views of the "color planet" of primary yellow (Gericke and Schöne 1973).

Figure 11.60.
The 24-grade hue circle of the Gericke and Schöne (1973) system with the colors of eight selected color planets projected onto a plane.

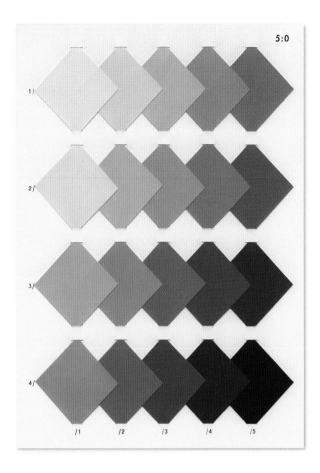

Figure 11.61.
Atlas page from the Farbenatlas für Gestaltung *(Gericke and Schöne 1973).*

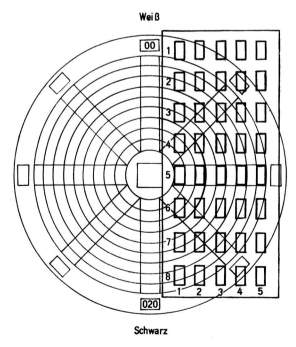

Figure 11.62.
Schematic view of the projection of the color of a color planet onto a plane and the translation of the colors onto the atlas pages (Gericke and Schöne 1973).

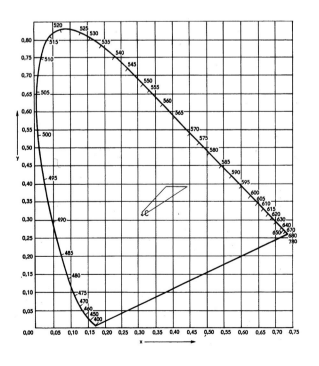

Figure 11.63.
CIE chromaticity diagram with the area encompassed by the samples of figure 11.59 Gericke and Schöne 1973).

HARALD KÜPPERS   1971

H. Küppers, *Das Rhomboeder-System*, 1971
Farbe – Ursprung, Systematik, Anwendung, 1972
Der DuMont Farbenatlas, 1978
Der Große Küppers-Farbenatlas, 1987

German printing technologist **Harald Küppers** (1928-) created a rhombohedric color order system based on the printing primaries yellow, cyan, and magenta, or their perceptual ideals. He commented, "[I]t was not my ambition to invent an additional color system" and mentioned "certain shortcomings in all known systems in regard to clarity, uniformity [in the sense of being isotropic], and logic" (Küppers 1971, p. 103, trans. by R.G.K.) as the impetus for his effort.

Küppers believes that "spectral colors" can not be fundamental color stimuli in additive color mixture. He believes that fundamental colors must have a broadband spectral signature to be properly distinguished in the eye's retina. Fundamental colors must occupy a spectral range comparable to those of the three cone types: "Each one of these must generate in a given cone type the maximal amount of decomposition product and in the other two types minimal amounts" (p. 105). Thus, he divides the spectral range into the three sectors (far from resulting in equal cone absorptions): 380-480 nm, 480-580 nm, and 580-680 nm.

Küppers says that he deliberately did not consider brightness/lightness sensations in his system. As a result, his system cannot be perceptually uniform. But, he says, it "is nearly uniform . . . much improved compared to the systems of Ostwald and Hickethier" (p. 106). Küppers's rhombohedron is a parallelepiped, as described in 1935 by Neugebauer (see entry in chapter 6). It is stretched to extend the distance between white and black (figures 11.64 and 11.65). The 90° angles in a cube are changed to 60° and 120° for reasons Küppers has explained only by saying that it improved uniform scaling of the space. In his view, his system is much more comprehensible and systematic than the systems of CIE, DIN 6164, or the Munsell Renotations (see entries in chapters 6 and 7).

Küppers takes the rhombohedron to be the basis of both the additive and the subtractive color-mixture systems, in what he calls "integrated mixture." The rhombohedron can be divided into "three clearly separated sections," each representing colors with a given idealized absorption range (figure 11.66). Constant hue planes are located on scalene triangular planes sharing the achromatic axis. Opposing planes represent complementary colors (figure 11.67). All colors on horizontal sections of the rhombohedron have identical sums of basic color components. Colors of constant primary color content are located on slanted, parallel rhombic planes (figure 11.68).

Küppers says he finds geometric definition of *Buntheit* (chromatic color content) to be impossible in his system, and instead defines its opposite *Unbuntheit* (achromatic content, figure 11.69). He identified colors with six digit numbers

(expanded from Alfred Hickethier's three digits; see entry in chapter 9), which allows identification of one million different colors. In *Das Grundgesetz der Farbenlehre* (The fundamental law of color theory, 1978a), Küppers acknowledged the value of a tilted cube: Its sections can be quadratic and thereby easier to use for color tables. Unlike Neugebauer's system, Küppers's has not been expressed in terms of cone absorption data or the CIE colorimetric system.

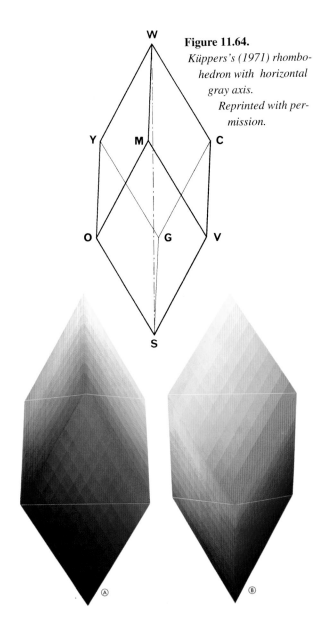

Figure 11.64.
Küppers's (1971) rhombohedron with horizontal gray axis.
Reprinted with permission.

Figure 11.65.
Colored illustration of two views of the surface of the color rhombohedron (Küppers 1977). Reprinted with permission.

In 1978, Küppers published a Hickethier-type cubic atlas (unrelated to the rhombohedric system; figure 11.70). Printed in the halftone process, it illustrates more than 5,500 colors with their numerical identification. In 1987, he published an atlas with 25,000 samples based on eight inks (figure 11.71), long before that of Michael and Pat Rogondino (see entry in chapter 9).

His rhombohedron and associated claims have resulted in considerable criticism (see, e.g., Scheidt et al. 1974). The rhombohedron is neither complete nor isotropic, and its derivation from broadband spectral functions involves misunderstandings about cone-sensitivity functions.

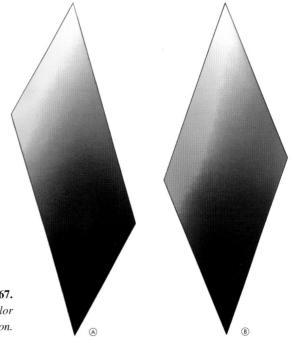

Figure 11.67.
Colored illustration of two cross sections through the color rhombohedron (Küppers 1977). Reprinted with permission.

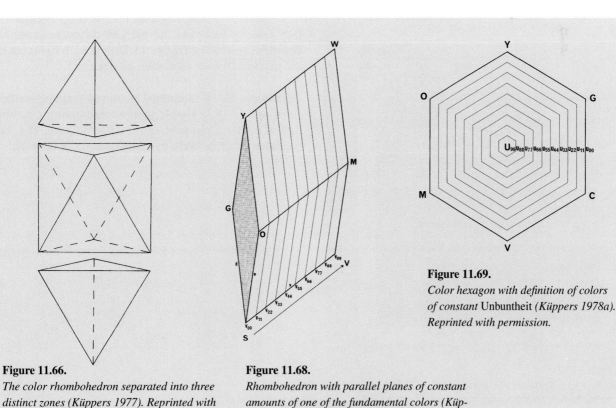

Figure 11.66.
The color rhombohedron separated into three distinct zones (Küppers 1977). Reprinted with permission.

Figure 11.68.
Rhombohedron with parallel planes of constant amounts of one of the fundamental colors (Küppers 1978a). Reprinted with permission.

Figure 11.69.
Color hexagon with definition of colors of constant Unbuntheit (Küppers 1978a). Reprinted with permission.

F. Gerritsen, *Farbe: optische Erscheinung, physikalisches Phänomen und künstlerische Ausdrucksweise*, 1972

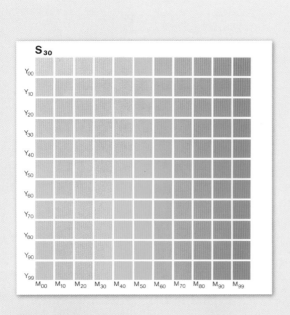

Figure 11.70.

A page of the printing ink atlas of 1978 (Küppers 1978b). Reprinted with permission.

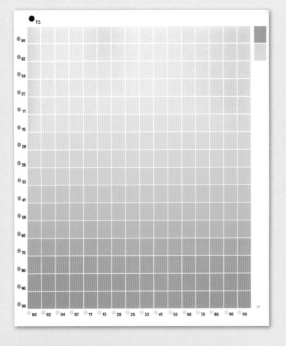

Figure 11.71.

A page of the printing ink atlas of 1987 (Küppers 1987). Reprinted with permission.

In 1974, the Dutch color theorist **Frans Gerritsen** published a color system that he claimed to be based on laws of perception. Like Küppers at about the same time (see entry in this chapter), Gerritsen separated the spectrum in three sectors, approximately aligned with the spectral spread of the CIE color-matching functions (see entry CIE X, Y, Z Color Stimulus Space, chapter 6). To each sector, he assigned a color perception: violet-blue, yellow-green, and scarlet. He then schematized reflectance curves of saturated object colors based on the three sectors. The perceived color is the additive mixture of the reflection of one, two, or all three sectors (figure 11.72).

Using this methodology, Gerritsen created an 18-hue color circle derived from additive mixture of the three primaries, with the printing primaries yellow, cyan, and magenta appearing as the initial secondary colors in the central hexagon (figure 11.73). The lightness dimension is represented with a 10-grade scale. He developed a color solid in the form of a deformed double cone by giving all saturated colors the same degree of saturation but assigning their location according to their lightness level.

The classical central circular color chart (e.g., the equatorial section of Runge's 1810 *Farben-Kugel*; see entry in chapter 4) is deformed to take account of lightness, thus making lightness the vertical dimension applicable throughout the solid (figure 11.74). At each lightness level are maximally four saturation grades.

Gerritsen's system attempted to convey the complexities of color order with a simple model representing a combination of additive and subtractive color mixture. The model is neither quantitatively psychophysical nor isotropic.

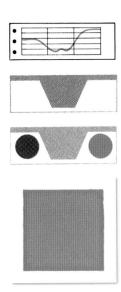

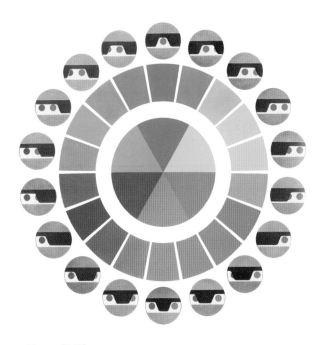

Figure 11.72.
Example of the reflectance function of a magenta pigment, its schematized form with three sectors, the two additive primaries attached to the reflecting sectors, and the resulting object color perception (Gerritsen 1975).

Figure 11.73.
Color circle with 18 hue grades derived from the additive primaries at 4, 8, and 12 o'clock positions in the center. Symbols on the periphery convey the conceptual reflectance curve and the mixture components for each grade (Gerritsen 1975).

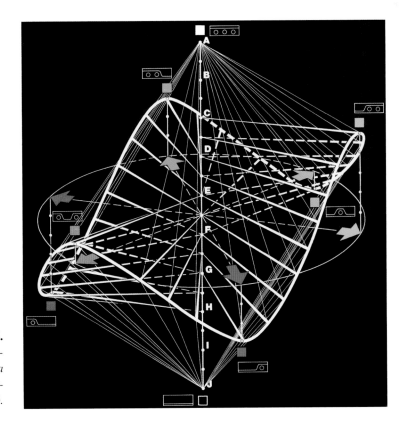

Figure 11.74.
Color solid of the Gerritsen system in the form of a distorted double cone (Gerritsen 1975).

MICHEL ALBERT-VANEL   1983

M. Albert-Vanel, *Le planetes-couleur-systeme*, 1983

The French color theorist, artist, and educator **Michel Albert-Vanel** (1935–) is aware of the many color systems. He says that he created a new one because the very existence of so many systems shows that the problem is far from resolved. The image of a galaxy imposed itself on his mind: "[A]round the original colored image in the center gravitate some principal planets with their own variants or satellites" (Albert-Vanel, 1983, p. 3, trans. by R.G.K.; see figure 11.75). According to him,

> The Planets-Color-System is a new approach to color. It concerns a space of colored groups, in which the color is no longer looked as individually [*sic*], but in relation to other colors, it then becomes possible to join together theory and practice and to offer the colorists a mean of getting closer to visual perception (Albert-Vanel 1990, p. 23).

Albert-Vanel sees the following as the four grand axes of this system: polychromy, based on yellowness, redness, blueness and greenness; tonality, for example, the difference between red and green, conventionally illustrated by the chromatic circle; saturation, the distance separating lively colors from the neutrals white, gray, or black; and lightness (*clarté*), separating light from dark colors.

All optical mixtures can be obtained by combining six colors: the pairs yellow-blue, green-red, and white-black (figure 11.76). Here, Albert-Vanel acknowledges that his model conforms to that of Hering (see entry in chapter 5). Every color can be described as more or less polychrome or monochrome, more or less warm or cold, more or less colored or de-colored, and more or less light or dark. In the "Hue-King" (figure 11.77), Albert-Vanel illustrates 63 color combinations in four zones.

Following Renaissance writers, Albert-Vanel also establishes concordances between colors and other properties; for example, greenness corresponds to renewal, natural, and cadaverous. Finally, Albert-Vanel describes successive abstractions that take place when moving from reality to isolated color: reality, space–time; three-dimensionality, volume–matter; figurative, realism–narration; abstract, forms; ensembles, associated colors; colorimetry, isolated colors.

Albert-Vanel (1990) wrote that the most complete color system "would have to incorporate not only scales showing hue, value, and saturation . . . , but also scales showing the oppositions of pigment and light, opacity and transparency, and matte and glossy surfaces" (p. 287). To illustrate these would require several diagrams.

Figure 11.75.
Planetary system (Albert-Vanel 1983). Reprinted with permission.

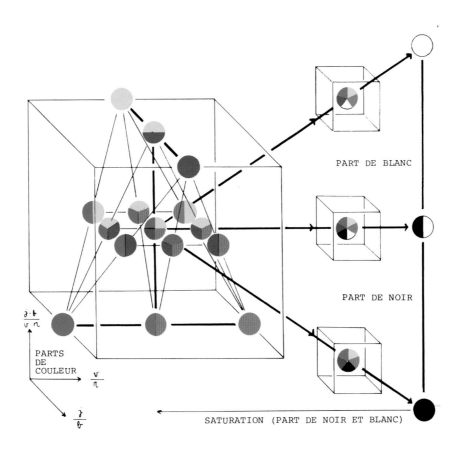

Figure 11.76.
Color mixture scheme (Albert-Vanel 1983). Reprinted with permission.

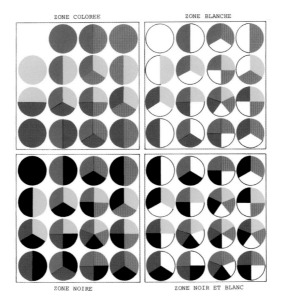

Figure 11.77.
"Hue-King." Colored zone: primary chromatic colors and binary, ternary and quaternary combinations. White zone: combinations of white with one to four primary chromatic colors. Black zone: combinations of black with one to four primary chromatic colors. Black and white zone: combinations of black, white and the four primary chromatic colors (Albert-Vanel 1983). Reprinted with permission.

CHRISTIAN RICHARDIÈRE   1987

C. Richardière, *Harmonies des couleurs*, 1987

Christian Richardière is a French color-harmony consultant. He names Lambert (see entry in chapter 4) and Wilhelm von Bezold and Wilhelm Wundt (see entries in chapter 10) as having influenced his thinking on color order. But his key influence is Hickethier (see entry in chapter 9). Richardière combines the simplicity of Hickethier's color-naming method with a form of color solid he finds more suitable than a cube for identifying harmonious and disharmonious color combinations. As a result, he translates the 1,000 colors of the Hickethier cube into the form of a cone. As in Hickethier's cube, the colors are illustrated using the halftone-printing process.

Richardière uses the modern printing primaries yellow, magenta, and cyan as his basic colorants. The zigzag line formed by the hue circuit on a tilted cube is flattened and expanded to a circle, with black at the center. White forms the top of the cone, and the tint colors form the cone mantle. A gray scale with 10 grades from white to black forms the vertical central cone axis. Black is located in the center of the basis circle. On the periphery of the basis plane are 54 hue steps, reduced to six on the innermost circle. Each color sample is named by a three-number code where the first number identifies the amount of yellow, the second number, the amount of

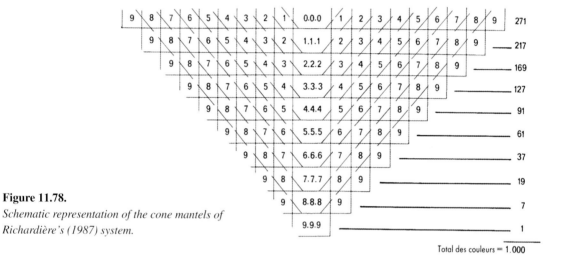

Figure 11.78.

Schematic representation of the cone mantels of Richardière's (1987) system.

Total des couleurs = 1.000

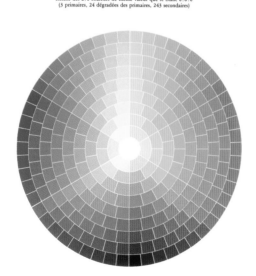

Rosace des 270 couleurs de même valeur que le blanc 0.0.0
(3 primaires, 24 dégradées des primaires, 243 secondaires)

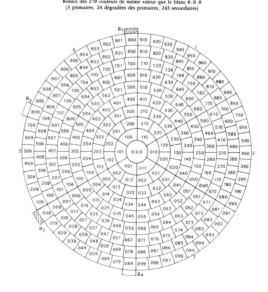

Rosace des 270 couleurs de même valeur que le blanc 0.0.0
(3 primaires, 24 dégradées des primaires, 243 secondaires)

Figure 11.79.

Top cone mantel with the most saturated colors at each lightness, Richardière's figure 4. On the right are the color field designations (Richardière 1987).

magenta, and the third number, that of cyan in all samples. Achromatic samples have three identical numbers.

The 1,000 samples are arranged in two ways. The first arrangement (Richardière's plates 4 and 15–23) shows the samples as a series of nine overlapping cone mantels (figure 11.78), with the first mantel showing the upper cone surface as projected on a plane (figure 11.79). The other eight mantels are concentric but smaller, as sketched in figure 11.78. Plate 23 has black only. The second arrangement (plates 23 bis to 32) shows the same samples as horizontal slices through the

cone (figures 11.80 and 11.81). Each chart has an associated transparent map identifying the color samples.

Richardière offered few technical details on the printing process. He does say that portions of the colors of plate 4 in the first series have been uniformly over-printed with the gray visible in the surround. His cone is limited to colors achievable with the pigments he used and is obviously not isotropic.

						0.0.0							
					1	1.1.1	1						
				2	2	2.2.2	2	2					
			3	3	3	3.3.3	3	3	3				
		4	4	4	4	4.4.4	4	4	4	4			
	5	5	5	5	5	5.5.5	5	5	5	5	5		
6	6	6	6	6	6	6.6.6	6	6	6	6	6	6	
7	7	7	7	7	7	7.7.7	7	7	7	7	7	7	7
8	8	8	8	8	8	8.8.8	8	8	8	8	8	8	8
9	9	9	9	9	9	9.9.9	9	9	9	9	9	9	9

Figure 11.80.
Schematic representation of the horizontal slices through the cone (Richardière 1987).

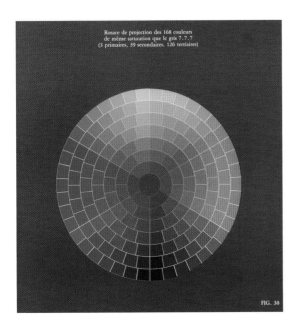

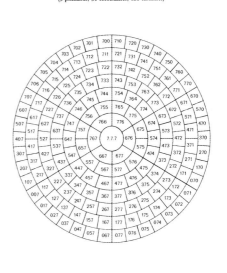

Figure 11.81.
Gray level 7 (Richardière's figure 30) of the horizontal sections shown against a gray surround of the same lightness with color designations on the right (Richardière 1987).

GERRIET HELLWIG   1990

G. Hellwig, *Zur Ordnung der Farben*, 1990
Color standard arrangement, US Patent 5,026 286, 1991

In 1990, German software developer **Gerriet Hellwig** (1951-2006) offered an unusual qualitative conceptual proposal for a perceptually uniform (isotropic) color order system. Surprisingly, Hellwig concluded that conventional geometries of color order systems fail because they have more perceptually uniform hue steps of unit magnitude at low saturation levels than at high. And a spherical system has fewer geometric unit distances near the center than near the periphery of the sphere.

As a result, Hellwig proposed to replace the sphere with a torus (donut-shaped) as an appropriate geometric model with pure colors along the inner circular contour and desaturated colors along the outer. Light colors are placed higher in the torus than are dark colors (figure 11.82). The full color yellow is placed high, and blue low, with red and green near the center, requiring a stretching of the torus (figure 11.83).

However, empirical findings complicate matters: In light regions, the attribute lightness is differentiated least; in dark regions, it is differentiated in first or second place. As a result, the inner walls of the stretched torus must be conical, instead of parallel. The solid takes on funnel shape with the funnel ending in black at the bottom (figure 11.84).

Hellwig thought that in the absence of a pure, ideal black, the black point theoretically can be located infinitely distant from the blue point. The funnel-shaped opening of the torus shows that colors of varying purity are located on a given circle in the torus. Arguing that there is no light bluish violet color, he sees the need to further deform the interior funnel shape. As a result, the funnel makes a complete turn during its progress, as sketched in figure 11.85. In 1991 Hellwig received a US patent for his invention.

Hellwig used arguments involving video displays as proof for the need to invert conventional color order systems. In his experience, an incremental step in an (undefined) graphics program results in several just noticeable differences (JNDs) if the reference color is desaturated, while several steps are required to produce a single JND if the reference color is pure. Some of his arguments involve a misunderstanding of the relationship between changes in color stimulus and resulting perceptions. The basic argument is contrary to the empirical data of psychological scaling.

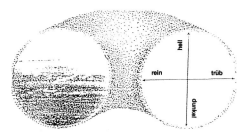

Figure 11.82.
Cross section through Hellwig's (1990) color torus. Light colors are located in the top; dark ones, in the bottom half. Bright colors are located near the inside; dull colors, near the outside.

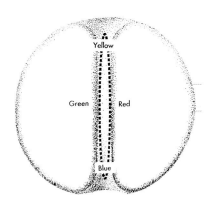

Figure 11.83.
Location of the full-color hue circle (in form of an elongated ellipse) on the inner surface of the torus (Hellwig 1990).

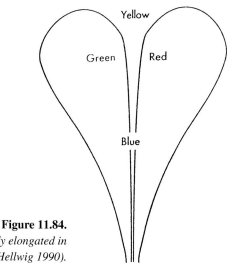

Figure 11.84.
To accommodate color order, the torus, already elongated in figure 11.83, needs to be elongated further (Hellwig 1990).

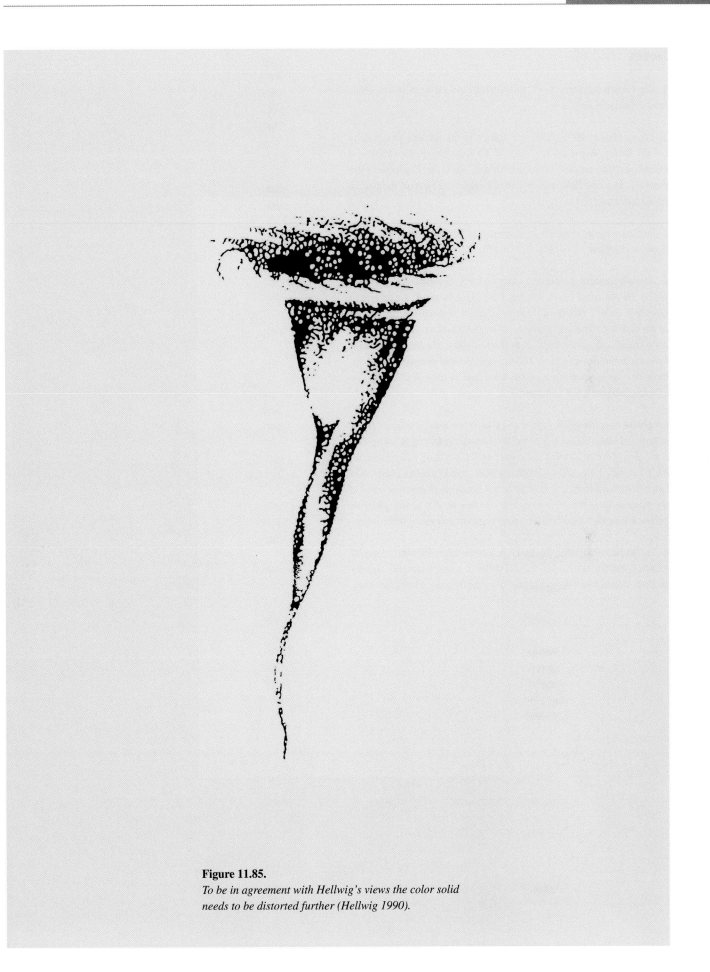

Figure 11.85.
*To be in agreement with Hellwig's views the color solid
needs to be distorted further (Hellwig 1990).*

Notes

1. The French bibliographer Quérard attributed the work to one Guillaume-Germain Guyot.

2. Young-Helmholtz trichromatic theory is the modern designation for the theory, based on the work of Thomas Young, Maxwell, and Helmholtz (see entries in chapter 6), that our color experiences are derivable from the light absorbed in the three cone types of the human retina (see chapter 1).

3. Barthel's figure contains an extensive explanatory note, translated here into English:

1. Complementary colors are diametrically opposed. 2. The four colors on the axial cross, j r and t m, are generated, according to Goethe's correct statement, along the right, respectively left border of the beam. Consult my corresponding publications. 3. Yellow is the specifically lightest, dark violet the darkest color. The tension quotient between lightness and darkness (numerator and denominator principle in color) passes through the hue circle in a cyclical process as already Schopenhauer indicated.

j = yellow or sodium line [of the spectrum]; o = orange; z = tile red; r = ruby red (spectral end); p = purple (complementary of grass green); v = color of violet blossoms or purple-violet; t = dark violet (spectral end); i = indigo blue; b = sky blue; m = sea green (spectral elementary color, complementary to rubine red); g = grass green (median color of the positive spectrum (complementary to purple; f = spring green or yellow green; O = place of passage of the gray axis through the plane.

4. The Russian language has two basic color terms for blue, *sinij* and *goluboy*, the former having a focal color of (comparatively) slightly reddish, darker blue, the latter one of a slightly greenish, lighter blue.

References

Albert-Vanel, M. 1983. *Le Planetes-Couleur-Systeme*, Paris: Exposition ENSAD.

Albert-Vanel, M. 1990. Systems: planetary color, in *Color Compendium*, Hope, A., Walch, M., eds., New York: Van Nostrand Reinhold.

Barthel, E. 1933. Die Farbenschraube. Vorschlag einer internationalen Farbennormung, *Zeitschrift für Sinnesphysiologie* 63:215–227.

Carpenter, J. M., and **H. T. Fisher**. 1974. *Color in art, a tribute to Arthur Pope*, exhibition catalogue, Cambridge, MA: Fogg Art Museum.

Forbes, J. D. 1849. Hints toward a classification of colours, *The London, Edinburgh and Dublin Philosophical Magazine and Journal of Science* 34:161–178.

Frieling, H. 1939. *Die Sprache der Farben*, München: Oldenbourg.

Frieling, H. 1968. *Gesetz der Farbe*, Göttingen: Musterschmidt.

Gericke, L., and **K. Schöne**. 1970. *Das Phänomen Farbe*. Berlin: Henschel; 2nd ed., 1973.

Gericke, L., **R. Schumitz, O. Richter**, and **K. Schöne**. 1969. *Farbenkatalog für Gestaltung*, Berlin: Zentralinstitut für Gestaltung; Ergänzungsteil, 1978.

Gerritsen, F. 1975. *Farbe: Optische Erscheinung, physikalisches Phänomen und künstlerische Ausdrucksweise* (Theory and practice of color), English ed., Ravensburg, Germany: Maier; originally published 1972.

[Guyot, E.-G.]. 1770. *Nouvelles récréations physiques et mathematiques*, Paris: Gueffier.

Heimendahl, E. 1961. *Licht und Farbe: Ordnung und Funktion der Farbwelt*, Berlin: Walter de Gruyter.

Hellwig, G. 1990. Zur Ordnung der Farben, *Farbe und Design* 49/50:57–62.

Hellwig, G. 1991. *Color standard arrangement*, US patent 5,026,286.

Hensel, C. ca. 1915. *Farben – Farbensehen*, Stuttgart: Franck'sche Verlagshandlung.

Klincksieck, P., and **T. Valette**. 1908. *Code des couleurs, à l'usage des naturalists, artistes, commerçants et industriels*, Paris: Klincksieck.

Küppers, H. 1971. Das Rhomboeder-System, *Die Farbe* 20:103–113.

Küppers, H. 1977. *Farbe – Ursprung, Systematik, Anwendung*, München: Callwey, 1st ed. 1972,.

Küppers, H. 1978a. *Das Grundgesetz der Farbenlehre*, Köln: Du Mont.

Küppers, H. 1978b. *Der Du Mont Farbenatlas*, Köln: Du Mont.

Küppers, H. 1987. *Der Große Küppers-Farbenatlas*, München: Callwey.

Oberthür, R., and **H. Dauthenay**. 1905. *Répertoire de couleurs pour aider à la détermination des couleurs des fleurs, des feuillages et des fruits*, Paris: Librairie Horticole.

Paclt, J. 1958. *Farbenbestimmung in der Biologie*, Jena, Germany: Fischer.

Pope, A. 1922. *Tone relations in painting*, Cambridge, MA: Harvard University Press.

Pope, A. 1929. *An introduction to the language of drawing and painting*, Cambridge, MA: Harvard University Press.

Rabkin, E. B. 1956. *Atlas cvetov*, Moscow: Megdiz.

Richardière, C. 1987. *Harmonies des couleurs*, Paris: Dessin et Tolra.

Ross, D. W. 1907. *A theory of pure design*, Boston: Houghton, Mifflin.

Scheffler, H. 1883. *Die Theorie des Lichtes*, Leipzig: Foerster.

Scheidt, P., **R. Boss**, and **K. H. Schirmer**. 1974. Kritische Betrachtungen zu Harald Küppers' Buch "Farbe," *Die Farbe* 23:339–345.

Schöne, W. 1954. *Über das Licht in der Malerei*, Berlin: Mann.

Séguy, E. 1936. *Code universel des couleurs*, Paris: Lechevalier.

Setarti AG. 1947. *Farbentafel*, Swiss Patent Nr. 243891, Bern: Swiss Patent Office.

Sol, J.-M.-C. 1849. *La palette théorique ou classification des couleurs*, Vannes, France: Lamarzelle.

Wilson, R. F. 1938. *Horticultural colour chart*, vol. 1, Banbury, England: Stone.

Wilson, R. F. 1957. *The practical Wilson colour system*, Göttingen: Musterschmidt.

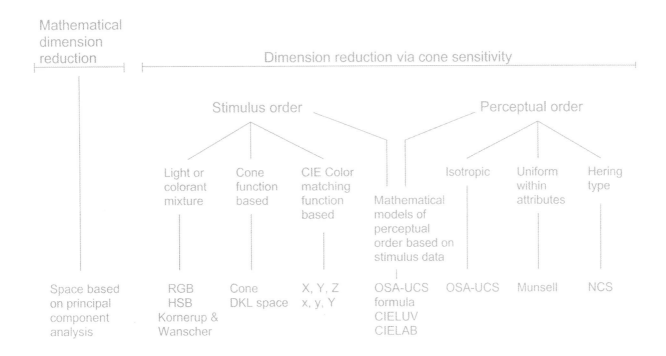

| Mathematical dimension reduction | Dimension reduction via cone sensitivity | | | | | | | |

Stimulus order   Perceptual order

Light or colorant mixture   Cone function based   CIE Color matching function based   Isotropic   Uniform within attributes   Hering type

Mathematical models of perceptual order based on stimulus data

Space based on principal component analysis   RGB HSB Kornerup & Wanscher   Cone DKL space   X, Y, Z x, y, Y   OSA-UCS formula CIELUV CIELAB   OSA-UCS   Munsell   NCS

CHAPTER 12

COLOR ORDER SYSTEMS
Categorization, Color Mixture, and Perceptual Experience

A synthesis of conclusions drawn from preceding chapters. Color categorization and order are shown as related. A classification of ordering systems is offered.

Color order can be seen as a categorization effort. Given the millions of distinguishable color experiences, one cannot practically communicate about colors without grouping certain of its experiences with others like them. Categorization is a key cognitive process. As Harnad (1996) pointed out, a large majority of words in a dictionary can be seen as the result of efforts to categorize related mental concepts. This also applies to color words. The cognitive process of categorization seems to require three tasks: (1) generate the concept of a category from related perceptual experiences (e.g., "red"), (2) determine the range of experiences included in the category, and (3) name the category. All three tasks are fluid processes, and the second and third tasks can overlap in time.

Color categorization

Categorization is a productivity tool that arose during early development of language (at least 50,000 years ago). Placing the endless number of objects, experiences, and abstract ideas into named categories allowed simplified, more productive communication about them. Once a category is established, at least in one mind, and a term is created for it, such a word becomes a unit of cultural information (a meme in the terminology of Dawkins 1976). If the category represents a cognitive achievement of importance (e.g., easier object identification) or if it proves fashionable, it may sweep the world, or at least a cultural group.

It seems likely that some kinds of categorization are aided by neurobiological processes other than cognition's specific "categorizer" apparatus (very likely those related to the senses). Other kinds, including meta-categories (e.g., "news media" being used for newspaper, radio, television, internet news, and commentary) may be the result of activity of the categorizer alone. In that case, the results of categorization influence, not just are influenced by, the group's culture.

In 1969, American linguists Brent Berlin and Paul Kay published a controversial hypothesis stating that a group of 11 simple color terms is universal, and the terms' development in languages follows a standard pattern. (The terms are black, white, gray, red, yellow, green, blue, orange, purple, pink, and brown.) Despite their conjectures, comparatively little is known about how the color-categorization process developed.

There is little doubt that the current human color vision apparatus (cones, the associated comparator system[1], and additional mechanisms in the cortex that support color vision) were available to the genus *Homo* from its beginning. With appropriate stimuli, hominids were capable of all color experiences we can have. Some 30,000 years ago, Cro-Ma-

gnon cave painters likely had names for the dyes and pigments that may have come to refer to their colors.[2]

With the development of writing and the resulting records, speculation gives way to facts. However, color terminology is sparse in the earliest writings in the Middle East. In the first 11 sections of the Sumerian *Gilgamesh* epic, dated to about 2000 B.C., no words are consistently interpreted as color terms. Section 12, believed to have been added much later, contains one usage each of terms interpreted as having the meanings of red and purple. The Egyptian dictionary of the middle period (ca. 2200 B.C. to ca. 1300 B.C.) has terms interpreted as yellow, red, blue, green, white, and black (Faulkner 1981). Wider use of color terms is also apparent in the early Greek language, such as in the Homeric epics (mentioned in chapter 1).

Colored artworks in Egyptian temples and graves, Minoan palaces, Greek temples, and mosaics in Sumerian temples and palaces required colorant technology and efficient communication about the colorants and their effects. This makes it likely that some form of color naming and categorization was well established in the first millennium B.C., at least among artists, craftsmen, and engineers. At the same time, accidental or purposeful paint mixture began to disclose aspects of the underlying rules of color mixture: Some color experiences are obtainable by mixing certain colorants, whereas others require pure colorants, facts providing glimpses of an underlying order.

Greek pre-Socratic philosophers were among the first to put theories of the world and the universe in writing, some of them specifically mentioning colors. Their lists of simple colors reflect early categorization as well as ideas regarding the generation of colors. The importance of the number 4 to Pythagoras and his followers, as expressed in the *tetractys*, is reflected in their four fundamental elements and basic colors associated with them: white, black, yellow, and red.[3]

This short list was slightly varied and maintained by Empedocles in the fifth century B.C., and Democritus in the fourth century B.C.. The changes may have been related to color categorization. For example, for the color of the element earth, Empedocles used the word *ochron* (yellow-brown), while Democritus used the term *khloron* (interpreted as yellow-green).

Aristotle demonstrated his interest in categorization in writings later collected under the name *Libri logicorum*. He proposed a list of eight simple colors (including gray, which he classified with black) that can be considered to represent categories: white, black, gray, yellow, red, violet, green, and blue (see entry in chapter 2). Only orange is missing to complete the Berlin and Kay (1969) basic hue-related categories. Orange was first included in a printed list of simple colors in Girolamo Cardano's (see entry in chapter 2) 1550 reinterpretation of Aristotle's list. Aristotle's list was strongly influential for the next 2,000 years, perhaps in part because it matched many people's categorical knowledge.

A likely key factor of color terminology was colorant technology. As the archaeological record shows, beginning with the Cro-Magnon, colorant technology steadily advanced in the fields of pigments and dyes as well as ceramic glazes and colored glass. By the end of the Roman Empire, it had progressed to a sophisticated state. Many colorants were available from where they occurred naturally, their names often reflecting those places. Other colorants, such as lead white, minium, viridian, or colored glasses crushed for use as pigments, were manufactured.

The *Papyrus Holmiensis* (Halleux 1981), written in Greek and believed to date from the third century A.D., contains a number of recipes for producing and applying dyes and pigments. Throughout the Middle Ages, alchemists shared and supplemented such knowledge. We find more knowledge about colorant mixture effects in the writings of Eraclius (ninth or tenth century A.D.; Merrifield 1849).

It is not known whether such knowledge was behind Chalcidius's fourth century A.D. reduction of Aristotle's eight basic colors to five (see entry in chapter 2). Aside from mentioning white and black, he listed what may have meant pale yellow (*pallidus*), red (*rubeus*), and blue (*cyaneus*). Chalcidius is the first author to posit three basic chromatic colors. This change from Aristotle's five chromatic color species represents the derivation of more fundamental categories based on knowledge of results of colorant mixture. Roger Bacon, in the twelfth century, declared whiteness, blackness, yellowness (*glauciditas*), redness (*rubedo*), and blue-greenness (*viriditas*) to be the fundamental color genera (Parkhurst 1990).

In the seventeenth century, Isaac Newton (see entry in chapter 6) decided to recognize seven different chromatic color categories in the spectrum. With an extraspectral category (purple, not considered by Newton, who was concerned with spectral colors only), the resulting number of chromatic categories had risen to eight.

The preeminence of yellow, red, and blue as simple chromatic colors was supported by Jakob Christof Le Blon's eighteenth-century achievements in color printing from three (or four, including black) corresponding plates (see figure 9.1). Those colors reigned as chromatic primaries

until the beginning of the nineteenth century, when Thomas Young (see entry in chapter 6) replaced them with red, green, and violet.[4]

Without standardization of exemplars, color categorization is a comparatively crude exercise. With about one million distinguishable object color stimuli, each of Berlin and Kay's 11 basic terms encompasses a large number of color experiences. But which stimuli and which resulting experiences? Not all possible experiences fall into the 11 categories. No complete set of experiences was available for the categorization efforts, and few desaturated colors were distinguishable enough as groups to result in category formation.

The number of categories can be enlarged at any time. But although the normal trichromat can distinguish one or two dozen categories without exemplars, it is not possible to distinguish several dozens or hundreds of categories.

Historically, extended categorization proceeded in the form of fixing stimuli that are the best representatives of the category (e.g., see Robert Ridgway, chapter 10). The selection of a given stimulus sample as representing the best exemplar for a given named category significantly varies by individual within a cultural group, as the World Color Survey[5] indicates. As a result, the fixing of category boundaries depends on the individuals involved in the process.

Extended categorization efforts have been systematic at times, resulting in color order systems (e.g., the Villalobos atlas; see Càndido and Julio Villalobos, chapter 10), and unsystematic at other times (color-chip collections such as the *Wiener Farbenkabinet* [Anonymous 1794] or Maerz and Paul's *Dictionary of Color* [1930]). Only in the mid-twentieth century was there a concerted effort to establish formal (American language) category boundaries in three dimensions, devised by the Inter-Society Color Council (ISCC) and the U.S. National Bureau of Standards (NBS; National Bureau of Standards 1955). The ISCC-NBS system shows the fluid transition from color categorization to color order. It is based on the Munsell Renotations (see entry in chapter 7) but divides the Munsell solid into subspaces that are not uniform in extension. In the Munsell solid, distances are presumed to have perceptual meaning, not linearly related to categorical meaning.

Color order

Pushing color categorization toward smaller and smaller categories requires systematic ordering of all possible color stimuli or experiences. But color order remains an abstract, mental exercise until it is demonstrated in form of a stimulus collection, typically an atlas of samples. Atlases require an ordering principle, a related geometric model, and different kinds of standardization. We discuss the last issue first.

Standardization

Sample chips of a color order system require standardization of methods and materials for generating them, including substrates (paper), colorants, application methods, and viewing methods. Such standardization was slow in developing. Standardization of colorants, as pursued by August Ludewig Pfannenschmid (see entry in chapter 4), is only one aspect.

Applying watercolors is intrinsically problematic because the amount of pigment applied per unit area of paper is not easily controlled, and it varies considerably in different copies of the same illustration. Only opaque paints avoided this problem. As a consequence, as Schwarz (2004) has shown, one must distinguish between illustrating color order systems (see, e.g., Phillip Otto Runge, chapter 4) and coloring them: Different copies of the same atlas that is exemplified with standardized color samples vary minimally.

Standardization is not limited to the manufacture of color chips. It should also involve viewing conditions. Their standardization is substantially lacking even today, in part because of practical difficulties. It is well known that the appearance of color chips can change significantly as a function of surround lightness and color. As discussed in chapter 1, appearance can also change substantially as a function of the light source in which the chips are viewed. Fixing a particular experience for a particular observer requires specifying the observer, the object, the surround, and the light. This is often difficult or impractical to achieve in conditions not involving research. Standardization, therefore, usually has meaning limited to an aspect of the stimulus, the color chip. As a result, color order systems expressed in atlases are essentially only color-stimulus order systems, regardless of their internal organization.

Ordering principles

The idea that mapping the multitude of color experiences requires at least three dimensions developed in the eighteenth century and is now well established.[6] There is less agreement about the attributes representing the three dimensions of the implicit Euclidean space. The habit of placing full object colors (colors of highest chroma of a given hue) in a circle on a common horizontal plane regardless of their lightness held sway in Europe from the eighteenth to deep into the twentieth century (e.g., in the DIN 6164 system; see Manfred Richter and DIN 6164, chapter 7).

At the beginning of the nineteenth century, Gaspard Grégoire and Matthias Klotz (see entries in chapter 4) proposed Newton's three parameters of color experience for object colors: hue (*colour*[7]), brightness (*luminosity*), and saturation (*intenseness*), the latter in a relative sense. Emphasized by James Clerk Maxwell and Hermann von Helmholtz (see entries in chapter 6), these attributes later became part of the colorimetric canon. Ewald Hering proposed the alternative idea that the total experience is the sum of separate experiences of full color, whiteness, and blackness (the two combining to account for brightness and saturation; see entry in chapter 5).

Geometry of color space

Spaces have geometric properties. In case of color spaces, there is the question what the color-order–related meaning of geometrical parameters is. On a fundamental basis, it is an open question whether the use of geometry to express psychological continua implies that all the burdens of geometry must be considered and met. Such a duty has rarely been observed in color order. The cone and tristimulus spaces presented in chapter 6 (except for Jozef B. Cohen's fundamental space) are not mathematically coherent because the cone or tristimulus functions are not orthonormal (see glossary). Whether there are advantages to orthonormalization in attempts to model psychological scaling data from cone response data remains to be seen, however. Cohen's fundamental space does not indicate so.

Classification of color order systems

There are two basic approaches in color order: systems based on uniform stimulus sampling of some kind, and systems based on uniform perceptual scales of some kind. Each of these two categories has several subcategories. There is also a mixed subcategory in which stimulus order has been modified so that the result resembles more closely perceptual order. The classification is illustrated in figure 12.1.

Colorimetric (cone or other color-mixture tristimulus function-derived) systems are useful for determining metamerism as related to a standard observer and standard illuminants. The International Commission on Illumination (CIE) version is generally in ordinal agreement with the

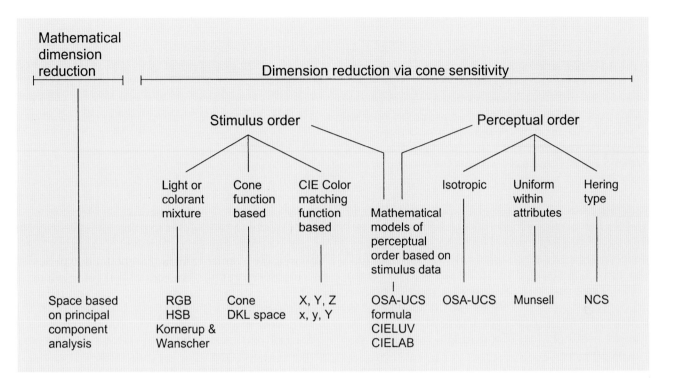

Figure 12.1.
*Schematic representation of the structure of different kinds
of color order systems, with some typical examples.*

three Newtonian attributes. Such systems have the advantage of relatively high internal precision due to the nature of the physical measurements involved. The precision is reduced in the case of colorant mixture or printing processes, such as halftone, because the number of variables is substantially higher than in case of stimulus measurement only. If the systems are based on three primary colorants and equal intervals of weights, volumes, or coverage of substrate (halftone), the results, generally, are no longer in even qualitative attribute order.

The elusive goal of uniformity

Uniformity of sampling according to perceptual scales has been a desired goal since Tobias Mayer (see entry in chapter 4). However, neither Mayer nor his followers clearly understood the consequences of this goal in terms of the interior structure of the solids involved. As a result, only the hue difference steps between primary colorants were made to look even, and as Johann Heinrich Lambert (see entry in chapter 4) noted, these vary in perceptual magnitude from one primary pair to the next. Although perceptual uniformity was among the goals of many of the systems in chapters 10 and 11, detailed descriptions of how it would be achieved are

usually absent. It is a psychological fact that many kinds of color charts appear at first glance to be uniform. When looking at such charts, the visual system is overwhelmed by the information presented. Judgments regarding the degree of uniformity require concentrated and extended attention. At the same time, variation in individual color vision systems, alluded to in chapter 1, results in disagreements about the perceived uniformity of systems.

The Munsell system attempted step uniformity also within the value and chroma attributes. Dorothy Nickerson (see entry in chapter 7) established the approximate relationship in the system between step sizes in the three attributes (in case of hue at a given chroma level) for establishing a color-difference formula based on the system. Deane Brewster Judd's geometric analysis of the formula disclosed the hue superimportance effect (see entry in chapter 7). One of its consequences is the impossibility of representing a perceptually uniform color space in Euclidean geometry.

Geometry also prevents a color solid from uniformity in all directions. Twelve is the maximum number of simultaneously uniform scales in various directions from a central color in such a solid, as exemplified in the OSA Uniform

Color Scales system (see entry in chapter 7). Also in this case, hue superimportance prevents accurate representation in a Euclidean space.

These facts show that isotropic Euclidean color solids are not natural kinds. The geometric complications are, in part, due to the fact that perceiving small spectral differences resulting in hue differences was evolutionarily important. Hue-difference perception was more important for survival than was chroma- or lightness-difference perception. The corresponding machinery appears to be implemented at a higher level in the visual system, not at the cone level, presumably as a result of an evolutionary adaptation.

The question arises of why the human visual system should have strived to agree with a Euclidean (or even non-Euclidean) space, or should have favored the idea of isotropy. This appears to be a cultural idea not met in reality. Instead, species survival has favored hue-difference detection as the most important aspect, resulting in the hue superimportance effect. Species survival has not pushed for an adaptation providing the ability to make accurate perceptual distance evaluations, which has been found useful only in the advanced industrial age.

Our visual system's technique of reducing spectral stimuli to only three dimensions has resulted in loss of varying amounts of detailed spectral information arriving at the eye. This loss and an endless variety of lighting and surround conditions have resulted in adaptations that (incompletely) normalize a number of such variables for conditions that our forebears encountered in the jungle and savanna. These effects and individual variations based on genetics act against a simple relationship between stimulus and percept. In this manner, they also act against any kind of universal, natural color order on more than the ordinal level.

Conventionally and historically, interobserver variability in color judgments is being treated within standard distribution statistics. The implication is that there is one kind of observer, the mean observer, perceiving the colors of objects accurately, and that all others do so with a smaller or larger error. The standard observer (in terms of color matching functions) defined by the CIE, is the identical twin of the "standard difference observer," as defined in the latest CIE color difference formula (see CIEDE2000, chapter 7). But the perceptual reality, as demonstrated by interobserver variability in unique hue stimulus selection and color difference judgment data, is much different. This fact is further strong evidence that a universally valid perceptual color space cannot exist.

A given atlas can be valid only for given specific conditions and the subgroup of observers that has the same percept-distance scaling pattern, not only an identical (or near-identical) choice pattern of unique hue stimuli. Color space atlases based on psychological data are therefore properly seen as color stimulus collections and not as representative of a (nonexisting) universal perceptual color solid.

The impossibility of an isotropic Euclidean color space means nonexistence of a widely comprehensible color space with maximal information content and spatial parameters with clear, mathematically consistent meaning. The absence of such a space leaves room for an infinite number of arrangements with more or less reduced information content (i.e., at the ordinal level only), all having a smaller or larger degree of arbitrariness. This book shows dozens of such arrangements. Arguments about their relative value lack fundamentality and take place in the realms of aesthetic or technical discourse.

These systems must be considered human cultural achievements, works of artifice of some kind. As such, they are emblematic of human creativity, within our natural boundaries. Despite these limitations, color atlases are useful tools for many applications, including design, architecture, graphics, and painting, coloration and display technologies, and their contemplation provides insight into the large variety of color experiences available to us.

Notes

1. "Comparator system" refers to the apparatus in the human visual system that compares the outputs from the different cone types to obtain chromatic distinction, and adds the outputs of two cone types to obtain brightness distinction. The exact operation of the comparator system is not yet known in detail.

2. Alternatively, they may have referred to locations where they are found.

3. The minimum number of color categories recorded in unwritten languages of the World Color Survey (see note 5) is two. In the single language with two color terms (among 111 in the survey), the terms mean red and grue (a combination of blue and green). The four basic terms of the pre-Socratic Greek philosophers thus represent an advance in categorization. However, as the Homeric epics show, several more color terms were in common use in Greece at the time.

4. Red, green, and violet are the spectral colors located at the beginning, middle, and end of the spectrum, making it possible to generate by additive mixture the maximal degree of saturation for all spectral and extraspectral hues, as mentioned in chapter 1. In the nineteenth century, James Clerk Maxwell identified the ideal colorants for subtractive mixture. The comparable modern colorants used for the purpose in printing or color photography are termed cyan, magenta, and yellow (CMY).

5. The World Color Survey is an anthropological study of the meaning of color terms in 111 unwritten languages around the world. Usually, 25 native observers were shown a total of 320 individual Munsell color chips and asked to name them. In a second step, the observers were shown a display of all 320 chips and were asked to indicate the best example for a given color name they used in the first step. Information on the World Color Survey is available at www.icsi.berkeley.edu/wcs/.

6. The three-dimensionality of color stimulus space is implicit in the fact of three cone types.

7. Newton used the general term "colour," together with hue names, to designate hue. The term "hue" originally had the meaning of form, appearance, complexion.

References

Anonymous. 1794. *Wiener Farbenkabinet*, 2 vols., 5,000 color samples, Vienna and Prague: Schönfeld.

Berlin, B., and **P. Kay**. 1969. *Basic color terms*, Berkley, CA: University of California Press.

Cardano, H. 1550. *Hieronimi Cardani medici mediolanensis de subtilitate libri XXI*, published simultaneously in Nürnberg, Germany, and Lyon and Paris, France.

Dawkins, R. 1976. *The selfish gene*, Oxford: Oxford University Press.

Faulkner, R. O. 1981. *A concise dictionary of Middle Egyptian*, Oxford: Griffith Institute, Ashmolean Museum.

Halleux, P. 1981. *Les alchimistes grecs*, vol. 1, Paris: Société d'Édition "Les Belles Lettres."

Harnad, S. 1996. The origin of words: a psychophysical hypothesis, in *Communicating meaning: evolution and development of language*, **Velichowsky, B.**, **Rumbaugh, D.**, eds., Hillsdale, NJ: Erlbaum.

Maerz, A., and **M. R. Paul**. 1930. *The dictionary of color*, New York: McGraw-Hill; 2nd ed., 1950.

Merrifield, M. P. 1849. *Medieval and Renaissance treatises on the arts of painting*, London: Murray; reprint, Mineola, NY: Dover, 1967.

National Bureau of Standards. 1955. *The ISCC-NBS method of designating colors and a dictionary of color names*, NBS Circular 553, Washington, DC: National Bureau of Standards.

Parkhurst, C. 1990. Roger Bacon on color: sources, theories and influence, in *The verbal and the visual*, Selig, K.-L., Heckscher, W. S., eds., New York: Italica Press.

Schwarz, A. 2004. *Farbsysteme und Farbmuster*, Hannover: BDK Verlag.

APPENDIX

The Appendix contains enlarged images of figures
found originally in the following system entries

Year	System	Page
Mediaeval	Urine Circle	33
1686	Waller	56
1708	Anonymous	57
1772	Schiffermüller	59
1772	Harris	61
1772	Lambert	75
1810	Runge	80
1816	Klotz	81
ca. 1820	Grégoire	82
1826	Hayter	65
1845	Hay	77
1871	Benson	254
1874	Bezold	234
1886	Ridgway	243
1890	Lacouture	188
1896	Steinheil	193
1901	Pilgrim	136
1907	Munsell	114
1905	Hering	101
1911	Höfler	102
1912	Chavkin	195
1917	Ostwald	247
1922	Fellowes	260
1922	Prase	259
1936	Painter	197
1947	Villalobos	265
1947	Rabkin	112
1956	Barding	308
1988	Colorcurve	221

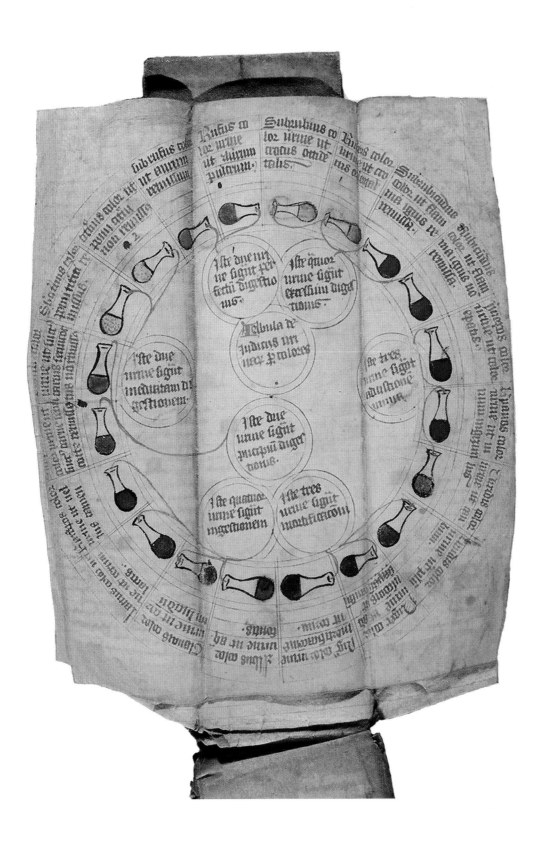

1. Figure 2.3.
Urine Circle, Mediaeval.
see page 33

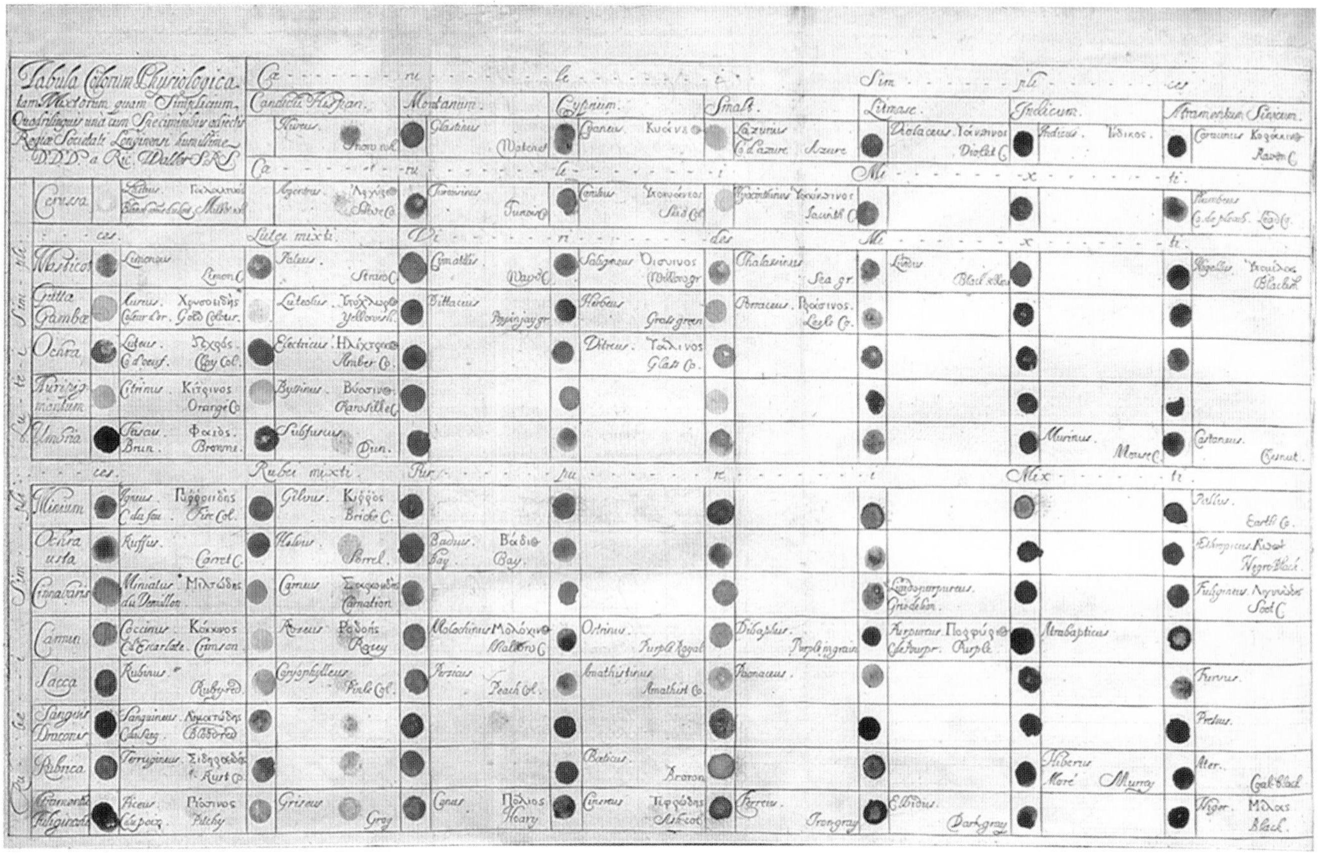

2. Figure 3.1.
Waller, 1686.
see page 56

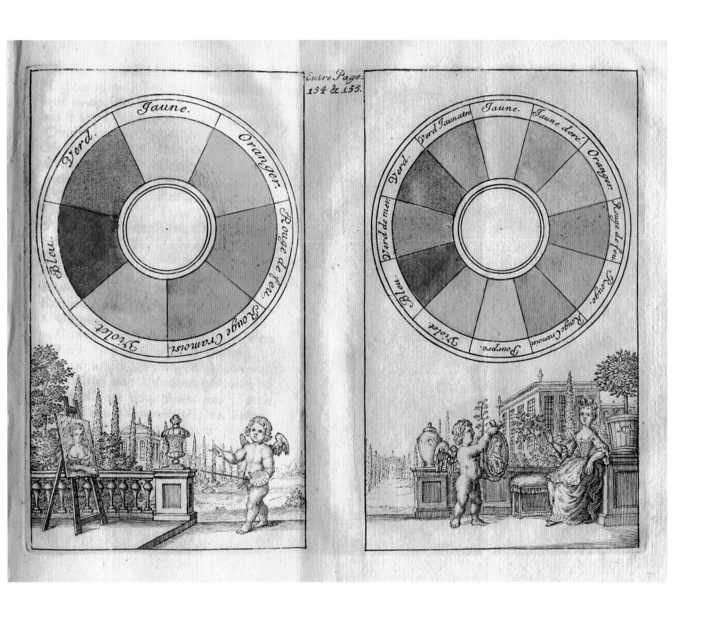

3. Figure 3.2.

Anonymous, 1708.

see page 57

I. Ordnung der Farbenclasse.

XII. *g.* Feuerblau.   I. *gatt.* Blau.

XI. *g.* Veilenblau.   II. *g.* Meergrün.

X. *g.* Veilen roth.   III. *g.* Grün.

IX. *g.* Karmasinroth.   IV. *g.* Oliven Grün.

VIII. *g.* Roth.   V. *g.* Gelb.

VII. *g.* Feuerroth.   VI. *g.* Oranien gelb.

Die Blühenden Farben.

4. Figure 3.5.
Schiffermüller, 1772.
see page 59

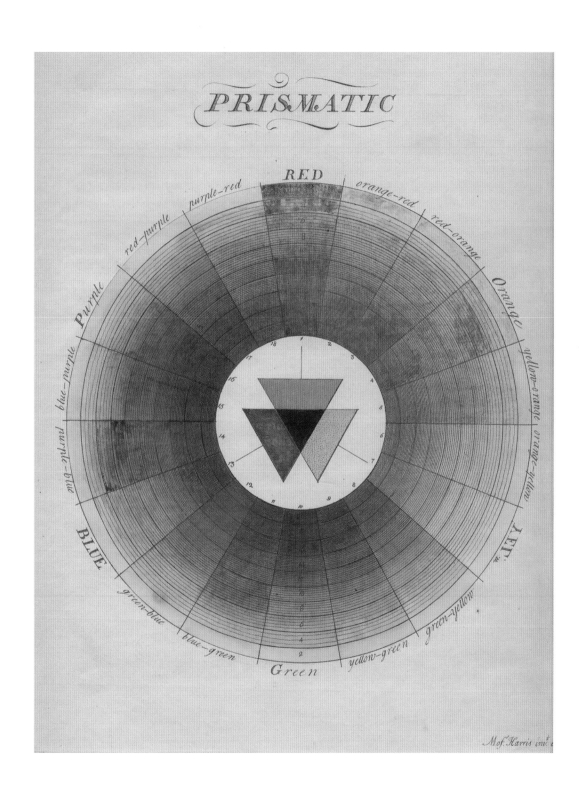

5. Figure 3.7.
Harris, 1772.
see page 61

6. Figure 4.6.
Lambert, 1772.
see page 75

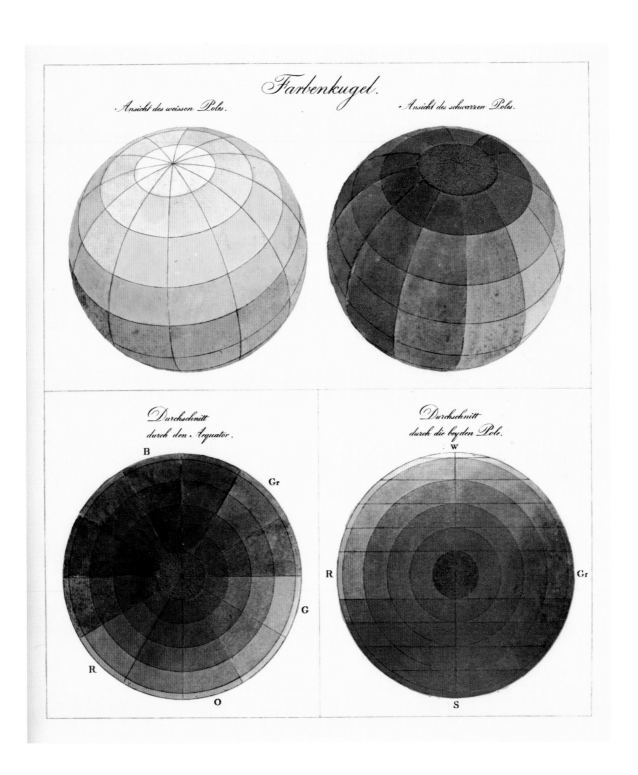

7. Figure 4.16.

Runge, 1810.

see page 80

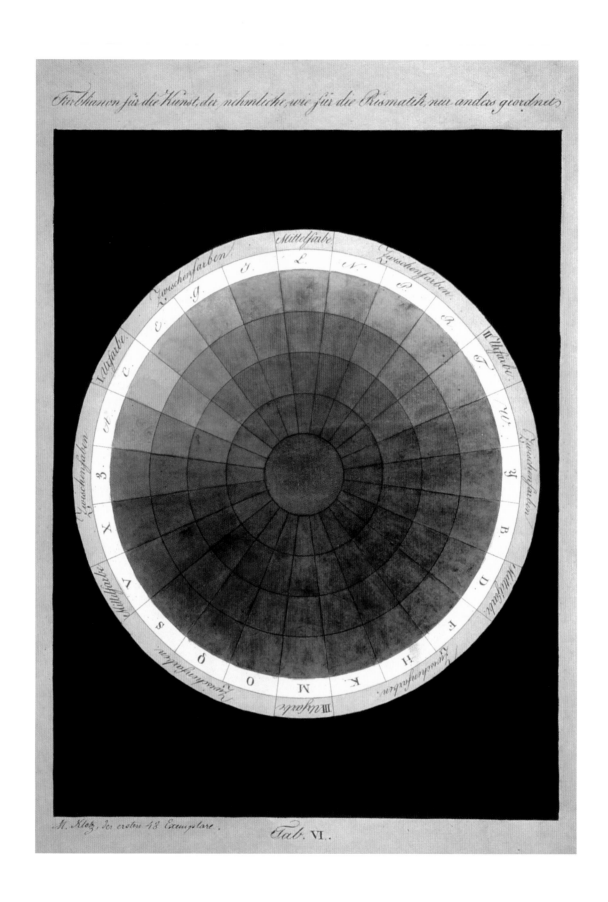

8. Figure 4.19.
Klotz, 1816.
see page 81

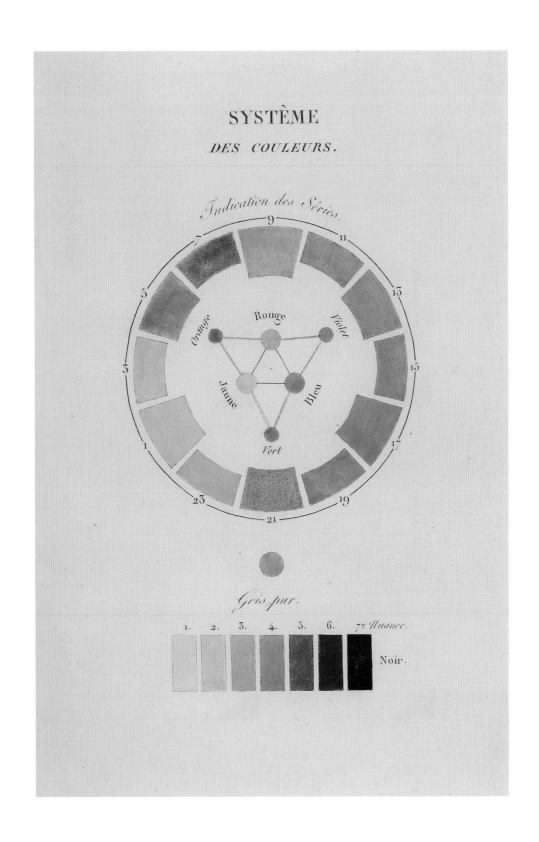

9. Figure 4.21.
Grégoire, ca. 1820.
see page 82

10. Figure 3.16.
Hayter, 1826.
see page 65

PLATE 13.

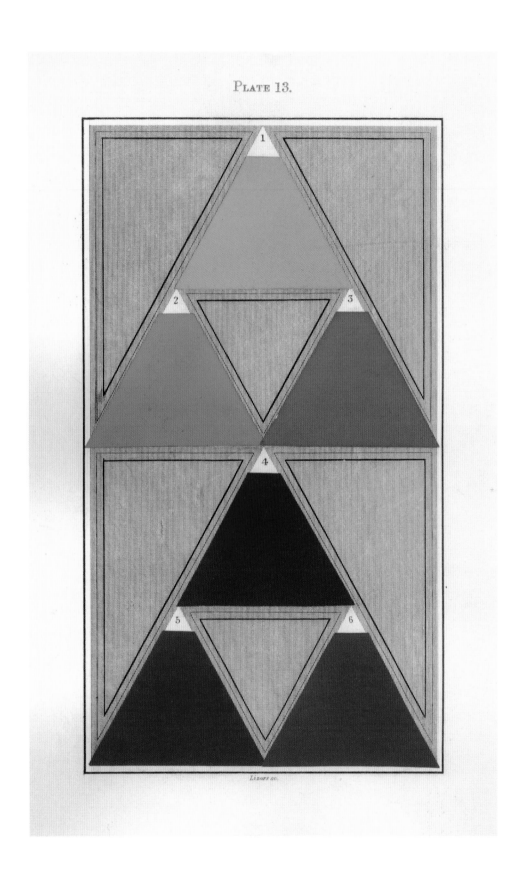

Lizars sc.

11. Figure 4.9.
Hay, 1845.
see page 77

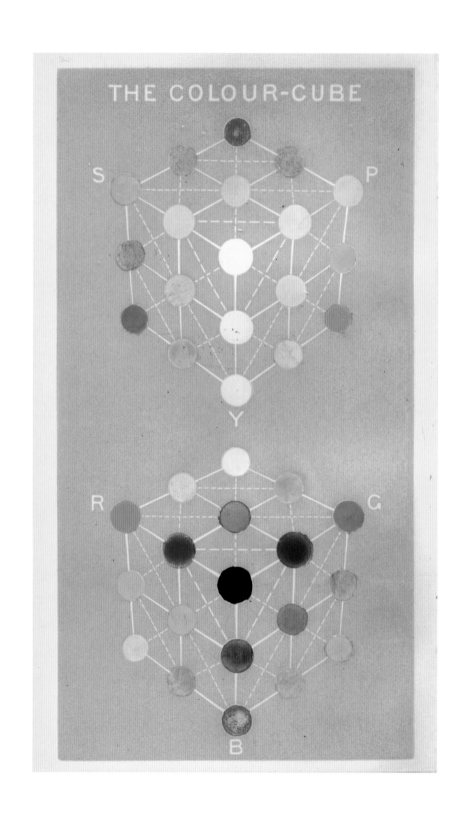

12. Figure 10.59.
Benson, 1871.
see page254

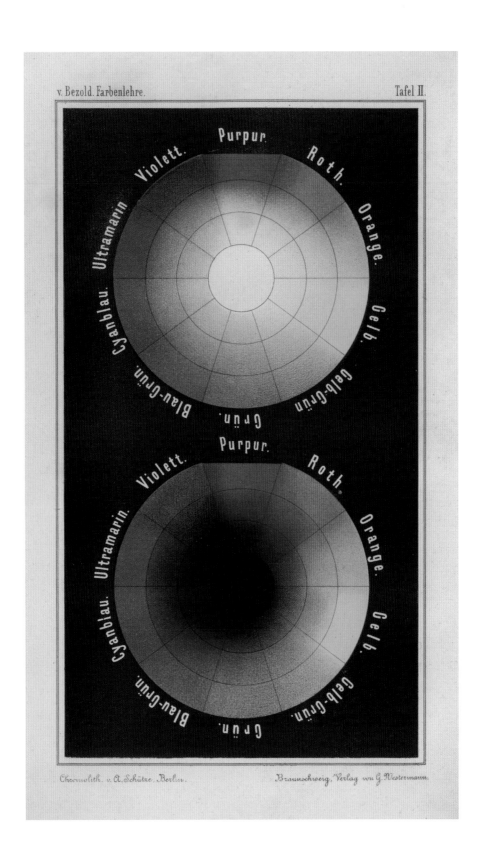

13. Figure 10.12.
Bezold, 1874.
see page 234

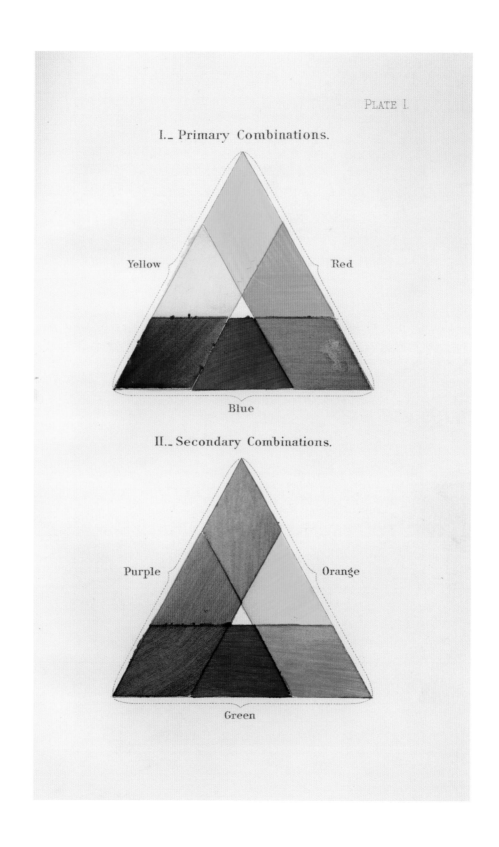

PLATE I

I._ Primary Combinations.

Yellow

Red

Blue

II._ Secondary Combinations.

Purple

Orange

Green

14. Figure 10.33.
Ridgway, 1886.
see page 243

TRILOBE SYNOPTIQUE

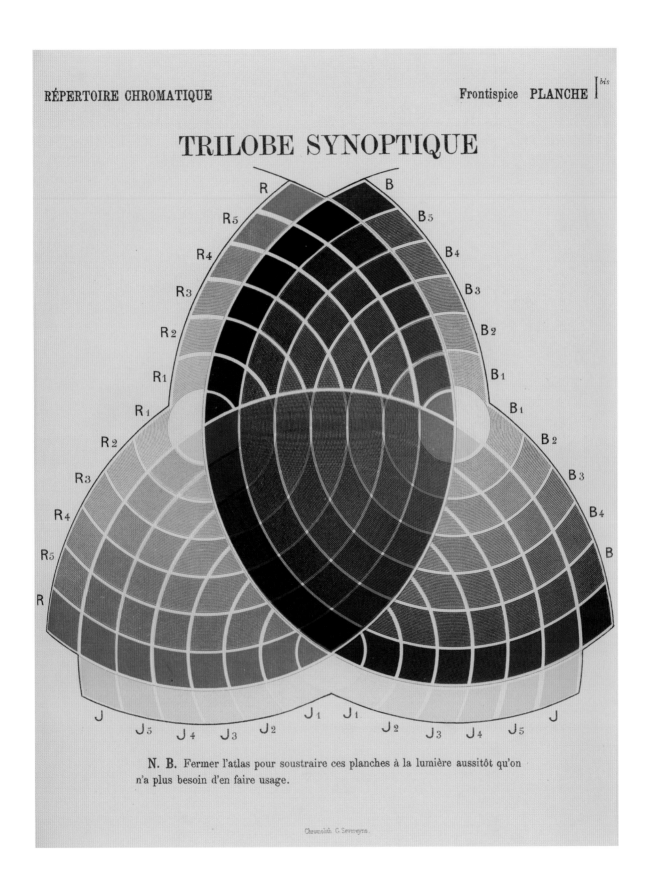

N. B. Fermer l'atlas pour soustraire ces planches à la lumière aussitôt qu'on n'a plus besoin d'en faire usage.

Chromolith. G. Severeyns.

15. Figure 9.17.
Lacouture, 1890.
see page 188

Pl. 82

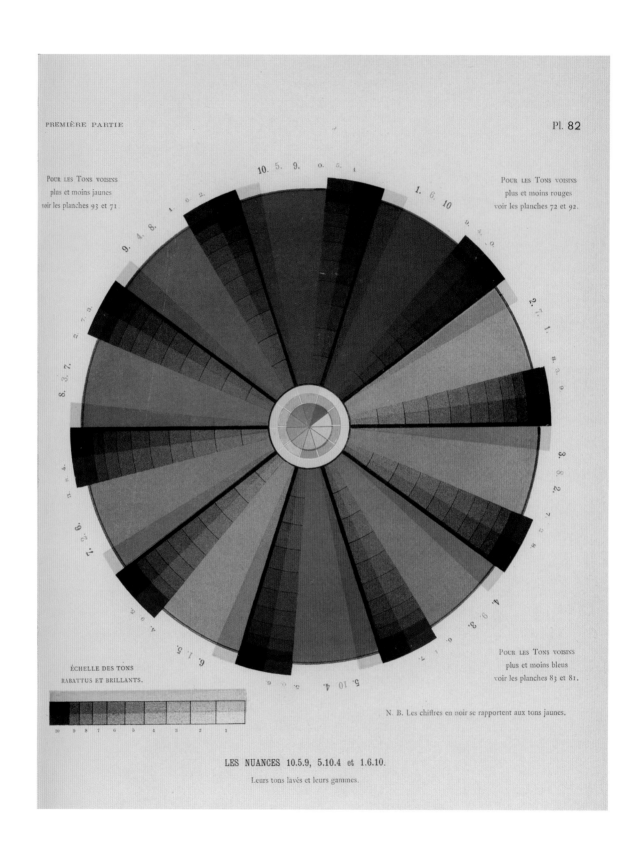

POUR LES TONS VOISINS
plus et moins jaunes
voir les planches 93 et 71.

POUR LES TONS VOISINS
plus et moins rouges
voir les planches 72 et 92.

POUR LES TONS VOISINS
plus et moins bleus
voir les planches 83 et 81.

ÉCHELLE DES TONS
RABATTUS ET BRILLANTS.

N. B. Les chiffres en noir se rapportent aux tons jaunes.

LES NUANCES 10.5.9, 5.10.4 et 1.6.10.

Leurs tons lavés et leurs gammes.

16. Figure 9.30.
Steinheil, 1896.
see page 193

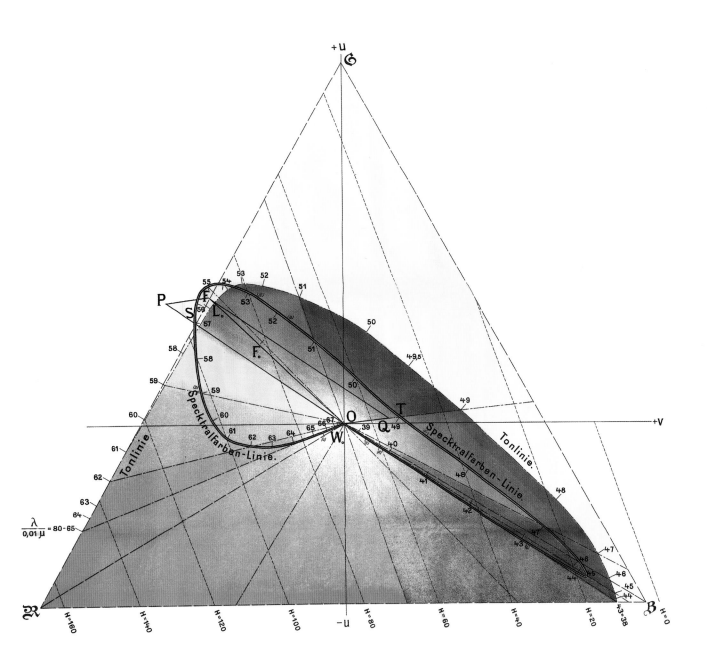

17. Figure 6.14.
Pilgrim, 1901.
see page 136

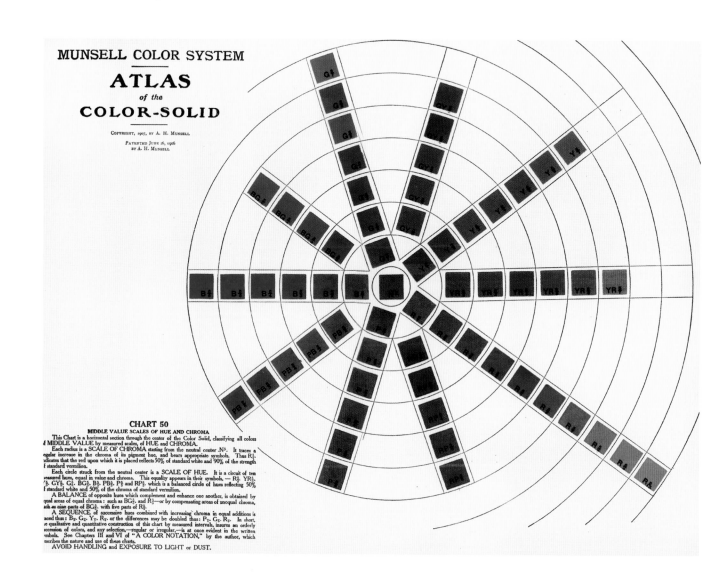

18. **Figure 5.57.**
Munsell, 1907.
see page 114

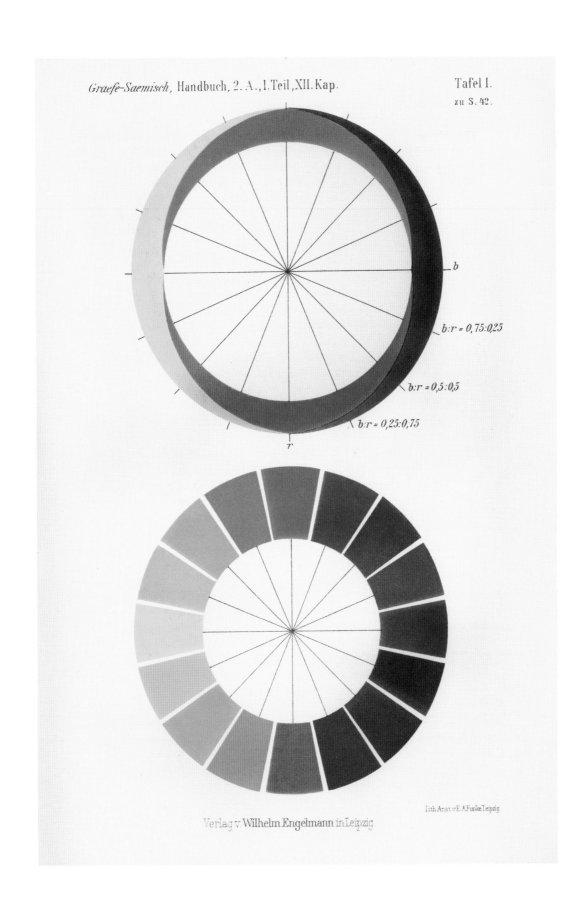

19. Figure 5.14.
Hering, 1905.
see page 101

ZWEI MODELLE PSYCHOLOGISCHER FARBENKÖRPER
(nach Alois Höfler)

blau

Farbenoktaeder

Farbendoppeltetraeder

20. Figure 5.18.
Höfler, 1911.
see page 102

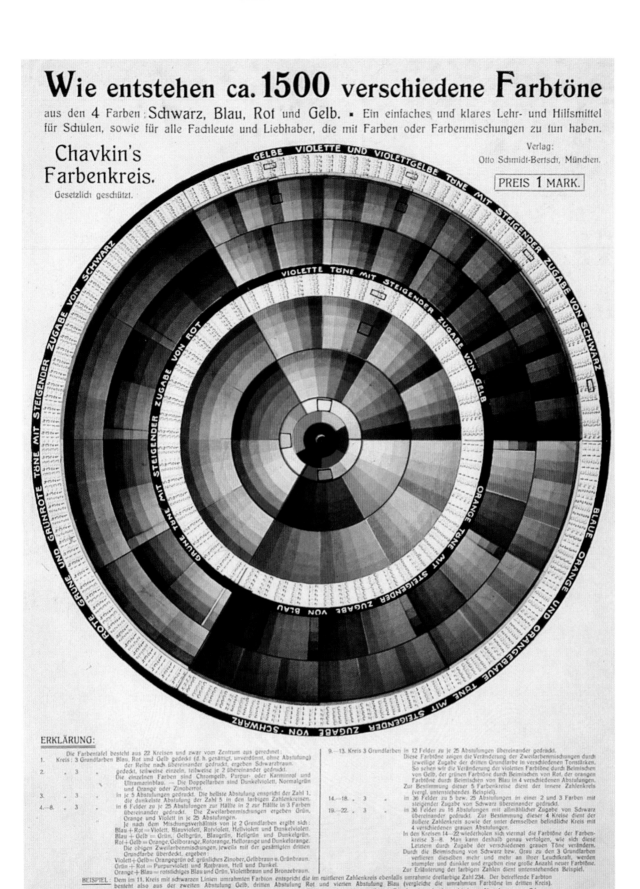

21. Figure 9.33.
Chavkin, Farbenkreis, 1912.
see page 195

Wilhelm Ostwald, Der Farbkörper. 2. A.

Verlag Unesma G. m. b. H. in Leipzig.

22. Figure 10.43.
Ostwald, 1919.
see page 246

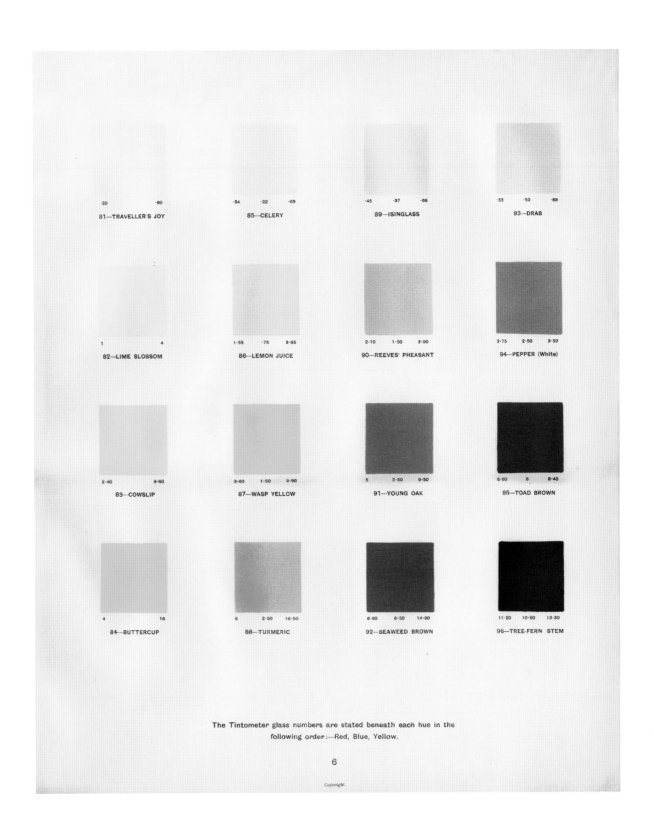

81—TRAVELLER'S JOY · ·20 ·80

85—CELERY · ·34 ·22 ·69

89—ISINGLASS · ·45 ·37 ·68

93—DRAB · ·55 ·52 ·68

82—LIME BLOSSOM · 1 4

86—LEMON JUICE · 1·55 ·75 3·95

90—REEVES' PHEASANT · 2·10 1·50 3·90

94—PEPPER (White) · 2·75 2·50 3·50

83—COWSLIP · 2·40 9·60

87—WASP YELLOW · 3·60 1·50 9·90

91—YOUNG OAK · 5 3·50 9·50

95—TOAD BROWN · 6·50 6 8·40

84—BUTTERCUP · 4 16

88—TURMERIC · 6 2·50 16·50

92—SEAWEED BROWN · 8·60 6·50 14·90

96—TREE-FERN STEM · 11·20 10·50 13·30

The Tintometer glass numbers are stated beneath each hue in the following order:—Red, Blue, Yellow.

6

23. Figure 10.74.
Fellowes, 1922.
see page 260

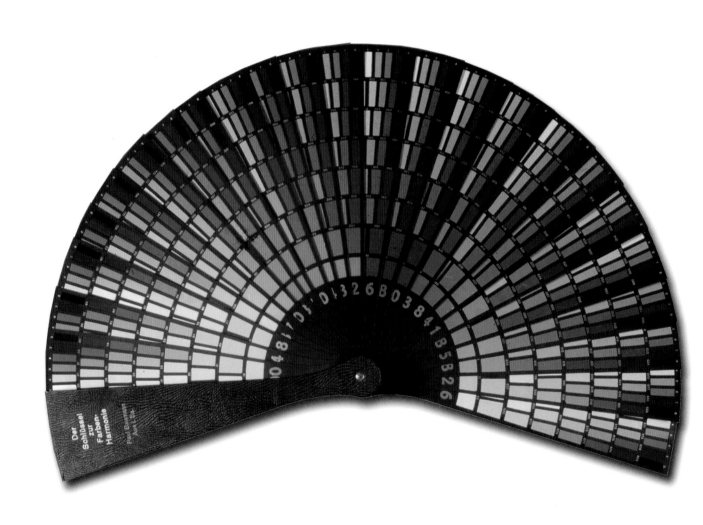

24. Figure 10.72.
Prase, 1924.
see page 259

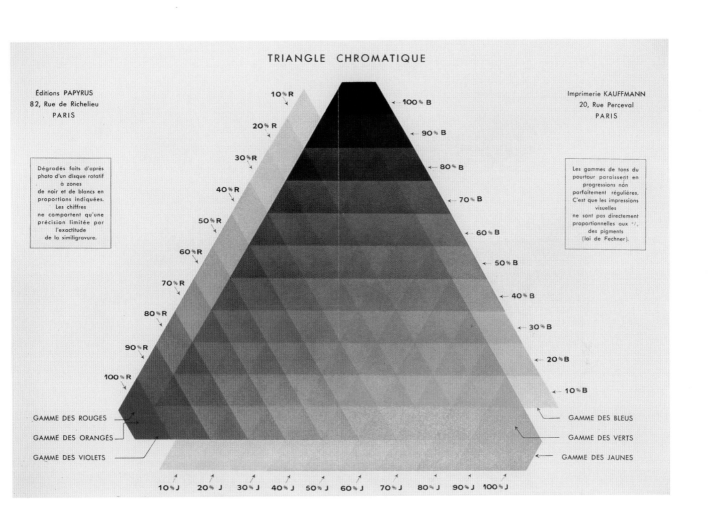

TRIANGLE CHROMATIQUE

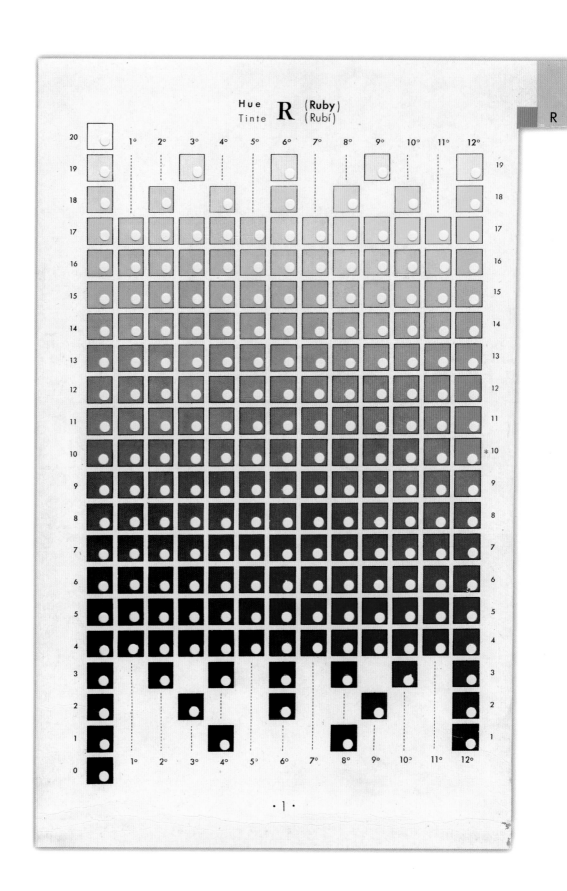

26. Figure 10.87.
Villalobos, 1947.
see page 265

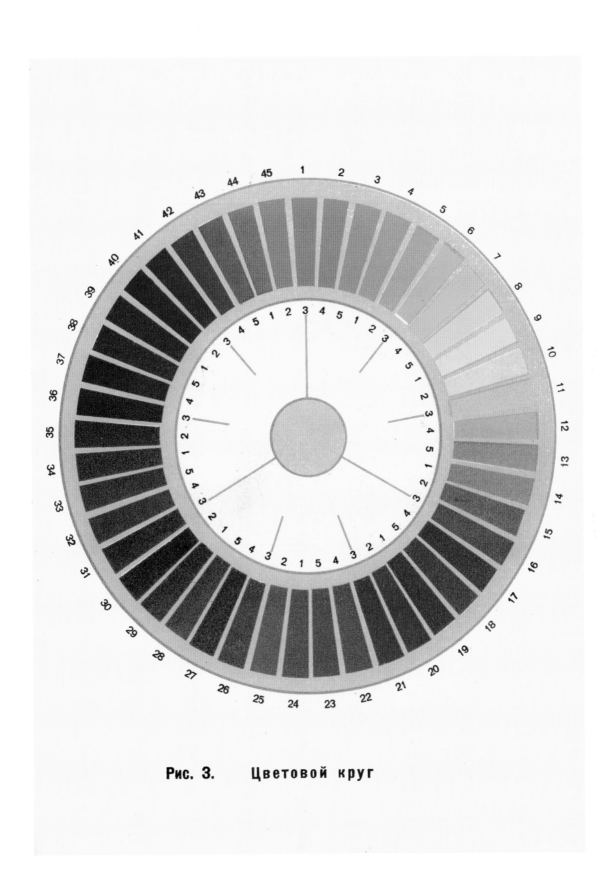

Рис. 3.   Цветовой круг

27. Figure 11.40.
Rabkin, 1956.
see page 308

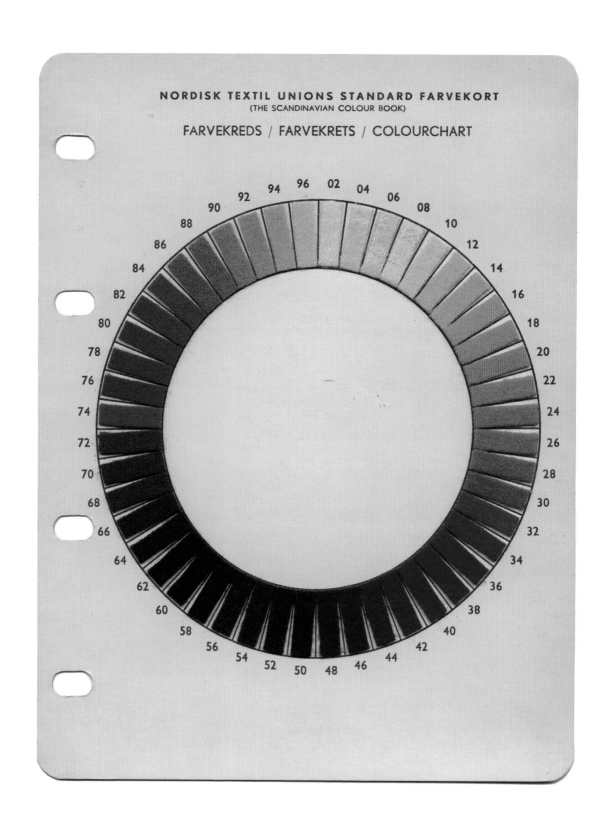

28. Figure 5.45.

Barding, 1956.

see page 112

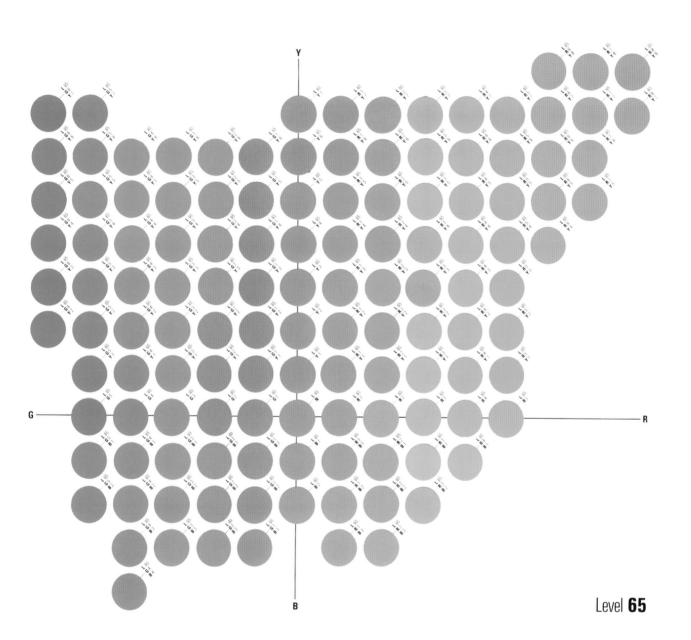

Level **65**

29. Figure 9.93.
Colorcurve , 1988.
see page 221

ALPHABETICAL AND CHRONOLOGICAL SEQUENCE OF ENTRIES

Alphabetical Sequence

Chronological Sequence

NAME	YEAR	CHAPTER	PAGE
Aristotle	ca. 330 B.C.	2	31
Urine color scales	Mediaeval	2	33
Chalcidius	ca. 325	2	32
Avicenna	ca. 1015	2	34
Theophilus	ca. 1120	2	35
Grossteste, Robert	ca. 1230	2	35
Theodoric of Freiberg	ca. 1310	2	36
Alberti, Leon Battista	ca. 1435	2	37
Ficino, Marsilio	ca. 1480	2	40
Leonardo da Vinci	ca. 1550	2	37
Cardano, Girolamo	1550	2	39
Dolce, Lodovico	1565	2	38
Lomazzo, Giovanni Paolo	1584	2	39
Forsius, Sigfridus Aronus	1611	2	45
d'Aguilon, Francois	1613	2	40
Fludd, Robert	1629	2	42
Kircher, Athanasius	1646	2	41
Traber, Zacharias	1675	2	43
Glisson, Francis	1677	2	46
Zahn, Johannes	1685	2	43
Waller, Richard	1686	3	56
Newton, Isaac	1704	6	128
Anonymous	1708	3	56
Castel, Louis-Bertrand	1740	3	58
Mayer, Tobias	1758	4	72
Guyot, Edmé-Gilles	1770	11	288
Lambert, Johann Heinrich	1772	4	74
Schiffermüller, Ignaz	1772	3	59
Harris, Moses	ca. 1772	3	59
Pfannenschmid, August Ludewig	1781	4	76
Frisch, Johann Christoph	1788	3	62
Sampayo, Diogo de Carvalho e Sampayo	1788	2	47
Young, Thomas	1807	6	129
Sowerby, James	1809	3	63
Goethe, Johann Wolfgang von	1810	3	62
Runge, Phillip Otto	1810	4	78
Grégoire, Gaspard	ca. 1810	4	82
Klotz, Matthias	1816	4	80
Hayter, Charles	1826	3	64
Nobili, Leopoldo	1830	2	49
Chevreul, Michel-Eugène	1839	4	84
Hay, David Ramsay	1845	4	76
Doppler, Christian	1847	10	230
Zheng, Funguang	1847	2	48
Forbes, James David	1849	11	290
Sol, J.-C.-M.	1849	11	291
Grassmann, Hermann Günter	1853	6	134
Helmholtz, Hermann von	1856	6	132

NAME	YEAR	CHAPTER	PAGE
Maxwell, James Clerk	1857	6	130
Brücke, Ernst Wilhelm von	1866	10	232
Benson, William	1868	10	252
Bezold, Wilhelm von	1873	10	233
Wundt, Wilhelm	1874	10	234
Hering, Ewald	1878	5	100
Radde, Otto	1878	4	86
Rood, Ogden Nicholas	1879	10	237
Scheffler, Hermann	1883	11	302
Carpentier, Jules	1885	10	254
Henry, Charles	1888	4	86
Lacouture, Charles	1890	9	187
Ziehen, Georg Theodor	1891	10	236
Hoffmann, Hermann	1892	9	190
Lovibond, Joseph Williams	1893	9	212
Kirschmann, August	1895	10	238
Steinheil, Robert	1896	9	192
Ebbinghaus, Hermann	1897	5	102
Höfler, Alois	1897	5	101
Warburg, John Cimon	1899	9	194
Pilgrim, Ludwig	1901	6	135
Titchener, Edward Bradford	1901	5	103
d'Udine, Jean	1903	10	239
Detlefsen, Emil	1905	9	213
Munsell, Albert Henry	1905	5	114
Oberthür, René, and Dauthenay, Henri	1905	11	293
Ross, Denman Waldo	1907	11	304
Klincksieck, Paul, and Valette, Theodore	1908	11	294
Rosenstiehl, Daniel-Auguste	1910	10	241
Ames, Adelbert	ca. 1910	5	113
Chavkin, Nathan	1912	9	195
Prase, Otto	1912	10	256
Ridgway, Robert	1912	10	243
Bloch, Leopold	1915	9	215
Ostwald, Wilhelm	1917	10	245
Schrödinger, Erwin	1920	6	137
Fellowes, E.	1922	10	260
Pope, Arthur	1922	11	305
Becke, Max	1923	10	255
Luther, Robert	1927	6	139
Vogel, Joseph Marcel, and Carl Sali Plaut	1927	10	261
Nyberg, Nikolaus D.	1928	6	139
Hillebrand, Franz	1929	5	103
Rösch, Sigfried	1929	6	140
Podestà, Hans	1930	5	103
CIE X,Y,C color stimulus space	1931	6	141
Judd, Dean Brewster	1932	7	154
Barthel, Ernst	1933	11	307
Gerstacker, Ludwig	1934	9	198
Ives, Herbert E.	1935	9	204

NAME	YEAR	CHAPTER	PAGE
MacAdam Limits	1935	6	144
Neugebauer, Hans	1935	6	143
Nickerson, Dorothy	1936	7	156
Painter, A.	1936	9	196
Séguy, Eugène	1936	11	295
Bernays, Adolphe	1937	5	104
Hiler, Hilaire	1937	10	262
Johansson, Tryggve	1937	5	107
Wilson, Robert Francis	1938	11	297
Adams, Elliot Q.	1942	7	155
MacAdam line element	1942	7	158
Blecher, Carl	1943	9	199
Hickethier, Alfred	1943	9	201
Munsell Renotations	1943	7	160
Absorption Space	1944	6	145
Müller, Aemilius	1945	5	115
NuHue Custom Color System	1946	10	263
Richter, Manfred	1947	7	162
Setarti	1947	11	299
Villalobos, Candidò, and Julio	1947	10	264
Boring, Langfeld, Weld	1948	5	105
Plochere, Gustave, and Gladys	1948	10	265
Cheskin, Louis	1949	10	268
Colorizer System	1949	10	269
DIN 6164	1952	7	162
Hesselgren, Sven	1953	5	108
Barding, Sven A.	1956	5	110
Braun, Frantz	1956	10	270
Hurvich, Leo M., and Jameson, Dorothea	1956	5	106
Rabkin, E. B.	1956	11	308
LTF/GATF Chart	1957	9	205
Paclt, Jiri	1958	11	300
Wezel, H.	1959	9	206
FDGB Farbsystem	late 1950s	9	207
Heimendahl, Eckart	1961	11	310
Kornerup, Andreas, and Wanscher, Johan H.	1961	10	272
DDR-Standard TGL 21 579	1965	10	248
Advanced Ink Mixing System (AIMS)	1968	10	273
Frieling, Heinrich	1968	11	312
Lemonnier, André	1968	10	250
Gericke, Lothar, and Schöne, Klaus	1969	11	314
Wedlake Cyril J.	1969	10	275
Color display solids (RGB, HSB, CMYK)	1970s	9	211
Imperial Chemical Industries Colour Atlas	1971	10	276
Küppers, Harald	1971	11	316
Gerritsen, Frans	1972	11	318
Foss, Carl E.	1973	9	208
Marthin, Perry	1974	5	112
Eucolor System	1975	10	277
CIELAB, CIELUV	1976	7	167

GENERAL LITERATURE ON THE SUBJECT OF COLOR ORDER: A SELECT BIBLIOGRAPHY

Agoston, G. A. 1987. *Color theory and its application in art and design*, chapters 8 – 10, Berlin: Springer.

Billmeyer, F. W. Jr. 1985. *AIC annotated bibliography on color order systems*, Beltsville, Maryland: Mimeoform Services (microfiche).

Birren, F. 1965. Color systems and color standards, Fort Wayne, IN.: General Color Cards.

Bradley, M. C., Jr, 1938. Systems of color classification, *Technical Studies* IV: 240-275.

Derefeldt, G. 1991. Colour appearance systems, in *The perception of colour* (Gouras, P., ed.), Boca Raton, FL: CRC Press.

Döring, G., 1980. Vergleich von Farbensammlungen mit Farbenkarten aus Farbsystemen, *Farbe + Design* 15/16: 2-12.

Foss, C. E. 1947. Color order systems, *Paper Trade Journal* 125: 443-447.

Gericke, L. and **K. Schöne**. 1973. *Das Phänomen Farbe. Zur Geschichte und Theorie ihrer Anwendung*, Berlin: Henschelverlag.

Gerritsen, F. 1984. *Entwicklung der Farbenlehre*, Göttingen, Germany: Muster-Schmidt. Translated as *Evolution in color*, Atglen, PA: Schiffer.

Herbert, R. L. 1974. The Faber Birren collection on color in the art library, *The Yale University Library Gazette* 49 (1) 1-49.

Indow, T. 1974. Colour Atlas and Colour Scaling, in *Colour '73, Proceedings of the 2nd Congress of the Association Internationale de la Couleur*, London: Hilger

Judd, D. B. and **G. Wyszecki**. 1967. *Color in business, science and industry*, 3rd ed., New York: Wiley.

Kuehni, R. G. 2003. *Color space and its divisions*, chap. 2, Historical development of color order systems, Hoboken, NJ: Wiley.

Lersch, T. 1974. Farbenlehre, in *Reallexikon zur Deutschen Kunstgeschichte*, vol. 7, München: Zentralinstitut für Kunstgeschichte.

Paclt, J. 1958. *Farbenbestimmung in der Biologie*, Jena: VEB Gustav Fischer.

Pander, H. 1938. Zur Geschichte der Farbtafeln, *Photographische Korrespondenz* 74: 184-204.

Richter, M. 1943. Der innere Aufbau einiger bekannter Farbensysteme, *Film und Farbe* 9: 113-118.

Richter, M. 1967. Gedanken über Farbsysteme, *Die Farbe* 16: 121-130.

Robertson, A. R. 1984. Colour order systems: An introductory review, *Color Research and Application* 9: 234-240.

Rösch, S. 1941. Farbenkarten und Farbenkörper, *Licht* 11: 71.

Silvestrini, N. and **E. P. Fischer**. 1998. *Farbsysteme in Kunst und Wissenschaft*, Cologne: Du Mont.

Schmuck, F. 1980. Farbordnungen, in *Farbe im Stadtbild* (Düttmann, M., F. Schmuck, and J. Uhl, eds.), Berlin: Archibook.

Schmuck, F., 1983. Farbsysteme und Farbordnungen, *Kunstforum International* 57: 163-180.

Schwarz, A. 1992. Psychologische Farbsysteme von Hering bis NCS, *Die Farbe* 38: 141-177.

Schwarz, A. 1995. Physikalisch begründete Farbsysteme, *Die Farbe* 41: 31-60.

Schwarz, A. 2004. *Farbsysteme und Farbmuster. Die Rolle der Ausfärbung in der historischen Entwicklung der Farbsysteme*, Hannover: BDK-Verlag.

Simon, F. T. 1997. Color order, in *Color technology in the textile industry*, 2nd ed. Research Triangle Park, North Carolina: American Association of Textile Chemists and Colorists.

Spillmann, W. 2000. Farbskalen – Farbkreise – Farbsysteme, *applica* (Sonderdruck): 4-23.

Wright, W. D. 1984. The basic concepts and attributes of colour order systems, *Color Research and Application* 9: 229-233.

Wyszecki, G. 1960. *Farbsysteme*, Göttingen: Musterschmidt.

Wyszecki, G. 1981a. Uniform color spaces, in *Golden jubilee of colour in the CIE* (Society of Dyers and Colourists, ed.), Bradford, England: Society of Dyers and Colourists.

Wyszecki, G. 1981b. Color-order systems, in *Proceedings of the 4th Congress of the International Color Association*, (Richter, M., ed.), Berlin: Deutsche Farbwissenschaftliche Gesellschaft.

GLOSSARY

absorption, the transfer of energy from photons to matter.

achromatic, neutral, possessing no hue and chroma.

adaptation, visual, modification of the visual response to stimuli due to the effects of the immediate surround and the total visual field of simultaneous or preceding stimuli, includes brightness adaptation and chromatic adaptation.

additive color mixture, produced by addition; specifically that the physical sum of two visual stimuli is seen as the psychological sum in the sense of matching color perceptions.

afterimage, a visual image experienced after its original stimulus has ended.

aim color, a psychophysical color specification for a color chip; typically in a systematic collection.

antagonistic, opposition in physiological action, specifically referring to neurons with opponent color character.

attention, whatever occupies the conscious mind at any given moment.

attribute, an inherent characteristic; there are two sets of widely accepted primary color attributes for object colors: (1) hue, chroma and lightness; (2) hue, whiteness and blackness.

attribute measurement, process of assigning numbers or other symbols to things in a manner that their relationship reflects the relationships of the attribute being measured.

Bezold-Brücke effect, a sensory effect named after German scientists, according to which the hue sensation caused under normalized viewing conditions by light of all but three wavelengths changes with changing intensity.

blackness, degree of resemblance of a visual field to the fundamental color contrast perception of black; a fundamental color attribute in the Hering system.

brightness, attribute of a visual perception according to which an area appears to emit, or reflect, more or less light; differences in brightness range from bright to dim.

category, a fundamental and distinct class to which concepts, or entities such as colors, belong. Categorization is the process of placing concepts or entities into categories.

centroid, a point representing the center of mass of a uniform body; in relation to color categories, the stimulus representing the color most representative of a given category.

chroma, the attribute of a visual sensation permitting the judgment of the degree to which a chromatic, related color differs from the achromatic color of the same lightness.

chromaticity diagram, a two-dimensional psychophysical diagram in which color stimuli are ordered according to hue and saturation.

chromaticness, attribute of a visual sensation according to which the perceived color of an area appears to be more or less chromatic.

chromatic plane; a plane in which all color perceptions, systematically ordered, of colors seen as equally bright or light are located.

CIE colorimetric system, a color specification system developed by the Commission Internationale de l'Éclairage (CIE, International Commission on Illumination.

CIE standard observers, two sets of three spectral functions defining the color matching properties of trichromatic observers believed average, one for a field of view of 2° (1931 standard observer), the other for a field of view of 10° (1964 standard observer).

cleavage plane, cleavage is the tendency of crystalline materials to break under strain along defined lines; a cleavage plane is a surface in a crystalline structure revealed after an actual or imagined break; in a color solid it contains colors that stand in simple mathematically definable relationship to each other.

cognition, the process of knowing with its aspects awareness and judgment.

color, dictionary definitions of color are typically circular. In the belief of one of the authors (RGK) colors are symbolic entities the human brain at an as yet unknown location assigns to electrochemical signals derived in the retina from electromagnetic radiation entering the eye, then variously modified to result in the most useful interpretation of local areas of light radiation in the complete array of light radiation arriving at the retina.

color, complementary, a color stimulus that when added to another one in appropriate ratio results in an achromatic perception; located on a straight line passing from one stimulus to the other through the achromatic point of the CIE chromaticity diagram. Colloquially also used for ranges of chromatic color percepts in the vicinity of colorimetrically complementary colors.

color, full, translation of Hering's term *Vollfarbe*, the mental image of a color at its highest chromaticness; the color with a particular hue at the highest level of chroma on the MacAdam limit.

color, primary, colloquial term used in different circumstances: 1. One of three lights the color appearance of which cannot be matched by the other two used with the other two to match the appearance of any other light. 2. One of three colorants used in color order systems or in color reproduction, e.g., yellow, red and blue, or yellow, magenta and cyan. 3. One of the four Hering *Urfarben* or fundamental hue perceptions: yellow, red, blue, and green.

color, related, color perception caused by light reflected from an object in the presence of other objects. The perceived color depends on the perceived color of surrounding objects.

color, spectral, refers to a color experience resulting from viewing light of a single wavelength or a narrow contiguous band.

color, tint or shade, color perceptions having chromatic and whiteness (tint), or blackness (shade) components, but not both.

color, tonal, for the purposes of this text defined as a veiled color containing both whiteness and blackness. Historically, definitions vary, at times being synonymous with "veiled."

color, unrelated, color perceived to belong to an area seen in isolation from other areas.

color, veiled, translation of Hering's term *verhüllte Farbe*: all color perceptions that have a white and/or black content.

colorant, a material that changes the absorption characteristics of another material: dyes, pigments, or dissolved metal salts.

colorant formulation, generation of colorant recipes that result in the desired coloration of the substrate involved; historically, colorant formulation was achieved by trial and error. Since the mid-twentieth century mathematical models run on computers have been increasingly used as tools to aid in the formulation process.

colorant strength, coloristic strength, the relative ability of a unit amount of colorant to impart coloration onto a substrate.

colorant trace, the trace in a geometric color space, such as the CIE x, y, Y space, resulting from the connection with a line of individual stimulus loci in that space representing application of the colorant in multiple concentrations.

color appearance, appearance is the sense impression or aspect of a thing; color appearance is the aspect of a colored field that distinguishes it from the comparable aspect of another field that has a different color appearance; visual appearance includes visual aspects other than color, such as glossiness, transparency, opacity, and so forth.

color assimilation, perceptual effect where colored lines superimposed on a field of a different color shift the appearance of the field color toward that of the lines.

coloration, in this text, term refers to exactly separated areas painted with given standardized colorants, that is, color chips or samples that have a clearly specified place in a color order system and can be used for comparison.

color atlas, a systematically arranged collection of colored chips that are symbols of the colors of a color solid; the chips illustrate the intended space only when viewed under prescribed conditions by an average color normal observer.

color attributes, fundamental aspects or components of color perception, for example, hue, chroma, and lightness, or hue, whiteness, and blackness.

color circle, a circular arrangement of hues in their spectral order, with nonspectral purple colors connecting the short-wavelength and the long-wavelength ends of the spectrum; usually illustrated with high chroma pigment colorations; color circles are a natural result of dimension reduction of reflectance functions to three dimensions.

color constancy, lack of change in the apparent color of an object regardless of quality or quantity of the illuminating light; natural objects tend to be reasonably constant in appearance when viewed after adaptation to various phases of daylight.

color difference, perceived difference between to nonidentical fields of color.

color difference formula, a psychophysical mathematical formula that allows the calculation of the approximate average perceived difference between two stimuli.

colorimeter, optical instrument for the investigation of color vision; in technology also an instrument that measures the reflectance of materials through three filters duplicating the color matching functions of a standard observer.

colorimetry, the branch of color science concerned with the psychophysical numerical specification of color stimuli.

colorimetric purity, a measure of saturation related to color stimuli and expressed in the CIE chromaticity diagram. Its relationship to perceived saturation in some standard conditions is complex.

color matching error, stimulus variability in repeated matches of a standard color.

color matching functions, three spectral functions describing the amounts of three primary lights required to result in color perceptions matching those obtained from spectral lights.

color metric, a metric describes the mathematical structure of a geometric space; a color metric applies to a color space.

color mixture diagram, diagram with which the chromatic result of mixing lights can be quantitatively predicted, for example, the CIE chromaticity diagram.

color moment, a mathematical term based on the view, introduced by Newton, that a quantitative chromaticity diagram of color stimuli is comparable to a force table; specifically, it denotes the three dimensions of the Luther space.

color order, systematic arrangement of color perceptions in terms of attributes and geometric or mathematical models thereof.

color order system, conceptual, a qualitative or semiquantitative color order system where color stimuli in the form of samples have been fitted into a predetermined geometric solid without consideration of the relationship of geometric distances to perceptual distances of some kind; the term "conceptual" is here applied in the sense that the system is purely qualitative and not populated with samples.

color rendering, the fidelity with which artificial light sources render the appearance of colored objects in comparison with a standard daylight.

color scale, a scale in which perceived colors change in a systematic manner, usually in one attribute.

color solid, subset of color space containing, in a given experimental situation, all possible color experiences of the observer under consideration.

color space, a conjectural geometric space in which all possible human color experiences are arranged in some kind of rational order, forming a color solid.

color specification, definition of a color stimulus in a particular reference system; typical reference systems are CIE tristimulus values and relative monitor primary light values.

color stimulus, a stimulus is something that excites an organism, or one of its components to functional activity; an external color stimulus normally consists of light of one or more wavelengths, viewed against a surround of different spectral composition.

cone, cone-shaped light-sensitive cells in the retina; the normal human retina has three types of cones that differ in spectral sensitivity.

cone sensitivity functions, spectral functions that describe the response of the three cone types to light energy arriving at the surface of the retina.

contrast, the difference between things having similar nature; specifically, the degree of difference between two adjacent fields of color; perceptually, contrast is expressed in terms of perceived difference; psychophysically, in cone activation terms (in a cone contrast diagram) or in colorimetric terms.

contrast, simultaneous, a change in apparent lightness, hue, and/or chroma of a colored field caused by an adjacent or surrounding field of different lightness, hue, and/or chroma; both fields change appearance in a direction away from the color of the other field; for example, alight gray field surrounded by a deep red field looks greenish and lighter than when viewed in a white surround.

contrast, successive, an imaginary colored field perceived in a location in the visual space where previously a colored object was located and viewed attentively for a time; the afterimage is either negative (say, red after viewing a green object) or positive and can be experienced with open or closed eyes.

correlated color temperature, absolute temperature on the Kelvin scale of a black body emitting light that gives rise to the same color perception as a given test light; used to describe an aspect of the quality of lamp light with spectral power distributions different from those of an emitting black body.

crispening, refers to changes in stimulus increments necessary for a criterion perceptual difference response; the increment is smallest if the surround color is intermediate to the colors of the two fields compared, in both luminance and chromaticity.

detection, discovery or determination of the existence or presence of something, specifically, for example, the determination of presence of redness in a perceived color.

diapason, the entire compass of musical notes, colloquially also used for the totality of colors.

dichromacy, property of a vision system based on two cone types only.

diffraction, the modulation of a wave passing the edge of an opaque material, resulting in a redistribution of energy due to bending of waves.

dimension reduction, natural or physical/mathematical reduction of the dimensionality of a complex stimulus from many to few, for example, the reduction of 31 individual spectral data points to three data points, the tristimulus values.

discrimination, the process by which two stimuli differing in some aspect result in different responses of some sort.

disk mixture, technique first mentioned by the Greek astronomer and geographer Claudius Ptolemy (first century A. D.) where differently colored sectors of a disk, on rapid spinning of the disk, are additively mixed because the visual system cannot resolve the colors of the individual sectors. The results are dependent on the quality and quantity of the colorants used to color the sectors.

dominant wavelength (of a color stimulus), wavelength of the monochromatic stimulus that, when additively mixed with the appropriate amount of achromatic stimulus, results in a color match with the test stimulus.

dyes, natural or artificial colorants that absorb but usually do not scatter light and are soluble in the substrate or that go through a solution stage in their application to the substrate.

electromagnetic radiation, transport of electromagnetic energy through space; electromagnetic radiation has a wide spectrum, from X-rays to radio and television transmission rays.

empirical, originating in observation or experience.

equal energy light source, a theoretical light source that has a relative spectral power distribution of 1.0 across the spectrum.

Euclidean space, space in which Euclid's axioms of straight and parallel lines and angles of plane figures apply, for example, a cube.

exemplar, a typical or standard model.

extraspectral, outside the spectrum; refers to stimuli that do not result in perceptions obtained from spectral stimuli, such as bluish red, purple, and bluish purple colors.

field of view, the size of the retinal image expressed in solid angle; the CIE has specified a 2° and a 10° standard observer.

flicker, variation in brightness or hue perceived upon stimulation by intermittent or temporally nonuniform light.

fovea, a small depression in the retina containing mainly cones at high density; it is the most sensitive area of the retina, that on which the optical image at the center of our gaze is focused.

Fraunhofer lines, a set of hundreds of dark lines appearing against the bright background of the continuous solar spectrum and produced by absorption of light by gases in the sun's atmosphere (or of another star). Named after the German physicist Joseph von Fraunhofer.

gamut, a region of color space occupied by stimuli resulting from colorations of two or more colorants, for example, three dyes in color photography.

geodesic, the shortest line between two points on a given surface; the curvature of the line depends on the geometry of the space.

grade, a position in a scale of ranks or qualities; specifically, a fixed point in a color scale.

gray scale, a series of grades representing an achromatic color scale, usually with visually equidistant steps between neighboring grades.

halftone printing, a printing process where the image is rendered in smooth variations of dots of the same or varying size of the four (or more) process inks yellow, magenta, cyan, and black.

Helmholtz-Kohlrausch effect, in heterochromatic brightness matching, the perception of chromatic colors as brighter than achromatic colors of the same luminance; the effect is dependent on the dominant wavelength of the color.

hue, attribute of a visual perception according to which an area appears to be similar to one of the colors yellow, red, blue, or green or to a combination of adjacent pairs of these colors considered to be arranged in a closed ring.

hue superimportance, a visual effect by which a smaller stimulus increment is required for a unit difference response if it represents a hue difference than if it represents a chroma or saturation difference of the same perceived magnitude.

hues, unique, the four hues of the color circle that can not be matched with colors other than themselves; the psychological primary hues yellow, red, blue, and green; for example, unique red is a red hue that is neither yellowish nor bluish, for example.

illuminant, an illuminating device; technically a set of numbers representing the spectral power distribution of a light source; several illuminants have been standardized by the CIE, such as daylight illuminant C and D65.

illustration, in this text, more or less continuous painted color transitions covering extended areas of the system without having locally homogeneous and spatially defined color fields suitable for comparison.

incandescence, the emission of visible light from a body at high temperature (above ca. 1,500 K).

interference colors, color stimuli generated as the result of addition or cancellation, of two light waves depending on whether they are in or out of phase when combined.

isotropic, exhibiting properties with the same values when measured along axes in all directions; in connection with color space; refers to a space in which distances in all directions are commensurate with the size of perceived distances; generally, a uniform color space; see also *uniform*.

just noticeable difference (JND), threshold difference, the initial perceptual difference that can be seen when one of two originally identical fields of color changes in any given direction.

Kelvin scale, temperature scale with Celsius scale units but beginning at absolute zero (minus 273.2° C).

Kubelka-Munk relationship, relates the reflectance of a partly absorbing and partly scattering object to absorption and scattering, for example, for a layer of paint or a dyed textile material.

lake, a kind of pigment produced by insolubilizing dyes through precipitation with an agent with low water solubility, such as aluminum or calcium compounds; the lightfastness of lakes is usually lower than that of regular pigment.

lateral geniculate nucleus, a mass of cells in the brain along the visual passageway between the retina and the visual area at the back of the brain.

lattice, a regular geometric arrangement of points in space.

lightness, perceptual attribute of related colors according to which a color field appears to emit equal or less light compared to a white field; can be understood as relative brightness; differences in lightness range from dark to light.

line element, the first fundamental form of a regular surface, defined by the Riemannian metric, in connection with color, the term describes a certain kind of color space defined by (weighted) increments of color fundamentals.

luminance, luminous flux of a light beam emanating from a surface in a given direction, per unit solid angle.

luminous reflectance, luminance of the surface of an object compared to the luminance of the surface of a perfectly reflecting diffuser, illuminated with the same light source and viewed at the same angle; in the CIE colorimetric system known as luminance factor Y.

magnocellular, layers in the lateral geniculate nucleus in which relatively large cells are located, believed to relay information necessary for motion perception.

matching, manipulation of color stimuli or colorants so that the combined stimulus or colorants result in a perception identical to that of a standard stimulus or coloration.

mathematical transformation, to change the form or direction of a point, line, area or solid from one set of coordinates to another set without changing the intrinsic information; based on Grassmann's laws, color matching functions can be transformed linearly to change their shapes without loosing the intrinsic information contained in them.

matrix algebra, algebra related to rectangular sets of transformations of coordinate spaces, or comparable data.

metamerism, a visual phenomenon according to which certain stimuli that are spectrally different have, for a given observer, identical color appearance.

metamers, two or more differing spectral power distributions resulting in identical color perceptions for an observer; also used for objects with different reflectance functions seen as having identical color when viewed in standard conditions under a given light source; a consequence of dimension reduction of spectra.

mezzotint, printing technology invented in the seventeenth century by Ludwig von Siegen; metal plates are roughened evenly on the surface and the image is developed by locally smoothing the surface to create an image, with the smooth surface printing dark and the rough one light.

moiré effect, a kind of usually unintended pattern formed by two overlapping geometric patterns, for example two halftone screens at certain angles that depend on the geometric patterns.

monochromatic, light of a single wavelength or a very narrow band of wavelengths seen as having identical color.

nanometer (nm), metric unit of distance measurement, a nanometer is a billionth of a meter; there are approximately 25 million nanometers in an inch.

neural network, refers to a structurally integrated unit of the brain, as well as to a mathematical procedure imitating its presumed activity on a digital computer; neural networks learn to extract fundamental information from input data and use it to generate or predict outputs.

neuron, cell in the nervous system specialized in the transmission of electrical signals.

neurophysiology, the physiology (science of the organic processes of living systems) of the nervous system.

object color, apparent color of an object; the color of an object can vary depending on the surround and contextual conditions in which it is viewed.

opaque, property of a material through which light cannot pass; not transparent or translucent.

opponent color theory, a theory according to which color perception is based on unique hues forming opposing pairs: red-green, yellow-blue, and the nonhued pair black and white; also the theory of how the outputs of cones are connected in the retina and the brain.

opsin, a kind of colorless protein molecule that together with the retinal molecule forms a spectrally tuned visual compound, different cone types have different opsins.

order, having elements arranged or identified according to a rule.

orthogonal, to intersect or lie at right angles.

orthonormal, a mathematical concept according to which the related space must be orthogonal (see previous entry) and the square of each function forming an axis of the space must, over a specified interval, equal to one.

parallelepiped, is a three-dimensional figure like a cube, except that its faces are not squares but parallelograms.

partitive color mixture, a special type of additive color mixture where the effect of addition of light is obtained by having very small stimuli (for example, pixels in video displays) that cannot by perceptually resolved and thus blend, or by rapidly exchanging one stimulus with another, such as in disk mixture.

parvocellular, layers in the lateral geniculate nucleus in which relatively small cells are located, believed to relay information necessary for brightness and color perception.

perception, the subjective, conscious awareness of any aspect of the external or internal environment.

photometer, instrument for the measurement of light intensity.

photometry, the measurement of light as related to the average human observer.

photon, unit or quantum of electromagnetic radiation.

photopic vision, vision mediated by cones active at daylight levels of light, and resulting in brightness and color perceptions

pigment, natural or artificial colorant that not only absorbs but also scatters light on the surface of its molecular aggregates; pigments are insoluble in the application medium or the substrate.

psychophysics, the study of mental processes by quantitative methods; specifically, the reports of human subjects of the perceptions resulting from carefully measured light stimuli.

qualia, singular *quale*: qualitative fundamentals of brain/mind activity, comparable to quanta as quantitative fundamentals of electromagnetic energy, for example, the redness of red, the sweetness of the taste of sugar; there is ongoing discussion concerning the reality of qualia.

quantum, (plural quanta), the smallest increment of electromagnetic radiation (see also photon).

reflection, the process by which a smooth surface returns electromagnetic radiation, specifically light, according to a simple optical law: the angle of reflection equals the angle of incidence.

refraction, change in direction of a beam of electromagnetic radiation due to change from one medium (for example, air) into another (for example, glass) in which its speed of propagation differs; refraction is the optical principle on which the refracting or lens telescope is based.

retina, a layer coating the inside of the camera-type eye, containing the light-sensitive rod and cone cells and cells connected to them; continuous with the optical nerve.

retinal, light-sensitive molecule derived from vitamin A.

Riemannian geometry, non-Euclidean geometry with positive curvature in which the parallel line postulate is replaced by the postulate that every pair of straight lines intersect.

rods, rod-shaped light-sensitive cells in the retina, specialized to operate primarily at low light levels, resulting in brightness perception only.

saturation, attribute of a visual perception that permits a judgment of the degree to which a chromatic stimulus differs in appearance from that of an achromatic stimulus, regardless of their brightness; colloquially designates the third attribute of color percepts regardless of their cause.

scattering, the process of returning electromagnetic radiation by a scattering surface; light is returned in all directions, resulting in diffusion of the light beam.

sensation, mental process due to bodily stimulation, now distinguished from awareness of the result of the mental process.

slit width, the width of the slit through which light passes on the way to a prism or grating that refracts or diffracts it according to wavelength; the wider the slit width, the more desaturated the resulting color when the light is reflected from a (white) surface.

spectral power distribution, the relative amount of light at different wavelengths of a polychromatic light with (usually) light at 555 nm = 1.0.

spectral spaces, spaces created from reflectance or spectral power distribution data by dimension reduction techniques other than those involving color matching or cone sensitivity functions.

spectrophotometer, instrument for measuring the spectral reflectance of objects.

standardization, to bring in conformance with a standard.

stimulus, an agent that directly influences the activity of a living organism or one of its parts; specifically, electromagnetic radiation within the visible band.

subtractive color mixture, the result of the combined effect of light absorption from two or more colorants; light reflected from the mixture is reduced compared to that reflected from either of the two individual colorants.

suprathreshold, exceeding the threshold; specifically, a difference that is larger than a threshold difference.

tetrachromacy, property of a visual system based on four different cone types.

tetrahedron, the first of the Platonic solids, having four identical triangular faces.

threshold, visual, the lowest level or increment of stimulus resulting in a visual perception or a difference perception.

tint-shade scale, a scale of colors of constant hue but varying lightness and chroma, beginning at white, passing through mixtures with increasing amounts of a chromatic pigment to the full color (color of highest chroma), and passing through mixtures with increasing amounts of black to black.

tonal scale, in this text defined as a constant-hue, constant-lightness scale varying in chroma.

translucent, attribute of a material that is neither transparent nor opaque, such as a frosted glass panel.

transmission, passage of light through a transparent medium, for example through glass or a liquid.

transparent, attribute of a material through which light of some or all wavelengths can pass.

trichromacy, property of a color vision system based on the activity of three cone types.

tristimulus values, the scalar values of the amounts of three primary lights required to match a given light; in the CIE system the tristimulus values *X, Y* and *Z* refer to nonreal lights.

uniform color space, a three-dimensional geometrical space in which geometric distances between points representing color perceptions correspond to their perceptual distances; see also *isotropic*.

value, Munsell's term for the grades of a perceptually uniform gray scale.

vector, a quantity with magnitude and direction, usually represented by an arrow with its length representing magnitude and its position in space representing direction.

vision, process by which the extended visual system extracts information from light energy to help generate appropriate response behavior.

wavelength, distance in the direction of propagation between two peaks of a wave.

Weber-Fechner law, colloquial term used to express a version of the nonlinear relationship between stimuli and resulting perceptions, such as between luminance and perceived brightness or lightness. Named after nineteenth century German physiologist Ernst Heinrich Weber and psychophysicist Gustav Theodor Fechner, the former stating the fact of the compression of the stimulus in the response, the latter proposing that the compression is of logarithmic nature.

whiteness, attribute of a diffusing surface permitting, when viewed under a standard light source, the judgment of similarity to a standard white surface viewed in the same light.

INDEX